CLASSICAL
SCIENTIFIC
PAPERS

CHEMISTRY
Second Series

CLASSICAL
SCIENTIFIC
PAPERS

CHEMISTRY
Second Series

Papers on the Nature and Arrangement
of the Chemical Elements
Arranged and Introduced by
DAVID M. KNIGHT, MA, D.Phil.
Department of Philosophy, University of Durham

Mills & Boon Limited
17–19 Foley Street
London W1A 1DR

American Elsevier
Publishing Company Inc.
New York

First published in 1970 Great Britain by Mills & Boon Limited
17–19 Foley Street, London W1A 1DR

First published in 1970 U.S.A. by American Elsevier Publishing Co, Inc.
52 Vanderbilt Avenue, New York, New York 10017

© Mills and Boon Limited 1970

British ISBN 0 263 51476 4

American ISBN 0 444 19646 3

Library of Congress Catalog Number 74 122441

Printed in Great Britain by Jarrold and Sons Ltd, Norwich

Contents

Acknowledgements

The editor and publishers of this collection of papers would like to express their thanks to the many individuals, libraries and publishing houses who have helped to bring the volume into being.

Dr W. H. Brock of Leicester University, Mr Frank Greenaway of the Science Museum, South Kensington, and Dr C. A. Russell of Harris College, Preston, have offered suggestions for papers which might be included. The editor is grateful for their helpful advice, at the same time assuming responsibility for the final selection.

Mr R. G. Griffin, Librarian of the Chemical Society, has been most generous of his time in making available the majority of the original volumes from which the required pages have been photographed.

We should also like to thank the Fellows of the Chemical Society for allowing us to use their library. Mr S. Gillam of the London Library, Mr J. Weston of the Royal Institution, and the Librarian of the National Lending Library for Science and Technology are others who have helped us in collecting the necessary material.

We acknowledge permission given to reproduce articles originally published by the following: Messrs Taylor and Francis (*Philosophical Magazine*), Messrs MacMillan (*Nature*), the Royal Society, the British Association for the Advancement of Science.

Finally, we should like to express our thanks to Messrs Jarrold and Sons for their enthusiasm and expertise in dealing with the technical side of this publication.

Foreword

This is the third of a series of volumes which contain, reprinted in facsimile, a selection of related papers on important topics in the development of modern science. *Classical Scientific Papers, Physics*, was devoted to the rise of atomic physics during the first decades of the twentieth century, and the first volume of *Classical Scientific Papers, Chemistry*, to the progress and vicissitudes of the atomic theory in chemistry from Dalton's book of 1808 to Perrin's paper of 1910 which finally convinced sceptics. The present volume traces ideas on the nature and arrangement of the chemical elements, beginning with Prout's hypothesis that all the different elements are not irreducibly different; showing the various attempts to classify the elements in families, which culminated in the Periodic Table of Mendeléeff; and finally arriving, by way of theories of inorganic evolution, at the first observation of a transmutation in Ramsay's laboratory at the beginning of the present century.

The three volumes are independent but are designed to complement one another. The object is to put the reader, as he follows the various controversies and developments, in the position of a participant or contemporary spectator; and thus to cast light on the way science develops. A general Introduction, and Introductions to the various sections, provide background and link the various phases of the story. It is hoped that this volume will be, as its predecessors seem to have been, useful to students, scholars and librarians.

D. M. Knight

General Introduction

When, in 1897, J. J. Thomson believed that he had proved the cathode rays to be composed of corpuscles, he wrote: 'The explanation which seems to me to account in the most simple and straightforward manner for the facts is founded on a view of the constitution of the chemical elements which has been favourably entertained by many chemists: this view is that the atoms of the different chemical elements are different aggregations of atoms of the same kind. In the form in which this hypothesis was enunciated by Prout, the atoms of the different elements were hydrogen atoms; in this precise form the hypothesis is not tenable, but if we substitute for hydrogen some unknown primordial substance X, there is nothing known which is inconsistent with this hypothesis, which is one that has recently been supported by Sir Norman Lockyer for reasons derived from the study of the stellar spectra.' Thomson went on to suggest that the chemical atom was an aggregation of a number of primordial atoms, which were probably the corpuscles composing the cathode rays; matter in these rays being 'subdivided very much more than in the ordinary gaseous state.'

This paper was the earliest reprinted in *Classical Scientific Papers—Physics*, which documented the rise of modern atomic physics; and in the first series of *Classical Scientific Papers—Chemistry* the appearance of chemical atomic theory, and its ultimate fusion with atomic and molecular theories derived from physics, was set out. The present volume, while it can stand on its own, complements both its predecessors. It contains in facsimile papers illustrating the tradition, stemming from Prout, according to which the elements were not irreducibly simple bodies, composed of *atoms* in the strict sense. Rather, they were stable configurations of indivisible particles of the same basic matter. We shall, in short, be preoccupied with the problem of the nature and arrangement of the chemical elements, from the time when Lavoisier defined the term 'element' until the first transmutation, the spontaneous radio-active decay of radon, was followed in William Ramsay's laboratory.

The plan of the volume is similar to that of its predecessors; but because more articles come from *Nature* and *Chemical News*, with their double columns, small type, and quarto format, it has been possible to reproduce a greater number of papers. These journals broke up long papers into short sections appearing over several weeks; Victorians thus often had to read their science, like their novels, in parts. An extreme example is Mendeléeff's long paper of 1879–80, which was spread over eighteen weeks; it is, therefore, tiresome to consult, and must, one would expect, have made less impact than if it had not been so fragmented. Even for those with access to the original, it will be easier to follow such papers here. This present volume has an index, and carries the author's name at the top of each paper; this should increase the usefulness of the collection.

The selection of papers for an anthology such as this poses certain problems. The chosen topic is one in which English-speaking chemists distinguished themselves, although in many respects British chemists played a peripheral role in the development of the science between about 1820 and 1900. Almost one third of the papers reprinted here were written by Americans or by Continental Europeans; in any attempt to display the whole history of chemistry in the nineteenth century, this ratio would have to be very different. But this collection, like its predecessors, does not set out to reprint the most important papers published in the whole field of chemistry, but to illustrate the progress of science in a series of connected papers on relatively narrow, but important, themes; reprinted in facsimile so that the reader can feel himself involved in the discussions. Richard Gough set out, not without success, to delineate human nature in his parish pump *History of Myddle;* an aim to which chroniclers of the rise and fall of empires do not as a rule aspire. The progress of science is not an affair of men of genius making breathtaking discoveries, and announcing truths which others had been too slothful to perceive, or too bigoted to acknowledge; the labours of more ordinary men, and of important scientists not on their main theme, deserve to be remembered, and their 'mistakes' are often instructive.

The literature of nineteenth century chemistry is like a great tumulus, through which any number of trenches may be dug. No one trench will reveal the whole truth, but it may bring out truths concealed from those who rely upon an aerial survey of the whole mound. The cut which we shall follow in this volume, besides showing how theories have led to discoveries which have in turn led to modifications in the theory, raises interesting philosophical questions: including the problems of accuracy, simplicity and falsifiability; the survival of exploded theories; the role of analogy in science; the difficulties of taxonomy, of drawing up artificial or natural systems of classification; and the carrying-over of ideas from a successful and prestigious science into one which is relatively backward. And by sticking in the main to papers written by natives of one country, one may gain an insight into general intellectual history; in this case, for example, the continuing vitality of Newtonian concepts far into the nineteenth century, and the development of doctrines of inorganic evolution parallel to those which revolutionised biology, and which in late Victorian Britain swept all before them.

Although it is possible thus to justify a parochial history of science, and to suggest that it may be a mistake to be too cosmopolitan in outlook, without the chauvinism of suggesting that all important discoveries have been made in one country: this collection, aiming at the best of both worlds, does in fact contain a higher proportion of papers written by non-British scientists than either of its predecessors. Berzelius was the most distinguished chemical analyst—indeed, the most distinguished chemist—of his epoch; the theory of isomerism and the radical theory were developed abroad; and in the field of classification the crucial contributions were not made by British chemists. Hence papers translated from foreign languages play an essential part in our story; and, in a new departure, we include two contributions not in English: an abstract in German of Mendeléeff's announcement of his Periodic Table, and an extract from a book in French by Marcelin Berthelot comparing Proutian speculation and alchemy. The former shows the first form of the Periodic Table, the consequences of which are amplified in Mendeléeff's long paper published ten years later; the latter is amusing and spirited, and was assailed by Mendeléeff in his Faraday Lecture to the Chemical Society in London in 1889; an English summary has been provided.

It was a dictum of the atomism, or 'corpuscular philosophy', of Boyle and Newton and their followers that everything was composed of particles of the same matter. Different agglomerations of these corpuscles constituted the various substances found in the world. Chemical reactions took place between these congeries of ultimate particles; but there was no reason in principle why, under more rigorous conditions, the ultimate particles might not be rearranged and one chemical element changed into another. The alchemists had been mistaken in the details of their theory and practice; but in principle they had been right. The alchemist Peter Woulfe, who invented a wash-bottle for gases, was a Fellow of the Royal Society in the last years of the eighteenth century; and the gap between the old alchemy and the new version of Ramsay and Rutherford is bridged by Davy, a student of alchemical literature, Faraday, Berthelot, Crookes and Lockyer. Davy was careful to separate this 'sublime chemical speculation', sanctioned by the authority of Newton and others and supported by chemical analogies, from the dreams of those deluded by charlatans and impostors; for a Newtonian, transmutation was not a mysterious process involving the philosopher's stone, but a deeper-level chemical change producing a more radical rearrangement of particles.

Another authority to whom Davy appealed was Roger Boscovich, who had in the late eighteenth century proposed a dynamical atomic theory. His atoms degenerated into mere points, and he was only interested in the forces between them. For Newtonians the world was composed of atoms and the void; but the void so predominated that it was even suggested that all the actual solid matter in the

3

solar system might be contained within a nutshell. Boscovich's atoms were mathematical points, and therefore took up no space at all; but associated with each point was a short range repulsive force rising to infinity, which prevented two centres occupying the same place. For Boscovich, whose theory was made known in Britain by Joseph Priestley and in a long article in the 1801 Supplement to the *Encyclopædia Britannica*, the points were identical; he compared them to minute black dots which form the different letters from which books on various subjects and in different languages are composed.

A more thoroughgoing dynamical theory, derived from Kant, made force the ground of matter, and denied the existence of atoms of any kind. On this view, there were no such things as chemical elements; all substances were on the same level, and all chemical reactions involved the generation of new qualities. In this world of flux, in which substances endured only as waterfalls or columns of smoke endure, all chemical reactions were in a sense transmutations. This world-view guided Johann Wilhelm Ritter and Hans Christian Oersted in their electro-chemical and electromagnetic researches; but in the early years of the nineteenth century it was only to be found in English in the *Principles of Modern Chemistry* of F. C. Gren, translated in 1800. Indeed it was confused with the theory of phlogiston; but, if we remember that it is possible to see phlogiston as a crude groping towards a conception of chemical energy, this will not surprise us. This Kantian dynamical science became important when, after a long interval, the eighteenth century interest in chemical energetics was revived about 1840 by workers such as P. L. Dulong and G. H. Hess.

When William Prout was a medical student at Edinburgh in the first decade of the nineteenth century, therefore, it was natural to believe that all matter was composed ultimately of similar particles. Davy provided further chemical arguments in the final chapter of his *Elements of Chemical Philosophy* of 1812, calling attention to the family resemblances among the elements and remarking on the similarity between derivatives of the element potassium and the radical ammonium, which was known to be composed of hydrogen and nitrogen; he suggested that this analogy in chemical behaviour might indicate some analogy of structure. This passage is reprinted in the first volume of *Classical Scientific Papers—Chemistry*; where can also be found the writings in which the new atomic theory of John Dalton was brought before the world.

Dalton's atomic theory differed from those of Newton and Boscovich in that he postulated distinct kinds of atom for each different chemical element. Atoms of, for example, hydrogen were for Dalton indivisible and irreducibly different from those of oxygen or iron; although in using the term 'atom' in writing of compound bodies like carbon dioxide he left open the possibility that some substances believed elementary might also turn out to be compound. Dalton did not think that all the

4

elements would be found to be constituted of one prime matter; but Thomas Thomson, the most important populariser of Dalton's theory, did, and it was in his journal, *Annals of Philosophy*, that Prout's hypothesis was published. For Thomson or Prout, the atoms of the chemical elements were not indivisible in principle; they were simply particles which were not divided in chemical reactions. If a piece of iron, or indeed of a compound body, were divided again and again, then at last the chemical atom of the substance would be reached; if this were divided, the products would be qualitatively different from the original substance. In the case of carbon dioxide, for example, the products would be carbon and oxygen; a chemical element would yield hydrogen or some simpler prime matter. This is not far from the view of the corpuscularians, discussed above.

For Newton or Davy, the arguments were analogical and qualitative; and if we follow philosophers of science who declare that a proposition is not scientific if it cannot be empirically falsified, then this corpuscularian belief in prime matter becomes a matter of metaphysics rather than of science. It was Prout who gave hostages to fortune by proposing a quantitative and testable form of the doctrine; and, as J. J. Thomson pointed out, this falsifiable part of his hypothesis was indeed falsified. Prout suggested that the atomic weights of all the elements would be found to be whole number multiples of that of hydrogen; and that hydrogen, or a component of it, might be the prime matter. Eighteenth century attempts to make chemistry a quantitative science based upon the forces of affinity between atoms— on the model of Newton's physics—had foundered on the inaccuracy of calorimetric and electrical measurements; and Proust and Dalton opened another road in insisting upon the definite proportions in which elements combine, and urging that atomic weights be determined. Prout's hypothesis gave a further impetus to this version of a quantitative chemistry, since an experimental test seemed not too difficult, and the question on which it might cast light was one which had puzzled so many.

The test turned out to be less straightforward than one might have expected. Prout himself turned towards physiological studies, identifying hydrochloric acid in the stomach; and it was Thomson who performed the analyses. He was a great teacher of chemistry, whose *System of Chemistry* became a standard textbook; and he was one of the first to introduce laboratory work for his students. But whereas Liebig at Giessen later achieved enormous success by encouraging his students to get on with their own projects, Thomson used his as research assistants, and published their results as his own, with vague acknowledgements. This practice is alleged not to be wholly extinct; in Thomson's case, it is one of the reasons why his analyses, which seemed to verify Prout's hypothesis, were not as accurate as those of Berzelius or Turner. But more interesting general points are raised by Thomson's analyses, which appeared first in *Annals of Philosophy* over a number of years, and were then collected into a book, with the splendid title *An Attempt to Establish the First Principles of Chemistry upon Experiment.*

Rather oddly, Thomson did not use the atomic weight scale, on which the relative weight of hydrogen is unity, and which on Prout's hypothesis would have given integer values for all the elements; instead he employed a scale in which oxygen had unit weight, and hydrogen became 0.125. All the other atomic weights for which he produced values were indeed integer multiples of that of hydrogen; to the modern eye the list of figures all ending in .125, .250, and so on, cannot but bring to mind the discussion of 'cooking' and 'trimming' in Babbage's *Decline of Science in England*. Thomson had an answer for this innuendo ready in advance. Earlier observers had produced complicated numbers because they had had no principle to guide them in selecting their results. Indeed, anybody who has verified Boyle's Law or Ohm's Law will know that the results of the observations do not accord exactly with the predicted values. The divergencies are put down to experimental error; which is just what Thomson did, arguing that Nature was not wont to indulge in complex relations and that the apparent precision of Berzelius' results, with their long chains of decimals, was sufficient to render their accuracy suspicious. Berzelius had only the test of consistency; Thomson had a theory to confirm. We should remember that at this date there was no generally agreed technique for estimating accuracy or error; the least squares method of averaging observations was introduced later by Gauss, and scientists tended to use those measurements which seemed best for one reason or another. Thomson was not perpetrating a fraud like the discoverers of the Piltdown Man; he was simply following canons of science which were becoming out of date.

The reviews of the *First Principles* were somewhat mixed; that in Benjamin Silliman's *American Journal of Science* ends on a note which sounds ironical but was meant seriously: 'The consummate skill discovered in devising and executing the experiments, and the surprising coincidence of the results of analysis with the deductions of theory, excite our astonishment, and prove beyond a question that chemistry, if not founded on intuitive, is built on demonstrative truth.' In the Royal Institution's *Quarterly Journal of Science*, Thomson's enemy Andrew Ure wrote that some of the analyses seemed cooked, but chiefly deplored the manner in which chemistry was passing from an investigation of the powers that modify matter to the mere measurement of atomic weights, which could hardly be called an intellectual activity. The really damaging blows came from Berzelius, in a slashing attack on the book; Thomson replied in equally good Billingsgate; but the analyses of Berzelius were confirmed by Edward Turner, and by 1845 Berzelius could write as though Prout's hypothesis were done for. As Babbage declared, 'the character of an observer, as of a woman, if doubted is destroyed'; and while Thomson published a

6

useful *History of Chemistry* in 1830–1, his main activity after this seems to have been playing academic politics in the University of Glasgow.

Newtonian atomism contributed to our story in another way, as the section on the Fourth State of Matter indicates. Newton had suggested that matter and light might be interconvertible; and the young Faraday echoed this doctrine in one of his earliest lectures. He remarked that the gaseous state is simpler than the solid or liquid state in that all gases obey, for example, the same simple laws of expansion on heating, whereas solids and liquids all expand differently, and in accordance with complex laws. Faraday suggested that 'radiant matter', the corpuscles composing light and the recently discovered infrared and ultraviolet radiations, might be a fourth state into which any substance might be transformed. Faraday and Davy experimented on the relations between the states of matter and succeeded in liquefying some gases, notably chlorine, by generating them in sealed tubes. These researches, which made it clear that there were no such things as really 'permanent gases', and that all substances could exist in all three states, are reprinted in an *Alembic Club Reprint;* they were followed by the classic work of Thomas Andrews on the continuity of the gaseous and liquid states. Andrews' long papers were summarised in *Nature* by James Thomson, the brother of Lord Kelvin and himself a famous engineer.

Faraday's lecture was published in the standard biography which appeared soon after his death; and William Crookes, not recognising a survival of the corpuscular theory of light, seems to have supposed that the lecture represented a new insight by a man of genius. He applied the theory to explain the cathode rays which he had recently discovered. These rays, which proceed in straight lines from the cathode in a highly evacuated tube, are identical, whatever gas was originally present, and whatever metal the cathode is made from; and Crookes identified them with the radiant matter of Faraday. Crookes therefore greatly overestimated the size of the particles; but, as his papers show, he was able to explain many of the phenomena of the discharge tube by his theory, and his experiments formed the basis of those of J. J. Thomson. A different interpretation was favoured by German physicists, who believed that the cathode rays were, like light, to be explained in terms of waves rather than particles or corpuscles. This alternative explanation of the facts was proposed by George Fitzgerald in an address to the British Association, in which he announced Heinrich Hertz's detection of radio waves. This splendid confirmation of the field physics of Maxwell seemed to indicate the fruitfulness of continuum rather than atomic theories; in the nineteenth century it seemed that one had to choose between wave and particle explanations, whereas today we have to be content with the two conceptions, recognising them as complementary.

We can now return to more strictly chemical questions. Davy had pointed out the similarity in chemical behaviour of potassium and the compound radical ammonium; and between chlorine and cyanogen, composed of carbon and nitrogen. He suggested that these analogies made it not improbable that potassium and chlorine, and indeed all the elements, were really radicals which were so stable that hitherto they had resisted all attempts to decompose them. Faraday, in another of his early lectures, argued that all the metals could not be believed irreducibly distinct; many of them, particularly the then recently discovered platinum group, were so similar in properties as to make it most improbable that they were separate creations. Then Dumas and Liebig made an important step towards organising the vast and previously inchoate field of organic chemistry, when they followed certain groups of atoms, or radicals, through a series of reactions. The substances studied in organic chemistry are for the most part composed of carbon, oxygen, nitrogen and hydrogen only; and Dumas and Liebig suggested that radicals such as benzoyl, C_6H_5CO, played in the organic realm the role played by elements in inorganic chemistry. In organic chemistry, they declared, the elements are compound; in inorganic chemistry they are simple; and that is the only difference between these great branches of the science.

Soon this difference was denied by those in search of analogical arguments for believing in the unity of matter. And at about the same time the phenomenon of isomerism was named by Berzelius; various chemists had noticed that various different substances seemed to be composed of the same elements in the same proportions, and Berzelius explained this in terms of the atoms being differently arranged in the different isomers. In Berzelius' laboratory, Eilhard Mitscherlich discovered the phenomenon of isomorphism; that is, the possession of identical crystalline form by two different compounds. This should have helped in the determination of chemical formulae, for isomorphism indicates similarity of chemical structure: but this was not apparent, for the isomorphic substances calcium carbonate and potassium nitrate were written, for example by J. F. W. Johnston in a Report to the British Association in 1837, $CaO + CO_2$ and $KO + NO_5$; our formulae are $CaCO_3$ and KNO_3, and the relationship between formula and structure is therefore clear. But ammonium and potassium salts were found to be isomorphous; KO could be replaced by $H_3N + HO$, and this gave a further reason for supposing that potassium might really be a compound radical.

The idea that the elements might be compound radicals or isomers received considerable attention; Liebig concluded that the families of elements, like chlorine, bromine and iodine, were too much alike to be isomers, which usually differ widely in properties. Various workers in Britain were nevertheless carried away by such analogies, and a transmutation, of carbon into silicon, was even reported by Samuel Brown, who was a candidate for the Chair of Chemistry at Edinburgh. George Wilson, who later became the first Professor of Technology at Edinburgh, was impressed by the arguments, but not by the experiments of Brown and others,

8

which he could not repeat. His review of the various speculations and purported experiments on the nature of the elements makes amusing reading. Berzelius' paper reprinted in this section shows that he was not averse to speculation, but only to poor quantitative forms of it, like Thomson's. Edward Frankland's contribution can remind us that it was from a concern with radicals, and in particular from attempts to prepare free radicals, that he derived his idea that, associated with each element, there was a fixed combining power or valency; an idea which in the hands of Kekulé and his school revolutionised chemistry. Hermann Kopp introduced a further reason for believing the elements to be complex, when he measured atomic heats (the product of atomic weight and specific heat). For metals, as Dulong and Petit had shown, the value was always about 6.4; but for non-metals, it was less, and the 'atomic heat' of hydrogen peroxide, for example, was the same as that of elementary metals. The metals might well therefore be also compounds of non-metallic substances. Roscoe's address to the British Association in 1884 is a most useful review of the state of chemistry, and of these controversies, at the time.

These arguments from analogy became compelling when, after 1869, the elements were arranged in the Periodic Table which exhibited at a glance the family resemblances and gradations of properties between them, and made it seem very unlikely that they were all independent and permanent creations. Newtonians had hoped that chemistry would become a quantitative science of forces, like astronomy; this did not happen, and there are ways in which chemistry more profitably followed biology. Linnaeus had from 1735 brought order into botany by his artificial classification, based upon the sexual parts of the plant. This enabled the scientist to place a specimen unambiguously within its class; and as travellers brought back great numbers of new plants, the system proved invaluable. But by about 1800 botanists were beginning to feel that this artificial system, while convenient, was not adequate. Certain plants were put into groups where the botanist guided by intuition would not have put them because they did not seem really to resemble others in the group. Antoine de Jussieu and others therefore urged a return to the natural system of Aristotle and John Ray, in which one must consider the whole range of characters and class the specimen with those with which it had most in common. A practical problem of this sort faced Davy when he discovered potassium; it had been a defining character of metals, laid down for example by Joseph Black, that they were dense. Potassium floats on water, but Davy after hesitation put it among the metals because 'in the philosophical division of the classes of bodies, the analogy between the greater number of properties must always be the foundation of arrangement'. Chemists insisted that it was not enough to order the elements conveniently; the ordering must also be in accord with chemical behaviour, and unlike elements must not be put together.

Lavoisier's classification was purely chemical, and based particularly upon reactions with oxygen. Light and the caloric fluid responsible for heat appear on the list, because for Lavoisier these substances, although imponderable, reacted chemically like ordinary elements. The table of Thomas Young represents the last attempt to quantify chemical affinities, and arrange substances in accordance with their affinities; it is perhaps closer to being an ancestor of our qualitative analysis tables than of our Periodic Table. In Ampère, who was a great classifier, we find the arguments in favour of a natural system; but Ampère's actual system, which took rather few chemical properties into account, was not a success. The discovery by Döbereiner that many elements could be viewed as members of triads, groups of three in which *either* all had about the same atomic weight (like iron, cobalt and nickel) *or* the atomic weight of the middle member was halfway between those of the others (like chlorine, bromine and iodine), promised a different way of arranging elements. In 1850 Max Pettenkofer tried to relate the atomic weights of different elements in the same family by algebraic series; and a more detailed attempt to do this was made by Professor Cooke of Harvard. His tables became quite widely known; an odd feature of them is that oxygen appears in a number of the series, and that the radical cyanogen appears among the elements which resemble it. Cooke, in the tables, assumed whole number atomic weights, except that a few end in .5; this is of course necessary if one is going to fit the numbers into series of the $(a+n.b)$ type. Dumas, who noticed such relations among the homologous series of radicals (such as methyl, ethyl and propyl), tried independently of Cooke to fit the elements into such groups; and set off further experiments to test Prout's hypothesis. This time the part of Thomson was played by C. Marignac, and that of Berzelius by J. S. Stas; Marignac was eventually forced towards the unfalsifiable hypothesis that the units of prime matter were so extremely small that large numbers of them could add up to the most complicated atomic weights, or that Prout's hypothesis was true but masked by other effects. Their papers are translated in the *Alembic Club Reprint* on Prout's hypothesis.

Another feature of Cooke's table is that his numerical values are not atomic weights at all, but equivalents. In the first series of *Classical Scientific Papers— Chemistry*, we saw how the atomic weights of Dalton were arrived at by the arbitrary assumption that where only one compound of two elements existed, its formula must be AB; and that if there were two, then one must be AB and the other AB_2 or A_2B; and so on; which did not inspire confidence. Sound men preferred not to speculate about atomic formulae or structures, and instead of atoms wrote of equivalents. The equivalent weight of an element or compound is that weight of it which will combine with or displace unit weight of hydrogen. The use of equivalent weights seemed to have all the advantages of Dalton's theory without introducing hypotheses. Two things changed this; one was the success of Kekulé and his followers in explaining in detail numerous facts of organic chemistry, which had hitherto seemed mysterious, in terms of the actual arrangement in space of atoms; the other was the introduction of the Periodic Table, which depended upon having atomic weights rather than equivalents, as its proposers realised.

At the Karlsruhe Conference of 1860, Cannizzaro reminded the chemical world of the hypothesis of Avogadro and Ampère, that equal volumes of all gases contain equal numbers of molecules; and showed how it could be used to determine atomic weights from equivalents. Various workers found that if the elements were put in order of increasing atomic weight, but not equivalent weight, then similar elements recurred at regular intervals. De Chancourtois proposed a scale on which elements of the same family fell below one another; Newlands announced his Law of Octaves; and Odling, and then Mendeléeff, published Periodic Tables which begin to look like those we are familiar with. De Chancourtois' paper was full of numerological speculations, which were prudently omitted by the translator of twenty years later; he cannot count as a serious precursor of Mendeléeff. Newlands energetically pushed his claims later; though his tables are clearly inferior to Mendeléeff's, he appreciated the principle and sometimes left gaps for elements as yet undiscovered. Because the Chemical Society was terrified of mere speculation, its *Journal* would not, as a matter of policy, accept papers of a purely theoretical nature; and Newlands therefore published his papers in Crookes' *Chemical News*, and Odling in the *Quarterly Journal of Science*, of which Crookes was also an editor. When Newlands read his paper, Carey Foster asked him whether he had tried, instead, arranging the elements in alphabetical order; this anecdote is sometimes told as an example of obscurantism, but Foster was really making the point that Newlands' groupings sometimes defied chemical analogies; they were not all natural groups, as we can see.

Odling's Table was a closer precursor of Mendeléeff's; but he seems to have spent little energy in forcing it upon the attention of chemists. Mendeléeff, on the other hand, with his very detailed predictions of undiscovered elements, had no doubt of the importance of his discovery, and staked his reputation upon it. It is remarkable that the Periodic Table seems to have aroused little interest in England for a decade, although Englishmen had been working along the same lines; perhaps Crookes was by 1869 too preoccupied with spiritualism to notice Mendeléeff's paper. Berthelot considered, quite unfairly, that Mendeléeff had been simply playing with numbers; but Mendeléeff had no difficulty in escaping this charge. When Ramsay and Rayleigh discovered argon, they were faced with the difficult problem of fitting it into the Table and predicting its congeners; a problem which was more acute because, since it apparently formed no compounds, it was difficult to determine its atomic weight from its vapour density; this was eventually done by computation from its specific heats, using the kinetic theory of gases, which indicated that there was but one atom in its molecule.

Chemists faced with the Table found themselves in the position of biologists looking at the Linnaean system. It was difficult to believe, particularly after the *Origin of Species* appeared in 1859, that all these families of related, and often closely related, elements were separate and immutable creations. Dumas had revived Prout's hypothesis before the British Association in 1851, encouraged by Faraday; J. H. Gladstone, in co-operation with the logician Augustus de Morgan, showed statistically that it was extremely unlikely that so many atomic weights would be nearly whole numbers as a result of chance. Then Crookes and the American geochemist Sterry Hunt proposed schemes of inorganic evolution; to be followed by Norman Lockyer, who believed that he had detected the synthesis of elements in the sun by means of the spectroscope. Lord Rayleigh's determination to put Prout's hypothesis to a final test by measuring the density of nitrogen led to the discovery that atmospheric and chemical nitrogen differed; in collaboration with Ramsay, Rayleigh showed that atmospheric nitrogen contained the gas argon. Ramsay soon afterwards discovered all the other members of the family, including helium which Lockyer had previously identified on the Sun.

The prevalence of the doctrine of inorganic evolution, sketched out in some detail by Crookes who even regarded the Rare Earth elements as the marsupials of the inorganic realm, prepared the way for the acceptance of Rutherford and Soddy's theory of radioactive disintegration; they saw radioactivity as the effect of subatomic chemical change. In Ramsay's laboratory at University College, London, the first transmutation in the annals of modern science was followed, as the emanation from radium—recognised as chemically an analogue of argon, and named radon—decayed to yield helium. In 1911, before the British Association, Ramsay predicted that radioactivity would provide a source of energy.

It has been suggested that Prout's fame is largely due to the shortness and memorability of his name, which formed a convenient label for all theories of the unity of matter; and indeed we can see that the new and quantitative part of his theory was soon falsified, whereas the general view of matter inherited from the corpuscularians persisted and is at least as much an ancestor of the atomism of our century as is the Daltonian chemical atomic theory. But Prout was historically important in that he did set off controversy, and he did attempt to provide a strictly testable version of the theory of the unity of matter. A feature of nineteenth century scientists which this story seems to reveal is their refusal to abandon a view of the world which was satisfactory on metaphysical grounds merely because a few chemical analyses seemed inconsistent with it. We are sometimes taught to regard such behaviour as unscientific; but in this case it led to the researches of J. J. Thomson on the electron, and to explanations of radioactivity.

BIBLIOGRAPHY

Alembic Club Reprints: XII, Faraday on the Liquefaction of Gases; XVIII, Cannizzaro, Sketch of a Course of Chemical Philosophy; XX, Prout's Hypothesis; all available in paperback.

C. Babbage, *Reflections on the Decline of Science in England*, London, 1830.

R. J. Boscovich, *A Theory of Natural Philosophy*, trans. J. M. Child, Cambridge, Mass., 1966

Robert Boyle on Natural Philosophy, ed. M. B. Hall, Bloomington, Indiana, 1965.

D. S. L. Cardwell (ed.), *John Dalton and the Progress of Science*, Manchester, 1968.

M. P. Crosland, *The Society of Arcueil*, London, 1967.

B. Z. Jones (ed.), *The Golden Age of Science*, New York, 1966.

D. M. Knight, *Atoms and Elements*, London, 1967; corrected reprint in press.
    Natural Science in English, 1600–1900, London, in press.

W. G. Palmer, *A History of the Concept of Valency*, Cambridge, 1965.

J. R. Partington, *A History of Chemistry*, III, London, 1962.

Royal Institute Library of Science, Physical Sciences, 10 vols. and index, London 1970.

C. A. Russell, *The History of Valency*, Leicester, in press.

R. E. Schofield, *A Scientific Autobiography of Joseph Priestley*, London, 1966.

B. Schonland, *The Atomists*, Oxford, 1968.

F. Szabadváry, *History of Analytical Chemistry*, trans. G. Svehla, Oxford, 1966.

T. I. Williams (ed.), *A Biographical Dictionary of Scientists*, London, 1969.

Section I

THE RISE AND FALL OF PROUT'S HYPOTHESIS

Thomas Thomson's *Annals of Philosophy* was one of the most important chemical journals of the second and third decades of the nineteenth century. In its pages Thomson pressed Dalton's atomic theory upon a world which was not at first very interested; and Dalton and Berzelius had an exchange over whether it was necessary to believe in atoms, in the course of which Berzelius proposed the notation which we still use in chemical equations and formulae. Dalton and Proust had convinced chemists that chemical compounds were the result of the combination of elements in fixed proportions by weight. Gay-Lussac then showed that gases combine in fixed and simple ratios by volume. Dalton did not see how this could be reconciled with his atomic theory; it seemed necessary to suppose that equal volumes of gases contained equal numbers of atoms, but then, if one volume of hydrogen and one of chlorine, for example, combined, there should be one volume of product: but there are two. So he rejected Gay-Lussac's experimental results, and maintained that there was no relation between atoms and volumes. Avogadro and Ampère saw the way out, and suggested that the hydrogen and chlorine existed in the form of molecules containing two—or an even number of—atoms; that equal volumes of gases contained equal numbers of molecules; and that the molecules of hydrogen and chlorine were divided into two when they combined.

For various good and bad reasons, their views, although widely known, were not generally accepted until after Cannizzaro presented them with great force and clarity at the Karlsruhe Conference of 1860. Instead, chemists quarrelled over whether to employ atomic weights or equivalents; and in dealing with gases whether weight or volume changes were more fundamental. It was against a background of debate on whether to use atoms or volumes that Prout's hypothesis was advanced; which explains its curious form, in which the hypothesis is almost submerged in measurements of specific gravity, and questions as to whether one volume of hydrogen corresponds to one atom or two.

15

Thomson's analyses were first published in *Annals of Philosophy*; by 1825 when the *First Principles* were published he was at the apex of his career; Professor at Glasgow, and the author of a textbook which had received the ultimate accolade for that date of being translated into French, the language of Lavoisier and his associates. His book received quite good reviews until, in 1828, Berzelius' violent onslaught was published. Thomson replied in kind to Berzelius; these were times of plain speaking. But Berzelius was superior to Thomson as an analyst both in technique and judgment; and when Edward Turner, first professor of Chemistry at University College, London, confirmed Berzelius' results, the whole edifice which Thomson had constructed in the *First Principles* fell down.

Prout himself backed away from this hypothesis in a letter written to Charles Daubeny, the Professor of Chemistry at Oxford, who in 1831 published a book on the Atomic Theory. Prout wrote that 'there seems to be no reason why bodies still lower in the scale than hydrogen . . . may not exist, of which other bodies may be multiples, without being actually multiples, of the intermediate hydrogen'; though he admitted that his earlier speculations had gone further. In his Bridgewater Treatise of 1834, on *Chemistry, Meteorology, and the Function of Digestion*, as proving the existence of God, he elaborated somewhat this idea of sub-molecules.

Berzelius and Turner had published their refutations in the *Philosophical Magazine*, which is a very valuable journal for the historian of the period. It contains articles written for a specialist and lay audience; reports of the meetings of scientific societies; reprints and translations of articles which had appeared elsewhere; and reviews of books. In America a similar function was fulfilled by Benjamin Silliman's *American Journal of Science*, which by the end of the nineteenth century was chiefly devoted to biological science, but earlier was very catholic in its coverage. Silliman was Professor of Chemistry at Yale, and, while undistinguished as a scientist, he thus played an important part in establishing science in the USA.

FURTHER READING

W. H. Brock, 'Prout's Hypothesis', *Annals of Science*, XXV (1969), 49–80, 127–37.

M. Daumas, 'Precision of Measurement . . . in the Eighteenth Century', in A. C. Crombie (ed.), *Scientific Change*, London, 1963, pp. 418–30.

W. V. Farrar, 'Nineteenth century Speculations on the Complexity of the Chemical Elements', *British Journal for the History of Science*, II (1965), 297–323; IV (1968), 65–7.

H. Hartley, *Studies of the History of Chemistry*, Oxford, in press.

J. E. Jorpes, *Jac. Berzelius, His Life and Work*, trans. B. Steele, Stockholm, 1966.

J. B. Morrell, 'Practical Chemistry in the University of Edinburgh, 1799–1843', *Ambix*, XVI (1969), 66–80; article on Thomson at Glasgow, forthcoming in the *British Journal for the History of Science*.

N. Reingold, *Science in Nineteenth Century America*, London, 1966.

ANNALS

OF

PHILOSOPHY.

NOVEMBER, 1815.

ARTICLE I.

On the Relation between the Specific Gravities of Bodies in their Gaseous State and the Weights of their Atoms.

THE author of the following essay submits it to the public with the greatest diffidence; for though he has taken the utmost pains to arrive at the truth, yet he has not that confidence in his abilities as an experimentalist as to induce him to dictate to others far superior to himself in chemical acquirements and fame. He trusts, however, that its importance will be seen, and that some one will undertake to examine it, and thus verify or refute its conclusions. If these should be proved erroneous, still new facts may be brought to light, or old ones better established, by the investigation; but if they should be verified, a new and interesting light will be thrown upon the whole science of chemistry.

It will perhaps be necessary to premise that the observations about to be offered are chiefly founded on the doctrine of volumes as first generalized by M. Gay-Lussac; and which, as far as the author is aware at least, is now universally admitted by chemists.

On the Specific Gravities of the Elementary Gases.

1. *Oxygen and Azote.*—Chemists do not appear to have considered atmospheric air in the light of a compound formed upon chemical principles, or at least little stress has been laid upon this circumstance. It has, however, been long known to be constituted by bulk of four volumes of azote and one volume of oxygen; and if we consider the atom of oxygen as 10, and the atom of azote as 17·5, it will be found by weight to consist of one atom of oxygen and two atoms of azote, or per cent. of

VOL. VI. N° V.   X

Oxygen .... 22·22

Azote .... 77·77

Hence, then, it must be considered in the light of a pure chemical compound; and indeed nothing but this supposition will account for its uniformity all over the world, as demonstrated by numerous experiments. From these data the specific gravities of oxygen and azote (atmospheric air being 1·000) will be found to be,*

Oxygen .... 1·1111

Azote .... ·9722

2. *Hydrogen.*—The specific gravity of hydrogen, on account of its great levity, and the obstinacy with which it retains water, has always been considered as the most difficult to take of any other gas. These obstacles made me (to speak in the first person) despair of arriving at a more just conclusion than had been before obtained by the usual process of weighing; and it occurred to me that its specific gravity might be much more accurately obtained by calculation from the specific gravity of a denser compound into which it entered in a known proportion. Ammoniacal gas appeared to be the best suited to my purpose, as its specific gravity had been taken with great care by Sir H. Davy, and the chance of error had been much diminished from the slight difference between its sp. gr. and that of steam. Moreover, Biot and Arrago had obtained almost precisely the same result as Sir H. Davy. The sp. gr. of ammonia, according to Sir H. Davy, is ·590164, atmospheric air being 1·000. We shall consider it as ·5902; and this we are authorized in doing, as Biot and Arrago state it somewhat higher than Sir H. Davy. Now ammonia consists of three volumes of hydrogen and one volume of azote condensed into two volumes. Hence the sp. gr. of hydrogen will be found to be ·0694,† atmospheric air being 1·0000. It will be also observed that the sp. gr. of oxygen as obtained above is just 16 times that of hydrogen as now ascertained, and the sp. gr. of azote just 14 times.‡

3. *Chlorine.*—The specific gravity of muriatic acid, according to Sir H. Davy's experiments, which coincide exactly with those of

---

* Let x = sp. gr. of oxygen. 22·22 = a

  y = sp. gr. of azote. 77·77 = b

Then $\dfrac{x + 4y}{5} = 1$.

And $x : 4y :: a : b$.

Hence $5 - 4y = \dfrac{4ay}{b}$

And $y = \dfrac{5b}{4a + 4b} = \cdot9722$. And $x = 5 - 4y = 1\cdot111\text{m}$.

† Let x = sp. gr. of hydrogen.

Then $\dfrac{3x + \cdot9722}{2} = \cdot5902$.

Hence $x = \dfrac{1\cdot1804 - \cdot9722}{3} = \cdot0694$.

‡ $1\cdot11111 \div \cdot0694 = 16$. And $\cdot9722 \div \cdot0694 = 14$.

Biot and Arrago, is 1·278. Now if we suppose this sp. gr. to be erroneous in the same proportion that we found the sp. gr. of oxygen and azote to be above, (which, though not rigidly accurate, may yet be fairly done, since the experiments were conducted in a similar manner), the sp. gr. of this gas will come out about 1·2845;* and since it is a compound of one volume chlorine and one volume hydrogen, the specific gravity of chlorine will be found by calculation to be 2·5.† Dr. Thomson states, that he has found 2·483 to be near the truth, ‡ and Gay-Lussac almost coincides with him. § Hence there is every reason for concluding that the sp. gr. of chlorine does not differ much from 2·5. On this supposition, the sp. gr. of chlorine will be found exactly 36 times that of hydrogen.

On the Specific Gravities of Elementary Substances in a Gaseous State that do not at ordinary Temperatures exist in that State.

1. *Iodine.*—I had some reason to suspect that M. Gay-Lussac had in his excellent memoir rated the weight of an atom of this substance somewhat too high; and in order to prove this 50 grains of iodine, which had been distilled from lime, were digested with 30 grs. of very pure lamellated zinc. The solution formed was transparent and colourless; and it was found that 12·9 grains of zinc had been dissolved. 100 parts of iodine, therefore, according to this experiment, will combine with 25·8 parts of zinc, and the weight of an atom of iodine will be 155,‖ zinc being supposed to be 40. From these data, the sp. gr. of iodine in a state of gas will be found by calculation to be 8·611111, or exactly 124 times that of hydrogen.**

2. *Carbon.*—I assume the weight of an atom of carbon at 7·5. Hence the sp. gr. of a volume of it in a state of gas will be found by calculation to be ·4166, or exactly 12 times that of hydrogen.

3. *Sulphur.*—The weight of an atom of sulphur is 20. Hence the specific gravity of its gas is the same as that of oxygen, or 1·1111, and consequently just 16 times that of hydrogen.

---

* As 1·104 : 1·11111 :: 1·278 : 1·286.
And as ·969 : ·9722 :: 1·278 : 1·283. The mean of these is 1·2845.
† Let x = sp. gr. of chlorine.
Then $\dfrac{x + ·0694}{2} = 1·2845$.
And $x = 2·569 - ·0694 = 2·5$ very nearly.
‡ *Annals of Philosophy*, vol. iv. p. 13.
§ Ditto, vol. vi. p. 126.
‖ As 25·8 : 100 :: 40 : 155. According to experiment 8th, stated below, the weight of an atom of zinc is 40. Dr. Thomson makes it 40·9, which differs very little. See *Annals of Philosophy*, vol. iv. p. 94.
** One volume of hydrogen combines with only half a volume of oxygen, but with a whole volume of gaseous iodine, according to M. Gay-Lussac. The ratio in volume, therefore, between oxygen and iodine is as ½ to 1, and the ratio in weight is as 1 to 15·5. Now ·5555, the density of half a volume of oxygen, multiplied by 15·5, gives 8·61111, and 8·61111 ÷ ·06944 = 124. Or generally, to find the sp. gr. of any substance in a state of gas, we have only to multiply half the sp. gr. of oxygen by the weight of the atom of the substances with respect to oxygen. See *Annals of Philosophy*, vol. v. p. 105.

x 2

4. *Phosphorus.*—I have made many experiments in order to ascertain the weight of an atom of this substance; but, after all, have not been able to satisfy myself, and want of leisure will not permit me to pursue the subject further at present. The results I have obtained approached nearly to those given by Dr. Wollaston, which I am therefore satisfied are correct, or nearly so, and which fix phosphorus at about 17·5, and phosphoric acid at 37·5, * and these numbers at present I adopt.

5. *Calcium.*—Dr. Marcet found carbonate of lime composed of 43·9 carbonic acid and 56·1 lime. † Hence as 43·9 : 56·1 :: 27·5 : 35·1, or 35 very nearly; and 35 — 10 = 25, for the atom of calcium. The sp. gr. of a volume of its gas will therefore be 1·3888, or exactly 20 times that of hydrogen.

6. *Sodium.*—100 grains of dilute muriatic acid dissolved 18·6 grs. of carbonate of lime, and the same quantity of the same dilute acid dissolved only 8·2 grs. of carbonate of lime, after there had been previously added 30 grs. of a very pure crystallized subcarbonate of soda. Hence 30 grs. of crystallized subcarbonate of soda are equivalent to 10·4 grs. of carbonate of lime, and as 10·4 : 30 :: 62·5 : 180. Now 100 grs. of crystallized subcarbonate of soda were found by application of heat to lose 62·5 of water. Hence 180 grs. of the same salt contain 112·5 water, equal to 10 atoms, and 67·5 dry subcarbonate of soda, and 67·5 — 27·5 = 40 for the atom of soda, and 40 — 10 = 30 for the atom of sodium. Hence a volume of it in a gaseous state will weigh 1·6666, or exactly 24 times that of hydrogen.

7. *Iron.*—100 grs. of dilute muriatic acid dissolved as before 18·6 grs. of carbonate of lime, and the same quantity of the same acid dissolved 10·45 of iron. Hence as 18·6 : 10·45 :: 62·5 : 35·1, or for the sake of analogy, 35, the weight of an atom of iron. The sp. gr. of a volume of this metal in a gaseous state will be 1·9444, or exactly 28 times that of hydrogen.

8. *Zinc.*—100 grs. of the same dilute acid dissolved, as before, 18·6 of carbonate of lime and 11·85 of zinc. Hence as 18·6 : 11·85 :: 62·5 : 39·82, the weight of the atom of zinc, considered from analogy to be 40. Hence the sp. gr. of a volume of it in a gaseous state will be 2·222, or exactly 32 times that of hydrogen.

9. *Potassium.*—100 grs. of the same dilute acid dissolved, as before, 18·6 carbonate of lime; but after the addition of 20 grs. of super-carbonate of potash, only 8·7 carbonate of lime. Hence 20 grs. of super-carbonate of potash are equivalent to 9·9 carbonate of lime; and as 9·9 : 20 :: 62·5 : 126·26, the weight of the atom of super-carbonate of potash. Now 126·26 — 55 + 11·25 = 60, the

---

* Some of my experiments approached nearer to 20 phosphorus and 40 phosphoric acid.

† I quote on the authority of Dr. Thomson, *Annals of Philosophy.* vol. iii. p. 376. Dr. Wollaston makes it somewhat different, or that carbonate of lime consists of 43·7 acid and 56·3 lime. Phil. Trans. vol. civ. p. 8.

8

weight of the atom of potash, and 60 — 10 = 50, the weight of the atom of potassium.   Hence a volume of it in a state of gas will weigh 2·7777, or exactly 40 times as much as hydrogen.

10. *Barytium.*—100 grs. of the same dilute acid dissolved exactly as much again of carbonate of barytes as of carbonate of lime. Hence the weight of the atom of carbonate of barytes is 125; and 125 — 27·5 = 97·5, the weight of the atom of barytes, and 97·5 — 10 = 87·5, the weight of the atom of barytium.   The sp. gr. therefore, of a volume of its gas will be 4·8611, or exactly 70 times that of hydrogen.

With respect to the above experiments, I may add, that they were made with the greatest possible attention to accuracy, and most of them were many times repeated with almost precisely the same results.

The following tables exhibit a general view of the above results, and at the same time the proportions, both in volume and weight, in which they unite with oxygen and hydrogen : also the weights of other substances, which have not been rigidly examined, are here stated from analogy.

TABLE I.—Elementary Substances.

Name.	Sp. gr. hydr. being 1.	Wt. of atom, 2 vols. hydr. being 1.	Wt. of atom, oxygen being 10.	Wt. of atom, oxygen being 10, from experiment.	Sp. gr. atmospheric air being 1.	Sp. gr. atmospheric air being 1, from experiment.	Wt. in grs. of 100 cub. inches. Barom. 30, Therm. 60.	Wt. in grs. of 100 cub. in. from exper.
Hydrogen	1	1	1·25	1·32 [2]	·06944	·073 [1]	2·118	2·23
Carbon	6	6	7·5	7·54 [2]	·4166	—	12·708	—
Azote	14	14	17·5	17·54	·9722	·969 [3]	29·652	29·56
Phosphorus	14	14	17·5	17·4 [4]	·9722	—	29·652	—
Oxygen	16	8	10	10	1·1111	1·104 [5]	33·888	33·612
Sulphur	16	16	20	20 [6]	1·1111	—	33·888	—
Calcium	20	20	25	25·467	1·3888	—	42·36	—
Sodium	24	24	30	29·1 [8]	1·6666	—	50·892	—
Iron	28	28	35	34·5 [9]	1·9444	—	59·302	—
Zinc	32	32	40	41 [10]	2·222	—	67·777	—
Chlorine	36	36	45	44·1 [11]	2·5	2·483 [12]	76·248	—
Potassium	40	40	50	49·1 [13]	2·7777	—	84·72	—
Barytium	70	70	87·5	87 [14]	4·8611	—	148·26	—
Iodine	124	124	155	156·21 [15]	8·6111	—	262·632	—

Observations.

[1] Dr. Thomson. See *Annals of Philosophy*, i, 177.
[2] Dr. Wollaston, from Biot and Arrago. Phil. Trans. civ. 20. Dr. Thomson makes it 7·51. *Annals of Philosophy*, ii. 42.
[3] Dr. W. from Biot and Arrago.
[4] Dr. W. from Berzelius and Rose.
[5] Dr. Thomson, from a mean of several experiments.
[6] Dr. W. from Berzelius.
[7] Dr. W. from experiment.
[8] Dr. W. from Davy.
[9] Dr. W. from Thenard and Berzelius.
[10] Dr. W. from Gay-Lussac.
[11] Dr. W. from Berzelius.
[12] Quoted from Dr. Thomson, *Annals of Philosophy*, iv. 13.
[13] Dr. W. from Berzelius.
[14] Dr. W. from Berzelius and Klaproth.
[15] Gay-Lussac. Ann. de Chim. xci. 5.

TABLE II.—Combinations with Oxygen.

Name.	Sp. Gr. hydro. being 1.	Wt. of atom, 2 vol. hydro. being 1.	Wt. of atom, ox. being 16.	Wt. of atom, ox. being 10, from exper.	Sp. Gr. air being 1, atmos.	Sp. Gr. atmos. air being 1, from exper.	Wt. of 100 cu. in. Bar. 30, Ther. 60.	Wt. of 100 cu. in. from exp.	Elements by volume.	No. of vol. after combination.	Elements by weight.	Observations.
Water	9	9	11·25	11·32	·625	·6896 [1]	19·062	21·033	·5 ox + 1 hyd	1	1 ox + 1 hy	[1] Trales, Dr. Thomson, *Annals*, i. 177.
Carbonic oxyde	14	14	17·5	17·54	·9722	·956 [2]	29·652	29·16	·5 ox + 1 ca	1	1 ox + 1 car	[2] Cruikshanks, quoted by Thomson.
Nitrous oxyde	22	22	27·5	—	1·5277	1·614 [3]	46·596	49·227	·5 ox + 1 az	1	1 ox + 1 az	[3] Sir H. Davy.
Atmospheric air	14·4	36	45	—	1·000	1·000	30·5	30·5 [4]	·5 ox + 2 az	2·5	1 ox + 2 az	[4] Sir G. S. Evelyn.
Phosphorous acid									·5 ox + 1 ph?		1 ox + 1 ph?	
Oxyde of sulphur?									·5 ox + 1 sul?		1 ox + 1 sul?	
Euchlorine	44	44	55	35·46	3·0555	2·409 [5]	93·192	73·474	·5 ox + 1 ch	1 ?	1 ox + 1 ch	[5] Sir H. Davy.
									·5 ox + 1 iod		1 ox + 1 iod	
Lime	28	28	35	—	1·9444	—	59·304	—	·5 ox + 1 cal	1 ?	1 ox + 1 cal	
									·5 ox + &c.		&c.	
Carbonic acid	22	22	27·5	27·54	1·5277	1·518 [7]	46·596	46·313	ox + 1 hy [6]	1	2 ox + 1 hy	[6] This and all higher combinations of hydrogen with oxygen are unknown.
Nitrous gas	15	30	37·5	—	1·0416	1·0388 [8]	31·77	31·684	ox + 1 car	2	2 ox + 1 car	[7] Saussure.
Phosphoric acid	30	30	37·5	37·4	2·0832	—	63·54	—	ox + 1 az		2 ox + 1 az	[8] Berard.
Sulphurous acid	32	32	40	—	2·2222	2·193 [9]	67·777	66·89	ox + 1 ph		2 ox + 1 ph	[9] Sir H. Davy.
									ox + 1 sul		2 ox + 1 sul	
									ox + 1 ch		2 ox + 1 ch	
									ox + 1 iod		2 ox + 1 iod	
									ox + &c.		&c.	
Nitrous acid	38	38	47·5	—	2·6388	2·427 [10]	80·484	74·0234	1·5 ox + 1 car	1	3 ox + 1 car	[10] Sir H. Davy.
Sulphuric acid	40	40	50	50	2·7777	—	84·72	—	1·5 ox + 1 az	1	3 ox + 1 az	
									1·5 ox + 1 ph		3 ox + 1 ph	
									1·5 ox + 1 sul		3 ox + 1 sul	
									1·5 ox + 1 ch		3 ox + 1 ch	
									1·5 ox + 1 iod		3 ox + 1 iod	
									&c.		&c.	
Nitric acid	54	54	67·5	67·51	3·75	—	114·372	—	2·5 ox + 1 car	1	5 ox + 1 car	See Gay-Lussac's memoir on iodine above referred to.
Chloric acid	76	76	95	—	5·2777	—	160·968	—	2·5 ox + 1 az	1	5 ox + 1 az	
Iodic acid	164	164	205	—	11·3883	—	347·352	—	2·5 ox + 1 ph		5 ox + 1 ph	
									2·5 ox + 1 sul		5 ox + 1 sul	
									2·5 ox + 1 ch		5 ox + 1 ch	
									2·5 ox + 1 iod		5 ox + 1 iod	
									&c.		&c.	

TABLE III.—Compounds with Hydrogen.

Name.	Sp. gr. hydro. being 1.	Wt. of atom, 2 vol. hydr. being 1.	Wt. of atom, oxygen being 10.	Wt. of atom, oxygen being 10, from experiment.	Sp. gr. atmospheric air being 1.	Sp. gr. atmospheric air being 1, from experiment.	Wt. of 100 cub. inch, Bar. 30. Ther. 60.	Wt. of 100 cub. inch, from exper.	Elements by volume.	No. of vol. after combination.	Elements by weight.	Observations.
Carbureted hydrogen	8	7	8·75	8·86	·5555	·5555 [1]	16·999	16·999	2 hy + 1 car	1	1 hy + 1 car	[1] Dr. Thomson.
Olefiant gas	14	13	16·25	16·4	·9722	·974 [2]	29·652	29·72	2 hy + 2 car	1	1 hy + 2 car	[2] Ditto.
Hydro-phosphorus gas									1 hy + 1 az		·5 hy + 1 az	⎫ I have omitted these from the uncertainty that still hangs over phosphorus.
Phosphoreted hydrogen												⎭ This compound is at present unknown, but it probably exists in fulminating gold, silver, &c. united to these metals.
Ammonia	8·5	15·5	19·375	21·5 [3]	·5902	·59 [3]	18·003	18·00	3 hy + 1 az	2	1·5 hy + 1 az	[3] Dr. Wollaston.
Sulphureted hydrogen	17	16·5	20·625	20·66	1·1805	1·177 [4]	36·006	35·89	1 hy + 1 sul	1	·5 hy + 1 sul	[4] Sir H. Davy.
Muriatic acid	18·5	36·5	45·625	45·66	1·284	1·278 [5]	39·183	38·979	1 hy + 1 ch	2	·5 hy + 1 ch	[5] Ditto.
Hydriodic acid	62·5	124·5	155·625	155·66	4·3402	4·3463 [6]	132·375		1 hy + 1 iode	2	·5 hy + 1 iod	[6] Gay-Lussac.

TABLE IV.—*Substances stated from Analogy, but of which we are yet uncertain.*

Name.	Sp. gr. hydr. being 1.	Wt. of atom, 2 vol. hydr. being 1.	Wt. of atom, oxygen being 10.	Wt. of atom, ox. being 10, from exper.	Observations.
Aluminum	8	8	10	10·68 [1]	[1] Berzelius.
Magnesium	12	12	15	14·6 [2]	[2] Henry. Berzelius makes it 15·77.
Chromium	18	18	22·5	23·6 [3]	[3] Berzelius.
Nickel	28	28	35	36·5 [4]	[4] Ditto.
Cobalt	28	28	35	36·6 [5]	[5] Rolhoff.
Tellurium......	32	32	40	40·27 [6]	[6] Berzelius.
Copper.........	32	32	40	40 [7]	[7] As deduced by Dr. Thomson.
Strontium......	48	48	60	59 [8]	[8] Klaproth.
Arsenic........	48	48	60	60 [9]	[9] Berzelius.
Molybdenum ..	48	48	60	60·13 [10]	[10] Bucholz and Berzelius.
Manganese	56	56	70	71·15 [11]	[11] Berzelius.
Tin............	60	60	75	73·5 [12]	[12] Ditto.
Bismuth	72	72	90	89·94 [13]	[13] Ditto.
Antimony......	88	88	110	111·11 [14]	[14] Ditto. Dr. Thomson makes it 112·49.
Cerium	92	92	115	114·87 [15]	[15] Hisinger.
Uranium	96	96	120	120 [16]	[16] Bucholz.
Tungsten	96	96	120	121·21 [17]	[17] Berzelius.
Platinum	96	96	120	121·66 [18]	[18] Ditto.
Mercury	100	100	125	125 [19]	[19] Fourcroy and Thenard.
Lead	104	104	130	129·5 [20]	[20] Berzelius.
Silver..........	108	108	135	135 [21]	[21] Wenzel and Davy.
Rhodium	120	120	150	149·03 [22]	[22] Berzelius.
Titanium	144	144	180	180·1 [23]	[23] Ditto.
Gold	200	200	250	249·68 [24]	[24] Ditto.

Observations.

Table I.—This, as well as the other tables, will be easily understood. In the first column we have the specific gravities of the different substances in a gaseous state, hydrogen being 1 : and if we suppose the volume to be 47·21435 cubic inches, the numbers will at the same time represent the number of grains which this quantity of each gas will weigh. In the third column are the corrected numbers, the atom of oxygen being supposed, according to Dr. Thomson, Dr. Wollaston, &c. to be 10 : and in the fourth, the same, as obtained by experiment, are stated, to show how nearly they coincide. Of the individual substances mentioned, I have no remark to make, except with respect to iodine. I made but one experiment to ascertain the weight of the atom of this substance, and therefore the results stated may be justly considered as deserving but little confidence ; and indeed this would be the case, did not all the experiments of Gay-Lussac nearly coincide in the same.

Table II.—This table exhibits many striking instances of the near coincidence of theory and experiment. It will be seen that Gay-Lussac's views are adopted, or rather indeed anticipated, as a good deal of this table was drawn up before I had an opportunity of seeing the latter part of that chemist's memoir on iodine. That table also exhibits one or two striking examples of the errors that have arisen from not clearly understanding the relation between the doctrine of volumes and of atoms. Thus ammonia has been stated to be composed of one atom of azote and three of hydrogen, whereas it is evidently composed of one atom of azote and only 1·5 of hydrogen, which are condensed into two volumes, equal therefore to one atom; and this is the reason why this substance, like some others, apparently combine in double proportions. *

Table III.—This table likewise exhibits some striking examples of the coincidence above noticed. Indeed, I had often observed the near approach to round numbers of many of the weights of the atoms, before I was led to investigate the subject. Dr. Thomson appears also to have made the same remark. It is also worthy of observation, that the three magnetic metals, as noticed by Dr. Thomson, have the same weight, which is exactly double that of azote. Substances in general of the same weight appear to combine readily, and somewhat resemble one another in their nature.

On a general review of the tables, we may notice,

1. That all the elementary numbers, hydrogen being considered as 1, are divisible by 4, except carbon, azote, and barytium, and these are divisible by 2, appearing therefore to indicate that they are modified by a higher number than that of unity or hydrogen. Is the other number 16, or oxygen? And are all substances compounded of these two elements?

2. That oxygen does not appear to enter into a compound in the ratio of two volumes or four atoms.

3. That all the gases, after having been dried as much as possible, still contain water, the quantity of which, supposing the present views are correct, may be ascertained with the greatest accuracy.

Others might doubtless be mentioned; but I submit the matter for the present to the consideration of the chemical world.

---

* See Gay-Lussac's memoir on iodine, *Annals of Philosophy*, vi. 189.

ARTICLE VII.

Correction of a Mistake in the Essay on the Relation between the Specific Gravities of Bodies in their Gaseous State and the Weights of their Atoms.

THE author of the essay On the Relation between the Specific Gravities of Bodies in their Gaseous State and the Weights of their Atoms is anxious to correct an oversight which influences some of the numbers in the third table given in that essay (vol. vi. p. 328). This oversight will be found in the head or title of the third column in each table, and consists in the statement of the atom of hydrogen being composed of two volumes instead of one, upon which latter supposition the tables are actually constructed, except in the instances corrected in the third table as follows, and in a sentence in the first paragraph on p. 330, beginning "This table also exhibits," &c. which is to be expunged.

Name.	Sp. gr. hydro, being 1.	Wt. of atom, hydrogen being 1.	Wt. of atom, oxygen being 10.	Wt. of atom, oxygen being 10, from experiment.	Sp. gr. atmospheric air being 1.	Sp. gr. atmospheric air being 1, from experiment.	Wt. of 100 cub. inch, Bar. 30, Ther. 60.	Wt. of 100 cub. inch, from exper.	Elements by volume.	No. of vol. after condensation.	Elements by weight.
Carbureted hydrogen	8	4	5	5·09	·5555	·5555 [1]	16·999	16·999	1 hyd + ·5car	·5	1 hyd + ·5car
Olefiant gas	14	7	8·75	8·86	·9722	·9740 [1]	29·652	29·72	1 hyd + 1 car	·5	1 hyd + 1 car
Sulphureted hydrogen	17	17	21·25	21·32	1·1805	1·177	36·006	35·89	1 hyd + 1 sul	1	1 hyd + 1 sul
Muriatic acid	18·5	37	46·25	45·42	1·284	1·278	39·183	38·979	1 hyd + 1 chl	2	1 hyd + 1 chl
Hydriodic acid	62·5	125	156·25	157·53 [3]	4·3402	4·3463 [3]	132·375	—	1 hyd + 1 iod	2	1 hyd + 1 iod
Ammonia	8·5	17	21·25	21·5 [3]	·59024	·5900	18·003	18·000	3 hyd + 1 az	2	3 hyd + 1 az
Cyanogen	26	26	32·5	32·52	1·8055	1·8064 [5]	55·068		2 car + 1 az	1	2 car + 1 az
Hydro-cyanic acid	13·5	27	33·75	33·846	·9374	·9360 [5]	28·593		1 cya + 1 hy	2	1 cya + 1 az
Chloro-cyanic acid	31	62	77·5	—	2·1527	2·1111 [5]	65·659		1 cya + 1 chl	2	1 cya + 1 chl

Observations.
[1] Dr. Thomson.
[2] Gay-Lussac.
[3] Dr. Wollaston.
[4] Sir H. Davy.
[5] Gay-Lussac. Ann. de Chim. Aug. 1815.

In this table it will be also observed that the new determinations of Gay-Lussac respecting the prussic acid, &c. are inserted, to show that they correspond with, and further corroborate, the views which have been brought forward in the essay above referred to.

There is an advantage in considering the volume of hydrogen equal to the atom, as in this case the specific gravities of most, or perhaps all, elementary substances (hydrogen being 1) will either exactly coincide with, or be some multiple of, the weights of their atoms; whereas if we make the volume of oxygen unity, the weights of the atoms of most elementary substances, except oxygen, will be double that of their specific gravities with respect to hydrogen. The assumption of the volume of hydrogen being equal to the atom will also enable us to find more readily the specific gravities of bodies in their gaseous state (either with respect to hydrogen or atmospheric air), by means of Dr. Wollaston's logometric scale.

If the views we have ventured to advance be correct, we may almost consider the πρώτη ὕλη of the ancients to be realised in hydrogen; an opinion, by the by, not altogether new. If we actually consider this to be the case, and further consider the specific gravities of bodies in their gaseous state to represent the number of volumes condensed into one; or, in other words, the number of the absolute weight of a single volume of the first matter (πρώτη ὕλη) which they contain, which is extremely probable, multiples in weight must always indicate multiples in volume, and *vice versâ*; and the specific gravities, or absolute weights of all bodies in a gaseous state, must be multiples of the specific gravity or absolute weight of the first matter (πρώτη ὕλη), because all bodies in a gaseous state which unite with one another unite with reference to their volume.

CHAP. XIX.

SOME GENERAL OBSERVATIONS ON THE ATOMIC WEIGHTS OF CHEMICAL BODIES.

1. HYDROGEN is the lightest of all known bodies. Its atomic weight is 0·125.

2. The atomic weights of all the other simple bodies are multiples of 0·25 = 2 atoms of hydrogen. Consequently, the number denoting the atomic weight of each is either a whole number, or a mixed number, ending with one or other of these decimals :—0·25, 0·5, 0·75.

3. The atomic weights of the supporters are as follows :

Oxygen	1
Chlorine	4·5
Iodine	15·5

They are all multiples of 0·5, or of 4 atoms of hydrogen.

4. The atomic weights of the acidifiable combustibles are as follows (omitting hydrogen) :

Acidifiable combustibles.

1. Carbon	.	.	0·75	6. Sulphur .	2
2. Boron	.	.	1	7. Tellurium .	4
3. Silicon	.	.	1	8. Arsenic .	4·75
4. Phosphorus	.	1·5		9. Selenium .	5
5. Azote	.	.	1·75		

9

They are all whole numbers, or terminate in 0·5 or 0·75—they never terminate in 0·25.

> 5 are multiples of 8 atoms of hydrogen.
> 1 is a multiple of 4 — —
> 3 are multiples of 2 — —

Intermediate combustibles.

5. The atomic weights of the intermediate combustibles are as follows:

1. Chromium	.	3·5	5. Tungsten	.	15·75
2. Titanium	.	4	6. Columbium	.	18
3. Antimony	.	5·5	7. Uranium	.	26
4. Molybdenum	.	6			

These, like the former set, are either whole numbers, or they end in 0·5 or 0·75, and never in 0·25.

> 4 are multiples of 8 atoms hydrogen.
> 2 are multiples of 4 — —
> 1 is a multiple of 2 — —

Alkalifiable combustibles.

6. The atomic weights of the alkalifiable combustibles are as follows:

1. Lithium	.	1·25	10. Manganese	.	3·5
2. Aluminum	.	1·25	11. Iridium	.	3·75
3. Magnesium	.	1·5	12. Copper	.	4
4. Glucinum	.	2·25	13. Yttrium	.	4·25
5. Calcium	.	2·5	14. Zinc	.	4·25
6. Sodium	.	3	15. Potassium	.	5
7. Nickel	.	3·25	16. Zirconium	.	5
8. Cobalt	.	3·25	17. Strontium	.	5·5
9. Iron	.	3·5	18. Rhodium	.	5·5

19.	Cerium	6·25	25.	Platinum	12
20.	Cadmium	7	26.	Lead	13
21.	Palladium	7	27.	Silver	13·75
22.	Tin	7·25	28.	Mercury	25
23.	Barium	8·75	29.	Gold	25
24.	Bismuth	9			

They are either whole numbers, or they terminate in 0·25 or 0·5 ; and very rarely in 0·75.

11 are multiples of 8 atoms of hydrogen,
6 are multiples of 4 — —
12 are multiples of 2 — —

Only 3 of them end in 0·75; namely, iridium, barium, and silver.

7. The simple combustibles (excluding hydrogen) amount to 44.   Of these

20 (or almost the half) are multiples of oxygen, and conseqently whole numbers.

9 are multiples of 4 atoms of hydrogen, and consequently terminate in 0·5.

9 terminate in 0·25, being multiples of 2 atoms of hydrogen.

6 terminate in 0·75, being likewise multiples of 2 atoms hydrogen.

8. When an acidifiable combustible terminates in 0·75, the acid, which it forms, does not act with much energy; but when an alkalifiable combustible terminates in 0·75 the base formed is usually energetic.   How far this applies to iridium is uncertain ; but it applies well both to barytes and oxide of silver.

16

Acids.

9. The acids are all compound bodies. They may, with reference to their atomic weights, be divided into three classes: 1. Combinations of a simple acidifiable or intermediate combustible with oxygen. 2. Acids, consisting of oxygen united to two bases at once, or undecomposed acids. 3. Hydracids, consisting of a simple or compound supporter united to an atom of hydrogen. The following table exhibits the atomic weights of all these acids:

1. OXYGEN UNITED TO A SINGLE BASE.

1. Silicic	.	2	14. Chromic	.	6·5
2. Phosphorous	.	2·5	15. Arsenious	.	6·75
3. Carbonic	.	2·75	16. Nitric	.	6·75
4. Boracic	.	3	17. Selenic	.	7
5. Hyposulphurous		3	18. Antimonic	.	7·5
6. Phosphoric	.	3·5	19. Manganesic	.	7·5
7. Sulphurous	.	4	20. Arsenic	.	7·75
8. Oxalic	.	4·5	21. Molybdous	.	8
9. Hyponitrous acid		4·75	22. Hyposulphuric		9
10. Sulphuric	.	5	23. Molybdic	.	9
11. Oxide of tellurium		5	24. Tungstic	.	18·75
12. Nitrous acid	.	5·75	25. Columbic	.	19
13. Titanic	.	6	26. Uranitic	.	28

2. OXYGEN UNITED TO A DOUBLE BASE—AND UNDECOMPOSED ACID.

1. Fluoric	.	1·25	6. Citric	.	7·25
2. Fluosilicic	.	·3·25	7. Tartaric	.	8·25
3. Fluoboric	.	4·25	8. Uric	.	9
4. Acetic	.	6·25	9. Saclactic	.	13
5. Succinic	.	6·25	10. Benzoic	.	15

3. HYDRACIDS.

1. Hydrocianic .	3·375	3. Muriatic .	4·625
2. Formic .	4·625	5. Hydriodic .	15·625

Of the first set 12 are whole numbers ; 6 end in 0·5 and 4 in 0·75. Or, in other words,

> 12 are multiples of oxygen,
> 6 are multiples of 4 atoms of hydrogen,
> 4 are multiples of 2 atoms of hydrogen.

Of the second set 3 are whole numbers, and the other 7 end in 0·25. Or, in other words,

> 3 are multiples of 1 atom of oxygen,
> 7 are multiples of 2 atoms of hydrogen.

The third set containing an atom of hydrogen united to an atom of a simple or compound supporter or acid, it is obvious that they can only be multiples of 1 atom of hydrogen. Accordingly, they end either in 0·625 or in 0·375.

10. The atomic weights of the bases (excluding ammonia) are as follows : Bases.

1. Lithia	.	2·25	10. Protoxide of man-	
2. Alumina	.	2·25	ganese .	4·5
3. Magnesia	.	2·5	11. Protoxide of chro-	
4. Glucina	.	3·25	mium .	4·5
5. Lime	.	3·5	12. Peroxide of iron	5
6. Soda	.	4	13. Deutoxide of man-	
7. Protoxide of nickel		4·25	ganese .	5
8. Protoxide of cobalt		4·25	14. Oxide of copper	5
9. Protoxide of iron		4·5	15. Yttria .	5·25

16.	Oxide of zinc	5·25	27.	Barytes .	9·75
17.	Potash .	6	28.	Oxide of bismuth	10
18.	Zirconia .	6	29.	Peroxide of plati-num .	14
19.	Strontian .	6·5			
20.	Protoxide of anti-mony .	6·5	30.	Protoxide of lead	14
			31.	Oxide of silver	14·75
21.	Protoxide of cerium	7·25	32.	Protoxide of ☿	26
22.	Deutoxide of rhodi-um .	7·5	33.	Peroxide of ☿	27
			34.	Protoxide of ura-nium .	27
23.	Peroxide of cerium	7·75			
24.	Oxide of palladium	8	35.	Peroxide of urani-um .	28
25.	Protoxide of tin	8·25			
26.	Peroxide of tin	9·25			

14 of these are whole numbers; 8 end in 0·5, 10 in 0·25, and 3 in 0·75.   Or, in other words,

> 14 are multiples of an atom of oxygen,
> 8 are multiples of 4 atoms of hydrogen,
> 13 are multiples of 2 atoms of hydrogen.

Ammonia being a compound of hydrogen, and containing 3 atoms of it, is obviously only a multiple of 1 atom of hydrogen.

11. Thus it appears, that out of 117 bodies, the atomic weights of which have been accurately determined in this work, there are only 5 which are merely multiples of 1 atom of hydrogen,—these are the four hydracids and ammonia. The four acids contain each 1 atom of hydrogen, and ammonia contains 3 atoms of hydrogen. 37 are multiples of 2 atoms of hydrogen, 11 of which are acids and 11 bases. The rest simple bodies.

25 are multiples of 4 atoms of hydrogen. Of these 6 are acids, 8 bases; 2 supporters, 3 acidifiable combustibles, and 6 alkalifiable.

50 are multiples of an atom of oxygen, and consequently whole numbers. Of these 15 are acids, 14 bases, 1 supporter, 9 acidifiable combustibles, and 11 alkalifiable combustibles.

12. It is obvious that all those bodies whose atomic weights are represented by whole numbers are multiples of the atomic weight of oxygen.

18 simple bodies are in this predicament, namely,

Boron,	Columbium,	Cadmium,
Silicon,	Uranium,	Palladium,
Sulphur,	Sodium,	Bismuth,
Tellurium,	Copper,	Platinum,
Selenium,	Potassium,	Lead,
Molybdenum,	Zirconium,	Titanium.

Most of the compounds of these bodies (to the amount of 28) are also whole numbers, and of course multiples of oxygen. This is the case also with three combustible acids; namely, uric, saclactic, and benzoic.

13. There are five or six of the simple bodies which we have found to combine both with 1 atom and with $1\frac{1}{2}$ atom of oxygen. This anomaly, I have no doubt, will startle many of my readers, and may even induce several persons to reject the whole system without farther exami-

Whether $1\frac{1}{2}$ atom of oxygen unite to bodies.

nation.   I have sometimes thought that the anomaly might be obviated by admitting, that oxygen in reality has an atomic weight amounting
to 0·5 instead of 1 ;—on that supposition, it will
have the property of usually entering into combinations by 2 atoms at a time.   All those compounds which I have considered as containing
only 1 atom of oxygen, will in reality contain 2 ;
and those which I suppose to contain $1\frac{1}{2}$ atom
will contain 3 atoms.   This supposition, if admitted, will make no alteration in the atomic
weights given in this work; it will only alter
our way of viewing them, and 1, instead of denoting the weight of 1 atom oxygen will represent 2 atoms.   It will be more convenient to
retain these atomic weights even if this new view
of the subject should be ultimately adopted ;
because, it is but very seldom that a less quantity of oxygen enters into combination with other
bodies than what we have denoted by 1.   One,
then, may still be considered as the representative of a compound atom of oxygen, because it
is the common multiple of it which enters into
combinations; though in a few rare instances we
find that 0·5, or a simple atom of it, also unites
with other bodies.   Such combinations constitute the peroxides of iron, nickel, cobalt, cerium, and sodium, and likewise the deutoxide of
manganese.

If we embrace this view of the subject, which

in the present state of our knowledge is by far
the most probable, then, not only those simple
bodies whose atomic weights are whole numbers,
but those likewise, whose weights end in 0·5,
are multiples of oxygen. These constitute the
whole of the supporters of combustion, and all
the acidifiable and intermediate combustibles,
except three; namely, carbon, arsenic, and
tungsten. Of the alkalifiable combustibles,
amounting to 29, 17 would be multiples of
oxygen, leaving altogether, 15 simple bodies,
which are multiples of 2 atoms of hydrogen.
Thus 31 simple bodies would be multiples of
oxygen, and 15 multiples of 2 atoms of hydrogen.

14. There are only 5 simple bodies which are
multiples of 2 atoms of water, or 2·25. These
are

Chlorine	4·5	or	4	water.
Tungsten	15·75	-	14	-
Columbium	18	-	16	-
Glucinum	2·25	-	2	-
Bismuth	9	-	8	-

But there are 12 compound bodies in the same
predicament: viz.

Oxalic acid	.	4·5	or	4 water.
Arsenious acid	.	6·75	-	6 -
Hyposulphuric	.	9	-	8 -
Molybdic	. .	9	-	8 -
Uric acid	. .	9	-	8 -

Lithia	.	.	2·25	or	2 water.
Alumina	.	.	2·25	-	2 -
Protoxide of iron			4·5	-	4 -
Protoxide of manganese			4·5	-	4 -
Peroxide of mercury			27	-	24 -
Protoxide of uranium			27	-	24 -

These comparisons might be carried a good deal farther. A bare inspection of the tables at the beginning of this chapter will enable the reader to perceive the relations which exist between the atomic weights of different bodies. Thus

The atomic weight of carbon is		0·75
That of phosphorus is	$0·75 \times 2 =$	1·5
sodium	$1·5 \times 2 =$	3
molybdenum	$3 \times 2 =$	6
platinum	$6 \times 2 =$	12

Or, in other words, the atom of

Phosphorus is	= 2 carbon.
Sodium	= 4 -
Molybdenum	= 8 -
Platinum	= 16 -

In like manner, the atom of

Glucinum is	= 3 carbon.
Oxalic acid	
Protoxide of iron	} = 6 -
Protoxide of manganese	
Bismuth	= 12 -

16

These, and many other similar relations between the atomic weights of the different bodies, will be obvious by barely inspecting the tables, and need not therefore be pointed out.

15. Though the seventeenth and eighteenth chapters of this work exhibit the composition of 307 salts, a greater number than has ever before been subjected to actual analysis by any individual, yet they do not enable us to detect any law with respect to the number of atoms of water which exist in the salts in the state of water of crystallization. Of these 307 species of salts, eighty are anhydrous—thirty-one contain 1 atom of water—fifty contain 2 atoms—twenty-seven 3 atoms—eighteen 4 atoms—twelve 5 atoms—nineteen 6 atoms—eighteen 7 atoms—ten 8 atoms—four 9 atoms; all of which are compound salts—six contain 10 atoms—none contain 11 atoms—and two contain 12 atoms.

No simple salt has been met with containing more than 12 atoms, except the sesquinitrate of uranium, which seems to contain 17 atoms. But the compound salts are found with a much greater quantity of water. Thus, one contains 16 atoms of water—five contain 25 atoms—and one contains 55 atoms. Next to the anhydrous salts, by far the most abundant are the salts which contain 2 atoms of water. The salts containing 1, 2, and 3 atoms, added together,

Water of crystallization of the salt.

G g 2

constitute about the half of the whole, if we exclude the anhydrous salts.

If we except acetate of ammonia and quadroxalate of potash, none of the salts of ammonia, potash, soda, barytes, strontian, or lime, contain 7 atoms of water. And, except phosphate of magnesia, all the salts which contain 7 atoms water are either sulphates or nitrates. The metalline sulphates seem to affect 7 atoms water.

No salt of barytes, strontian, iron, manganese, bismuth, or mercury, contains 6 atoms of water; and only 14 simple salts occur containing that quantity of water. The other five are compound salts.

Salts containing 5, 8, 10, and 12 atoms are very rare. No salt has been met with containing 11 atoms.

The oxalates are either anhydrous, or they contain 1, 2, or 3 atoms of water. Only 2 salts contain more: namely, the oxalate of nickel, which contains 4, and the quadroxalate of potash, which contains 7 atoms.

The tartrates are either anhydrous, or they contain 1, 2, or 3 atoms of water. One salt contains 4 atoms, the tartrate of lime, and one contains 5 atoms, the tartrate of bismuth.

The acetates vary much more in their composition. They occur with 1, 2, 3, 4, 6, and 7 atoms water; but very rarely anhydrous.

16. Before concluding these general observations, I may say a few words respecting Berzelius' law, that "in all salts the atoms of oxygen
in the acid constitute a multiple by a whole
number of the atoms of oxygen in the base."
This law was founded upon the first set of exact
analyses of neutral salts which Berzelius made.
Now, as neutral salts in general are combinations
of an atom of a protoxide with an atom of an
acid, it is obvious that the atoms of oxygen in the
acid must in all such salts be multiples of the
atom of oxygen in the base; because every
whole number is a multiple of unity. Neutral
salts, therefore, are not the kind of salts by means
of which the precision of this supposed law can
be put to the test.

Even in the subsalts, composed of 1 atom of
acid united to 2 atoms of base, it is obvious
enough, that the law will hold whenever the
acid combined with the base happens to contain
2 or 4, or any even number of atoms; because
all even numbers are multiples of 2. Now, this
is the case with the following acids:

Phosphoric,	Nitrous,	Antimonic,	Citric,
Carbonic,	Titanic,	Manganesic,	Saclactic,
Boracic,	Arsenious,	Molybdous,	Chromous.
Sulphurous,	Selenic,	Uranitic,	

Consequently, the law must hold good in all

G g 3

combinations of 1 atom of these acids with 2 atoms of base.

In the case of all those acids which contain only 1 atom of oxygen, all the subsalts composed of 1 atom of the acid united to 2 atoms of base, the law will also in some sort hold; for the atoms of the oxygen in such acids being 1, this number will always be a submultiple of 2, the number of atoms of oxygen in 2 atoms of base. This is the case with the following acids:

Silicic,	Hyposulphurous,
Phosphorous,	Oxide of tellurium.

It is only in the subsalts of acids containing an odd number of atoms of oxygen, that exceptions to the law can exist. It is to them, therefore, that we must have recourse when we wish to determine whether this empyrical law of Berzelius be founded in nature or not. Now, there are thirteen acids, the integrant particles of which contain an odd number of atoms of oxygen. The following table exhibits the names of these acids, together with the number of atoms of oxygen in each.

	ATOMS OF OXYGEN.		ATOMS OF OXYGEN.
Sulphuric acid	3	Acetic acid	3
Arsenic	3	Succinic	3
Chromic	3	Benzoic	3
Molybdic	3	Nitric	5
Tungstic	3	Tartaric	5
Oxalic	3	Hyposulphuric	$2\frac{1}{2}$
Formic	3		

Now, although the number of subsalts which I have examined is exceedingly small, because my object was not to investigate the truth of Berzelius' law, but to determine the quantity of water of crystallization which the salts contain, yet there occur several which are inconsistent with Berzelius' law. This is the case, for example, with the disulphate of alumina, the atoms of oxygen in the base being 2, and those in the acid 3. The following subsalts are precisely in the same predicament :

	ATOMS OF OXYGEN IN BASE.	DITTO IN ACID.
Dinitrate of alumina	2	5
Trisnitrate of alumina	3	5
Diprotarseniate of iron	2	3
Dinitrate of lead	2	5
Diacetate of lead	2	3
Diacetate of copper (verdigris)	2	3
Dinitrate of bismuth	2	5

These examples comprehend not only nitric acid, which Berzelius has recognised as an exception to his law; but likewise, sulphuric acid, arsenic acid, and acetic acid.

It would certainly be a most remarkable circumstance if 2 atoms of any protoxide were incapable of combining with 1 atom of any of the 13 acids in the preceding list. I have given seven examples of such combinations; and am persuaded that many more will be discovered

Gg4

whenever the attention of chemists is particularly turned to the subsalts.

There is another kind of saline combination in which exceptions to the law of Berzelius may also be looked for; I mean those salts which I have distinguished by the epithet *sesquisalts* or *subsesquisalts*.   In the sesquisalts, $1\frac{1}{2}$ atom of acid unite with 1 atom of base; or, which comes to the same thing, 3 atoms of acid unite with 2 atoms of base.   In the subsesquisalts, $1\frac{1}{2}$ atom of the base unite with 1 atom of the acid; for example, the *sesquicolumbate* of barytes is composed of

3 atoms columbic acid, containing   3 atoms oxygen.
2 atoms barytes   ———   2 —— ————

Here we see, that the oxygen of the acid is not a multiple of that in the base.

When the acid contains 2 atoms of oxygen, and the base 1 atom, it is plain that the sesquisalts must all come under Berzelius' law; because $1\frac{1}{2}$ atom of acid will contain 3 atoms of oxygen, and 3 is, of course, a multiple of 1; but in acids containing 1 or 3 atoms of oxygen, the law of Berzelius cannot hold.

With respect to the subsesquisalts they will all come under Berzelius' law when the acid happens to contain 3 atoms oxygen, and the base only 1 atom; but they will deviate from it whenever the acid contains 1 or 2 atoms of oxygen.

9

Upon the whole, though the subsalts and sesquisalts have not been sufficiently investigated
to enable us to decide upon the point with
perfect certainty; yet from what we do know,
there appears sufficient evidence that Berzelius'
rule cannot be considered as a general chemical
law; and that we run the risk of falling into
most egregious mistakes, if we make use of such
a law in calculating the atomic weight and chemical constitution of the acids or bases. I
pointed out some remarkable examples of this
error when treating of uranium, to which it is
merely necessary to refer the reader.

LXXV. *Attack of* Berzelius *on* Dr. Thomson's "*Attempt to establish the First Principles of Chemistry by Experiment.*"

OUR scientific readers need not be reminded, that in the work above quoted, Dr. Thomson has endeavoured to fix the combining equivalents of chemical substances.

---

* Ægeria, Sect. B. b. *Steph.*

† *Ses. Cephiformis,* alis hyalinis, anticis marginibus fasciâque nigro-cæruleis; abdomine barbato flavo, cingulis tribus flavis.—*Ochs.* II. p. 169. *Sp.* 22.

‡ *Ses. Euceræformis,* alis anticis fuscis apice inauratis, maculis duabus hyalinis; abdomine barbato nigro, strigâ medii interruptâ flavâ.—*Ochs.* IV. p. 171. No. 22.

§ Ægeria, Sect. B. a. *Steph.*

‖ *Ses. Masariformis,* alis hyalinis, anticis apice flavo irroratis, marginibus fasciâque nigris: abdomine nigro, cingulis tribus flavis; barbâ terminali flavâ.—*Ochs.* II. p. 173. *Sp.* 24.

Dr.

Dr. Thomson has found reason to adopt an idea, suggested some years ago by Dr. Prout, that the numbers which express the atomic weight of bodies are multiples by a whole number of the atomic weight of hydrogen; and his favourite object, visible in almost every page, is to prove the coincidence to be perfect. In this attempt he has been so successful, that the correspondence between his hypothesis and the result of his experiments is startlingly precise. As the accuracy of his results, which, if true, are very important, can be duly estimated only by an analyst of extensive experience, we looked forward with impatience to hear the opinion of Berzelius. His opinion has at length reached us; and as it is expressed in language extremely strong and extremely unusual, we think it necessary to employ his own words. We have accordingly translated a few passages from the *Yahres-Bericht* for 1827 (Woehler's Translation).

" This work belongs to those few productions from which science will derive no advantage whatever. Much of the experimental part, even of the fundamental experiments, appears to have been made at the writing-desk; and the greatest civility which his contemporaries can show its author, is to forget that it was ever published." (page 77.)

" Thomson has published an essay ' On the method of analysing sulphate of zinc;' a subject scarcely requiring, one would think, a separate essay, since the composition of this salt is known with considerable certainty. The great importance attached to it is owing to the circumstance, that in his large work on the atomic weights and chemical proportions, the analysis of this salt is the basis on which the whole superstructure is founded. In describing this analysis, Thomson states that the oxide of zinc was precipitated by carbonate of soda, and that 18·125 grains of crystallized sulphate of zinc yielded 8 grains of anhydrous neutral carbonate of zinc. In this fundamental analysis are two errors;—errors of such a nature as it is difficult to commit, and which appear to prove that the results were invented. Some one had told Thomson that his whole work was of little value, because, in the fundamental experiment, the zinc was precipitated by carbonate of soda in the cold. To this privately communicated remark, Thomson openly replied, That he had supposed chemists would have given him credit for a knowledge of the mode of separating oxide of zinc from acids, and had therefore omitted details; but as he found this opinion erroneous, it became necessary to publish a full account of his process. Ninety grains (five atoms) of sulphate of zinc were precipitated by carbonate of soda, and yielded from 29·3 to 31·03 grains

3 M 2

of

of carbonate of zinc dried at the temperature of 212° Fahr. As this carbonate, when heated to redness, yielded 20·37 of oxide, it was anhydrous and neutral. The filtered solution was boiled, and the oxide of zinc which subsided, after being collected and ignited, weighed 4·54 grains. The solution was then evaporated to dryness in a porcelain vessel, and the salt again dissolved in water; when some oxide of zinc remained, which, when dried at a red heat, weighed 0·431. The solution was then again evaporated to dryness, and the residue ignited; and on being dissolved in water a little silicate of zinc was left, which after decomposition yielded 0·22 grains of oxide of zinc. To the remaining alkaline solution, after being neutralized by muriatic acid, a few drops of hydro-sulphuret of ammonia were added, which threw down a quantity of sulphuret corresponding to 0·65 grains of oxide of zinc. All these five portions together make up the sum of 26·211 grains, the weight of five atoms of oxide of zinc; so that 5·245 is the weight of one atom. Thomson, on this occasion, does not appear to have reflected on the fact, that neutral carbonate of zinc is never obtained by precipitation from an alkaline carbonate; and he has left this difficult point unexplained. The reason why all the oxide of zinc contained in a solution cannot be precipitated in the cold, is, that a portion is dissolved in the form of bi-carbonate, while that portion of oxide which loses its carbonic acid, subsides in the form of a subcarbonate. We have seen in the experiments of Boussingault, that even the sesquicarbonate of soda precipitates a subsalt. Consequently, the statement as to the nature of the first and largest quantity of the precipitated oxide of zinc is obviously erroneous. And such is the method and result of one of the fundamental experiments on which Thomson's whole system stands, by which he obtained more accurate results than any preceding chemist, and through which he established for ever the atomic weights of bodies. The character of this work of Thomson's ought to exclude it from notice here; but it appears to me, that love for the real progress of science makes it imperative to detect quackery, and expose it to the judgement of every one as it merits." (page 181.)

We have thought it right to give as nearly as possible the very words of this critique, that scientific men may judge of its tone and merits. With respect to the former, we regret to see the dignity of science sacrificed by the intemperance of those who profess to be her advocates. It well becomes Berzelius to expose fallacy in argument, or detect error in analysis; but let him not pass beyond the limits of fair criticism: let him not arraign the character of an individual, who may be
actuated

actuated by motives and principles as pure as his own. Intemperate attacks, such as this, reflect back upon their author, and indicate a mind inflamed by pique, jealousy, or some unworthy passion. We know not whether any cause for such feelings may exist in the present case, nor does it concern us to inquire; but we know Dr. Thomson to be devotedly attached to his profession, and we believe him to be sincere and honourable in his transactions. If deception exists at all, we are satisfied that Dr. Thomson himself is more deceived than any one. It is possible that, misled by a favourite hypothesis, he may, like many before him, have been too eager in seizing facts favourable to his views, and too tardy in perceiving those that are unfavourable. On this we offer no opinion at present; but must confess that several circumstances concur in shaking that confidence in the accuracy of his results which we once entertained. Dr. Thomson must be aware that the composition of the chloride of barium, as stated by him, has been declared by Berzelius to be erroneous; and that this error, if such, will vitiate many of his analyses. Would it not be prudent in Dr. Thomson to come forward and correct any mistake which he may have committed, rather than by delay allow others to do so for him? Does not the deference which British chemists have of late paid to him in adopting his atomic weights, impose on him the duty of admission or defence?

XXXIII. *Reply to* Berzelius's *Attack on* Dr. Thomson's *"Attempt to establish the First Principles of Chemistry by Experiment;"* noticed in the Philosophical Magazine and Annals, vol. iv. p. 450. By THOMAS THOMSON, *M.D. F.R.S. Regius Professor of Chemistry in the University of Glasgow.*

To the Editors of the Philosophical Magazine and Annals.

Gentlemen,   Glasgow, February 6, 1829.

YOUR December Number, though published I presume more than a month, has only reached me about half an hour ago. It contains Berzelius's attack upon my character inserted in his *Arberetälre,* for 1827. I had not seen this attack before; but I had heard of it, and been informed of its nature and spirit by several foreign gentlemen, whom I have the pleasure of reckoning among the number of my friends. I had resolved to take no notice of it whatever, being perfectly aware that, as far as my reputation and character are concerned, it would do me no injury. My character and reputation are too well established in my own country, where I am best known, to run any risk from the foul aspersions of the Stockholm Professor. I could only have told him that my feelings were at least as high, and my conduct through life at least as honourable, as his own. I could only have thrown back his foul aspersions with the contempt which they deserved, and demanded that satisfaction which every gentleman feels himself entitled to, when his character has been unjustly traduced. The question was not whether my experiments were accurate or inaccurate; but whether I was an honest man or a scoundrel. Such a question I might surely be pardoned for not thinking it necessary to discuss. My experiments were all made in the laboratory within the walls of the College of Glasgow, and there was scarcely one of them that was not witnessed by more than one competent judge. Indeed more than one-fourth of the salts whose composition I have given in my *First Principles,* were analysed by my pupils. Ample testimony might therefore be produced to authenticate the actual performance of all my experiments. But surely that man must be wofully ignorant of the state of moral feeling in Great Britain, who could allow himself to suppose that a chemical Professor could exist in one of its most celebrated medical schools, capable of setting honour and honesty at defiance. So certain indeed did I feel that not one of my countrymen could for a moment adopt such an idea, that I read the tirade of Berzelius with comparative indifference. And nothing would have induced me to have noticed it at all, but the remarks which you have attached to it. Had I continued silent after these remarks, it occurred to me that your readers would have supposed me conscious of inaccuracies which I do not believe to exist, and of defects which I had not the spirit to acknowledge.

With respect to Berzelius's observations on my analysis of sulphate of zinc, it is only necessary to state a few facts to enable the reader to appreciate their justice. I sent a copy of my *First Principles* to Berzelius, because I had combated many of his opinions in that work, and thought it right that he should have an opportunity of vindicating himself, if he thought himself unjustly treated. He wrote me some months after that he *could not credit the experiments of a man who did not know that zinc cannot be precipitated from its acid solutions in the cold.* This letter was obviously intended to hurt my feelings, but it was at the same time so foolish that it only excited a smile. It was impossible that he could believe that one who had been actively engaged in chemical investigations for almost thirty years, and who had perused every chemical tract of any value that appeared during that long and most momentous period, could be ignorant of one of the most elementary parts of the science. I had been engaged for years in teaching practical chemistry; and there was at the time a manuscript treatise on analysis written by me, lying in my laboratory, which was open to all my practical students, many copies of which had been taken and dispersed through the country. In that book the most minute directions are given how to separate the constituents of minerals; and oxide of zinc is not forgotten.

On reading Berzelius's letter, I thought that it might silence his malignity if I published a single analysis of sulphate of zinc. I transcribed out of the book where I register my experiments, the first accurate analysis of this salt that I had made. This book still exists: and should any person have the least doubts about the fact, it is open to his inspection. As this analysis had been made without any view to publication, but merely for my own private satisfaction, I cannot conceive any motive that could induce me to falsify the register, unless my object had been to impose upon myself. The weights which I have given, and the quantities of reagents used, are precisely those which I found in my register. The analysis had been made probably a couple of years before I published it, though I do not recollect precisely how long. As for Berzelius's hypotheses about subcarbonates and supercarbonates, I have nothing to do with them. I had only to state

state exactly what I got, and what I doubt not I should get again were I to repeat the analysis.

But this solitary analysis was not the only one from which I deduced the atomic weight of oxide of zinc, though I thought at the time, and still think, that it affords sufficient data for the purpose. I may mention another here, which I made about a year ago, and which was witnessed by one of my practical pupils. 5·25 grains of pure oxide of zinc were mixed with their own weight of flowers of sulphur, and heated in a covered porcelain crucible over a spirit-lamp till the crucible was made red hot. It was kept at that temperature till all sulphur fumes had ceased to exhale. The crucible was then allowed to cool. By this process the oxide was converted into sulphuret of zinc. Its weight was 6·25 grains very nearly. It rather exceeded 6·25, but was not so much as 6·26 grains. Now the atom of oxygen is 1, and that of sulphur 2. It is obvious that the Oxide must have been a compound of

$$
\begin{array}{ll}
\text{Zinc} & 4\cdot25 \\
\text{Oxygen} & \underline{1} \\
& 5\cdot25
\end{array}
$$

And the Sulphuret, of Zinc

$$
\begin{array}{ll}
\text{Zinc} & 4\cdot25 \\
\text{Sulphur} & \underline{2} \\
& 6\cdot25
\end{array}
$$

When this sulphuret was dissolved in muriatic acid it left a trace of sulphur too small to be weighed, but visible to the eye, and giving out a sensible odour of sulphurous acid when heated. This slight surplus of sulphur was doubtless the cause of the slight additional weight above 6·25 grains.

Such an experiment could leave no doubt about the accuracy of my analysis of sulphate of zinc. The analysis of *blende* which I made last year with great care, and repeated four times, tends still further to corroborate the same thing. As I have sent the result of my investigation of this mineral to the Royal Society of Edinburgh, I do not consider myself at liberty to detail it here.

I had seen from the new edition of Dr. Turner's First Principles of Chemistry, that Berzelius had announced my number for barytes to be erroneous. But I have not yet seen the paper in which this announcement is made, and do not know what the alleged inaccuracy amounts to. I had found the atomic weight of the 4 alkaline earths to be

$$
\begin{array}{ll}
\text{Magnesia} & 2\cdot5 \\
\text{Lime} & 3\cdot5 \\
\text{Strontian} & 5\cdot5 \\
\text{Barytes} & 9\cdot75
\end{array}
$$

2 F 2   Had

Had the number for barytes been 9·5 instead of 9·75, there would have existed a very obvious analogy among them all. They would all have terminated in 0·5, or they would all have been multiples of 4 hydrogen. This analogy struck me at an early period of my investigations, and I was anxious to find the weight of barytes only 9·5. The experiments of Berzelius rather favoured the idea; according to his analysis the constituents of sulphate of barytes are

Sulphuric acid 5
Barytes . . . . 9·55

But Klaproth's analysis, made with great care, gave

Sulphuric acid 5
Barytes . . . . 10·01

But when I mixed together sulphate of potash and chloride of barium, I found in many trials, that 11 of the former and 13·25 of the latter were the weights which decomposed each other completely. When I employed only 13 of chloride of barium (the weight, if barytes be only 9·5), there was always a residue of sulphuric acid in the solution. Even 13·125 chloride left a residue of sulphuric acid; showing clearly that the weight of barytes is more than 9·625.

Berzelius, in the French edition of his tables published in 1819, gives for the weight of barium 17·1386; and under the name oxidum baryticum, we have the four following numbers, which are all obviously multiples of the first.

19·1386
38·2772
57·4158
76·5544

In his new table, published since he had an opportunity of seeing my *First Principles*, I observe a vast number of changes. He has abandoned a great deal for which he had formerly stickled; and though he has not had the candour to acknowledge as much, I see the great impression which my views have made upon him. I am uncharitable enough to believe, that it was in order to prevent his countrymen and the Germans from being aware of the benefit which he derived from my labours, that his attack upon me was made. I had touched his selfish feelings, and disturbed those dreams of chemical sovereignty in which he has been evidently indulging. In his new table he gives the atom of barytes

9·5688

This is about $\frac{1}{34}$th part less than my determination. It was impossible that my error could have amounted to 2 per cent. It could not have been greater than $\frac{1}{1000}$dth part at the utmost.

But there is a circumstance of which I was not aware when
I de-

I determined the atomic weight of barytes. The muriate of barytes of commerce always contains lead. The reason I take to be, that it is manufactured from the carbonate of barytes of Anglesark, which is probably mixed with some carbonate of lead. I do not recollect whether the chloride of barium which I employed was prepared by myself, or purchased. Supposing it purchased, it was possible that my number might have been affected by the lead present, which would undoubtedly tend to increase the apparent weight of the atom of barytes. To obviate this uncertainty, I purified a quantity of muriate of barytes by passing a current of sulphuretted hydrogen through its solution. It was then crystallized and ignited. With this purified chloride I made several of my practical pupils in succession, at least as many as six of them, make the following experiment: 11 grains of sulphate of potash and 13·25 grains of chloride of barium were dissolved each in a minimum of water. The solutions were mixed, and after standing for twenty-hours were tested for sulphuric acid and barytes, and in no one case was the least trace of either found. I consider these experiments as more satisfactory than if I had made them myself, because the experimenters could have no undue leaning to my numbers. When I see Berzelius's observations, I shall be able to judge whether any additional experiments are necessary.

The only atomic weights given in my *First Principles*, which I have since found to be inaccurate, are the following:

Chromium I stated to have an atomic weight of 3·5. This was merely from analogy. I had determined the atomic weight of chromic acid to be 6·5; and as there were three compounds of chromium and oxygen, I was led to consider them as composed of 1 atom chromium, and 1, 2, and 3 atoms oxygen respectively, which would make the atoms of chromium 3·5. Since that time I have examined the atomic weight of chromium and its oxides with much care. The reader will find the result of this investigation in the Philosophical Transactions for 1827. I found the supposed deutoxide of chromium to be merely the protoxide contaminated with a little chromic acid. The atom of chromium I found to be 4, that of protoxide 5, and that of chromic acid 6·5.

I find the atomic weight of the phosphoric acid which exists in earth of bones and in the phosphate of soda of commerce to be 4·5, and not 3·5 as I state it in my *First Principles*. My number 3·5 was obtained from the analysis of a phosphate of soda which I had prepared myself many years ago in Edinburgh. My stock of this phosphate was considerable, and it was only exhausted in the summer of 1825. On using some
phosphate

phosphate of soda from the Apothecaries'-hall for a particular purpose, I was astonished to find that when I mixed 7·5 grains of the ignited phosphate with a muriatic solution of 6·25 grains of calcareous spar and evaporated the mixture to dryness, and digested the dry mass in water, this water contained a quantity of unprecipitated lime. I found that to precipitate the whole lime, it was necessary to employ 8·5 grains of anhydrous phosphate of soda instead of 7·5. From this it is obvious that the acid in the phosphate weighed 4·5 and not 3·5. I extracted a quantity of phosphoric acid from earth of bones and combined it with soda. 8·5 grains of this salt when anhydrous were still necessary to throw down all the lime from the muriatic solution of 6·25 grains of calcareous spar. I made a quantity of phosphoric acid by the slow combustion of phosphorus and subsequent digestion in nitric acid. The atomic weight of this acid was also 4·5.

I think that there exists two different phosphoric acids which have not hitherto been distinguished from each other, one weighing 3·5 and the other 4·5. Stromeyer seems to have encountered the former in his analysis of Cornish hydrous phosphate of iron (*Unternuhunger*, p. 274); and I found it in the phosphate of soda prepared by me many years ago in Edinburgh. I made my phosphoric acid, if I remember right, by dissolving phosphorus in nitric acid; but the atomic weight of the most common phosphoric acid is 4·5.

I have read over carefully the experiments of Rose on phosphuretted hydrogen gas, and have found nothing in them in the least inconsistent with my experiments on the same gas, which were made so carefully that I cannot doubt their accuracy. Rose's conclusions indeed are inconsistent with mine. But I still think my number for phosphorus, viz. 1·5, right. There are undoubtedly three acids of phosphorus, which must weigh respectively 2·5, 3·5, and 4·5.

There is a circumstance connected with the water of crystallization in oxalic acid which I find myself unable to account for. I find that 9 grains of the crystals of this acid saturate 6 grains of potash and precipitate 6.25 grains of calcareous spar dissolved in muriatic acid without leaving any residue. Hence I conclude, as I have stated in my *First Principles*, that these crystals contain half their weight of water. Dr. Prout wrote me, before the publication of my *First Principles*, that he had uniformly found the crystals of oxalic acid composed of

Acid . . . . 4·5
Water . . . . . . 3·375
———
7·875

This

This information induced me to make the experiments stated in the note (vol. ii. p. 103). Dr. Prout wrote me after the publication of my work, that he still found the crystallized oxalic acid as he had stated. On receiving this letter I requested my assistant, Mr. Andrew Steel, a chemist of much practical experience, to repeat my experiments and give me the result in writing. His experiments agreed exactly with mine to the hundredth of a grain. I am unable to account for this circumstance, and wish much that some other individual would repeat this experiment, and tell us on which side the error lies. Is it possible that two varieties of crystals of oxalic acid occur in commerce? Those that I first tried I had prepared myself; but I afterwards bought acid, and found its composition just the same.

I have been long aware of the malignant feeling which Berzelius harboured with respect to me, and had even got notice of some attacks which he had sent to certain foreign journals; but which the editors had refused to insert. Neither am I ignorant of the origin of this malignant feeling; though I do not pretend to be less of the *genus irritabile* than other people, I must acknowledge that I have viewed the conduct and the attacks of Berzelius with great indifference. I never had the pleasure of meeting with him, and was thoroughly satisfied that he had formed a very erroneous idea both of my character and conduct. It was not against me, but against a man of straw of his own creation, that the attacks were made. I formed a very early resolution not to retaliate, and I still intend not to deviate from that resolution. I shall continue to avail myself of all Berzelius's experiments, and still use the privilege of calling in question his theories and hypotheses when I think them erroneous. But I shall continue to speak of him, as I have always done, with that respect for his talents and industry which I feel; and allow no improper conduct on his part to drag me into any thing which would derogate from the rank which I am conscious of holding as a man of science and of upright conduct.

I am, Gentlemen, your humble servant,

THOMAS THOMSON.

XXIII. *On some Atomic Weights.* By EDWARD TURNER, *M.D. F.R.S. Lond. & Ed., Sec. G.S., Professor of Chemistry in the University of London.**

THE adoption by British chemists of the opinion that atomic weights are multiples by whole numbers of the atomic weight of hydrogen, and the experimental contradiction given to that opinion by so distinguished an analyst as Berzelius, induced me about three years ago to undertake an inquiry into the subject. As nearly the sole evidence in proof of the multiple theory is embodied in the First Principles of Chemistry, published by Dr. Thomson, I turned to that work with the view of putting some of the statements, contained in it, to the test of careful experiment. I commenced with investigating the composition of the chloride of barium, because Dr. Thomson had employed it as a means of obtaining a considerable number of his results. My inquiry, published in the Philosophical Transactions for 1829†, proved the existence of a material error; and Dr. Thomson has since acknowledged it by changing the equivalent of barium from 70 to 68 ‡. It is obvious that this error vitiates many of his other equivalents; and that as so great a mistake has been committed in a fundamental question, an inquiry into the accuracy of minor points is superfluous.

I apprehend, therefore, that the atomic weights at present employed by British chemists are unsupported by satisfactory experiments, and that those who adopt the multiple theory cannot adduce exact analyses in defence of the practice. With this feeling I have occupied my leisure for some time past in examining the equivalents of several important substances, endeavouring to ascertain the value of the numbers adopted in this country compared with those of Berzelius. I shall confine myself entirely to results, partly because some of the points are not yet settled to my satisfaction, and partly because I hope early in the ensuing winter to lay the details in a more perfect form before the Royal Society.

Lead.—The equivalent of lead is frequently employed as the basis of calculation in chemistry. The number adopted in this country, on the authority of Dr. Thomson, is 104. Berzelius has lately repeated his earlier experiments on the

---

* Read before the Chemical Section of the British Association at Oxford, June 27, 1832; and communicated by the Author.

† Dr. Turner's paper " *On the Composition of Chloride of Barium,*" will be found in Phil. Mag. and Annals, N.S. vol. viii. p. 180.—EDIT.

‡ Dr. Thomson's correction will be found in Phil. Mag. and Annals, N.S. vol. x. p. 392.—EDIT.

subject, by reducing oxide of lead to the state of metal by means of hydrogen gas. Taking his two most widely differing results, the equivalent of lead, oxygen being 8, will be 103·42 in the one case, and 103·64 in the other. His mode of analysis, though apparently easy and simple, is by no means free from practical difficulty. My experiments were made by converting the oxide into sulphate of lead, a method, I believe, susceptible, with the requisite precautions, of greater accuracy than that employed by Berzelius. After many trials I feel certain that the equivalent of lead is not higher than 103·6. It is probably somewhat lower; so that 103·5, nearly the mean of Berzelius's experiments, is very near the truth. This point I hope to clear up by renewed experiments, which are rendered necessary by the extreme difficulty of getting oxide of lead in a state of adequate purity. In the mean time 103·5 is the nearest approximation which experiment justifies: it is useless to go beyond the first decimal, because we are ignorant whether 103·5 is greater or less than the real number.

The following experiment will test the value of this estimate:—If the equivalent of lead be 103·5, then 100 parts of metallic lead should yield 146·38 parts of sulphate of lead. The mean of several closely corresponding experiments by Berzelius is 146·419; and the mean of my own is 146·401. If 104 were the equivalent of lead, 100 parts ought to yield 146·16 of the sulphate,—a number differing widely from the result of experiment, and much beyond the errors of manipulation.

Chlorine.— The most satisfactory experiments I have met with respecting the equivalent of chlorine are those of Berzelius. He obtained from 100 of chlorate of potash 39·15 of oxygen, and 60·85 of chloride of potassium; and found that 100 of chloride of potassium correspond to 192·4 of chloride of silver. According to my own experiments, 100 parts of silver give 132·8 of chloride of silver,—an estimate extremely close to that of Berzelius. From these data it follows that the equivalent of chlorine is 35·45.

To compare with this number the equivalent of chlorine determined in a totally different manner, I prepared some very pure chloride of lead, and separated its chlorine by means of nitrate of silver. From the best experiments I could make, 100 of chloride of lead correspond to 103·24 of chloride of silver. Now, even taking as the equivalent of lead the theoretic number 104, the preceding analysis gives 35·578 as the equivalent of chlorine; and when we take the more correct equivalent of lead 103·5, that of chlorine is 35·45, identical with the number deduced from the experiments of Berzelius. The equivalent of chlorine commonly used by British chemists, namely 36, is therefore erroneous.

I may add that the preceding analysis agrees closely with that of Berzelius; but I prefer my own result, because my chloride of lead appears to have been purer than the specimen employed by him, dissolving in water without the slightest residue. It affords an instructive test of the value of the atomic weights current among us. For, supposing 104, 36, and 110 to be the respective equivalents of lead, chlorine, and silver, it follows that 100 of chloride of lead should yield 104·28 parts of chloride of silver, instead of 103·24 as given by experiment. In fact, as will immediately appear, the equivalent of silver is still more erroneous than those of lead and chlorine.

Silver.—My first attempts to determine the equivalent of silver were by means of the oxide; but different analyses disagreed so widely, that I was obliged to resort to another method. Knowing very nearly the equivalent of chlorine, that of silver may be inferred from the composition of the chloride. According to Dr. Thomson, 100 parts of silver correspond to 132·73, according to Berzelius to between 132·75 and 132·79, and by my experiments to 132·8. The coincidence is very close, and therefore the principal difference in the equivalent of silver will depend on that of chlorine. If 36 be assumed as the equivalent of chlorine, that of silver is 110; and it is 108·08 if 35·45 be chosen as the equivalent of chlorine. An extremely slight difference in the number for chlorine, such as lies entirely within the ordinary limits of error, would raise the equivalent of silver to 108·1 or rather higher, or depress it to 108. While the matter is uncertain it will be most convenient to employ the whole number.

In order, by an independent analysis, to ascertain which of the numbers above mentioned is the more accurate, I prepared some very pure nitrate of silver, kept it for some time in fusion, and converted it into chloride of silver. After repeated experiments, I find that 100 of the chloride of silver corresponds to a quantity of fused nitrate, varying from 118·544 to 118·50. But the theoretic quantity deduced by adopting 110 and 36 to represent silver and chlorine is 117·81, which differs widely from the result of actual experiment; whereas, supposing the equivalent of silver and chlorine to be represented by 108 and 35·45, 100 of chloride of silver should correspond to 118·51 of the fused nitrate. This, then, is additional evidence in favour of the atomic weight of chlorine as above stated.

Nitrogen.—I have endeavoured to ascertain the equivalent of nitrogen by the analysis of the nitrates of silver and lead.

1. From the analysis just stated, I consider 100 of the chloride of silver, containing 75·3012 silver, to be equivalent to 118·5 of

nitrate of silver.   Calculating from these elements, and with 108 as the equivalent of silver, we shall find 14·06 as the equivalent of nitrogen.   It will be 14·046 if the equivalent of silver be 108·1.

2. As a mean of three closely corresponding analyses, made by converting the nitrate into sulphate of lead, I find that 100 parts of sulphate of lead correspond to 109·307 of nitrate of lead.   Calculating the equivalent of nitrogen, on the presumption that 103·5 and 40 are the respective equivalents of lead and sulphuric acid, we shall find it to be 14·101.

Berzelius calculates his equivalent of nitrogen from an analysis of nitrate of lead, and estimates it at 14·18.   The difference between us principally depends on a different estimate of the composition of the oxide of lead; and until this point shall be settled with more precision than at present, no certain inference can be deduced from the analysis of the nitrate.   I have more confidence in the estimate from nitrate of silver, and feel little doubt that 14 is a very close approximation.   Some analyses of nitrate of baryta, but which are not fully in a state for publication, induce me to believe that the real equivalent of nitrogen is nearer 14 than 14·1.

Barium.—From the analysis of chloride of barium, published in my Essay on that compound, no inference could at first be drawn in consequence of the uncertainty respecting the equivalent of chlorine.   Now, however, that we have reason to take 35·45 as the equivalent of chlorine, it follows from my analysis that the equivalent of barium is 68·76; and according to the analysis of chloride of barium by Berzelius, it is 68·588. I believe the equivalent of barium is intermediate between 68·6 and 68·8, and in the absence of more exact knowledge 68·7 may be taken as a very good approximation.

The general conclusions which I deduce from the preceding account are the following:

1. The atomic weights commonly used by British chemists have been adopted without due inquiry, and several of the most important ones are erroneous.

2. The hypothesis, that all equivalents are multiples by a whole number of the equivalent of hydrogen, is inconsistent with the present state of chemical knowledge, being at variance with experiment.

3. The subjoined equivalents are very nearly correct:—

Lead	103·5
Silver	108
Barium	68·7
Chlorine	35·45
Nitrogen	14

APPENDIX.

———◆———

When the preceding sheets had gone through the press, they were submitted to Dr. Prout, who after perusing them favoured me with the following remarks, which I gladly insert, as serving to explain more fully those peculiar views of his to which I have alluded.

" Sackville-street, Sept. 12, 1831.
" DEAR SIR,

" I WAS much gratified by a perusal of your Essay on the Atomic Theory: there are, however, a few points in which I am more immediately concerned, apparently requiring some remarks, and which I shall consider in the order they occur.

In page 39 you observe, ' I believe, indeed, that I shall not be misrepresenting Dr. Prout's opinions, if I remark that in the paper alluded to he seems to have noticed the relation between the numbers......chiefly as a presumption in favour of the idea of their being possibly compounded of oxygen and hydrogen, of which they appear to be multiples.' The original opinion to which I was led by the observations of others, and innumerable experiments (never published) of my own, was, that the combining or atomic weights of bodies bear certain simple relations to one another, frequently by multiple, and consequently that many of them must necessarily be multiples of some one unit; but as the atom of hydrogen, the lowest body known, is frequently subdivided when in combination with oxygen, &c. there seems to be no reason why bodies still lower in the scale than hydrogen (similarly however related to one another, as well as to those above hydrogen) may not exist, of which

K

other bodies may be multiples, without being actually multiples of the intermediate hydrogen. Such was my *opinion* in general terms; my *speculations*, I confess, went further, and were indeed pretty much as you have stated them to be.

" Page 44 and 62, you remark, with respect to the general notion of atomic series rather than units, ' that you are not aware of any facts which do not equally admit of being referred to the theory more commonly adopted, and that you do not see the absurdity of supposing that in organic compounds where the terms of the series are, as is the case of water, represented as $3:6:9:12$, &c. the true relations may not be as $9:18:27:36$ corresponding to $1:2:3:4$ atoms of water;' and again, ' that you do not see that the theory of Dalton holds out any stronger temptation to fraud than the laws substituted for it by me.'

" To reply to the first of these remarks as it ought to be replied to, and indeed as I perhaps *could* reply to it, would lead me far beyond my present purpose; I shall therefore merely observe, that by adhering to a single term (with reference to which I am quite aware all others may be expressed) great difficulties often occur, and the real (often very simple) compositions of bodies are so masked and *apparently* misrepresented, that they cannot without difficulty be recognised in some instances.

" With respect to the second objection, I may remark, that my notions were not proposed with the expectation that they would make honest men of knaves: though it may be worth while to observe, that by diminishing the number, the *amount* of error is likely to be diminished.

" The series given for water, I wish it to be observed, applies to its combination with carbon, and perhaps some other bodies; but in uniting with bodies having different combining series, the aqueous series itself *may* become modified or different—and hereby hangs, if I am not much mistaken, a very curious tale, which I hope some one will tell ere long better than I am able to do.

" At page 62, you speak of the ' *censures* I have cast on
the atomic theory.' Now this is a much stronger term than
I am willing to allow.   There is no one can possibly have
greater respect for Mr. Dalton, and all that he has done,
than myself, and I am a firm believer in his principles *as
far as they go*, because I believe them to be founded
in truth.   What I meant to say was, that they do not
contain *all* the truth, and that consequently in *their present state* they are inadequate to explain the operations
of nature.   It is however my opinion, that the system of
Dalton, even in its present state, on account of its great
simplicity and convenience, never will nor ought to be
superseded; and that consequently it will continue to be
employed for these reasons, just as the Linnean system
continues to be employed for very similar reasons by
botanists; and here I may remark, by the by, that I referred to botany in my lectures rather for the sake of illustration, than from any close analogy between this science
and chemistry, which I was well aware did not exist.

" Pages 68 et seq. you speak of the doctrines of *isomorphism* and *isomerism;* and though I do not observe that
you allude to any thing that I have said on the subject, I
am anxious to make a few remarks on a passage in my
lectures, which from the terms employed may be liable to
be misunderstood. I have said that the continental chemists
have succeeded in establishing the curious and important
doctrines of isomorphism and isomerism—doctrines totally
inexplicable upon the principles of Dalton and Berzelius,
but which seem to me to flow necessarily in conjunction
with some others from the principles which I have long
considered as regulating the union of bodies in nature.' In
the lectures as delivered, these doctrines were very briefly
explained, and I wish here to remark, that I mean nothing
more by the above than that the doctrines in the abstract,
or generally speaking, are established, which I believe to
be the case; as for the details, I always considered many of
them exceedingly unsatisfactory. So long ago as 1815,

K 2

I was led to infer that relation in *weight* might indicate a relation also in *size* among the atoms of bodies *; and that many of those striking and curious analogies in property, form, &c. which I thought I observed among bodies atomically related, might depend upon one or other of these circumstances. But, as soon after this period I relinquished chemistry in general, I thought little more of the matter, till the doctrines of Mitscherlich were announced. I merely mention this, but without advancing the shadow of a claim to the honour of the discovery of isomorphism, which, as far as I know, is entirely due to the eminent philosopher above mentioned.   With respect to *isomerism*, in my lectures as originally written, I alluded to three varieties or modifications of this principle as existing in bodies—one in which the same elements are differently arranged ; a second, in which the arrangement (still crystalline) is different, but which difference depends upon the presence of minute quantities of foreign bodies; and a third, which I have provisionally termed *merorganization,* in which the general arrangement, besides being peculiar, may be also supposed to be subject to or influenced by the same causes which produce the peculiar arrangements in one or both the other two varieties. All these varieties, I believe, are inexplicable upon the principles of Dalton *as they at present stand*, but on these principles *as they may be extended*, I have strong hopes that one day or other the two first varieties at least will be explained.

" You allude to the speculations of M. Decandolle on the forms of plants, and I will amuse you with a speculation of mine on the same subject, viz. that these forms are somehow or other connected with the oxygen series 2 : 4 : 6 : 8, &c. and the isobaric series of carbon and water 3 :

* See Annals of Philosophy, VII. 113. where the general principles on which this notion was founded are briefly stated with another view, viz. that of explaining the relation between the doctrine of atoms and of volumes.

$6:9:12$, &c. I think I could bring forward many curious circumstances illustrative of this opinion.

" I remain, dear sir, Yours very truly,

"W. PROUT."

" P. S. I approve of your ingenious observations on mineral waters, which, I think, throw considerable light on their constitution. My professional pursuits keep me away so much from chemical details, that it is very possible I may have committed some errors in the preceding observations. If so, I beg you will correct them.   W. P."

ART. XIV.—*On the hypothesis of Mr. Prout, with regard to Atomic Weights;*—in a letter from BERZELIUS, dated Stockholm, Dec. 6, 1844.

Remark.—The following is the translation of a letter received from Berzelius, in reply to a request from the Junior Editor of this Journal, that he would favor its pages with an expression of his views on this much mooted question. The reader cannot fail to admire the candor of the distinguished author, and at the same time must admit the justness of his views.—*Eds.*

IN scientific questions, errors are avoided only by making experiment the basis of opinions. Mr. Prout, in bringing forward his views, based them, on the contrary, upon a presumed inaccuracy in experiments made for determining atomic weights. It seemed to him convenient and advantageous to science, that the atomic weights should be expressed by small numbers, and without fractions, and as the weight of hydrogen was relatively much below the others, he proposed to secure this simplicity by considering all a multiple of the atomic weight of hydrogen; and in carrying out these views, he found that it required but a small change of the ascertained weights, not greater than was deemed unavoidable errors in analyses.

The basis of Mr. Prout's hypothesis, then, is the supposed inexactness of experiment. The seeming correspondence with it which oxygen and carbon afforded, was thought to authorize a correction of the other atomic weights, so as to make them accord with his hypothesis, although but a mere hypothesis, not verified by observations in the great majority of instances.

Mr. Prout has given no explanation of the supposed fact on which his theory is based. Such a result could hardly proceed from any thing but the existence of only a single ponderable element—hydrogen—of atoms of which combined, the compound molecules of all the other so-called elements must consist—each of some definite number. In such a case, whatever be the attraction uniting the atoms in their compound molecules, we might have expected that in the course of the 1300 years since the alchymists commenced their experiment, or under the more skillful research of the modern chemist, some instance might have come to light, in which an atom should become detached from a given compound molecule, or another added, and in one way or the other a new element have come to light. But no instance of this is known: the elements have remained unchanged—incommutable.

Dr. Thomson endeavored to establish Mr. Prout's hypothesis, by a series of experiments; but his investigations will not bear even the most superficial examination; yet they contributed much to give popularity to Mr. Prout's ideas in England and also in other countries. Subsequent to this, Prof. Turner was deputed by the British Association, to determine the atomic weight of certain of the elements, in order to settle the question. You, perhaps, recollect that his labors gave but negative results, and this able chemist finally abandoned the hypothesis.

The discussions, however, did not terminate here. The atomic weight of carbon caused their renewal. I obtained for its equivalent, by chemical experiments, 75·33. The method by calculation from the specific weight of carbonic acid gas, I at that time rejected, as I had found that sulphurous acid gas, being condensable, gave the atomic weight of sulphur much too high. But some years afterward, on examining with Dulong the specific weight of carbonic acid gas, at that time not known to be capable of liquefaction, it seemed to us that the method by calculation, would lead to surer results than direct experiment. Admitting that oxygen gas alters not in volume, when converted into carbonic acid gas, we deduced for the atomic weight of carbon 76·438—a number soon very generally admitted. A few years subsequent to this, however, it was discovered that carbonic acid gas, far from being permanent, was condensable; but no one then dreamed of the influence which this would have upon the result obtained by calculation. But after a while it was found that the atomic weight thus deduced was much too high. Dumas undertook new experiments, and arrived quite exactly at the number 75, which, as it is just 6 times the weight of hydrogen, gave a new impulse to the hypothesis of Prout. He examined the atomic weight of hydrogen, and by experiments made apparently with much care, arrived at results a little varying, from which he adopted 12·5—the same that Prout had obtained by dividing the atomic weight of oxygen by 8. By new experiments upon oxygen, nitrogen and atmospheric air, he went on to show other coincidences with the hypothesis of Prout.

These researches engaged another chemist, M. Marignac, to undertake a series of experiments upon the atomic weights of chlorine, bromine, iodine, nitrogen, potassium and silver. His incomparable researches are a *chef-d'œuvre*, as much for conscientious exactness as for the judicious and varied methods employed. The results approach multiples of 12·5, but to make them precisely so, requires larger corrections than the greatest variations in the results of experiments. One alone, that of bromine, is very nearly 80 times 12·5. The atomic weight of chlorine, found to be 443·2, varies from 447·5, the nearest multiple, by too large a quantity, to be an error in experiment. Marignac was of the opinion that his experiments did not settle the question at issue, as the approximation to the hypothesis of multiples was still quite close ; but in the case of chlorine he admitted that there was an undeniable exception. He endeavored, however, to make an approximation to the hypothesis of Mr. Prout, by comparing the weight of two atoms of chlorine with one of hydrogen, in which he found the former to be almost exactly 73 times that of the latter.

But let us examine whether it is proper to view them in this relation. You know that nearly all English chemists, and also many others, consider water a compound of an atom of each element, notwithstanding the fact that the amount of hydrogen gas is in volume double that of oxygen ; and this view is adopted because it is the most convenient, although mere convenience cannot establish the truth in the case. It is well known that several simple bodies combine with oxygen in such a manner that the molecule of the oxyd sometimes contains one, sometimes two atoms of the radical, as an example of which I mention here the series, $Mn + O$, $2M + 3O$, $Mn + 2O$, $Mn + 3O$, $2Mn + 7O$. There is no doubt that the last, the hypermanganic acid, contains 2 per cent. of the radical. It is besides known that the *hypermanganate* and *hyperchlorate* of potassa are isomorphous. From this it follows conclusively that the hyperchloric acid contains also 2 atoms of chlorine for 7 atoms of oxygen. It must hence be an error to consider two volumes of chlorine as a single atom. You know also that chlorine combines itself with hydrogen only in equal volumes, and that in the great number of isomorphous compounds in which chlorine is substituted for hydrogen, without a change of crystalline form, this substitution takes place invariably under equal volumes. We cannot, therefore, avoid the conclusion that *two* volumes of hydrogen as well as of chlorine, represent *two* atoms, and consequently an atom of water contains two atoms of hydrogen, just as one atom of cuprous oxyd, mercurous oxyd, and hyposulphurous acid, &c. contain two atoms of the radical. It therefore follows—and not equivocally—that chlorine and hydrogen cannot be compared except in the relation of equal volumes, and consequently the atomic weight of chlorine cannot be a multiple of that of hydrogen.

In glancing over the table of atomic weights, you will find numerous exceptions to the hypothesis of Mr. Prout. It is true that the most of these numbers have been determined by experiments made 30 years since, and at the first development of chemical proportions. They cannot, therefore, have all the exactness which they might have with the new analytical processes since discovered. But I am persuaded that they will be found, with few exceptions, to approach quite near the truth, and not to admit of that latitude of correction demanded by Prout's hypothesis. I cite, for example, the atomic weight of lead, which has been the subject of reiterated experiments in the course of this discussion, and which has been fixed at 1294·5—also that of copper 395·7, both lately verified by Messrs. Erdmann and Marchand. To accord with Mr. Prout's hypothesis, the first ought to be 1287·5 or 1300, and the second 387·5 or 400.

Among the atomic weights there are those which are apparently equal, and others which approximate quite near to being double of one another in weight—from which it is probable that there is between them a certain relation; but these pertain only to particular groups of the elements—such as the metals which accompany platinum,—molybdenum and tungsten,—chrome, iron, manganese, &c. But this is a new question, to be discussed only after farther investigation.

Yours, &c.   JAC. BERZELIUS.

Section II

THE FOURTH STATE OF MATTER

While Faraday was still working in a junior capacity at the Royal Institution, he delivered some lectures to other young men at the City Philosophical Society in London; 'philosophical' at that date meant 'scientific'. Henry Bence Jones reprinted some of these lectures in his biography of Faraday. As we saw in the General Introduction, they are in the corpuscularian tradition of Newton and Boyle. In them we see already the polish which was to attract great audiences to Faraday's lectures at the Royal Institution; and the interest in hypotheses, coupled with a suspicion of them, which one would expect to find in a fertile scientist.

Faraday's experiments on the liquefaction of gases were followed by those of Thomas Andrews of Belfast; who demonstrated the continuity of the liquid and gaseous states, and showed that for each gas there is a critical temperature above which it cannot be liquefied by pressure alone. These researches made possible the liquefaction of gases which had previously resisted all attempts; and they helped to elucidate somewhat the relations between the three states of matter.

The fourth state, or radiant matter, was what Crookes believed that he had obtained in his discharge tubes. It is difficult for us not to identify the cathode rays with electrons; but Helmholtz's famous Faraday Lecture advocating the particulate nature of electricity was not delivered until 1881, and there was no reason why Crookes should have anticipated him. Whereas for us the phenomena are explicable in terms of electrons ejected from the cathode, for Crookes they arise from the gas being at such a low pressure that the mean free path of the molecules between collisions is comparable with the dimensions of the tube. The electric current does not bring about any change; it simply illustrates how differently gases behave at extremely low pressures, and leads us to assume that we are confronted by matter in a fourth state, as far removed from the gaseous as that state is from the liquid.

Crookes had in 1861 discovered thallium, identifying it from a green line in the spectrum of selenium. In 1863 he had been elected a Fellow of the Royal Society; and in 1875 he had constructed the radiometer. He was therefore an established figure when in 1879 he lectured on Radiant Matter to the British Association, and when he took up the subject again before the Institute of Electrical Engineers in 1891. In between had come in 1886 his Presidential Address to the Chemical Section of the British Association, reprinted in *Classical Scientific Papers—Chemistry*, first series, with its doctrine of inorganic evolution, touched on in 1891. Crookes, like Andrews and Faraday, was an extraordinarily able experimentalist, and his experiments on the cathode rays are classics. He had also a gift for language; perhaps his lectures are too florid for the modern taste, but the descriptions are vivid and the generalisations bold.

George Fitzgerald was a theoretical physicist, best known for his hypothesis of the 'Fitzgerald Contraction' to explain the negative result of Michelson and Morley's experiment. His published writings do not apparently indicate his importance, for he was very fertile in suggesting to others fruitful lines of research. He called attention at the British Association in 1888 to Hertz's detection of radio waves; and went on to deplore a physics of hard particles, urging his colleagues to base their theory of matter upon the ether.

The journal *Chemical News* had begun in 1860, and *Nature* in 1869; for the later nineteenth century they are the journals which the historian will look at first. Neither was published by a Society; their editors, Crookes and Lockyer, were both extremely able and dashing scientists. They carried full reports of the meetings of the British Association, and of interesting meetings of other societies. They reprinted, condensed, and translated articles from other journals; and both have also a considerable number of important original papers which first appeared in their pages.

FURTHER READING

T. Andrews, *Scientific Papers*, London, 1889; contains full texts.

M. Faraday, *Chemical Manipulation*, London, 1827; reprint of 3rd ed., London, in press, with introduction by D. M. Knight; *Experimental Researches in Chemistry and Physics*, London, 1859; reprint, Brussels, in press; *Diary*, ed. T. Martin, 7 vols. and index, London, 1932–6.

M. Hesse, *Forces and Fields*, London, 1961.

T. H. Levere, 'Faraday, Matter and Natural Theology', *British Journal for the History of Science*, IV (1968), 95–107.

J. E. McGuire, 'Force, Active Principles and Newton's Invisible Realm', *Ambix*, XVI (1969), 154–208.

R. Olson, 'The Reception of Boscovich's Ideas in Scotland', *Isis*, LX (1969), 91–103.

A. Thackray, 'Matter in a Nut-shell: Newton's *Opticks* and Eighteenth-Century Chemistry', *Ambix*, XV (1968), 29–53.

L. P. Williams, *Michael Faraday*, London, 1965.

This year a lecture, On the Forms of Matter, was given at the City Philosophical Society : in it he shows his views regarding matter and force at this time.

' In the constant investigation of nature pursued by curious and inquisitive man, some causes which retard his progress in no mean degree arise from the habits incurred by his exertions ; and it not unfrequently happens, that the man who is the most successful in his pursuit of one branch of philosophy thereby raises up difficulties to his advancement in another.

' Necessitated as we are, in our search after the laws impressed upon nature, to look for them in the effects which are their aim and end, and to read them in the abstracted and insulated phenomena which they govern, we gradually become accustomed to distinguish things with almost preternatural facility ; and induced by the ease which is found to be afforded to the memory and other faculties of the mind, division and subdivision, classification and arrangement, are eagerly adopted and strenuously retained.

'Much as the present stage of knowledge owes to this tendency of the human mind to methodise, and therefore to facilitate its labours, still it may complain that in some directions it has been opposed and held down to error by it. All method is artificial and all arrangement arbitrary. The distinction we make between classes, both of thoughts and things, are distinctions of our own; and though we mean to found them on nature, we are never certain we have actually done so. That which appears to us a very marked distinctive character may be really of very subordinate importance, and where we can perceive nothing but analogies and resemblances, may be concealed nature's greatest distinctions.

'The evil of method in philosophical pursuits is indeed only apparent, and has no real existence but in the abuse. But the system-maker is unwilling to believe that his explanations are not perfect, the theorist to allow that incertitude hovers about him. Each condemns what does not agree with *his method*, and consequently each departs from nature. And unfortunately, though no one can conceive why another should presume to bound the universe and its laws by his wild and fantastic imaginations, yet each has a reason for retaining and cherishing his own.

'The disagreeable and uneasy sensation produced by incertitude will always induce a man to sacrifice a slight degree of probability to the pleasure and ease of resting on a decided opinion; and where the evidence of a thing is not quite perfect, the deficiency will be easily supplied by desire and imagination.   The efforts a man makes to obtain a knowledge of nature's secrets merit, he thinks, their object for their reward; and though he may, and in many cases must, fail of obtaining his desire, he seldom thinks himself unsuccessful, but substitutes the whisperings of his own fancy for the revelations of the goddess.

'Thus the love a man has for his own opinion, his readiness to form it on uncertain grounds rather than remain in doubt, and the necessity he is under of referring to particular and individual examples in illustration of his views of nature, all tend to the production of habits of mind which are partial and warped.   These habits it is which give rise to the difference of opinion in men on every possible subject.   All parts of the system both of the moral and natural world are constant in their natures, presenting the same appearances at all times and to all men.   But we cannot perceive them in all their bearings and relations; we view them in different states and tempers of mind, and we hasten to decide upon them.   Hence it happens, that a judgment is made for future use which not only differs in different individuals, but, unfortunately, from truth itself.

'As it regards natural philosophy, these bad, but more or less inevitable, effects are perhaps best opposed by cautious but frequent generalisations.   It is true that with the candid man experience will do very much, and after having found in some instances the necessity of altering previous opinions, he will retain a degree of scepticism in future on all those points which are not proved to him.   But generalisations will aid the efforts a man makes to free himself from erroneous ideas and prejudices; for, presenting to us the immense family of facts ranged according to the relationships of the individuals, it makes evident many analogies and distinctions which escape the mind when engaged on each separately, and corrects those errors consequent on partial views.

'We are obliged, from the confined nature of our powers, to consider of but one thing at a time. Generalisation compensates in part the resulting inconveniences, and in an imperfect way places many things before us; and the more carefully this is done the more accurately our partial notions are corrected.

' Ultimately, however, facts are the only things which we are *sure* are worthy of trust. All our theories and explanations of the laws which govern them, whether particular or general, are necessarily deduced from insufficient data. They are probably most correct when they agree with the greatest number of phenomena, and when they do not appear incompatible with each other. The test of an opinion is its agreement in association with others, and we associate most when we generalise.

' Hence I should recommend the practice of generalising as a sort of parsing in philosophy. It occasions a review of single opinions, requires a distinct impression of each, and ascertains their connection and government. And it is on this idea of the important use that may be made of generalisation, that I venture to propose for this evening a lecture on the general states of matter.

' Matter defined—essential and secondary properties.

' Matter classed into four states—solid, liquid, gaseous,

and radiant—which depend upon differences in the essential properties.

. . . . . .

'Radiant state.—Purely hypothetical.   Distinctions.

'Reasons for belief in its existence.   Experimental evidence.   Kinds of radiant matter admitted.

'Such are the four states of matter most generally admitted.   They do not belong to particular and separate sets of bodies, but are taken by most kinds of matter ; and it will now be found necessary, to a clear comprehension of their nature, to notice the phenomena which cause and accompany their transition into each other.

'Some curious points arise respecting the changes in the forms of matter, which, though not immediately applicable to any convenient or important use, claim our respect as buddings of science which at some future period will be productive of much good to man. Of the bodies already taken and presented in various forms in illustration of this part of our subject, some have evinced their Protean nature in the production of striking effects ; others there are which, being more constant to the states they take on, suffer a conversion of form with greater difficulty ; and others, again, have as yet resisted the attempts made to change their state by the application of the usual agencies of heat and cold.   By the power of heat all solid bodies have been fused into fluids, and there are very few the conversion of which into a gaseous form is at all doubtful. In inverting the method, attempts have not been so successful.   Many gases refuse to resign their form, and some fluids have not been frozen. If, however, we adopt means which depend on the rearrangement of particles, then these refractory instances disappear, and by combining substances together we can make them take the solid, fluid, or gaseous form at pleasure.

'In these observations on the changes of state, I have purposely avoided mentioning the radiant state of matter, because, being purely hypothetical, it would not have been just to the demonstrated parts of the science to weaken the force of their laws by connecting them with what is undecided. I may now, however, notice a curious progression in physical properties accompanying changes of form, and which is perhaps sufficient to induce, in the inventive and sanguine philosopher, a considerable degree of belief in the association of the radiant form with the others in the set of changes I have mentioned.

'As we ascend from the solid to the fluid and gaseous states, physical properties diminish in number and variety, each state losing some of those which belonged to the preceding state. When solids are converted into fluids, all the varieties of hardness and softness are necessarily lost. Crystalline and other shapes are destroyed. Opacity and colour frequently give way to a colourless transparency, and a general mobility of particles is conferred.

'Passing onward to the gaseous state, still more of the evident characters of bodies are annihilated. The immense differences in their weights almost disappear; the remains of difference in colour that were left are lost. Transparency becomes universal, and they are all elastic. They now form but one set of substances, and the varieties of density, hardness, opacity, colour, elasticity and form, which render the number of solids and fluids almost infinite, are now supplied by a few slight variations in weight, and some unimportant shades of colour.

'To those, therefore, who admit the radiant form of matter, no difficulty exists in the simplicity of the properties it possesses, but rather an argument in their favour. These persons show you a gradual resignation of properties in the matter we can appreciate as the matter ascends in the scale of forms, and they would be surprised if that effect were to cease at the gaseous state. They point out the greater exertions which nature makes at each step of the change, and think that, consistently, it ought to be greatest in the passage from the gaseous to the radiant form; and thus a partial reconciliation is established to the belief that all the variety of this fair globe may be converted into three kinds of radiant matter.

'There are so many theoretical points connected with the states of matter that I might involve you in the discussions of philosophers through many lectures without doing justice to them. In the search after the *cause* of the changes of state of bodies, some have found it in one place, some in another; and nothing can be more opposite than the conclusions they come to. The old philosophers, and with them many of the highest of the modern, thought it to be occasioned by a change either in the motion of the particles or in their attractive power; whilst others account for it by the introduction of another kind of matter, called heat, or caloric, which dissolves all that we see changed. The one set assume a change in the *state* of the matter already existing, the other create a new kind for the same end.

'The nature of heat, electricity, &c., are unsettled points relating to the same subject. Some boldly assert them to be matter; others, more cautious, and not willing to admit the existence of matter without that evidence of the senses which applies to it, rank them as qualities. It is almost necessary that, in a lecture on matter and its states, I should give you my own opinion on this point, and it inclines to the immaterial nature of these agencies. One thing, however, is fortunate, which is, that whatever our opinions, they do not alter nor derange the laws of nature. We may think of heat as a property, or as matter: it will still be of the utmost benefit and importance to us. We may differ with respect to the way in which it acts: it will still act effectually, and for our good; and, after all, our differences are merely squabbles about words, since nature, our object, is one and the same.

'Nothing is more difficult and requires more care than philosophical deduction, nor is there anything more adverse to its accuracy than fixidity of opinion. The man who is certain he is right is almost sure to be wrong, and he has the additional misfortune of inevitably remaining so. All our theories are fixed upon uncertain data, and all of them want alteration and support. Ever since the world began, opinion has changed with the progress of things; and it is something more than absurd to suppose that we have a sure claim to perfection, or that we are in possession of the highest stretch of intellect which has or can result from human thought. Why our successors should not displace us in our opinions, as well as in our persons, it is difficult to say; it ever has been so, and from analogy would be supposed to continue so; and yet, with all this practical evidence of the fallibility of our opinions, all, and none more than philosophers, are ready to assert the real truth of their opinions.

'The history of the opinions on the general nature of matter would afford remarkable illustrations in support of what I have said, but it does not belong to my subject to *extend upon it*. All I wish to point out is, by a reference to light, heat, electricity, &c., and the opinions formed on them, the necessity of cautious and slow decision on philosophical points, the care with which evidence ought to be admitted, and the continual guard against philosophical prejudices which should be preserved in the mind. The man who wishes to advance in knowledge should never of himself fix obstacles in the way.'

1819.
Æt. 27.

THE CONTINUITY OF THE GASEOUS AND LIQUID STATES OF MATTER*

IT may be truly affirmed of Physical Science, that its history, for some generations at least, has been one of rapid progress and unceasing change, and that its most earnest promoters have not claimed infallibility for their opinions, nor finality for their results. Its advancing progress has been marked by eras when some long-accepted theory or hypothesis, which had appeared so closely in accordance with all known experiments and observations as to have been received as an obvious truth, has, by further experiments extending into regions previously unexplored, been found to be a faulty or incomplete representation of the phenomena.

Such an era has occurred in the discovery recently announced by Dr. Andrews of the Continuity of the Gaseous and Liquid States of Matter.

We have all been accustomed to consider matter as existing in one or other of three states,—the solid, liquid, and gaseous.

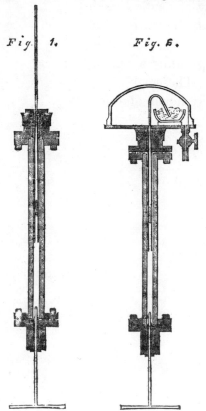

Fig. 1.   Fig. 2.

The transition, from any one of these states to another, has hitherto been regarded as necessarily abrupt ; at least, if we except the imperfectly understood conditions of softening or plasticity, assumed by such bodies as glass or iron, when gradually passing from the solid to the molten condition. The true state of the case is now found to be very different.

The memoir of Dr. Andrews, of which we propose to give an account in this article, opens with the following historical *résumé* of previous researches bearing more or less in the direction of his investigations :—" In 1822 M. Cagniard de la Tour observed that certain liquids, such as ether, alcohol, and water, when heated in hermetically sealed glass tubes, became apparently reduced to vapour in a space from twice to four times the original volume of the liquid. He also made a few numerical determinations of the pressures exerted in these experiments. In the following year Faraday succeeded in liquefying, by the aid of pressure alone, chlorine and several other bodies known before only in the gaseous form. A few years later Thilorier obtained solid carbonic acid, and observed that the coefficient of expansion of the liquid for heat is greater than that of any aëriform body.

*"The Bakerian Lecture for 1869." By Thomas Andrews, M.D., F.R.S. (Abridged from an Original Essay of Professor James Thomson, LL.D.)

A second memoir by Faraday, published in 1845, greatly extended our knowledge of the effects of cold and pressure on gases. Regnault has examined with care the absolute change of volume in a few gases when exposed to a pressure of twenty atmospheres,

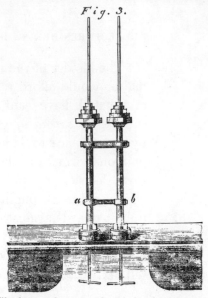

Fig. 3.

and Pouillet has made some observations on the same subject. The experiments of Natterer have carried this inquiry to the enormous pressure of 2,790 atmospheres ; and although his method is not altogether free from objection, the results he

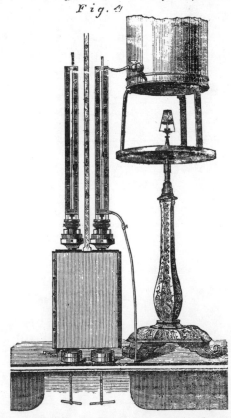

Fig. 4.

obtained are valuable, and deserve more attention than they have hitherto received."

In 1861 a brief notice appeared of some early experiments by Dr. Andrews in this direction. Oxygen, hydrogen, nitrogen, carbonic oxide, and nitric oxide were submitted to greater pressures than had previously been attained in glass tubes, and while under these pressures they were exposed to the cold of the carbonic acid and ether bath. None of these gases exhibited any appearance of liquefaction, although reduced to less than $\frac{1}{500}$ of their ordinary volume by the combined action of cold and pressure. Subsequently, in the third edition of Miller's "Chemical Physics," published in 1863, a short account, communicated by Dr. Andrews, appeared of some further results he had obtained, under certain fixed conditions of pressure and temperature, with carbonic acid. These results constitute the foundation of the researches which form the general subject of the present article, and the following extract from the original communication of Dr. Andrews to Dr. Miller may here be quoted :—"On partially liquefying carbonic acid by pressure alone, and gradually raising at the same time the temperature to 88° Fahr., the surface of demarcation between the liquid and the gas became fainter, lost its curvature, and at last disappeared. The space was then occupied by a homogeneous fluid, which exhibited, when the pressure was suddenly diminished or the temperature slightly lowered, a peculiar appearance of moving or flickering striæ throughout its entire mass. At temperatures

sure of 400 atmospheres or more. A section, exhibiting all the details, is given in Fig. 1. Before commencing an experiment the body of the apparatus was filled with water ; the upper end-piece, carrying the glass tube, in which was the gas to be operated on, was firmly secured in its place, and the pressure was obtained by screwing the steel screw into the water chamber. In Fig. 2 the same apparatus is shown with the modifications required when the gas or liquid is exposed to very low temperatures under high pressure. The end of the capillary tube dips into a bath of ether and solid carbonic acid, under a bell jar, from which the air may be exhausted.

In order to estimate the pressure exerted in these experiments, a duplex or compound form of the apparatus was employed, as shown in Fig. 3. The two sides of the apparatus freely communicate through $a\,b$, so that on turning either of the steel screws the pressure is immediately transmitted through the entire apparatus. In the second tube a known volume of air is confined, and the pressure is approximately estimated by its contraction.

Figure 4 exhibits the complete apparatus with the arrangements for maintaining the capillary tubes and the body of the apparatus itself at fixed temperatures. A rectangular brass case, closed before and behind with plate glass, surrounds each capil-

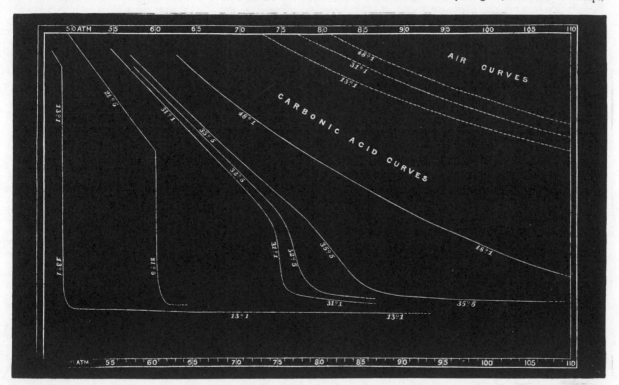

above 88°, no apparent liquefaction of carbonic acid or separation into two distinct forms of matter could be effected, even when a pressure of 300 or 400 atmospheres was applied. Nitrous oxide gave analogous results."

For his recent researches Dr. Andrews again selected carbonic acid as the substance for investigation. He devised for his experiments an apparatus, novel in construction, and well suited to exhibit the properties acquired by fluids under very varied conditions of pressure and temperature. The carbonic acid was contained in a glass tube, capillary in the upper and larger part of its length, and for the remainder, of the widest bore in which a column of mercury would remain without displacement when the tube was placed in a vertical position. A movable column or bar of mercury confined the gas to be operated on. This glass tube was secured by careful packing in a massive end-piece of brass, which carried a flange, by means of which a water-tight junction could be made with a corresponding flange, attached to a cold-drawn copper tube of great strength. To the other end of the copper tube a similar end-piece was firmly bolted. The latter carried a fine steel screw, 7 inches long, which was packed with such care that the packing was capable of resisting a pres-

lary tube, and allows it to be maintained at any required temperature by the flow of a stream of water. In the figure, the arrangement for obtaining a current of heated water in the case of the carbonic-acid tube is shown. The body of the apparatus itself, as is shown in the figure, is enclosed in an external vessel of copper, which is filled with water at the temperature of the apartment. This latter arrangement is essential when accurate observations are made.

The temperature of the water surrounding the air-tube was made to coincide, as closely as possible, with that of the apartment, while the temperature of the water surrounding the carbonic-acid tube varied in different experiments from 13° C. to 48° C. In the experiments as they were performed, the mercury did not come into view in the capillary part of the air-tube till the pressure amounted to about forty atmospheres. The volumes of the air and of the carbonic acid were carefully read by a cathetometer, and the results could be relied on with certainty to less than 1-20 of a millimetre or 1-500 of an inch. We must refer the reader to the original memoir for an account of the details of the experiments, and of the numerous precautions adopted to secure accuracy. The object of the present article is to place as clearly as possible before his mind the main results arrived at, and the general features of the apparatus employed.

In the above diagram, we have a graphical representation of the results of a large number of comparative experiments on air and carbonic acid, under pressures ranging from 48 to 107 atmospheres, and at temperatures for the carbonic acid varying from 13°·1 C. to 48°·1 C. The dotted lines (*Air Curves*) represent a portion of the curves of a perfect gas (assumed to have the same volume originally at 0° C., and under one atmosphere as the carbonic acid), for the temperatures of 13°·1 C., 31°·1 C., and 48°·1 C. The lines designated *Carbonic Acid Curves* show the volumes to which the carbonic acid is reduced at the temperatures marked on each curve, and under the approximate pressures indicated by the numbers at the top and bottom of the figure. Ordinates drawn from the inner horizontal line at the lower part of the figure to meet the curves, will represent the volume of the carbonic acid. These ordinates do not always refer to homogeneous matter, but sometimes to a mixture of gas and liquid.

It will be observed that in the curves for 13°·1 C. there occurs an abrupt, or almost quite abrupt, fall, when a pressure of about 49 atmospheres has been attained. The curve for 21°·5 C. exhibits a corresponding fall, but not till a higher pressure (about 60 atmospheres) has been reached. On close inspection of the figure, a slight deviation from perfect abruptness will be observed in the portion of the curves representing these falls, which Dr. Andrews showed to be due to a trace of air (about $\frac{1}{500}$ part) in the carbonic acid with which the experiments were made. Had the carbonic acid been absolutely pure, there can be no doubt that the fall would have been quite abrupt.

In the curve for 31°·1 C. there is no abrupt fall; but a rapid descent, indicating a corresponding diminution of volume, occurs between the pressures of 73 and 75 atmospheres. As the temperature rises this descent becomes gradually less marked, and when a temperature of 48°·1 C. has been attained, it has almost, if not altogether, disappeared.

At any temperature between −57°C., and 30°·92 C., carbonic acid, under the ordinary pressure of the atmosphere, is unquestionably in the state of a gas or vapour. If within these limits we take a given volume of carbonic acid, and gradually augment the pressure, the volume will steadily diminish, not however uniformly, but according to a more rapid rate than the law for a perfect gas, till we reach the point at which liquefaction begins. A sudden fall or diminution of volume will now take place, and with a little care it will be found easy so to arrange the experiment that part of the carbonic acid shall be in the liquid, and part of it in the gaseous state; the carbonic acid thus coexisting in two distinct physical conditions in the same tube, and under the same external pressure. But if the experiment be made at 30°·92 C., or any higher temperature, the result will be very different. At 30°·92 C., and under a pressure of about 74 atmospheres, the densities of liquid and gaseous carbonic acid, as well as all their other physical properties, become absolutely identical, and the most careful observation fails to discover any heterogeneity at this or higher temperatures in carbonic acid, when its volume is so reduced as to occupy a space in which, at lower temperatures, a mixture of gas and liquid would have been formed. In other words, all distinctions of state have disappeared, and the carbonic acid has become one homogeneous fluid, which cannot by change of pressure be separated into two distinct physical conditions. This temperature of 30°·92 is called by Dr. Andrews the *critical point* of carbonic acid. Other fluids which can be obtained in both the liquid and gaseous states have shown similar phenomena, and have each presented a critical point of temperature. The rapid changes of density which slight changes of temperature or pressure produce, when the gas is reduced at temperatures a little above the critical point, to the volume at which it might be expected to liquefy, account for the flickering movements referred to in the beginning of this article.

The general conclusions arrived at we give in the words of the original memoir. "I have frequently exposed carbonic acid," observes Dr. Andrews, "without making precise measurements, to much higher pressures than any marked in the tables, and have made it pass, without break or interruption, from what is regarded by every one as the gaseous state, to what is, in like manner, universally regarded as the liquid state. Take, for example, a given volume of carbonic acid gas at 50° C., or at a higher temperature, and expose it to increasing pressure till 150 atmospheres have been reached. In this process its volume will steadily diminish as the pressure augments, and no sudden diminution of volume, without the application of external pressure, will occur at any stage of it. When the full pressure has been applied, let the temperature be allowed to fall, till the carbonic acid has reached the ordinary temperature of the atmosphere. During the whole of this operation, no breach of continuity has occurred. It begins with a gas, and by series of gradual changes, presenting nowhere any abrupt alteration of volume or sudden evolution of heat, it ends with a liquid. The closest observation fails to discover anywhere indications of a change of condition in the carbonic acid, or evidence, at any period of the process, of part of it being in one physical state and part in another. That the gas has actually changed into a liquid would, indeed, never have been suspected, had it not shown itself to be so changed by entering into ebullition on the removal of the pressure. For convenience this process has been divided into two stages, the compression of the carbonic acid, and its subsequent cooling; but these operations might have been performed simultaneously, if care were taken so to arrange the application of the pressure and the rate of cooling that the pressure should not be less than 76 atmospheres when the carbonic acid had cooled to 31°.

"We are now prepared for the consideration of the following important question. What is the condition of carbonic acid when it passes, at temperatures above 31°, from the gaseous state down to the volume of the liquid, without giving evidence at any part of the process of liquefaction having occurred? Does it continue in the gaseous state, or does it liquefy, or have we to deal with a new condition of matter? If the experiment were made at 100°, or at a higher temperature, when all indications of a fall had disappeared, the probable answer which would be given to this question is that the gas preserves its gaseous condition during the compression; and few would hesitate to declare this statement to be true, if the pressure, as in Natterer's experiments, were applied to such gases as hydrogen or nitrogen. On the other hand, when the experiment is made with carbonic acid at temperatures a little above 31°, the great fall which occurs at one period of the process would lead to the conjecture that liquefaction had actually taken place, although optical tests carefully applied failed at any time to discover the presence of a liquid in contact with a gas. But against this view it may be urged, with great force, that the fact of additional pressure being always required for a further diminution of volume, is opposed to the known laws which hold in the change of bodies from the gaseous to the liquid state. Besides, the higher the temperature at which the gas is compressed, the less the fall becomes, and at last it disappears.

"The answer to the foregoing question, according to what appears to me to be the true interpretation of the experiments already described, is to be found in the close and intimate relations which subsist between the gaseous and liquid states of matter. The ordinary gaseous and ordinary liquid states are, in short, only widely separated forms of the same condition of matter, and may be made to pass into one another by a series of gradations so gentle that the passage shall nowhere present any interruption or breach of continuity. From carbonic acid as a perfect gas to carbonic acid as a perfect liquid, the transition we have seen, may be accomplished by a continuous process, and the gas and liquid are only distant stages of a long series of continuous physical changes. Under certain conditions of temperature and pressure, carbonic acid finds itself, it is true, in what may be described as a state of instability, and suddenly passes, with the evolution of heat, and without the application of additional pressure or change of temperature, to the volume which by the continuous process can only be reached through a long and circuitous route. In the abrupt change which here occurs, a marked difference is exhibited, while the process is going on, in the optical and other physical properties of the carbonic acid which has collapsed into the smaller volume, and of the carbonic acid not yet altered. There is no difficulty here, therefore, in distinguishing between the liquid and the gas. But in other cases the distinction cannot be made; and under many of the conditions I have described it would be vain to attempt to assign carbonic acid to the liquid rather than the gaseous state. Carbonic acid, at the temperature of 35°·5, and under a pressure of 108 atmospheres, is reduced to $\frac{1}{130}$ of the volume it occupied under a pressure of one atmosphere; but if any one ask whether it is now in the gaseous or liquid state, the question does not, I believe, admit of a positive reply. Carbonic acid at 35°·5, and under 108 atmospheres of pressure, stands nearly midway between the gas and the liquid ; and we have no valid grounds for assigning it to the one form of matter any more than to the other. The same observation would apply with even greater force to the state in which carbonic acid exists at higher temperatures and under greater pressures than those just mentioned. In the original experiment of Cagniard de la Tour, that distinguished physicist inferred that the liquid had disappeared, and had changed into a gas. A slight modification of the conditions of his experiment would have led him to the opposite conclusion, that what had been before a gas was changed into a liquid. These conditions are, in short, the intermediate states which matter assumes in passing, without sudden change of volume, or abrupt evolution of heat, from the ordinary liquid to the ordinary gaseous state.

"In the foregoing observations I have avoided all reference to the molecular forces brought into play in these experiments. The resistance of liquids and gases to external pressure tending to produce a diminution of volume proves the existence of an internal force of an expansive or resisting character. On the other hand, the sudden diminution of volume, without the application of additional pressure externally, which occurs when a gas is compressed, at any temperature below the critical point, to the volume at which liquefaction begins, can scarcely be explained without assuming that a molecular force of great attractive power comes here into operation, and overcomes the resistance to diminution of volume, which commonly requires the application of external force. When the passage from the gaseous to the liquid state is effected by the continuous process described in the foregoing pages, these molecular forces are so modified as to be unable at any stage of the process to overcome alone the resistance of the fluid to change of volume.

"The properties described in this communication, as exhibited by carbonic acid, are not peculiar to it, but are generally true of all bodies which can be obtained as gases and liquids. Nitrous oxide, hydrochloric acid, ammonia, sulphuric ether, and sulphuret of carbon, all exhibited, at fixed pressures and temperatures, critical points, and rapid changes of volume with flickering movements, when the temperature or pressure was changed in the neighbourhood of those points. The critical points of some of these bodies were above 100° ; and in order to make the observations, it was necessary to bend the capillary tube before the commencement of the experiment, and to heat it in a bath of paraffin or oil of vitriol.

"The distinction between a gas and vapour has hitherto been founded on principles which are altogether arbitrary. Ether in the state of gas is called a vapour, while sulphurous acid in the same state is called a gas, yet they are both vapours, the one derived from a liquid boiling at 35°, the other from a liquid boiling at − 10°. The distinction is thus determined by the trivial condition of the boiling-point of the liquid, under the ordinary pressure of the atmosphere, being higher or lower than the ordinary temperature of the atmosphere. Such a distinction may have some advantages for practical reference, but it has no scientific value. The critical point of temperature affords a criterion for distinguishing a vapour from a gas, if it be considered important to maintain the distinction at all. Many of the properties of vapours depend on the gas and liquid being present in contact with one another ; and this, we have seen, can only occur at temperatures below the critical point. We may accordingly define a vapour to be a gas at any temperature under its critical point. According to this definition, a vapour may, by pressure alone, be changed into a liquid, and may therefore exist in presence of its own liquid ; while a gas cannot be liquefied by pressure, that is, so changed by pressure as to become a visible liquid distinguished by a surface of demarcation from the gas. If this definition be accepted, carbonic acid will be a vapour below 31°, a gas above that temperature ; ether, a vapour below 200°, a gas above that temperature.

"We have seen that the gaseous and liquid states are only distant stages of the same condition of matter, and are capable of passing into one another by a process of continuous change. A problem of far greater difficulty yet remains to be solved, the possible continuity of the liquid and solid states of matter. But this must be a subject for future investigation ; and for the present I will not venture to go beyond the conclusion I have already drawn from direct experiment, that the gaseous and liquid forms of matter may be transformed into one another by a series of continuous and unbroken changes."

JAMES THOMSON

NOTES

AT last a sum of money has been voted for a new Natural History Museum. In introducing the vote the Chancellor of the Exchequer said the British Museum had long been suffering from repletion, and there were no means of exhibiting the valuable articles which, from time to time, were bought for the national collection. Five years ago the trustees resolved in favour of separating the collections, and it had been determined to separate the natural history department from the books and antiquities. For the natural history collection the typical mode of exhibition had been decided on, and the building required must cover at least four acres. Even the present collection would pretty well fill a building of these dimensions, and provision must be made for further extension. The question was, where should this building be situated? and after referring to possible sites he referred to the locality which we were enabled to state some time ago had been chosen—a plot of ground 16½ acres in extent, which the trustees of the Exhibition of 1851 sold to the Government at 7,000*l.* an acre. It therefore cost 120,000*l.*, but is now worth 100,000*l.* more. The sale was coupled with the condition that any building erected upon the land must be for purposes of science and art. For seven years the land had remained waste, a sort of Potter's field, and a scandal to that part of the metropolis. The Government now proposed to place on that piece of land the museum required for the natural history collection. It would occupy four acres ; there would be room for wings, and the outside estimate for the building was 350,000*l.*, not an unreasonable price, considering its extent. For the present, however, the Government merely asked for a small vote to enable them to clear the ground, and in order to take the opinion of the House. Railway communication had now made South Kensington easily accessible, and unless a more eligible, a more accessible, and a cheaper site could be suggested, he hoped the Committee would agree to the proposal. He might add that, if it were hereafter thought desirable to do so, there would be room enough on the same site for the Patent Museum, the necessity of which had been much insisted on. We trust that after the discussion which followed the introduction of the vote the scientific men will speak for themselves, and again let their wishes and opinions be heard.

THE American Association for the Advancement of Science met yesterday (Wednesday) at Troy. Professor W. Chauvenet is president for the year.

IT is gratifying to learn that some of the recommendations of the Royal Commission on Military Education, which were most inimical to the scientific instruction of the army, will not be carried out.

BY Imperial decree the *Association Scientifique de France* has been acknowledged to be an *établissement d'utilité publique*.

THE French observers are making preparations for a combined attack on the 10th of August meteors.

THE list of pensions granted during the year ended the 20th of June, 1870, and charged upon the civil list (presented pursuant to Act 1 Victoria, cap. 2, sec. 6) has been published this week. Among them we note the following :—Mr. Augustus De Morgan, 100*l.*, in consideration of his distinguished merits as a mathematician ; Mrs. Charlotte J. Thompson, 40*l.*, in consideration of the labours of her late husband, Mr. Thurston Thompson, as Official Photographer to the Science and Art Department, and of his personal services to the late Prince Consort ; Dame Henrietta Grace Baden Powell, 150*l.*, in consideration of the valuable services to science rendered by her husband during the 33 years he held the Savilian Professorship of Geometry and Astronomy at Oxford ;

ON THE GASEOUS AND LIQUID STATES OF MATTER

A DISCOURSE was delivered on Friday evening, June 2, at the Royal Institution in Albemarle Street, by Dr. Andrews on the " Gaseous and Liquid States of Matter," from which we make the following extracts :—" The liquid state of matter forms a link between the solid and gaseous states. This link is, however, often suppressed, and the solid passes directly into the gaseous or vaporous form. In the intense cold of an arctic winter, hard ice will gradually change into transparent vapour without previously assuming the form of water. Carbonic acid snow passes rapidly into gas when exposed to the air, and can with difficulty be liquefied in open tubes. Its boiling point, as Faraday has shown, presents the apparent anomaly of being lower in the thermometric scale than its melting point, a statement less paradoxical than it may at first appear, if we remember that water can exist as vapour at temperatures far lower than those at which it can exist as liquid. Whether the transition be directly from solid to gaseous, or from solid to liquid and from liquid to gaseous, a marked change of physical properties occurs at each step or break, and heat is absorbed, as was proved long ago by Black, without producing elevation of temperature. Many solids and liquids will for this reason maintain a low temperature, even when surrounded by a white hot atmosphere, and the remarkable experiment of solidifying water and even mercury on a red hot plate, finds thus an easy explanation. The term spheroidal state, when applied to water floating on a cushion of vapour over a red hot plate, is, however, apt to mislead. The water is not here in any peculiar state. It is simply water evaporating rapidly at a few degrees below its boiling point, and all its properties, even those of capillarity, are the properties of ordinary water at 96°·5C. The interesting phenomena

exhibited under these conditions are due to other causes, and not to any new or peculiar state of the liquid itself. The fine researches of Dalton upon vapours, and the memorable discovery by Faraday of the liquefaction of gases by pressure alone, finished the work which Black had begun. Our knowledge of the conditions under which matter passes abruptly from the gaseous to the liquid and from the liquid to the solid state may now be regarded as almost complete.

"In 1822 Cagniard de la Tour made some remarkable experiments, which still bear his name, and which may be regarded as the starting point of the investigations which form the chief subject of this address. Cagniard de la Tour's first experiments were made in a small Papin's digester constructed from the thick end of a gun barrel, into which he introduced a little alcohol and also a small quartz ball, and firmly closed the whole. On heating the gun barrel with its contents over an open fire, and observing from time to time the sound produced by the ball when the apparatus was shaken, he inferred that after a certain temperature was attained the liquid had disap-

Fig. 1—Cloud below critical point

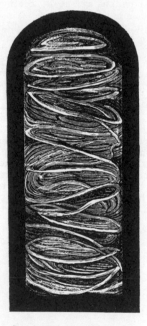

Fig. 2—Striæ above critical point

peared. He afterwards succeeded in repeating the experiment in glass tubes, and arrived at the following results. An hermetically sealed glass tube, containing sufficient alcohol to occupy two-fifths of its capacity, was gradually heated, when the liquid was seen to dilate, and its mobility at the same time to become gradually greater. After attaining to nearly twice its original volume, the liquid completely disappeared, and was converted into a vapour so transparent that the tube appeared to be quite empty. On allowing the tube to cool, a very thick cloud was formed, after which the liquid reappeared in its former state.

"It is singular that in this otherwise accurate description Cagniard de la Tour should have overlooked the most remarkable phenomenon of all—the moving or flickering striæ which fill the tube, when, after heating it above the *critical point*, the temperature is quickly lowered. This phenomenon was first observed by the lecturer in 1863, when experimenting with carbonic acid, and may be admirably seen by heating such liquids as ether or sulphurous acid in hermetically sealed tubes, of which when cold they occupy about one-third of the capacity. The

appearances exhibited by the ascending and descending sheets of matter of unequal density are most remarkable, but it is difficult to give an adequate description of them in words or even to delineate them.

"These striæ arise from the great changes of density which slight variations of temperature or pressure produce when liquids are heated in a confined space above the critical point already referred to; but they are not formed if the temperature and pressure are kept steady. When seen they are always a proof that the matter in the tube is homogeneous, and that we have not liquid and gas in presence of one another. They are, in short, an extraordinary development of the movements observed in ordinary liquids and gases when heated from below. The fact that at a temperature 0°·2 above its critical point carbonic acid diminishes to one-half its volume from an increase of only $\frac{1}{37}$ of the entire pressure is sufficient to account for the marked characters they exhibit.

"If the temperature is allowed to fall a little below the critical point, the formation of cloud shows that we have now heterogeneous matter in the tube, minute drops of liquid in presence of a gas. From the midst of this cloud (as shown in Fig. 1) a faint surface of demarcation appears, constituting the boundary between liquid and gas, but at first wholly devoid of curvature. We must, however, take care not to suppose that a cloud necessarily precedes the formation of true liquid. If the pressure be sufficiently great, no cloud of any kind will form."

After describing the results obtained by the lecturer with carbonic acid under varied conditions of temperature and pressure, of which a full account has already appeared in NATURE,* Dr. Andrews remarked that it would be erroneous to say that between liquid and gas there exists one intermediate state of matter, but that it is correct to say that between ordinary liquid and ordinary gas there is an infinite number of intermediate conditions of matter, establishing perfect continuity between the two states. Under great pressures the passage from the liquid to the gaseous state is effected on the application of heat without any break or breach of continuity. A solid model, constructed by Prof. J. Thomson, from the data furnished by the experiments of the lecturer, exhibited very clearly the different paths which connect the liquid and gaseous states, showing the ordinary passage by break from the liquid, as well as the continuous passages above the critical point.

After referring to the experiments of Frankland on the change produced by pressure in the spectrum of hydrogen, and to those of the same able chemist and Lockyer on the spectrum of the spark in compressed gases, Dr. Andrews described the remarkable change from a translucent to an opaque body, which occurs when bromine is heated above the critical point; and then drew attention to the general fact that when the critical point is reached, the density of the liquid and gas become identical.

In order to establish the continuity of the solid and liquid states, it would be necessary in like manner, by the combined action of heat and pressure, to obtain the solid and liquid of the same density and of like physical properties. To accomplish this result would probably require pressures far beyond any which can be reached in transparent tubes, but future experiment may show that the solid and liquid can be made to approach to the required conditions.

See NATURE, vol. ii. p. 278.

ON RADIANT MATTER [1]

TO throw light on the title of this lecture I must go back more than sixty years—to 1816. Faraday, then a mere student and ardent experimentalist, was twenty-four years old, and at this early period of his career he delivered a series of lectures on the general properties of matter, and one of them bore the remarkable title, "On Radiant Matter." The great philosopher's notes of this lecture are to be found in Dr. Bence Jones's "Life and Letters of Faraday," and I will here quote a passage in which he first employs the expression *Radiant Matter* :—

"If we conceive a change as far beyond vaporisation as that is above fluidity, and then take into account also the proportional increased extent of alteration as the changes rise, we shall perhaps, if we can form any conception at all, not fall far short of radiant matter ; and as in the last conversion many qualities were lost, so here also many more would disappear."

Faraday was evidently engrossed with this far-reaching speculation, for three years later—in 1819—we find him bringing fresh evidence and argument to strengthen his startling hypothesis. His notes are now more extended, and they show that in the intervening three years he had thought much and deeply on this higher form of matter. He first points out that matter may be classed into four states—solid, liquid, gaseous, and radiant—these modifications depending upon differences in their several essential properties. He admits that the existence of radiant matter is as yet unproved, and then proceeds, in a series of ingenious analogical arguments, to show the probability of its existence. [2]

If, in the beginning of this century, we had asked, What is a gas ? the answer then would have been that it is matter, expanded and rarefied to such an extent as to be impalpable, save when set in violent motion ; invisible, incapable of assuming or of being reduced into any definite form like solids, or of forming drops like liquids ; always ready to expand where no resistance is offered, and to contract on being subjected to pressure. Sixty years ago such were the chief attributes assigned to gases. Modern research, however, has greatly enlarged and modified our views on the constitution of these elastic fluids. Gases are now considered to be composed of an almost infinite number of small particles or molecules, which are constantly moving in every direction with velocities of all conceivable magnitudes. As these molecules are exceedingly numerous, it follows that no molecule can move far in any direction without coming in contact with some other molecule. But if we exhaust the air or gas contained in a closed vessel, the number of molecules becomes diminished, and the distance through which any one of them can move without coming in contact with another is increased, the length of the mean free path being inversely proportional to the number of molecules present. The further this process is carried the longer becomes the average distance a molecule can travel before entering into collision ; or, in other words, the longer its mean free path the more the physical properties of the gas or air are modified. Thus, at a certain point, the phenomena of the radiometer become possible, and on pushing the rarefaction still further, *i.e.*, decreasing the number of

molecules in a given space and lengthening their mean free path, the experimental results are obtainable to which I am now about to call your attention. So distinct are these phenomena from anything which occurs in air or gas at the ordinary tension, that we are led to assume that we are here brought face to face with matter in a fourth state or condition, a condition as far removed from the state of gas as a gas is from a liquid.

Mean Free Path. Radiant Matter

I have long believed that a well-known appearance observed in vacuum tubes is closely related to the phenomena of the mean free path of the molecules. When the negative pole is examined while the discharge from an induction-coil is passing through an exhausted tube, a dark space is seen to surround it. This dark space is found to increase and diminish as the vacuum is varied, in the same way that the mean free path of the molecules lengthens and contracts. As the one is perceived by the mind's eye to get greater, so the other is seen by the bodily eye to increase in size ; and if the vacuum is insufficient to permit much play of the molecules before they enter into collision, the passage of electricity shows that the "dark space" has shrunk to small dimensions. We naturally infer that the dark space is the mean free path of the molecules of the residual gas, an inference confirmed by experiment.

I will endeavour to render this "dark space" visible to all present. Here is a tube (Fig. 1) having a pole in the centre in

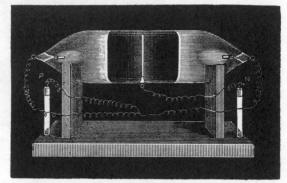

FIG. 1.

the form of a metal disk, and other poles at each end. The centre pole is made negative, and the two end poles connected together are made the positive terminal. The dark space will be in the centre. When the exhaustion is not very great the dark space extends only a little on each side of the negative pole in the centre. When the exhaustion is good, as in the tube before you, and I turn on the coil, the dark space is seen to extend for about an inch on each side of the pole.

Here, then, we see the induction spark actually illuminating the lines of molecular pressure caused by the excitement of the negative pole. The thickness of this dark space is the measure of the mean free path between successive collisions of the molecules of the residual gas. The extra velocity with which the negatively electrified molecules rebound from the excited pole, keeps back the more slowly moving molecules which are advancing towards that pole. A conflict occurs at the boundary of the dark space, where the luminous margin bears witness to the energy of the discharge.

Therefore the residual gas—or, as I prefer to call it, the gaseous residue—within the dark space, is in an entirely different state to that of the residual gas in vessels at a lower degree of exhaustion. To quote the words of our last year's President, in his address at Dublin :—

"In the exhausted column we have a vehicle for electricity not constant like an ordinary conductor, but itself modified by the passage of the discharge, and perhaps subject to laws differing materially from those which it obeys at atmospheric pressure."

In the vessels with the lower degree of exhaustion, the length of the mean free path of the molecules is exceedingly small as compared with the dimensions of the bulb, and the properties belonging to the ordinary gaseous state of matter, depending upon constant collisions, can be observed. But in the phenomena now about to be examined, so high is the exhaustion carried that the dark space around the negative pole has widened out till it entirely fills the tube. By great rarefaction the mean free path

[1] A lecture delivered to the British Association for the Advancement of Science, at Sheffield, Friday, August 22, 1879, by William Crookes, F.R.S.

[2] "I may now notice a curious progression in physical properties accompanying changes of form, and which is perhaps sufficient to induce, in the inventive and sanguine philosopher, a considerable degree of belief in the association of the radiant form with the others in the set of changes I have mentioned.

"As we ascend from the solid to the fluid and gaseous states, physical properties diminish in number and variety, each state losing some of those which belonged to the preceding state. When solids are converted into fluids, all the varieties of hardness and softness are necessarily lost. Crystalline and other shapes are destroyed. Opacity and colour frequently give way to a colourless transparency, and a general mobility of particles is conferred.

"Passing onward to the gaseous state, still more of the evident characters of bodies are annihilated. The immense differences in their weight almost disappear ; the remains of difference in colour that were left are lost. Transparency becomes universal, and they are all elastic. They now form but one set of substances, and the varieties of density, hardness, opacity, colour, elasticity, and form, which render the number of solids and fluids almost infinite, are now supplied by a few slight variations in weight, and some unimportant shades of colour.

"To those, therefore, who admit the radiant form of matter, no difficulty exists in the simplicity of the properties it possesses, but rather an argument in their favour. These persons show you a gradual resignation of properties in the matter we can appreciate as the matter ascends in the scale of forms, and they would be surprised if that effect were to cease at the gaseous state. They point out the greater exertions which nature makes at each step of the change, and think that, consistently, it ought to be greatest in the passage from the gaseous to the radiant form."—*Life and Letters of Faraday*, vol. i. p. 308.

has become so long that the hits in a given time in comparison to the misses may be disregarded, and the average molecule is now allowed to obey its own motions or laws without interference. The mean free path, in fact, is comparable to the dimensions of the vessel, and we have no longer to deal with a *continuous* portion of matter, as would be the case were the tubes less highly exhausted, but we must here contemplate the molecules *individually*. In these highly exhausted vessels the molecules of the gaseous residue are able to dart across the tube with comparatively few collisions, and radiating from the pole with enormous velocity, they assume properties so novel and so characteristic as to entirely justify the application of the term borrowed from Faraday, that of *Radiant Matter*.

Radiant Matter exerts powerful Phosphorogenic Action where it strikes

I have mentioned that the radiant matter within the dark space excites luminosity where its velocity is arrested by residual gas outside the dark space. But if no residual gas is left, the molecules will have their velocity arrested by the sides of the glass; and here we come to the first and one of the most noteworthy properties of radiant matter discharged from the negative pole—its power of exciting phosphorescence when it strikes

against solid matter. The number of bodies which respond luminously to this molecular bombardment is very great, and the resulting colours are of every variety. Glass, for instance, is highly phosphorescent when exposed to a stream of radiant matter. Here (Fig. 2) are three bulbs composed of different glass: one is uranium glass (*a*), which phosphoresces of a dark green colour; another is English glass (*b*), which phosphoresces of a blue colour; and the third (*c*) is soft German glass—of which most of the apparatus before you is made—which phosphoresces of a bright apple-green.

My earlier experiments were almost entirely carried on by the aid of the phosphorescence which glass takes up when it is under the influence of the radiant discharge; but many other substances possess this phosphorescent power in a still higher degree than glass. For instance, here is some of the luminous sulphide of calcium prepared according to M. Ed. Becquerel's description. When the sulphide is exposed to light—even candle-light—it phosphoresces for hours with a bluish-white colour. It is, however, much more strongly phosphorescent to the molecular discharge in a good vacuum, as you will see when I pass the discharge through this tube.

Other substances besides English, German, and uranium glass, and Becquerel's luminous sulphides, are also phosphorescent.

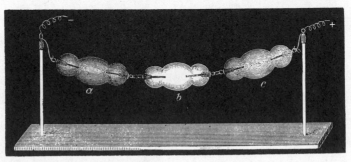

FIG. 2.

The rare mineral phenakite (aluminate of glucinum) phosphoresces blue; the mineral spodumene (a silicate of aluminium and lithium) phosphoresces a rich golden yellow; the emerald gives out a crimson light. But without exception, the diamond is the most sensitive substance I have yet met for ready and brilliant phosphorescence. Here is a very curious fluorescent diamond,

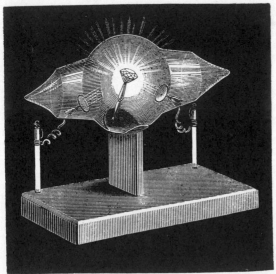

FIG. 3.

green by daylight, colourless by candle-light. It is mounted in the centre of an exhausted bulb (Fig. 3), and the molecular discharge will be directed on it from below upwards. On darkening the room you see the diamond shines with as much light as a candle, phosphorescing of a bright green.

Next to the diamond the ruby is one of the most remarkable stones for phosphorescing. In this tube (Fig. 4) is a fine collection of ruby pebbles. As soon as the induction-spark is turned on you will see these rubies shining with a brilliant rich red tone, as if they were glowing hot. It scarcely matters what colour the ruby is, to begin with. In this tube of natural rubies there are stones of all colours—the deep red and also the pale pink ruby. There are some so pale as to be almost colourless, and some of the highly-prized tint of pigeon's blood; but under the impact of radiant matter they all phosphoresce with about the same colour.

Now the ruby is nothing but crystallised alumina with a little

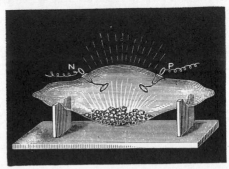

FIG. 4.

colouring-matter. In a paper by Ed. Becquerel,[1] published twenty years ago, he describes the appearance of alumina as glowing with a rich red colour in the phosphoroscope. Here is some precipitated alumina prepared in the most careful manner. It has been heated to whiteness, and you see it also glows under the molecular discharge with the same rich red colour.

The spectrum of the red light emitted by these varieties of alumina is the same as described by Becquerel twenty years ago. There is one intense red line, a little below the fixed line B in

[1] *Annales de Chimie et de Physique*, 3rd series, vol. lvii., p. 50, 1859.

the spectrum, having a wave-length of about 6895. There is a continuous spectrum beginning at about B, and a few fainter lines beyond it, but they are so faint in comparison with this red line that they may be neglected. This line is easily seen by examining with a small pocket spectroscope the light reflected from a good ruby.

There is one particular degree of exhaustion more favourable than any other for the development of the properties of radiant matter which are now under examination. Roughly speaking, it may be put at the millionth of an atmosphere.[1] At this degree of exhaustion the phosphorescence is very strong, and after that it begins to diminish until the spark refuses to pass.[2]

I have here a tube (Fig. 5) which will serve to illustrate the

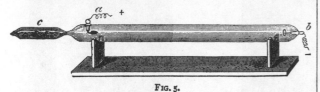

FIG. 5.

dependence of the phosphorescence of the glass on the degree of exhaustion. The two poles are at *a* and *b*, and at the end (*c*) is a small supplementary tube, connected with the other by a narrow aperture, and containing solid caustic potash. The tube has been exhausted to a very high point, and the potash heated so as to drive off moisture and injure the vacuum. Exhaustion has then been recommenced, and the alternate heating and exhaustion repeated until the tube has been brought to the state in which it now appears before you. When the induction-spark is first turned on nothing is visible—the vacuum is so high that the tube is non-conducting. I now warm the potash slightly, and liberate a trace of aqueous vapour. Instantly conduction commences, and the green phosphorescence flashes out along the length of the tube. I continue the heat, so as to drive off more gas from the potash. The green gets fainter, and now a wave of cloudy luminosity sweeps over the tube, and stratifications appear, which rapidly get narrower, until the spark passes along the tube in the form of a narrow purple line. I take the lamp away, and allow the potash to cool; as it cools, the aqueous vapour, which the heat had driven off, is re-absorbed. The purple line broadens out, and breaks up into fine stratifications; these get wider, and travel towards the potash tube. Now a wave of green light appears on the glass at the other end, sweeping on and driving the last pale stratification into the potash;

[1]

1·0 millionth of an atmosphere	=	0·00076 millim.	
1315·789 millionths of an atmosphere	=	1·0 millim.	
1,000,000· ,, ,, ,,	=	760·0 millims.	
,, ,, ,, ,,	=	1 atmosphere.	

[2] Nearly 100 years ago Mr. Wm. Morgan communicated to the Royal Society a paper entitled "Electrical Experiments made to ascertain the Non-conducting Power of a Perfect Vacuum, &c." The following extracts from this paper, which was published in the *Phil. Trans.* for 1785 (vol. lxxv. p. 272), will be read with interest:—

"A mercurial gage about 15 inches long, carefully and accurately boiled till every particle of air was expelled from the inside, was coated with tin-foil 5 inches down from its sealed end, and being inverted into mercury through a perforation in the brass cap which covered the mouth of the cistern; the whole was cemented together, and the air was exhausted from the inside of the cistern through a valve in the brass cap, which, producing a perfect vacuum in the gage, formed an instrument peculiarly well adapted for experiments of this kind. Things being thus adjusted (a small wire having been previously fixed on the inside of the cistern to form a communication between the brass cap and the mercury, into which the gage was inverted) the coated end was applied to the conductor of an electrical machine, and notwithstanding every effort, neither the smallest ray of light, nor the slightest charge, could ever be procured in this exhausted gage."

"If the mercury in the gage be imperfectly boiled, the experiment will not succeed; but the colour of the electric light, which, in air rarefied by an exhauster, is always violet or purple, appears in this case of a beautiful green, and, what is very curious, the degree of the air's rarefaction may be nearly determined by this means; for I have known instances, during the course of these experiments, where a small particle of air, having found its way into the tube, the electric light became visible, and, as usual, of a green colour; but the charge being often repeated, the gage has at length cracked at its sealed end, and in consequence the external air, by being admitted into the inside, has gradually produced a change in the electric light from green to blue, from blue to indigo, and so on to violet and purple, till the medium has at length become so dense as no longer to be a conductor of electricity. I think there can be little doubt, from the above experiments, of the non-conducting power of a perfect vacuum."

"This seems to prove that there is a limit even in the rarefaction of air, which sets bounds to its conducting power; or, in other words, that the particles of air may be so far separated from each other as no longer to be able to transmit the electric fluid; and if they are brought within a certain distance of each other their conducting power begins, and continually increases till their approach also arrives at its limit."

and now the tube glows over its whole length with the green phosphorescence. I might keep it before you, and show the green growing fainter and the vacuum becoming non-conducting, but I should detain you too long, as time is required for the absorption of the last traces of vapour by the potash, and I must pass on to the next subject.

Radiant Matter proceeds in straight Lines

The radiant matter whose impact on the glass causes an evolution of light, absolutely refuses to turn a corner. Here is a V-shaped tube (Fig. 6), a pole being at each extremity. The pole at the right side (*a*) being negative, you see that the whole of the right arm is flooded with green light, but at the bottom it stops sharply and will not turn the corner to get into the left side. When I reverse the current and make the left pole negative, the green changes to the left side, always following the negative pole and leaving the positive side with scarcely any luminosity.

In the ordinary phenomena exhibited by vacuum tubes—phenomena with which we are all familiar—it is customary, in order to bring out the striking contrasts of colour, to bend the tubes into very elaborate designs. The luminosity caused by the phosphorescence of the residual gas follows all the convolutions into which skilful glass-blowers can manage to twist the glass. The

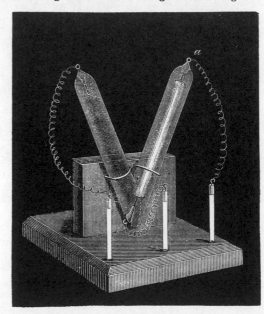

FIG. 6.

negative pole being at one end and the positive pole at the other, the luminous phenomena seem to depend more on the positive than on the negative at the ordinary exhaustion hitherto used to get the best phenomena of vacuum tubes. But at a very high exhaustion the phenomena noticed in ordinary vacuum tubes when the induction spark passes through them—an appearance of cloudy luminosity and of stratifications—disappear entirely. No cloud or fog whatever is seen in the body of the tube, and with such a vacuum as I am working with in these experiments, the only light observed is that from the phosphorescent surface of the glass. I have here two bulbs (Fig. 7), alike in shape and position of poles, the only difference being that one is at an exhaustion equal to a few millimetres of mercury—such a moderate exhaustion as will give the ordinary luminous phenomena—whilst the other is exhausted to about the millionth of an atmosphere. I will first connect the moderately exhausted bulb (A) with the induction-coil, and retaining the pole at one side (*a*) always negative, I will put the positive wire successively to the other poles with which the bulb is furnished. You see that as I change the position of the positive pole, the line of violet light joining the two poles changes, the electric current always choosing the shortest path between the two poles, and moving about the bulb as I alter the position of the wires.

This, then, is the kind of phenomenon we get in ordinary exhaustions. I will now try the same experiment with a bulb (B) that is very highly exhausted, and as before, will make the side pole (*a'*) the negative, the top pole (*b*) being positive. Notice how widely different is the appearance from that shown by the last bulb. The negative pole is in the form of a shallow cup. The molecular rays from the cup cross in the centre of the bulb, and thence diverging fall on the opposite side and produce a circular patch of green phosphorescent light. As I turn the bulb round you will all be able to see the green patch on the glass. Now observe, I remove the positive wire from the top,

and connect it with the side pole (*c*). The green patch from the divergent negative focus is there still. I now make the lowest pole (*d*) positive, and the green patch remains where it was at first, unchanged in position or intensity.

We have here another property of radiant matter. In the low vacuum the position of the positive pole is of every importance, whilst in a high vacuum the position of the positive pole scarcely matters at all; the phenomena seem to depend entirely on the negative pole. If the negative pole points in the direction of the

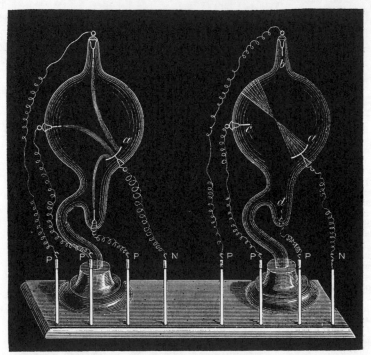

FIG. 7.

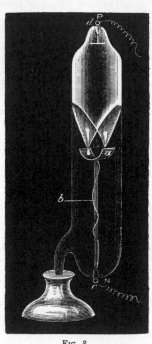

FIG. 8.

positive, all very well, but if the negative pole is entirely in the opposite direction it is of little consequence : the radiant matter darts all the same in a straight line from the negative.

If, instead of a flat disk, a hemi-cylinder is used for the negative pole, the matter still radiates normal to its surface. The tube before you (Fig. 8) illustrates this property. It contains, as a negative pole, a hemi-cylinder (*a*) of polished aluminium. This is connected with a fine copper wire, *b*, ending at the platinum terminal, *c*. At the upper end of the tube is another terminal, *d*. The induction-coil is connected so that the hemicylinder is negative and the upper pole positive, and when exhausted to a sufficient extent the projection of the molecular rays to a focus is very beautifully shown. The rays of matter being driven from the hemi-cylinder in a direction normal to its surface, come to a focus and then diverge, tracing their path in brilliant green phosphorescence on the surface of the glass.

Instead of receiving the molecular rays on the glass, I will show you another tube in which the focus falls on a phosphorescent screen. See how brilliantly the lines of discharge shine out, and how intensely the focal point is illuminated, lighting up the table.

Radiant Matter when intercepted by Solid Matter casts a Shadow

Radiant matter comes from the pole in straight lines, and does not merely permeate all parts of the tube and fill it with light, as would be the case were the exhaustion less good. Where there is nothing in the way the rays strike the screen and produce phosphorescence, and where solid matter intervenes they are obstructed by it, and a shadow is thrown on the screen. In this pear-shaped bulb (Fig. 9) the negative pole (*a*) is at the pointed end. In the middle is a cross (*b*) cut out of sheet aluminium, so that the rays from the negative pole projected along the tube will be partly intercepted by the aluminium cross, and will project an image of it on the hemispherical end of the tube which is phosphorescent. I turn on the coil, and you will all see the black

shadow of the cross on the luminous end of the bulb (*c*, *d*). Now, the radiant matter from the negative pole has been passing by the side of the aluminium cross to produce the shadow; the glass has been hammered and bombarded till it is appreciably warm, and at the same time another effect has been produced on the glass—its sensibility has been deadened. The glass has got tired, if I may use the expression, by the enforced phosphorescence. A change has been produced by this molecular bom-

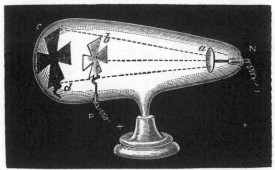

FIG. 9.

bardment which will prevent the glass from responding easily to additional excitement; but the part that the shadow has fallen on is not tired—it has not been phosphorescing at all and is perfectly fresh; therefore if I throw down this cross—I can easily do so by giving the apparatus a slight jerk, for it has been most ingeniously constructed with a hinge by Mr. Gimingham—and so allow the rays from the negative pole to fall uninterruptedly on to the end of the bulb, you will suddenly see the black cross (*c*, *d*, Fig. 10) change to a luminous one (*e*, *f*), because the background is now only capable of faintly phosphorescing, whilst the part which had the black shadow on it retains its full phosphorescent power. The stencilled image of the luminous cross unfortunately soon dies out. After a period of rest the glass partly recovers its power of phosphorescing, but it is never so good as it was at first.

Here, therefore, is another important property of radiant matter. It is projected with great velocity from the negative pole, and not only strikes the glass in such a way as to cause it to vibrate and become temporarily luminous while the discharge is going on, but the molecules hammer away with sufficient energy to produce a permanent impression upon the glass.

(*To be continued.*)

NOTES

In accordance with the resolution come to at the recent International Congress of Meteorology, the International Committee have issued circulars for a special Conference at the Deutsche Seewarte at Hamburg, on October 1, to consider the scheme of Count Wilczek and Lieut. Weyprecht for the establishment of circumpolar observing stations. The Conference will consider specially the following points :—1. The number of observatories and the most convenient places at which to establish them. The decision will depend on the number of co-operating states and the sums which they are willing to devote to this purpose. Count Wilczek and Lieut. Weyprecht have proposed the following places :—In the Northern Hemisphere : north coasts of Spitzbergen and of Novaya Zemlya, the neighbourhood of the North Cape, the mouth of the Lena, New Siberia, Point Barrow, on the north-east of Behring Strait, west coast of Greenland, east coast of Greenland, about 75° N. lat. In the Southern Hemisphere : the neighbourhood of Cape Horn, Kerguelen or Macdonald Islands, one of the groups south of the Auckland Islands. 2. There will be considered the exact epoch of the observations and their maximum duration. 3. Uniform instruction for observations, which will have to fix especially : (*a*) The minimum of elements to be observed at each station, both for meteorological phenomena and for those of terrestrial magnetism, as well as for other phenomena of terrestrial physics connected with them. (*b*) The minimum number of daily observations for the different elements. (*c*) The first meridian which will serve as basis for simultaneous observations. (*d*) Methods of observation for the different elements and methods of reduction. (*e*) Instruments of observation and their arrangement, as far as they may influence the comparability of the results.

At a recent meeting of the Committee of the Iron and Steel Institute in Liverpool it was arranged that this year's meeting should be held in Liverpool on September 24, 25, and 26. The use of St. George's Hall has been granted by the Corporation, and numerous places for inspection and excursion have been partly arranged for, including Messrs. Blundell's collieries, near Wigan, and the Tubular Bridge at Menai Straits. In addition to papers on the manufacture and application of steel and iron, papers on subjects of work more immediately connected with Liverpool have been promised.

M. Janssen, we are glad to see, has been appointed to represent the Paris Academy of Sciences, at the inauguration of the statue to Arago, at Perpignan.

The prizes instituted by Prof. Schäfli (Lausanne) for scientific works on Switzerland will now be awarded not only to Swiss naturalists, as hitherto, but also to foreign, a resolution in this sense having been accepted at the last meeting of Swiss naturalists.

We regret to hear of the death of Mr. Edward Edwards, late of Menai Bridge, Anglesey, at the age of seventy-five. For upwards of twenty years he had studied the habits and characters of marine animals in their native haunts, and his contrivance of the "dark chamber tank" was the first by which these animal, could be kept alive and healthy for an indefinite period in confinement, and the principle of which was afterwards carefully recognised in the construction of the Crystal Palace and other aquariums.

The *Times* Geneva correspondent writes, under date August 22 :—"On the evening of August 5, six persons who were standing in the gallery of a *châlet* in the Jura, above St. Cergues, witnessed an atmospheric phenomenon equally rare and curious. The aspect of the sky was dark and stormy. The air was thick with clouds, out of which darted at intervals bright flashes of lightning. At length one of these clouds, seeming to break loose from the mountains between Nyon and the Dôle, advanced in the direction of a storm which had, meanwhile, broken out over Morges. The sun was hidden and the country covered with thick darkness. At this moment the pine forest round St. Cergues was suddenly illuminated and shone with a light bearing a striking resemblance to the phosphorescence of the sea as seen in the tropics. The light disappeared with every clap of thunder, but only to re-appear with increased intensity until the subsidence of the tempest. M. Raoul Pictet, the eminent chemist, who was one of the witnesses of the phenomenon, thus explains it in the last number of the *Archives des Sciences Physiques et Naturelles* :— 'Before the appearance of this fire of St. Elmo, which covered the whole of the forest, it had rained several minutes during the first part of the storm. The rain had converted the trees into conductors of electricity. Then, when the cloud, strongly charged with the electric fluid, passed over this multitude of points, the discharges were sufficiently vivid to give rise to the luminous appearance. The effect was produced by the action of the electricity of the atmosphere on the electricity of the earth, an effect which, on the occasion in question, was considerably increased by the height of the locality, the proximity of a storm-cloud, and the action of the rain, which turned all the trees of the forest into conductors.' "

A young female gorilla is now being exhibited at the Crystal Palace.

At the last meeting of the Swiss Naturalists, Prof. Kollmann (Basel) presented a report of the Anthropological and Statistical Commission, appointed by the Swiss Natural History Society for the investigation of the distribution of the light-coloured and dark-coloured population in Switzerland. Thanks to the collaboration of many schoolmasters, no less than 250,000 children in twenty-one cantons were described as to the colour of the eyes, hair, and skin, and a very rich and reliable material was collected. It is proved that in Switzerland, as well as in all middle Europe, the light-coloured population decreases from north to south, while the dark-coloured increases, and that it reaches its greatest quantity in the Graubünden, sending a rather dense branch to the south-west. It may be concluded that a dark-coloured population immigrated in Switzerland from the south, having also a side-branch which followed the direction from the Rhone to the Rhine.

ON RADIANT MATTER[1]

II.

Radiant Matter exerts strong Mechanical Action where it Strikes

WE have seen, from the sharpness of the molecular shadows, that radiant matter is arrested by solid matter placed in its path.

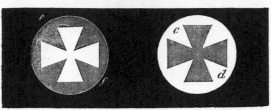

FIG. 10.

If this solid body is easily moved the impact of the molecules will reveal itself in strong mechanical action. Mr. Gimingham has constructed for me an ingenious piece of apparatus which when placed in the electric lantern will render this mechanical action visible to all present. It consists of a highly-exhausted glass tube (Fig. 11), having a little glass railway running along it from one end to the other. The axle of a small wheel revolves on the rails, the spokes of the wheel carrying wide mica paddles. At each end of the tube, and rather above the centre, is an aluminium pole, so that whichever pole is made negative the stream of radiant matter darts from it along the tube, and striking the upper vanes of the little paddle-wheel, causes it to turn round and travel along the railway. By reversing the poles I can arrest the wheel and send it the reverse way, and if I gently incline the tube the force of impact is observed to be sufficient even to drive the wheel up-hill.

This experiment therefore shows that the molecular stream

FIG. 11.

from the negative pole is able to move any light object in front of it.

The molecules being driven violently from the pole there should be a recoil of the pole from the molecules, and by arranging an apparatus so as to have the negative pole movable and the body receiving the impact of the radiant matter fixed, this recoil can be rendered sensible. In appearance the apparatus (Fig. 12) is not unlike an ordinary radiometer with aluminium disks for vanes, each disk coated on one side with a film of mica. The fly is supported by a hard steel instead of glass cup, and the needle-point on which it works is connected by means of a wire with a platinum terminal sealed into the glass. At the top of the radiometer bulb a second terminal is sealed in. The radiometer therefore can be connected with an induction-coil, the movable fly being made the negative pole.

For these mechanical effects the exhaustion need not be so high as when phosphorescence is produced. The best pressure

[1] A lecture delivered to the British Association for the Advancement of Science, at Sheffield, Friday, August 22, 1879, by William Crookes, F.R.S. Continued from p. 423.

for this electrical radiometer is a little beyond that at which the dark space round the negative pole extends to the sides of the glass bulb. When the pressure is only a few millims. of mercury, on passing the induction current a halo of velvety violet light forms on the metallic side of the vanes, the mica side remaining dark. As the pressure diminishes, a dark space is seen to separate the violet halo from the metal. At a pressure of half a millim. this dark space extends to the glass, and rotation commences. On continuing the exhaustion the dark space further widens out and appears to flatten itself against the glass, when the rotation becomes very rapid.

Here is another piece of apparatus (Fig. 13) which illustrates

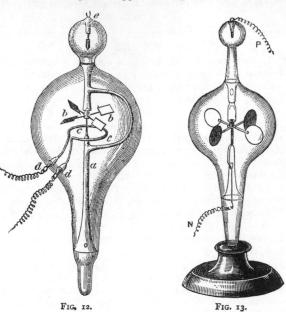

FIG. 12.   FIG. 13.

the mechanical force of the radiant matter from the negative pole. A stem (*a*) carries a needle-point in which revolves a light mica fly (*b b*). The fly consists of four square vanes of thin clear mica, supported on light aluminium arms, and in the centre is a small glass cap which rests on the needle-point. The vanes are inclined at an angle of 45° to the horizontal plane. Below the fly is a ring of fine platinum wire (*c c*), the ends of which pass through the glass at *d d*. An aluminium terminal (*e*) is sealed in at the top of the tube, and the whole is exhausted to a very high point.

By means of the electric lantern I project an image of the vanes on the screen. Wires from the induction-coil are attached,

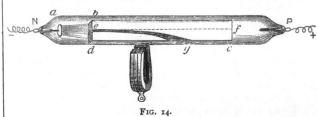

FIG. 14.

so that the platinum ring is made the negative pole, the aluminium wire (*e*) being positive. Instantly, owing to the projection of radiant matter from the platinum ring, the vanes rotate with extreme velocity. Thus far the apparatus has shown nothing more than the previous experiments have prepared us to expect; but observe what now happens. I disconnect the induction-coil altogether, and connect the two ends of the platinum wire with a small galvanic battery; this makes the ring *c c* red-hot, and under this influence you see that the vanes spin as fast as they did when the induction-coil was at work.

Here, then, is another most important fact. Radiant matter in these high vacua is not only excited by the negative pole of an induction-coil, but a hot wire will set it in motion with force sufficient to drive round the sloping vanes.

Radiant Matter is deflected by a Magnet

I now pass to another property of radiant matter. This long glass tube (Fig. 14), is very highly exhausted; it has a negative pole at one end (*a*) and a long phosphorescent screen (*b, c*) down the centre of the tube. In front of the negative pole is a plate of mica (*b, d*) with a hole (*e*) in it, and the result is, when I turn on the current, a line of phosphorescent light (*e, f*) is projected along the whole length of the tube. I now place beneath the tube a powerful horse-shoe magnet: observe how the line of light (*e, g*) becomes curved under the magnetic influence waving about like a flexible wand as I move the magnet to and fro.

This action of the magnet is very curious, and if carefully followed up will elucidate other properties of radiant matter. Here (Fig. 15) is an exactly similar tube, but having at one end a small potash tube, which if heated will slightly injure the vacuum. I turn on the induction current, and you see the ray of radiant matter tracing its trajectory in a curved line along the screen, under the influence of the horse-shoe magnet beneath. Observe the shape of the curve. The molecules shot from the negative pole may be likened to a discharge of iron bullets from a mitrailleuse, and the magnet beneath will represent the earth curving the trajectory of the shot by gravitation. Here on this luminous screen you see the curved trajectory of the shot accurately traced. Now suppose the deflecting force to remain constant, the curve traced by the projectile varies with the velocity. If I put more powder in the gun the velocity will be greater and the trajectory flatter, and if I interpose a denser resisting medium between the gun and the target, I diminish the velocity of the shot, and thereby cause it to move in a greater curve and come to the ground sooner. I cannot well increase before you the velocity of my stream of radiant molecules by putting more

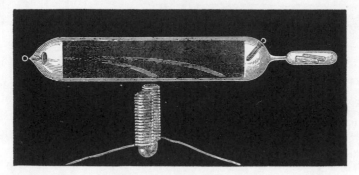

FIG. 15.

powder in my battery, but I will try and make them suffer greater resistance in their flight from one end of the tube to the other. I heat the caustic potash with a spirit-lamp and so throw in a trace more gas. Instantly the stream of radiant matter responds. Its velocity is impeded, the magnetism has longer time on which to act on the individual molecules, the trajectory gets more and more curved, until, instead of shooting nearly to the end of the tube, my molecular bullets fall to the bottom before they have got more than half-way.

It is of great interest to ascertain whether the law governing the magnetic deflection of the trajectory of radiant matter is the same as has been found to hold good at a lower vacuum. The experiments I have just shown you were with a very high vacuum. Here is a tube with a low vacuum (Fig. 16). When I turn on the induction spark, it passes as a narrow line of violet light

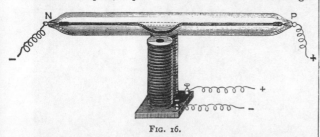

FIG. 16.

joining the two poles. Underneath I have a powerful electro-magnet. I make contact with the magnet, and the line of light dips in the centre towards the magnet. I reverse the poles, and the line is driven up to the top of the tube. Notice the difference between the two phenomena. Here the action is temporary. The dip takes place under the magnetic influence; the line of discharge then rises and pursues its path to the positive pole. In the high exhaustion, however, after the stream of radiant matter had dipped to the magnet, it did not recover itself, but continued its path in the altered direction.

By means of this little wheel, skilfully constructed by Mr. Gimingham, I am able to show the magnetic deflection in the electric lantern. The apparatus is shown in this diagram (Fig. 17). The negative pole (*a, b*) is in the form of a very shallow cup. In front of the cup is a mica screen (*c, d*), wide enough to intercept the radiant matter coming from the negative pole. Behind this screen is a mica wheel (*e, f*) with a series of vanes, making a sort of paddle-wheel. So arranged, the molecular rays from the pole *a b* will be cut off from the wheel, and will not produce any movement. I now put a magnet, *g*, over the tube, so as to deflect the stream over or under the obstacle *c, d*, and the result will be rapid motion in one or the other direction, according to the way the magnet is turned. I throw the image of the apparatus on the screen. The spiral lines painted on the wheel show which way it turns. I arrange the magnet to draw the molecular stream so as to beat against the upper vanes, and the wheel revolves rapidly as if it were an over-shot water-wheel. I turn the magnet so as to drive the radiant

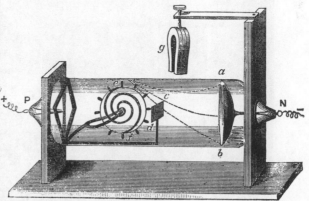

FIG. 17.

matter underneath; the wheel slackens speed, stops, and then begins to rotate the other way, like an under-shot water-wheel. This can be repeated as often as I reverse the position of the magnet.

I have mentioned that the molecules of the radiant matter discharged from the negative pole are negatively electrified. It is probable that their velocity is owing to the mutual repulsion between the similarly electrified pole and the molecules. In less high vacua, such as you saw a few minutes ago (Fig. 16), the discharge passes from one pole to another, carrying an electric current, as if it were a flexible wire. Now it is of great interest to ascertain if the stream of radiant matter from the negative pole also carries a current. Here (Fig. 18) is an apparatus which will decide the question at once. The tube contains two negative terminals (*a, b*) close together at one end, and one positive terminal (*c*) at the other. This enables me to send two streams of radiant matter side by side along the phosphorescent screen— or by disconnecting one negative pole, only one stream.

If the streams of radiant matter carry an electric current they will act like two parallel conducting wires and attract one

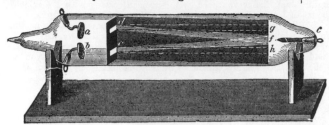

FIG. 18.

another; but if they are simply built up of negatively electrified molecules they will repel each other.

I will first connect the upper negative pole (*a*) with the coil, and you see the ray shooting along the line *d, f.* I now bring the lower negative pole (*b*) into play, and another line (*e, h*) darts along the screen. But notice the way the first line behaves; it jumps up from its first position, *d f*, to *d g*, showing that it is repelled, and if time permitted I could show you that the lower ray is also deflected from its normal direction: therefore the two

parallel streams of radiant matter exert mutual repulsion, acting not like current carriers, but merely as similarly electrified bodies.

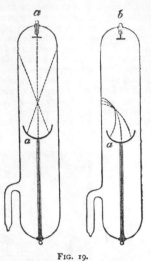

FIG. 19.

Radiant Matter produces Heat when its Motion is arrested

During these experiments another property of radiant matter has made itself evident, although I have not yet drawn attention

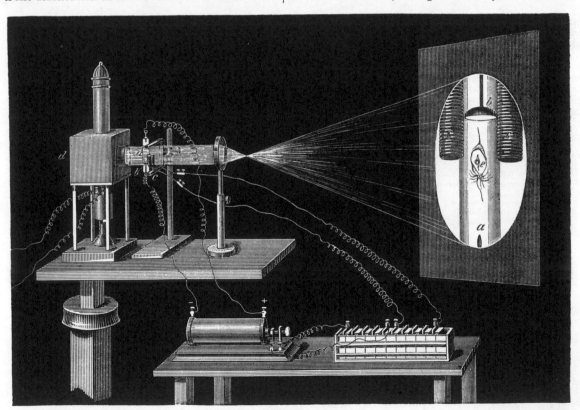

FIG. 20.

to it. The glass gets very warm where the green phosphorescence is strongest. The molecular focus on the tube, which we saw earlier in the evening (Fig. 8) is intensely hot, and I have prepared an apparatus by which this heat at the focus can be rendered apparent to all present.

I have here a small tube (Fig. 19, *a*) with a cup-shaped negative pole. This cup projects the rays to a focus in the middle of the tube. At the side of the tube is a small electro-magnet, which I can set in action by touching a key, and the focus is then drawn to the side of the glass tube (Fig. 19, *b*). To show the first action of the heat I have coated the tube with wax. I will put the apparatus in front of the electric lantern (Fig. 20, *d*), and throw a magnified image of the tube on the screen. The coil is now at work, and the focus of molecular rays is projected along the tube. I turn the magnetism on, and draw the focus to the side of the glass. The first thing you see is a small circular patch melted in the coating of wax. The glass soon begins to disintegrate, and cracks are shooting starwise from the centre of heat. The glass is softening. Now the atmospheric pressure forces it in, and now it melts. A hole (*e*) is perforated in the middle, the air rushes in, and the experiment is at an end.

I can render this focal heat more evident if I allow it to play on a piece of metal. The bulb (Fig. 21) is furnished with a negative pole in the form of a cup (*a*). The rays will therefore be projected to a focus on a piece of iridio-platinum (*b*) supported in the centre of the bulb.

I first turn on the induction-coil slightly, so as not to bring out its full power. The focus is now playing on the metal, raising it to a white heat. I bring a small magnet near, and you see I can deflect the focus of heat just as I did the luminous focus in the other tube. By shifting the magnet I can drive the focus up and down, or draw it completely away from the metal, and leave it non-luminous. I withdraw the magnet, and let the molecules have full play again; the metal is now white hot. I increase

FIG. 21.

the intensity of the spark. The iridio-platinum glows with almost insupportable brilliancy, and at last melts.

The Chemistry of Radiant Matter

As might be expected, the chemical distinctions between one kind of radiant matter and another at these high exhaustions are difficult to recognise. The physical properties I have been elucidating seem to be common to all matter at this low density. Whether the gas originally under experiment be hydrogen, carbonic acid, or atmospheric air, the phenomena of phosphorescence, shadows, magnetic deflection, &c., are identical, only they commence at different pressures. Other facts, however, show that at this low density the molecules retain their chemical characteristics. Thus by introducing into the tubes appropriate absorbents of residual gas, I can see that chemical attraction goes on long after the attenuation has reached the best stage for showing the phenomena now under illustration, and I am able by this means to carry the exhaustion to much higher degrees that I can get by mere pumping. Working with aqueous vapour I can use phosphoric anhydride as an absorbent; with carbonic acid, potash; with hydrogen, palladium; and with oxygen, carbon, and then potash. The highest vacuum I have yet succeeded in obtaining has been the 1-20,000,000th of an atmosphere, a degree which may be better understood if I say that it corresponds to about the hundredth of an inch in a barometric column three miles high.

It may be objected that it is hardly consistent to attach primary

importance to the presence of *Matter*, when I have taken extraordinary pains to remove as much matter as possible from these bulbs and these tubes, and have succeeded so far as to leave only about the one-millionth of an atmosphere in them. At its ordinary pressure the atmosphere is not very dense, and its recognition as a constituent of the world of matter is quite a modern notion. It would seem that when divided by a million, so little matter will necessarily be left that we may justifiably neglect the trifling residue and apply the term *vacuum* to space from which the air has been so nearly removed. To do so, however, would be a great error, attributable to our limited faculties being unable to grasp high numbers. It is generally taken for granted that when a number is divided by a million the quotient must necessarily be small, whereas it may happen that the original number is so large that its division by a million seems to make little impression on it. According to the best authorities, a bulb of the size of the one before you (13·5 centimetres in diameter) contains more than 1,000,000,000,000,000,000,000,000 (a quadrillion) molecules. Now, when exhausted to a millionth of an atmosphere we shall still have a trillion molecules left in the bulb—a number quite sufficient to justify me in speaking of the residue as *matter*.

To suggest some idea of this vast number I take the exhausted bulb, and perforate it by a spark from the induction-coil. The spark produces a hole of microscopical fineness, yet sufficient to allow molecules to penetrate and to destroy the vacuum. The inrush of air impinges against the vanes, and sets them rotating after the manner of a windmill. Let us suppose the molecules to be of such a size that at every second of time a hundred millions could enter. How long, think you, would it take for this small vessel to get full of air? An hour? A day? A year? A century? Nay, almost an eternity! A time so enormous that imagination itself cannot grasp the reality. Supposing this exhausted glass bulb, indued with indestructibility, had been pierced at the birth of the solar system; supposing it to have been present when the earth was without form and void; supposing it to have borne witness to all the stupendous changes evolved during the full cycles of geologic time, to have seen the first living creature appear, and the last man disappear; supposing it to survive until the fulfilment of the mathematician's prediction that the sun, the source of energy, four million centuries from its formation, will ultimately become a burnt-out cinder;[1] supposing all this—at the rate of filling I have just described, 100 million molecules a second—this little bulb even then would scarcely have admitted its full quadrillion of molecules.[2]

But what will you say if I tell you that all these molecules, this quadrillion of molecules, will enter through the microscopic hole before you leave this room? The hole being unaltered in size, the number of molecules undiminished, this apparent paradox can only be explained by again supposing the size of the molecules to be diminished almost infinitely—so that instead of entering at the rate of 100 millions every second, they troop in at a rate of something like 300 trillions a second. I have done the sum, but figures when they mount so high cease to have any meaning, and such calculations are as futile as trying to count the drops in the ocean.

In studying this fourth state of matter we seem at length to have within our grasp and obedient to our control the little indivisible particles which with good warrant are supposed to constitute the physical basis of the universe. We have seen that in some of its properties radiant matter is as material as this table, whilst in other properties it almost assumes the character of radiant energy. We have actually touched the borderland where matter and force seem to merge into one another, the shadowy realm between Known and Unknown which for me has always

---

[1] The possible duration of the sun from formation to extinction has been variously estimated by different authorities, at from 18 million years to 400 million years. For the purpose of this illustration I have taken the highest estimate.

[2] According to Mr. Johnstone Stoney (*Phil. Mag.*, vol. 36, p. 141), 1 c.c. of air contains about 1,000,000,000,000,000,000,000 molecules. Therefore a bulb 13·5 centims. diameter contains 13·5³ × 0·5236 × 1,000,000,000,000,000,000,000 or 1,288,252,350,000,000,000,000,000 molecules of air at the ordinary pressure. Therefore the bulb when exhausted to the millionth of an atmosphere, contains 1,288,252,350,000,000,000 molecules. leaving 1,288,251,061,747,650,000,000,000 molecules to enter through the perforation. At the rate of 100,000,000 molecules a second, the time required for them all to enter will be

12,882,510,617,476,500 seconds, or
214,708,510,291,275 minutes, or
3,578,475,171,521 hours, or
149,103,132,147 days, or
408,501,731 years.

had peculiar temptations. I venture to think that the greatest scientific problems of the future will find their solution in this Border Land, and even beyond ; here, it seems to me, lie Ultimate Realities, subtle, far-reaching, wonderful.

"Yet all these were, when no Man did them know,
　　Yet have from wisest Ages hidden beene ;
　And later Times thinges more unknowne shall show.
　Why then should witlesse Man so much misweene,
　That nothing is, but that which he hath seene ?"

THE BRITISH ASSOCIATION

GENERAL satisfaction is expressed with the Sheffield meeting. The people of the town and district did their best, amid many difficulties, to give the members of the Association a hearty reception, and they succeeded. The excursions on Thursday were well attended, and those who took part in them seem to have enjoyed themselves. At the meeting of the General Committee, Swansea was selected as next year's place of meeting, with Prof. A. R. Ramsay as president ; the date of meeting is August 25. A letter was read from the Archbishop of York, warmly urging upon the Association to meet in the archiepiscopal City in 1881, when, for some unaccountable reason, the jubilee is to be celebrated, as we have already said, in the fifty-first year of the Association's existence. As the result of the important discussion in Section F on science teaching in schools, a committee was appointed for the purpose of reporting, in addition to other matters, whether it is important that her Majesty's inspectors of elementary schools should be appointed with reference to their ability for examining on scientific specific subjects of the code, the committee to consist of Mr. Mundella, M.P., Mr. Shaw, Mr. Bourne, Mr. Jas. Heywood, Mr. Wilkinson, and Dr. J. H. Gladstone.

REPORTS

Report of the Committee on Erratic Blocks, presented by the Rev. H. W. Crosskey, F.G.S. (Abstract.)

Several contributions of interest and importance have been received respecting the position and distribution of erratic blocks.

A granite boulder $3 \times 2{\cdot}5 \times 2$ feet has been found by Mr. Hall, in the village of Bickington, parish of Fremington. There is no similar rock nearer than Lundy Island, twenty-five miles west-north-west from the boulder and Dartmoor, twenty-five miles south by east. Its height above the sea is 80 feet.

Among the most remarkable erratic blocks yet described in the midland district, are those reported upon Frankley Hill, at a height of 650 feet above the sea. They were examined by the writer in company with Prof. T. G. Bonney, and the following is a summary of the observations made :—

A section of drift beds is exposed in a cutting of the new Hales Owen Railway passing through Frankley Hill. The section is as follows :—Permian clay, sand of clayey texture, yellowish sand, greyish sandy clay with brinter pebbly clay, somewhat sandy. The heights of the clays and sands are very irregular throughout the section which is in itself about 60 feet in depth.

Fragments of permian sandstone (which is exposed in a part of the section) are scattered through the sands and clays, but erratic blocks are rare. Indeed, one only—a green-stone—was noticed in the cutting itself, although others doubtless occur.

No part of this section can be called a "boulder clay"—if by "boulder clay" be meant either a clay formed beneath land ice, or a clay carried away by an iceberg and deposited on the sea-bottom, as the berg melted or stranded.

The various sands and gravels have all the appearance of being a "wash" from older beds, effected during the depression and subsequent upheaval of the present land surface. They are neither compactly crowded with erratics, nor are fragments of local rocks heaped irregularly together, and grooved and striated. The way in which the pieces of native rock are scattered through the beds, does not indicate any other force than that which would be exerted by the ordinary "wash" of the waters during the movements just mentioned.

The presence of a few erratics shows that the *wash* must have taken place beneath the waters of a glacial sea, over which icebergs floated.

These beds appear to have been formed in the earlier rather than the later part of the glacial epoch. In a field *on the summit of the section* a large number of erratics are to be seen which have been taken from a recent surface-drain. Twenty of these boulders are felsite, two are basalt, one is a piece of vein-quartz, and one is a Welsh diabase. They constitute a group of allied rocks, evidently from one district. Probably they belong to the great Arenig dispersion. Two of the felsites close to the group are of considerable size, the larger being about $6 \times 4 \times 2$ feet. Similar blocks may be traced to the summit of the hill. One felsite boulder opposite the Yew Trees is about $4{\cdot}5 \times 3 \times 2$ feet, and is partly buried in the ground.

The height of the boulders above the sea is remarkable, their highest level being 650 feet.

This indicates a corresponding depression of the land, since no Welsh glacier could have travelled over hill and down dale to this summit-level. To render any such glacier work conceivable, the Welsh mountains must have stood at a height beyond any point for which there is the slightest evidence.

This group of boulders on Frankley Hill appears to have been dropped by an iceberg travelling from Wales upon the top of the clays and sands exposed in the railway cutting at a time when the land was depressed at least 700 feet. In the clays and sands upon which the summit group of erratics rests, we must have beds belonging to an earlier date than the *close* of the glacial epoch ; and the erratics in the cutting must be discriminated from those left at the higher level.

Some remarkable boulders were described from the neighbourhood of Wolverhampton : (1) a striated boulder of felsite $11 \times 3 \times 3$ feet ; (2) one of slate, broken into two parts, but which, when whole, measured $11{\cdot}25 \times 6{\cdot}25 \times 3{\cdot}5$ feet ; (3) one of granite about $4{\cdot}75$ feet in each dimension, and weighing about three tons.

Mr. D. Mackintosh traces the origin of the so-called "greenstone" boulders (more properly to be called diorites or dolerites) around the estuaries of the Mersey and the Dee.

The area in which they are very much concentrated is intensely striated, and nearly all the striæ point divergently to the south of Scotland, *i.e.*, between N. 15° W. and N. 45° W.

A large "greenstone" boulder has been found at Crosby, resting on a perfectly flat glaciated rock surface, with striæ pointing N. 40° W.

Additional presumptions in favour of the Scottish derivation of these boulders may be found (1) in the fact that nearly all these boulders consist of basic rocks similar to some found in the south of Scotland, and (2) in the extent to which they are locally concentrated on the peninsula of Wirral and the neighbouring part of Lancashire. Many fresh greenstone boulders have been lately exposed in the newest Bootle Dock excavation. The largest is $6 \times 4{\cdot}5 \times 3$ feet, and was found on the surface of the upper boulder clay. As a rule these boulders are excessively flattened and regularly grooved.

Mr. J. R. Dakyns describes the occurrence of Shap granite boulders on the Yorkshire coast. There are several at Long Nab on the north side of the Nab ; one of these measures 3 cubic feet. Others are on the north side of Cromer Point ; south of Cromer Point there are more till you come nearly to Filey. There is one measuring $3 \times 2{\cdot}5 \times 2$ feet on the top of the cliff about a mile from Filey. It is probably practically undisturbed, for the ground slopes inland from the cliff, and therefore, if it has been turned up in ploughing and moved, it cannot have been moved far, for no one would take the trouble to cart a huge boulder far up-hill.

There are several boulders of Shap granite on the shore along the north of Filey Bay, but none along the south till one reaches Flamborough Head. Several occur along the shore between Flamborough Head and Flamborough south landing ; one of these measures 36 cubic feet. One may be seen rather more than a mile south of Bridlington Quay, and doubtless they have travelled still further south, since there is one built into a wall at Hornsea.

The destruction of erratic blocks is going on so rapidly that the Committee invite continued contributions of information concerning them.

SECTION A.

MATHEMATICAL AND PHYSICAL SCIENCE.

OPENING ADDRESS BY PROF. G. F. FITZGERALD, M.A., F.R.S., PRESIDENT OF THE SECTION.

THE British Association in Bath, and especially we here in Section A, have to deplore a very great loss. We confidently anticipated profit and pleasure from the presence in this chair of one of the leading spirits of English science, Dr. Schuster. We deplore the loss, and we deplore the cause of it. It is always sad when want of strength makes the independent dependent, and it is doubly sad when a life's work is thereby delayed ; and to selfish humanity it is trebly sad when, as in this case, we ourselves are involved in the loss. And our loss is great. Dr. Schuster has been investigating some very important questions. He has been studying electric discharges in gases, and he has been investigating the probably allied question of the variations of terrestrial magnetism. We anticipated his matured pronouncements upon these subjects, and also the advantage of his very wide general information upon physical questions, and the benefit of his judicial mind while presiding here.

As to myself, his substitute, I cannot express how much gratified I feel at the distinguished honour done me in asking me to preside. It has been one of the ambitions of my life to be worthy of it, and I will do my best to deserve your confidence ; man can do no more, and upon such a subject "the less said the soonest mended."

I suppose most former occupants of this chair have looked over the addresses of their predecessors to see what sort of a thing was expected from them. I find that very few had the courage to deliver no address. Most have devoted themselves to broad general questions, such as the relations of mathematics to physics, or more generally deductive to inductive science. On the other hand, several have dealt each with his own specialty. On looking back over these addresses my attention was specially arrested by the first two past Presidents of this Section whose bodily presence we cannot have here. They were Presidents of Section A in consecutive years. In 1874, Provost Jellett occupied this chair ; and in 1875, Prof. Balfour Stewart occupied it. Both have gone from us since the last meeting of this Association. Each gave a characteristic address. The Provost, with the clearness and brilliancy that distinguished his great intellect, plunged through the deep and broad questions surrounding the mechanism of the universe, and with impassioned earnestness claimed on behalf of science the right to prosecute its investigations until it attains, if it ever does attain, to a mechanical explanation of all things. This intrepid honesty, to carry to their utmost the principles of whose truth he was convinced, the utter abhorrence of the shadow of double-dealing with truth, was eminently characteristic of one whom all, but especially we of Trinity College, Dublin, will long miss as a lofty example of the highest intellectual keenness and honesty, and mourn as the truest-hearted friend, full of sympathy and Christian charity. In 1875, Prof. Stewart gave us a striking example of the other class of address in a splendid exposition of the subject he did so much to advance—namely, solar physics. He brought together from the two great storehouses of his information and speculation a brilliant store, and displayed them here for the advancement of science. Him, too, all science mourns. Though, from want of personal acquaintance, I am unequal to the task of bringing before you his many abilities and great character, you can each compose a fitting epitaph for this well-known great one of British science. In this connection I am only expressing what we all feel when I say how well timed was the Royal bounty recently extended to his widow. At the same time, the niggardly recognition of science by the public is a disgrace to the enlightenment of the nineteenth century. What Chancellor or General with his tens of thousands has done that for his country and mankind that Faraday, Darwin, and Pasteur have done ? The "public" now are but the children of those who murdered Socrates, tolerated the persecution of Galileo, and deserted Columbus.

In a Presidential address on the borderlands of the known delivered from this chair the great Clerk Maxwell spoke of it as an undecided question whether electro-magnetic phenomena are due to a direct action at a distance or are due to the action of an intevening medium. The year 1888 will be ever memorable as the year in which this great question has been experimentally decided by Hertz in Germany, and, I hope, by others in England.

It has been decided in favour of the hypothesis that these actions take place by means of an intervening medium. Although there is nothing new about the question, and although most workers at it have long been practically satisfied that electro-magnetic actions are due to an intervening medium, I have thought it worth while to try and explain to others who may not have considered the problem, what the problem is and how it has been solved. A Presidential address such as this is not for specialists—it is for the whole Section; and I would not have thought of dealing with this subject, only that its immediate consequences reach to all the bounds of physical science, and are of interest to all its students.

We are all familiar with this, that when we do not know all about something there are generally a variety of explanations of what we do know. Whether there is anything of which there are in reality a variety of explanations is a deep question, which some have connected with the freedom of the will, but which I am not concerned with here. A notable example of the possibility of a variety of explanations for us is recorded in connection with an incident said to have occurred in the neighbouring town of Clifton, where a remarkable meteorological phenomenon, as it appeared to an observing scientist, was explained by others as a bull's-eye lantern in the hands of Mr. Pickwick. Another kind of example is the old explanation of water rising in a pump, that "Nature abhors a vacuum," as compared with the modern one. Nowadays, when we know as little about anything, we say, "It is the property of electricity to attract." This is really little or no advance on the old form, and is merely a way of stating that we know a fact but not its explanation. There are plenty of cases still where a variety of explanations are possible. For example, we know of no *experimentum crucis* to decide whether the people I see around me are conscious or are only automata. There are other questions which have existed, but which have been experimentally decided. The most celebrated of these are the questions between the caloric and kinetic theories of heat, and between the emission and undulatory theories of light. The classical experiments by which the case has been decided in favour of the kinetic theory of heat and the undulatory theory of light are some of the most important experiments that have ever been performed. When it was shown that heat disappeared whenever work appeared, and *vice versâ*, and so the caloric hypothesis was disproved; when it was shown that light was propagated more slowly in a dense medium than in a rare, the sciences of light and heat were revolutionized. Not but that most who studied the subject had given their adhesion to the true theory before it was finally decided in general estimation. In fact, Rumford's and Davy's experiments on heat, and Young and Fresnel's experiments on light, had really decided these questions long before the erroneous views were finally abandoned. I hope that science will not be so slow in accepting the results of experiment in respect of electro-magnetism as it was in the case of light and heat, and that no Carnot will throw back science by giving plausible explanations on a wrong hypothesis. Rowland's experiment proving an electro-magnetic action between electric charges depending on their absolute and not relative velocities has already proved the existence of a medium relative to which the motion must take place, but the connection is rather metaphysical, and is too indirect to attract general attention. The importance of these striking experiments was that they put the language of the wrong hypothesis out of fashion. Elementary text-books that halted between two opinions, and, after the manner of text-books, leant towards that enunciated in preceding text-books, had all perforce to give prominence to the true theory, and the whole rising generation began their researches from a firm and true stand-point. I anticipate the same results to follow Hertz's experimental demonstration of a medium by which electro-magnetic actions are produced. Text-books which have gradually been invoking lines of force, in some respects to the aid of learners and in others to their bewilderment, will now fearlessly discourse of the stresses in the ether that cause electric and magnetic force. The younger generation will see clearly in electro-magnetic phenomena the working of the all-pervading ether, and this will give them a firm and true stand-point for further advances.

And now I want to spend a short time in explaining to you how the question has been decided. An illustrative example may make the question itself clearer, and so lead you to understand the answer better. In colloquial language we say that balloons, hot air, &c., rise because they are light. In old times this was stated more explicitly, and therefore much more clearly. It was said that they possessed a quality called "levity." "Levity" was opposed to "heaviness." Heaviness made things tend downwards, levity made things tend upwards. It was a sort of action at a distance. At least, it would have required such an hypothesis if it had survived until it was known that heaviness was due to the action of the earth. I expect levity would have been attributed to the direct action of heaven. It was comparatively recently in the history of mankind that the rising of hot air, flames, &c., was attributed to the air. Everybody knew that there was air, but it was not supposed that the upward motion of flames was due to it. We now know that this and the rising of balloons are due to the difference of pressure at different levels in the air. In a similar way we have long known that there is an ether, an all-pervading medium, occupying all known space. Its existence is a necessary consequence of the undulatory theory of light. People who think a little, but not much, sometimes ask me, "Why do you believe in the ether? What's the good of it?" I ask them, "What becomes of light for the eight minutes after it has left the sun and before it reaches the earth?" When they consider that, they observe how necessary the ether is. If light took no time to come from the sun, there would be no need of the ether. That it is a vibratory phenomenon, that it is affected by matter it acts through—these could be explained by action at a distance very well. The phenomena of interference would, however, require such complicated and curious laws of action at a distance as practically to put such an hypothesis out of court, or else be purely mathematical expressions for wave propagation. In fact, anything except propagation in time is explicable by action at a distance. It is the same in the case of electro-magnetic actions. There were two hypotheses as to the causes of electro-magnetic actions. One attributed electric attraction to a property of a thing called electricity to attract at a distance, the other attributed it to a pull exerted by means of the ether, somewhat in the way that air pushes balloons up. We do not know what the structure of the ether is by means of which it can pull, but neither do we know what the structure of a piece of india-rubber is by means of which it can pull; and we might as well ignore the india-rubber, though we know a lot about the laws of its action, because we do not know its structure, as to ignore the ether because we do not know its structure. Anyway, what was wanted was an experiment to decide between the hypothesis of direct action at a distance and of action by means of a medium. At the time that Clerk Maxwell delivered his address no experiment was known that could decide between the two hypotheses. Specific inductive capacity, the action of intervening matter, the delay in telegraphing, the time propagation of electro-magnetic actions by means of conducting material—these were known, but he knew that they could be explained by means of action at a distance, and had been so explained. Waves in a conductor do not necessarily postulate action through a medium such as the ether. When we are dealing with a conductor and a thing called electricity running over its surface, we are, of course, postulating a medium on or in the conductor, but not outside it, which is the special point at issue. Clerk Maxwell *believed* that just as the same air that transmits sound is able by differences of pressure—*i.e.* by means of its energy per unit volume—to move bodies immersed in it, so the same ether that transmits light causes electrified bodies to move by means of its energy per unit volume. He believed this, but there was no experiment known then to decide between this hypothesis and that of direct action at a distance. As I have endeavoured to impress upon you, no *experimentum crucis* between the hypotheses is possible except an experiment proving propagation in time, either directly, or indirectly by an experiment exhibiting phenomena like those of the interference of light. A theorist may speak of propagation of actions in time without talking of a medium. This is all very well in mathematical formulæ, but, as in the case of light we must consider what becomes of it after it has left the sun and before it reaches the earth, so every hypothesis assuming action in time really postulates a medium whether we talk about it or not. There are some difficulties surrounding the complete interpretation of some of Hertz's experiments. The conditions are complicated, but I confidently expect that they will lead to a decision on most of the outstanding questions on the theory of electro-magnetic action. However, there is no doubt that he has observed the interference of electro-magnetic waves quite analogous to those of light, and that he has proved that electro-magnetic actions are propagated in air with the velocity of light. By a beautiful device Hertz has produced rapidly alternating currents of such frequency that their wavelength is only about 2 metres. I may pause for a minute to call your attention to what that means. These waves are propagated three hundred thousand kilometres in a second. If they vibrated three hundred thousand times a second, the waves would be each a kilometre long. This rate of vibration is much higher than the highest audible note, and yet the waves are much too long to be manageable. We want a vibration about a thousand times as fast again with waves about a metre long. Hertz produced such vibrations, vibrating more than a hundred million times a second. That is, there are as many vibrations in one second as there are seconds—in a day? No, far more. In a week? No, more even than that. The pendulum of a clock ticking seconds would have to vibrate for four months before it would vibrate as often as one of Hertz's vibrators vibrates in one second. And how did he detect the vibrations and their interference? He could not see them; they are much too slow for that; they should go about a million times as fast again to be visible. He could not hear them; they are much too quick for that. If they went a million times more slowly they would be well heard. He made use of the principle of resonance. You all understand how by a succession of well-timed small impulses a large vibration may be set up. It explains many things, from speech to spectrum analysis. It is related that a former Marquess of Waterford used the principle to overturn lamp-posts—his ambition soared above knocker-wrenching. So that it is a principle known to others besides scientific men. Hertz constructed a circuit whose period of vibration for electric currents was the same as that of his generating vibrator, and he was able to see sparks, due to the induced vibration, leaping across a small air-space in this resonant circuit. The well-timed electrical impulses broke down the air-resistance just as those of my Lord of Waterford broke down the lamp-post. The combination of a vibrating generating circuit with a resonant receiving circuit is one that I spoke of at the meeting of the British Association at Southport as one by which this very question might be studied. At the time I did not see any feasible way of detecting the induced resonance: I did not anticipate that it could produce sparks. By its means, however, Hertz has been able to observe the interference between waves incident on a wall and the reflected waves. He placed his generating vibrator several wave-lengths away from a wall, and placed the receiving resonant circuit between the generator and the wall, and in this air-space he was able to observe that at some points there were hardly any induced sparks, but at other and greater distances from his generator they reappeared, to disappear again in regular succession at equal intervals between his generator and the wall. It is exactly the same phenomenon as what are known as Lloyd's bands in optics, which are due to the interference between a direct and a reflected wave. It follows hence that, just as Young's and Fresnel's researches on the interference of light prove the undulatory theory of optics, so Hertz's experiment proves the ethereal theory of electro-magnetism. It is a splendid result. Henceforth I hope no learner will fail to be impressed with the theory—hypothesis no longer—that electro-magnetic actions are due to a medium pervading all known space, and that it is the same medium as the one by which light is propagated, that non-conductors can, and probably always do, as Prof. Poynting has taught us, transmit electro-magnetic energy. By means of variable currents energy *is* propagated into space with the velocity of light. The rotation of the earth is being slowly stopped by the diurnal rotation of its magnetic poles. This seems a hopeful direction in which to look for an explanation of the secular precession of terrestrial magnetism. It is quite different from Edlund's curious hypothesis that free space is a perfect conductor. If this were true, there would be a pair of great antipoles outside the air, and terrestrial magnetism would not be much like what it is, and I think the earth would have stopped rotating long ago. With alternating currents we *do* propagate energy through non-conductors. It seems almost as if our future telegraph-cables would be pipes. Just as the long sound-waves in speaking-tubes go round corners, so these electro-magnetic waves go round corners if they are not too sharp. Prof. Lodge will probably have something to tell us on this point in connection with lightning-conductors. The silvered glass-bars used by surgeons to conduct light are exactly what I am describing. They are a glass, a non-conducting, and therefore transparent,

bar surrounded by a conducting, and therefore opaque, silver sheath, and they transmit the rapidly alternating currents we call light. There would not be the same difficulty in utilizing the energy of these electro-magnetic waves as in utilizing radiant heat. Having all the vibrations of the same period we might utilize Hertz's resonating circuits, and in any case the second law of thermodynamics would not trouble us when we could practically attain to the absolute zero of these, as compared with heat, long-period vibrations.

We seem to be approaching a theory as to the structure of the ether. There are difficulties from diffusion in the simple theory that it is a fluid full of motion, a sort of vortex-sponge. There were similar difficulties in the wave theory of light owing to wave propagation round corners, and there is as great a difficulty in the jelly theory of the ether arising from the freedom of motion of matter through it. It may be found that there is diffusion, or it may be found that there are polarized distributions of fluid kinetic energy which are not unstable when the surfaces are fixed: more than one such is known. Osborne Reynolds has pointed out another, though in my opinion less hopeful, direction in which to look for a theory of the ether. Hard particles are abominations. Perhaps the impenetrability of a vortex would suffice. Oliver Lodge speaks confidently of a sort of chemical union of two opposite kinds of elements forming the ether. The opposite sides of a vortex-ring might perchance suit, or maybe the ether, after all, is but an atmosphere of some infra-hydrogen element: these two latter hypotheses may both come to the same thing. Anyway we are learning daily what sort of properties the ether must have. It must be the means of propagation of light; it must be the means by which electric and magnetic forces exist; it should explain chemical actions, and, if possible, gravity.

On the vortex-sponge theory of the ether there is no real difficulty by reason of complexity why it should not explain chemical actions. In fact, there is every reason to expect that very much more complex actions would take place at distances comparable with the size of the vortices than at the distances at which we study the simple phenomena of electro-magnetism. Indeed, if vortices can make a small piece of a strong elastic solid, we can make watches and build steam-engines and any amount of complex machinery, so that complexity can be no essential difficulty. Similarly the instantaneous propagation of gravity, if it exists, is not an essential difficulty, for vortices each occupy all space, and they act on one another simultaneously everywhere. The theory that material atoms are simple vortex-rings in a perfect liquid otherwise unmoving is insufficient, but with the innumerable possibilities of fluid motion it seems almost impossible but that an explanation of the properties of the universe will be found in this conception. Anything purporting to be an explanation founded on such ideas as "an inherent property of matter to attract," or building up big elastic solids out of little ones, is not of the nature of an ultimate explanation at all; it can only be a temporary stopping-place. There are metaphysical grounds, too, for reducing matter to motion and potential to kinetic energy.

These ideas are not new, but it is well to enunciate them from time to time, and a Presidential address in Section A is a fitting time. Besides all this, it has become the fashion to indulge in quaint cosmical theories and to dilate upon them before learned Societies and in learned journals. I would suggest, as one who has been bogged in this quagmire, that a successor in this chair might well devote himself to a review of the cosmical theories propounded within the last few years. The opportunities for piquant criticism would be splendid.

Returning to the sure ground of experimental research, let us for a moment contemplate what is betokened by this theory that in electro-magnetic engines we are using as our mechanism the ether, the medium that fills all known space. It was a great step in human progress when man learnt to make material machines, when he used the elasticity of his bow and the rigidity of his arrow to provide food and defeat his enemies. It was a great advance when he learnt to use the chemical action of fire; when he learnt to use water to float his boats and air to drive them; when he used artificial selection to provide himself with food and domestic animals. For two hundred years he has made heat his slave to drive his machinery. Fire, water, earth, and air have long been his slaves, but it is only within the last few years that man has won the battle lost by the giants of old, has snatched the thunderbolt from Jove himself, and enslaved the all-pervading ether.

ELECTRICITY
IN TRANSITU:
FROM PLENUM TO VACUUM.*

By WILLIAM CROOKES, F.R.S.,
President of the Institution of Electrical Engineers.

Introduction.

WHILST steadily bearing in mind that I have the honour to address a Society, not only of physicists, but of Electrical Engineers, I shall not, I hope, be out of order in venturing to call your attention to a purely abstract phase of Electrical Science. Numberless instances show that pure research is the abundant source from which spring endless streams of practical applications. We all know how speculative inquiry into the influence of electricity on the nervous system of animals led to knowledge of current electricity, and ultimately to the priceless possession of the telegraph and the telephone. The abstract study of certain microscopic forms of parasitic vegetable life has enabled us to give to fermented solutions of sugar the exact flavour and aroma of the most highly prized wines, and probably, ere long, will put us in a position to increase at will the fertility of the soil, In a different direction the same class of abstract researches applied to medical science has brought us within measurable distance of the final conquest over a large class of diseases hitherto incurable; and without egotism I may, perhaps, be allowed to say that my own researches into high vacua to some extent have contributed to the present degree of perfection of the incandescence lamp. Surely, therefore, whilst eagerly reaping and storing the harvest of practical benefits, we must not neglect to scatter more seed for future results, perchance not less wonderful and valuable.

In another respect I deviate to some extent from the course taken by many of my predecessors. I am about to treat electricity, not so much as an end in itself, but rather as a tool, by whose judicious use we may gain some addition to our scanty knowledge of the atoms and molecules of matter and of the forms of energy which by their mutual reactions constitute the universe as it is manifest to our five senses.

I will endeavour to explain what I mean by characterising electricity as a tool. When working as a chemist in the laboratory, I find the induction spark often of great service in discriminating one element from another, also in indicating the presence of hitherto unknown elements in other bodies in quantity far too minute to be recognisable by any other means. In this way, chemists have discovered thallium, gallium, germanium, and numerous other elements. On the other hand, when examining electrical reactions in high vacua, various rare chemical elements become in turn tests for recognising the intensity and character of electric energy. Electricity, positive and negative, effect respectively different movements and luminosities. Hence the behaviour of the substances upon which electricity acts may indicate with which of these two kinds we have to deal. In other physical researches both electricity and chemistry come into play simply as means of exploration.

In submitting to you certain researches in which electricity is used as a tool, or as a means of bringing within scope of our senses phenomena that otherwise would be unrevealed, I must for a moment recal to your minds the now generally accepted theory of the constitution of matter.

Kinetic Theory of Gases.

Matter, at its ultimate degree of extension, is conjectured to be not continuous, but granular. Maxwell illustrates this view as follows:— To a railway contractor driving a tunnel through a gravel-hill the gravel may be viewed as a continuous substance. To a worm wriggling through gravel, it makes all the difference whether the creature pushes against a piece of gravel or directs its course between the interstices. To the worm, therefore, gravel seems by no means homogeneous and continuous.

With speculations as to the constitution of liquid and solid matter I need not trouble you, but will proceed at once to the third or gaseous state of matter.

The kinetic theory of gases teaches that the constituent molecules dart in every possible direction with great but continually varying velocities, coming almost ceaselessly in mutual collision with each other. The distance each molecule traverses without hitting another molecule is known as its *free path*; the average distance traversed without collision by the whole number of molecules of a gas at any given pressure and temperature is called the *mean free path*. The molecules exert pressure in all directions, and are only restrained by gravitation from dissipating themselves into space. In ordinary gases, the length of the mean free path of the molecules is exceedingly small compared with the dimensions of the vessel, and the properties we then observe are such as constitute the ordinary gaseous state of matter, which depend upon constant collisions. But if we greatly reduce the number of molecules in a given extent of space the free path of the molecules under electric impulse is so long that the number of their mutual collisions in any given time in comparsion with the number of times they fail to collide may be disregarded. Hence, the average molecule can carry out its own motions without interference. When the mean free path becomes comparable to the dimensions of the containing vessel, the attributes which constitute gaseity shrink to a minimum, the matter attains the ultra-gaseous or "radiant" state, and we arrive at a condition where molecular motions under electrical impulse can easily be studied.

The mean free path of the molecules of a gas increases so rapidly with progressive exhaustion, that whilst that of the molecules of air at the ordinary pressure is only 1-10,000th of a millimetre, at an exhaustion of a hundred-millionth of an atmosphere—a point (which, with present appliances, is easy to attain) corresponding to the rarefaction of the air 90 miles above the earth's surface—the mean free path will be about 30 feet; whilst at 200 miles above the earth it will be 10,000,000 miles, and millions of miles out in the depth of space it will become practically infinite. I could go on speculating in spite of Aristotle, who said:—" Beyond the universe there are neither space, nor vacuum, nor time."

In discussing the motions of molecules we have to distinguish the *free path* from the *mean free path*. Nothing is yet known of the *absolute* length of the free path nor of the *absolute* velocity of a molecule. For anything we can prove to the contrary, these values may vary almost from zero to infinity. We can deal only with the *mean* free path and the *mean* velocity.

The Vacuum Pump.

As most of the experiments I put before you to-night are connected with high vacua, it is not out of place to refer to the pump by means of which these tubes are exhausted. Much has been said lately in recommendation of the Geissler pump and its many improvements, but I am still strongly in favour of the Sprengel, as with it I have obtained greater exhaustion than with any other. I should like to point out that the action does not stop when we cease to see air specks passing down the tubes but continues long after this point has been passed. Neither is the non-conducting vacuum, so easily obtained by the Sprengel pump, due in any way to the presence of mercury vapour, since non-conduction can be obtained just as rapidly when special precautions have been taken to keep mercury vapour out of the tubes.

One of the great advantages of the Sprengel pump over all others lies in the fact that its internal capacity need not exceed a few cubic centimetres, and there is, therefore much less wall surface for gases to condense upon. I have brought the very latest modification of this form

---

* Inaugural Address delivered January 15th, 1891.

of pump here to-night, and you will have an opportunity of seeing it in action and of measuring with the McLeod gauge the rarefaction it produces.*

* My measurements of high vacua have all been taken with the beautiful little gauge devised by Professor McLeod. Unmerited discredit has recently been cast on this gauge, the principal fault alleged being its inability to distinguish between the tension of the permanent gas and that of the mercury vapour present. Now it is evident that, under ordinary circumstances, the tension of mercury vapour may be disregarded, as it will be the same on both sides of the gauge ; and it will be only in cases where no mercury is present on one side of the gauge that a slight error is introduced. It is, however, very difficult to devise and successfully experiment with apparatus in which a trace of mercury vapour shall not enter, and it is not

The Passage of Electricity Through Rarefied Gas.

The various phenomena presented when an induction spark is made to pass through a gas at different degrees of exhaustion point to a modified condition of the matter at the highest exhaustions. Here are three exactly similar bulbs, the electrodes being aluminium balls, and the internal pressures being respectively 75 m.m., 2 m.m., and 0·1 m.m. If I pass the induction current in succession through the bulbs, you will perceive in each case a very different luminous phenomena. Here is a slightly exhausted tube (Fig. 1), like the first in the series just exhibited (75 m.m.), the induction spark passes from one

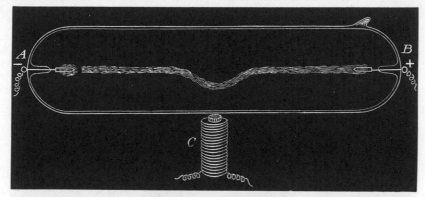

FIG. 1.—P. = 75 m.m.

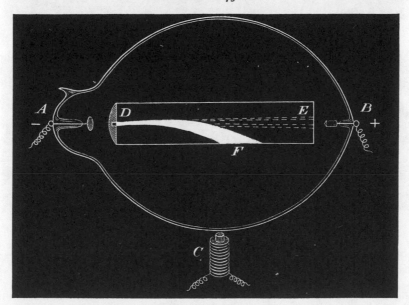

FIG. 2.—P. = 0·1 m.m.

likely that an experimentalist who would be working with such mercury-free apparatus would attempt to use the gauge without remembering that in this special case the indications would be incorrect. To use the McLeod gauge requires much patience and some amount of experience, but I have always found it trustworthy to register exhaustions far beyond the millionth of an atmosphere. I can adduce circumstantial evidence of the accuracy of its readings at these high vacua. In the year 1881 I read a paper before the Royal Society on " The Viscosity of Gases at High Exhaustions " (*Phil. Trans.*, 1881, p. 387), and illustrated my results in three large diagrams, on which I plotted the experimental results obtained at rarefactions up to 0·02 millionth of an atmosphere, giving curves comparing the decrease in viscosity with that of the repulsion resulting from radiation, at the different pressures. Now these curves, in the case of air for instance, are perfectly regular and uniform in their falling off, and it is evident that this could not have been the case unless the ordinates

end to the other, A, B, and the luminous discharge is seen as a line of light, acting as a flexible conductor. Under the tube I have an electro-magnet, C, and on making con-

representing viscosity and the abscissæ representing pressure were equally accurate. I am satisfied that, within narrow limits, the ordinates of viscosity are correct to the highest point, and the conformity of experiment to theory in the shape of these curves is a conclusive proof that, at as high an exhaustion as 0·02 M., the McLeod gauge is to be trusted to give accurate results within 2 per cent of the truth. To give some idea what these high exhaustions mean I may mention that the highest measured exhaustion—0·02 M.—bears the same proportion to the ordinary pressure of the atmosphere that a millimetre does to 30 miles, or in point of time, that one second bears to 20 months.

tact the line of light dips in the centre down to the poles of the magnet, and then rising again proceeds in a straight line. On reversing the current the line of light curves upwards. Notice that the action of the magnet in this case is only local.

In a highly exhausted tube the action is quite otherwise. Such a tube is before you (Fig. 2), and in it I have carried the exhaustion to a high point (0·1). I pass the induction current and you perceive the electrified molecules, like the line of light in the first tube, also move in straight lines, and make their path apparent by impinging on a phosphorescent screen, D E. If, however, I submit them to the action of a magnet, C, their behaviour is different. The line dips down to F, but does *not* recover itself. It seems that in the tube first shown we have to do with the average behaviour of the molecules of gas in its totality. In the second case, where the gas has been greatly attenuated, we are merely concerned with the be-

struction is created. The passengers behind catch up to the block and increase it, and those in front, passing on unchecked at their former rate, leave a comparatively vacant space. If a crowd is moving all in the same direction the formation of these groups becomes more distinct. With vehicles in crowded streets, the result, as everyone may have remarked, will be the same.

Hence mere differences in speed suffice to resolve a multitude of passengers into alternating gaps and knots.

Instead of observing moving men and women, suppose we experiment on little particles of some substance, such as sand, approximately equal in size. If we mix the particles with water in a horizontal tube and set them in rhythmical agitation, we shall see very similar results, the powder sorting itself with regularity into alternate heaps and blank spaces.

If we pass to yet more minute substances we observe the behaviour of the molecules of a rarefied gas, when

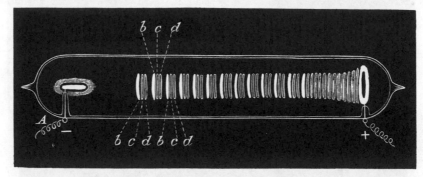

FIG. 3.—P. = 2 m.m.

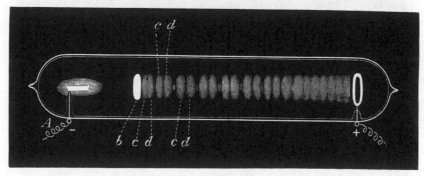

FIG. 4.—P. = 2 m.m.

haviour of the individual molecules of which it was originally composed.

The Stratified Discharge.

When the gas is rarer than is necessary to give the flexible line of light, as shown in the first experiment, the luminosity is plainly discontinuous, or, as it is termed, stratified.

A very good illustration of this fact may be taken from the moving crowd in any much frequented street—say Fleet Street. If at some time when the stream of traffic runs almost equally in both directions, we take our stand at a window from which we can overlook the passing crowd, we shall notice that the throng on the foot-way is not uniformly distributed, but is made up of knots—we might almost say blocks—interrupted by spaces which are comparatively open. We may easily conceive in what manner these knots and groups are formed. Some few persons walking rather more slowly than the average rate slightly retard the movements of others whether travelling in the same or in an opposite direction. Thus a temporary ob-

submitted to an induction current. The molecules here are free, of course, from any caprice, and simply follow the law I seek to illustrate, and though originally in a state of rampant disorder yet under the influence of the electric rhythm they arrange themselves into well-defined groups or stratifications; the luminosities show where arrested motion with concomitant friction occurs, and the dark intervals indicate where the molecules travel with comparatively few collisions.

Party-coloured Stratifications.

As another illustration of stratifications in a moderately exhausted tube (P = 2 m.m.), I will take the case of hydrogen prepared from zinc and sulphuric acid after being passed through various purifying agents, dried in the usual manner, and exhausted with a mercury pump (Fig. 3). I pass the induction current, and we see that the stratifications are tri-coloured, blue, pink, and grey. Next the negative pole A is a luminous layer, then comes a dark interval or Faraday's dark space (see below), and after this are the stratifications, the front component (*b*) of each group blue, the next (*c*) pink, and the third (*d*) grey. The blue disks are somewhat erratic. At a certain stage of exhaustion all the blue components of the stratifications suddenly migrate to the front, forming one bright blue disk, and leaving the pink and grey components by themselves. The tube before you (Fig. 4) is at this particular stage of exhaustion, and on passing the current you observe the blue disk only (*b*) is in front. When the tube contains a compound gaseous residue of this kind, the form of stratifications can be very considerably altered by varying the potential of the discharge. This alteration in the forms of stratification was first pointed out by Gassiot (1865, " B.A. Abstracts," p. 15), who gave very full descriptions and drawings of the alterations produced by putting in resistances of various lengths of distilled water. That the alteration depends simply upon the difference of potential the following experiment pretty clearly shows :—Here is a tube giving on my coil the coloured stratification usually attributed to the presence of residual hydrogen, but which I find is due to a mixture of hydrogen, mercury, and hydrocarbon vapours. Now by altering the brake so as to produce frequent discharges of lower potential, you see the stratifications gradually change in shape and become all pink ; again altering the brake so as to send less rapid discharges at a much higher potential, once more we get the coloured stratifications. When in this state, if we introduce a water resistance into the circuit so as to damp down the potential, exactly the same thing happens. The blue disk is caused by mercury ; its spectrum is that of mercury only, without even a trace of the bright red line of hydrogen. Experiments not yet finished make it very probable that the pink disks are due to hydrogen, and that the grey disks indicate carbon. The tube you have just seen contains nothing but hydrogen, mercury, and a minute trace of carbon ; but with all the resources at my command I have not been able to get hydrogen quite free from impurity. Indeed I do not think absolutely pure hydrogen has ever yet been obtained in a vacuum tube. I have so far succeeded as to completely eliminate the mercury, and almost completely to remove the trace of carbon. On the table is such a tube giving uniformly pink stratifications and showing no blue or grey disks with any potential of current.

(To be continued).

ELECTRICITY
IN TRANSITU:
FROM PLENUM TO VACUUM.*

By WILLIAM CROOKES, F.R.S.,
President of the Institution of Electrical Engineers.

(Continued from p. 56).

The Dark Space.

AFTER the stratification stage is passed we come to a very curious phenomenon, the so-called " Dark Space." Studying electrical phenomena in gases, in the year 1838, Faraday† pointed out a break in the continuity of the luminous discharge separating the glow of the positive slightly rarefied, as in this tube (Fig. 5, P. = 6 m.m.), where you will observe that the positive glow, extending as a pink streak from the positive electrode, B, ends about ten millimetres before the spot of blue light, C, representing the negative glow. This gap, or non-luminous hiatus, D, is Faraday's " dark space." Separating the negative glow from its electrode is another space. In this tube it is so small that the glow appears to be in actual contact with the electrode, but on exhausting a little further it rapidly separates; and in the next tube (Fig. 6), containing air at a little less pressure (P. = 3 m.m.), this dark space, E, has extended so as to remove the negative glow about four millimetres from the electrode, A. It is with this second dark space that I particularly wish to deal to-night. Therefore I shall refer to it as the " dark space," meaning always that in the negative glow.

In the experiments just shown with hydrogen stratifications the contents of the tube under the electric discharge still obey the laws following from the average properties of an immense number of molecules moving in every direction with velocities of all conceivable magnitudes. But if I continue exhausting, the dark space, E, round the negative pole, A, becomes visible, grows larger and larger, and at last fills up the entire tube. The molecules at this stage are in a condition different from those in a less highly exhausted tube. At low exhaustions they behave

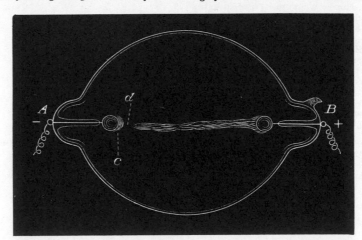

FIG. 5.—P. = 6 m.m.

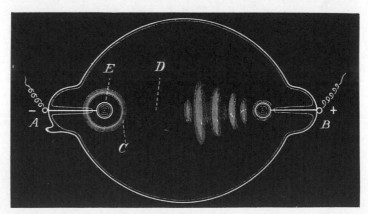

FIG. 6.—P. = 3 m.m.

electrode from that of the negative. This he called " the dark space." It is seen in tubes containing gas only as gas in the ordinary sense of the term, but at these high exhaustions, under electric stress, they have become exalted to an *ultra-gaseous* state, in which very decided properties, hitherto masked, come into play.

* Inaugural Address delivered January 15th, 1891.
‡ " Experimental Researches in Electricity," 1838, par. 1544.

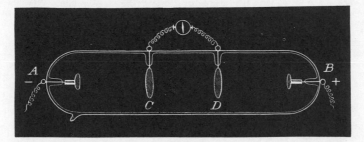

Fig. 7.

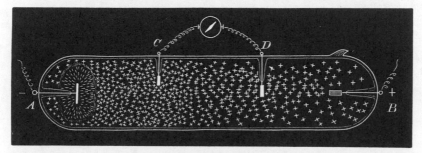

Fig. 8a.—P. = 0·25 m.m.

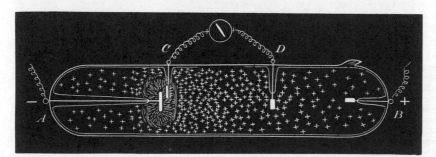

Fig. 8b.—P. = 0·25 m.m.

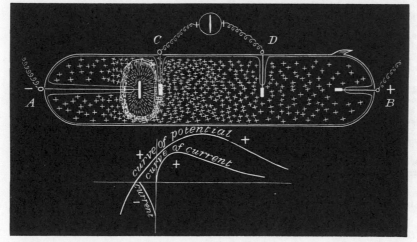

Fig. 8c.—P. = 0·25 m.m.

The radius of the dark space varies with the degree of exhaustion, with the kind of gas in which it has been produced, with the temperature of the negative pole, and to a less extent with the intensity of the spark.

It has been erroneously assumed that I ever said the thickness of the "dark space" represents the mean free path of the molecules in their ordinary condition, and it has been pointed out that the radius of the dark space is decidedly greater than the calculated mean free path of the molecules. I have taken accurate measurements of the radius of the dark space at different pressures, and compared it with the calculated mean free path of the gaseous molecules at corresponding pressures when not under the influence of electrical energy, and I find that they do not bear a constant relation one to the other. The length of the dark space is not 20 times the mean free path, as some have estimated, but a gradually increasing multiple must be taken as the exhaustion becomes greater.

Exploration with Idle Poles.

Wishing to learn something of the electrical condition of the matter within and without the dark space, I made a tube (Fig. 7), having besides the positive and negative terminals, A B, two extra intermediate poles, C and D; the tube showed that when the exhaustion was such that both the idle poles were outside the dark space, on passing the current through the tube, there was a considerable difference of potential between them when measured on the galvanometer. If the exhaustion was carried so high that one of the extra poles was just on the border of the dark space, then no current passed between them. When the exhaustion was still further increased so as to inclose one of the extra poles fully in the dark space, again there was a great difference of potential between them, but the direction was reversed, the pole at highest potential now being the one formerly at lowest potential.

When the dark space has been further explored by means of a movable negative pole, I found that the effects did not depend essentially on the exhaustion, and were really due to the position occupied by the extra pole with regard to the dark space.

These phenomena are difficult to understand from mere description, and the experiments themselves are not easy to carry out so as to be visible to many at a time. I have here, however, a working model of an apparatus which will make these puzzling indications clear to all.

A cylindrical tube (Fig. 8, a, b, and c, P.=0·25 m.m.), furnished with the usual poles, A, B, at the ends, has two extra or idle poles near together at C and D. The pole A is movable along the axis of the tube, so that when exhausted the dark space can be brought to any desired position with respect to the idle poles C and D. The shading and + and − marks roughly show the distribution of positive and negative electricity inside the tube. I start with the negative pole A as far as possible from either idle pole (Fig. 8, a). Turning on the coil you see the dark space surrounding the pole A, and the idle poles quite outside.

The shading shows that each idle pole is in the positive area, and on testing with a gold leaf electroscope it will be seen that each is charged with positive electricity. But the shading also shows more positive at C than at D, and on connecting C and D with a galvanometer the needle indicates a rush of current from C to D, D being negative to C.

The dark space is next brought to such a position that the pole C is well within it (Fig. 8, b). A change has now come over the indications. The galvanometer shows a reverse current to that which was seen on the former occasion. C is now negative and D positive, but the gold leaves still tell us both poles are positively electrified.

At a certain position of the dark space, when its edge is on the pole C (Fig. 8, c), a neutral state is found at which the gold leaves still show strongly positive electrifications, and no current is seen on the galvanometer. The curves below (Fig. 8, c), roughly show the rise and fall of negative and positive current at different parts of the tube, whilst the potential curve keeps positive.

When a substance that will phosphoresce under electrical excitement is introduced into the tube, the position of greatest luminosity is found to be at the border of the dark space, just where the two opposing armies of negative and positive atoms meet in battle array and re-combine. Later on I shall refer to this phenomenon in connection with the phosphorescence of yttria.

(To be continued).

KEEPING OF "NESSLER" STANDARDS.

By W. P. MASON.

STANDARD tubes containing 0·01, 0·02, 0·03, 0·04, 0·05, 0·06, 0·07, 0·08, and 0·09 m.grm. NH_3 in 50 c.c. water were prepared for use. Three days afterward they were compared with fresh standards of above values. It was observed that 0·03, 0·05, and 0·07 had not changed, and the remainder had darkened to a very slight degree—less than 0·0025 in each case.

Standards made in the morning may be thoroughly relied upon for working during the day.

ELECTRICITY
IN TRANSITU:
FROM PLENUM TO VACUUM.*

By WILLIAM CROOKES, F.R.S.,
President of the Institution of Electrical Engineers.

(Continued from p. 70).

Radiant Matter.

BY means of this tube (Fig. 9), I am able to show that a stream of ultra-gaseous particles, or Radiant Matter, does not carry a current of electricity, but consists of a succession of negatively electrified molecules whose electrostatic repulsion overbalances their electro-magnetic attraction, probably because their speed along the tube is less than the velocity of light. The tube has two negative terminals, A, A', close together at one end, enabling me to send along the tube two parallel streams of radiant matter, rendered visible by impinging them through holes in a mica diaphragm on a screen of phosphorescent substance. It is exhausted to a pressure of 0·1 m.m. I connect one of the negative poles, A, with the induction coil and the luminous stream darts along the tube from C to D parallel with the axis. I now connect the other negative pole, giving a second parallel stream of radiant matter. If these streams are in the nature of wires carrying a current they will attract each other, but if they are simply two streams of electrified molecules they will repel each other. As soon as the second stream is started you see the first stream jump away in the direction C, E, showing strong repulsion, proving that they do not act like current carriers, but merely like similarly electrified bodies. It is, however, probable that were the velocity of the streams of molecules greater than that of light, they would behave differently, and attract each other, like conductors carrying a current.

To ascertain the electrical state of the residual molecules in a highly exhausted tube such as you have just seen, I introduced an idle pole or exploring electrode between the positive and negative electrodes in such a manner that the molecular stream might play upon it. The intention was to ascertain whether the molecules on collision with an obstacle gave off any of their electrical charge. In this experiment (Fig. 10, P. = 0·0001 m.m., or 0·13 M.†) it was found that an idle pole, C, placed in the direct line between the positive and the negative poles, A, B, receiving in consequence the full impact of the molecules shot from the negative pole, manifested a strong *positive* charge. In a variety of other experiments made to decide this question, the electricity obtained was always found positive on testing with the gold leaf or Lippman's electrometer, and when the idle pole was connected to earth through a galvanometer a current passed as if this pole were the copper element of a copper-zinc cell, indicating leakage of a current to earth, the idle pole being positive. If, instead of sending this current to earth, the wire was connected to the negative pole of the tube, a much more powerful current passed in the same direction.

The Edison Effect.

An exactly parallel experiment has been made by Mr. Edison, Mr. Preece, F.R.S., and Prof. Fleming, using, instead of a vacuum tube, an incandescent lamp. They found that from an idle pole placed between the ends of the filament the electricity always flowed as if the pole were the zinc element of a copper-zinc cell; having repeated their experiments I entirely corroborate them. I get a powerful current in one direction from an idle pole placed between the limbs of an incandescent carbon filament, and one in the opposite direction from an idle pole in an highly exhausted vacuum tube. This discrepancy

---

* Inaugural Address delivered January 15th, 1891.
† M. = one-millionth of an atmosphere.
1,000,000 M. = 760 m.m.
,, = one atmosphere.

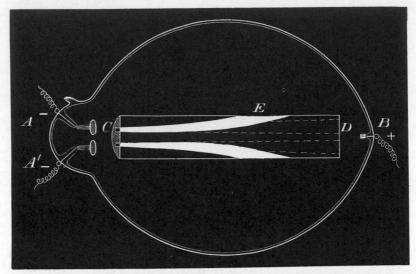

FIG. 9.—P. = 0·1 m.m.

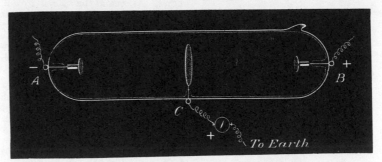

FIG. 10.—P. = 0·0001 m.m., or 0·13 M.

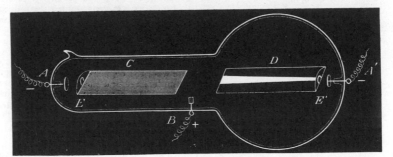

FIG. 11.—P. = 0·001 m.m., or 1·3 M.

was extremely puzzling, and I tested, with a similar result, very many experimental tubes made in different ways. The electricity obtained from an idle pole placed between the positive and negative terminals in a highly exhausted tube was always strongly *positive*, and it is only recently that continued experiment has cleared the matter up.

Some of the contradictory results are due to the exhaustion not being identical in all cases. In my vacuum tubes the directions of current between the idle pole and the earth changes from negative to positive as the exhaustion rises higher. Testing the current when exhaustion is proceeding, there is a point reached when the galvano-meter deflection—hitherto negative—becomes nil, showing that the potential at this point is zero. At this stage the passage of a few more drops of mercury down the pump tube renders the current positive. This change occurs at a pressure of about 2 m.m.

After this point is reached, when the induction current is passed through the tube, the walls rapidly become positively electrified, probably by the friction of the molecular stream against the glass, and this electrification extends over the surface of any object placed inside the tube. I will show you how this electrification of the inner walls of the tube acts on the molecular stream at high

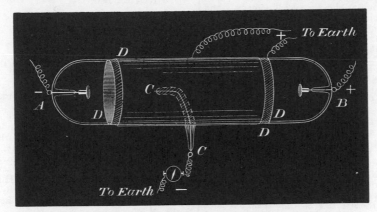

FIG. 12.—P. = 0·0001 m.m., or 0·13 M.

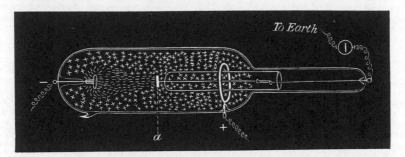

FIG 13*a*.—P. = 0·0001 m.m., or 0·13 M.

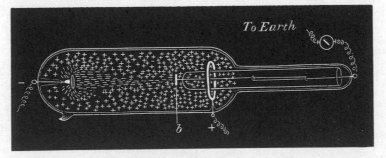

FIG. 13*b*.—P. = 0·0001 m.m., or 0·13 M.

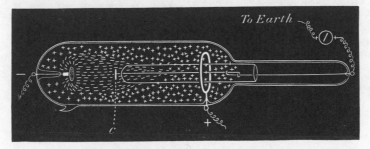

FIG. 13*c*.—P. = 0·0001 m.m., or 0·13 M.

vacua. In this tube, Fig. 11 (P.=0·001 m.m., or 1·3 M.), are fixed to exactly similar phosphorescent screens, c and D; at one end of each is a mica gate, E, E', with a negative pole, A, A', facing it. One of the screens, c, is in the cylindrical part of the tube and close to the walls; the other, D, is in the spherical portion, and therefore far removed from the walls. On passing the current the screen D in the globe shows a narrow sharp streak of phosphorescence, proving that here the molecules are free to follow their normal course straight from the negative pole. In the cylindrical part of the tube, however, so great is the attraction of the walls that the molecular stream is widened out sufficiently to make the whole surface of the screen, c, glow with phosphorescent light.

If an idle pole, c, c, Fig. 12 (P.=0·0001 m.m., or 0·13 M.), protected all but the point by a thick coating of glass, is brought into the centre of the molecular stream in front of the negative pole A, and the whole of the inside and outside of the tube walls are coated with metal, D, D, and "earthed," so as to carry away the positive electricity as rapidly as possible, then it is seen that the molecules leaving the negative pole and striking upon the idle pole c on their journey along the tube carry a negative charge, and communicate negative electricity to the idle pole.

This tube is of interest since it is the one in which I was first able to perceive how in my earlier results I always obtained a positive charge from an idle pole placed in the direct stream from the negative pole. Having got so far, it was easy to devise a form of apparatus that completely verified the theory, and at the same time threw considerably more light upon the subject. Fig. 13, a, b, c, is such a tube, and in this model I have endeavoured to show the electrical state of it at a high vacuum by marking a number of + and − signs. The exhaustion has been carried to 0·0001 m.m., or 0·13 M., and you see that in the neighbourhood of the positive pole, and extending almost to the negative, the tube is strongly electrified with positive electricity, the negative atoms shooting out from the negative pole in a rapidly diminishing cone. If an idle pole is placed in the position shown at Fig. 13 a, the impacts of positive and negative molecules are about equal, and no decided current will pass from it, through the galvanometer, to earth. This is the neutral point. But if we imagine the idle pole to be as at Fig. 13 b, then the positively electrified molecules greatly preponderate over the negative molecules, and positive electricity is shown. If the idle pole is now shifted as shown at Fig. 13 c, the negative molecules preponderate, and the pole will give negative electricity.

As the exhaustion proceeds, the positive charge in the tube increases, and the neutral point approaches closer to the negative pole, and at a point just short of non-conduction so greatly does the positive electrification preponderate that it is almost impossible to get negative electricity from the idle pole, unless it actually touches the negative pole. This tube is before you, and I will now proceed to show the change in direction of current by moving the idle pole.

I have not succeeded in getting the "Edison" current in incandescent lamps to change in direction at even the highest degree of exhaustion which my pump will produce. The subject requires further investigation, and like other residual phenomena these discrepancies promise a rich harvest of future discoveries to the experimental philosopher, just as the waste products of the chemist have often proved the source of new and valuable bodies.

(To be continued).

PROCEEDINGS OF SOCIETIES.

CHEMICAL SOCIETY.
Ordinary Meeting, January 15th, 1891.

Dr. W. J. RUSSELL, F.R.S., in the Chair.

CERTIFICATES were read for the first time in favour of Messrs. Henry Austin Appleton, 19, South Street, Middlesbro'; John Charles Aydans, 44, Crescent Road, Plumstead, S.E.; Clayton Beadle, Beadonwell, Belvedere, Kent; Thomas Byrne, Glenville, Dundrum, co. Dublin; Arthur Cole, B.A., Holmleigh, Charles Street, Berkhamsted; Reginald Lorn Marshall, 25, Lancaster Park, Richmond; Tom Kirke Rose, 9, Royal Mint, E.; R. Greig Smith, Springwells, New Street, Musselburgh; Howard C. Sucré, Breeze House, Higher Broughton, Manchester; Matthew Carrington Sykes, Sykeshurst, Barnsley, Yorks; W. Will, Ph.D., 1, Beethoven Strasse, Berlin, N.W.

The following papers were read:—

1. "*Magnetic Rotation.*" By W. OSTWALD.

The magnetic rotation of organic compounds, according to Perkin, is an additive function of their composition and equal to the sum of the rotations of the components, but this is not the case with the rotation of inorganic compounds, which is usually found greater than that calculated on such an assumption. In the case of hydrogen chloride, for instance, the calculated value is about 2·18, and as a matter of fact the value obtained for hydrogen chloride dissolved in an organic solvent, isoamyl oxide, is 2·24, but when dissolved in water the value found is from 4·05 to 4·42, increasing with the dilution. The author points out that these exceptional values are only obtained in the case of electrolytes, and that they must therefore be referred to a fundamental difference existing between the constitution of electrolytes and that of non-conductors. That such a difference exists has been already deduced from other considerations, and has led Arrhenius to the formulation of the theory of electrolytic dissociation. The author claims that the facts established with regard to magnetic rotation are in perfect accordance with this theory, and that any exceptional values in the magnetic rotations of electrolytes are due to the occurrence of electrolytic dissociation.

DISCUSSION.

Mr. PICKERING said that in Professor Ostwald's attempt to appropriate Dr. Perkin's results on the magnetic rotation of solutions of electrolytes in support of the dissociation theory, no attempt was made to explain what connection should exist between the magnetic rotation and the supposed dissociation into ions, but it was boldly stated that if, as in the case of hydrogen chloride, the magnetic rotation and dissociation both increased with dilution, the result proved the truth of the dissociation theory, while equal support to this latter theory was afforded if, as in the case of sulphuric acid, dilution diminished the rotation and increased the dissociation; and the most astonishing part of the argument appeared to be that the nearly double magnetic rotations obtained in some cases should be brought forward as a proof of dissociation, when the observations were made on solutions so strong that the dissociation theory represents them as containing hardly any dissociated substance at all. Surely the natural conclusion to draw from such a doubling when it occurs in the absence of dissociation would be that it could not be explained by dissociation.

ELECTRICITY
IN TRANSITU:
FROM PLENUM TO VACUUM.*

By WILLIAM CROOKES, F.R.S.,
President of the Institution of Electrical Engineers
(Continued from p. 80).

Properties of Radiant Matter.

ONE of the most characteristic attributes of Radiant
Matter—whence its name—is that it moves in approxi-

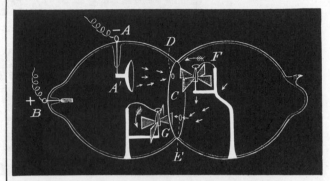

FIG. 14.—P. = 0·001 m.m., or 1·3 M.

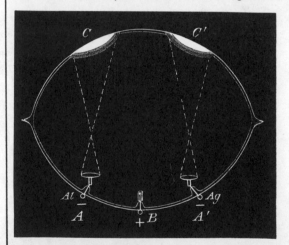

FIG. 15.—P. = 0·00068 m.m., or 0·9 M.

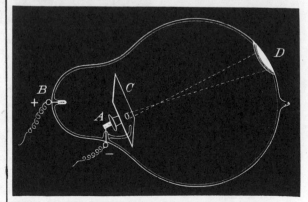

FIG. 16.—P. = 0·00068 m.m., or 0·9 M.

* Inaugural Address delivered January 15th, 1891.

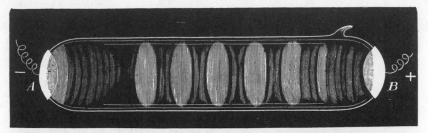

FIG. 17.—P. = 2 m.m.

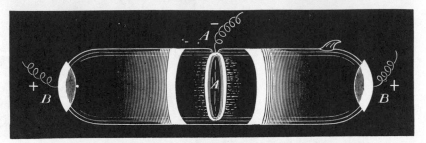

FIG. 18.—P. = 0.076 m.m., or 100 M.

FIG. 19.—P. = 0.00068 m.m., or 0.9 M.

mately straight lines and in a direction almost normal to the surface of the electrode. If we keep the induction current passing continuously through a vacuum tube in the same direction, we can imagine two ways in which the action proceeds ; either the supply of gaseous molecules at the surface of the negative pole must run short, and the phenomena come to an end, or the molecules must find some means of getting back. I will show you an experiment which reveals the molecules in the very act of returning. Here is a tube (Fig. 14) exhausted to a pressure of 0.001 m.m., or 1.3 M. In the middle of the tube is a thin glass diaphragm, C, pierced with two holes, D and E. At one part of the tube a concave pole, A', is focussed on the upper hole, D, in the diaphragm. Behind the upper hole and in front of the lower one are movable vanes, F and G, capable of rotation by the slightest current of gas through the holes.

On passing the current with the concave pole negative, the small vanes rotate in such a manner as to prove that at this high exhaustion a stream of molecules issues from the lower hole in the diaphragm, whilst at the same time a stream of freshly charged molecules is forced by the negative pole through the upper hole. The experiment speaks for itself, showing as forcibly as an experiment can show that so far the theory is right.

This view of the ultra-gaseous state of matter is advanced merely as a working hypothesis, which, in the present state of our knowledge, may be regarded as a necessary help to be retained only so long as it proves useful. In experimental research early hypotheses have necessarily to be modified, or adjusted, or perhaps entirely abandoned, in deference to more accurate observations. Dumas said, truly, that hypotheses were like crutches which we throw away when we are able to walk without them.

Radiant Matter and "Radiant Electrode Matter."

In recording my investigations on the subject of radiant matter and the state of gaseous residues in high vacua under electrical strain, I must refer to certain attacks on the views I have propounded.   The most important of

by Puluj on "Radiant Electrode Matter and the So-called Fourth State."   Dr. Puluj's paper concerns me most, as the author has set himself vigorously to the task of

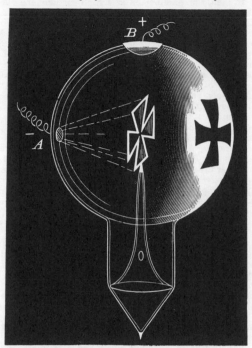

Fig. 20.—P. = 0·00068 m.m., or 0·9 M.

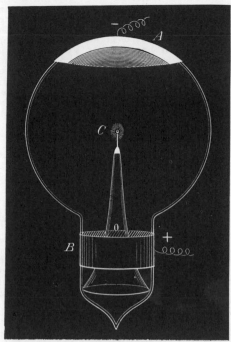

Fig. 22.   P. = 0·000076 m.m., or 0·1 M.

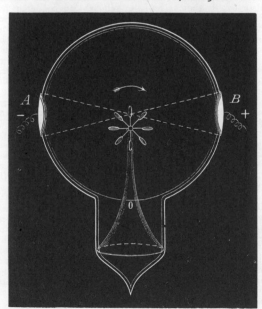

Fig. 21.—P. = 0·001 m.m., or 1·3 M.

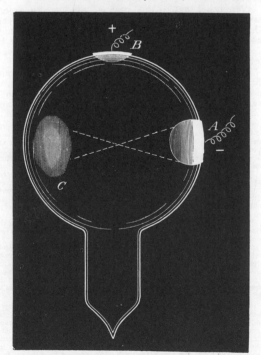

Fig. 23.—P. = 0·00068 m.m., or 0·9 M.

these questionings are contained in a volume of "Physical Memoirs," selected and translated from foreign sources under the direction of the Physical Society (vol. i., Part 2). This volume contains two memoirs, one by Hittorff on the "Conduction of Electricity in Gases," and the other

opposing my conclusions.   Apart from my desire to keep controversial matter out of an address of this sort, time would not permit me to discuss the points raised by my critic. I will therefore only observe in passing that Dr. Puluj has no authority for linking my theory of the fourth state of matter with the highly transcendental doctrine of four dimensional space.

Reference has already been made to the mistaken supposition that I have pronounced the thickness of the dark space in a highly-exhausted tube, through which an induction spark is passed, to be identical with the natural mean free path of the molecules of gas at that exhaustion. I could quote numerous passages from my writings to show that what I meant and said was the mean free path as amplified and modified by the electrification.* In this view I am supported by Professor Schuster,† who in a passage quoted below distinctly admits that the mean free path of an electrified molecule may differ from that of one in its ordinary state.

The great difference between Puluj and me lies in his statement that‡ "_the matter which fills the dark space consists of mechanically detached particles of the electrodes which are charged with statical negative electricity, and move progressively in a straight direction._"

To these mechanically detached particles of the electrodes, "of different sizes, often large lumps,"§ Puluj attributes all the phenomena of heat, force, and phosphorescence that I from time to time have described in my several papers.

Puluj objects energetically to my definition "Radiant Matter," and then proposes in its stead the misleading term, "Radiant Electrode Matter." I say "misleading," for while both his and my definitions equally admit the existence of "Radiant Matter," he drags in the hypothesis that the radiant matter is actually the disintegrated material of the poles.

Puluj declares that the phenomena I have described in high vacua are produced by his irregularly shaped lumps of Radiant Electrode Matter. My contention is that they are produced by Radiant Matter of the residual molecules of gas.

Were it not that in this case we can turn to experimental evidence I would not mention the subject to you. On such an occasion as this controversial matter must have no place; therefore I content myself, at present, by showing a few novel experiments which demonstratively prove my case.

Let me first deal with the Radiant Electrode hypothesis. Some metals, it is well known, such as silver, gold, or platinum, when used for the negative electrode in a vacuum tube, volatilise more or less rapidly, coating any

object in their neighbourhood with a very even film. On this depends the well-known method of electrically preparing small mirrors, &c. Aluminium, however, seems exempt from this volatility. Hence, and for other reasons, it is generally used for electrodes.

If, then, the phenomena in a high vacuum are due to the "electrode matter," the more volatile the metal used the greater should be the effect.*

Here is a tube (Fig. 15, P. = 0·00068 m.m., or 0·9 M.), with two negative electrodes, A, A', so placed as to project two luminous spots on the phosphorescent glass of the tube. One electrode, A', is of pure silver, a volatile metal, the other, A, is of aluminium, practically non-volatile. A quantity of "electrode matter" will be shot off from the silver pole, and practically none from the aluminium pole; but you see that in each case the phosphorescence, C, C', is identical. Had the Radiant Electrode Matter been the active agent the more intense phosphorescence would proceed from the more volatile pole.

A drawing of another experimental piece of apparatus is shown in Fig. 16. A pear-shaped bulb of German glass, has near the small end an inner concave negative pole, A, of pure silver, so mounted that its inverted image is thrown upon the opposite end of the tube. In front of this pole is a screen of mica, C, having a small hole in the centre, so that only a narrow pencil of rays from the silver pole can pass through, forming a bright spot, D, at the far end of the bulb. The exhaustion is about the same as in the previous tube, and the current has been allowed to pass continuously for many hours so as to drive off a certain portion of the silver electrode, and upon examination it is found that the silver has all been deposited in the immediate neighbourhood of the pole, whilst the spot, D, at the far end of the tube that has been continuously glowing with phosphorescent light is practically free from silver.

The experiment is too lengthy for me to repeat it here, so I shall not attempt it, but I have on the table the results for examination.

The identity of action of silver and aluminium in the first case, and the non-projection of silver in this second instance, in themselves are sufficient to condemn Dr. Puluj's hypothesis, since they prove that phosphorescence is independent of the material of the negative electrode. In front of me is a set of tubes that to my mind puts the matter wholly beyond doubt. The tubes contain no inside electrodes with the residual gaseous molecules, and with them I will proceed to give some of the most striking radiant matter experiments without any inner metallic poles at all.

In all these tubes the electrodes, which are of silver, are on the outside, the current acting through the body of the glass. The first tube contains gas only slightly rarefied and at the stratification stage. It is simply a closed glass cylinder, with a coat of silver deposited outside at each end, and exhausted to a pressure of 2 m.m. The outline of the tube is shown in Fig. 17. I pass a current and, as you see, the stratifications, though faint, are perfectly formed.

The next tube, seen in outline in Fig. 18, shows the dark space. Like the first it is a closed cylinder of glass, with a central indentation forming a kind of hanging pocket, and almost dividing the tube into two compartments. This pocket, silvered on the air side, forms a hollow glass diaphragm that can be connected electrically from the outside, forming the negative pole, A; the two ends of the tube, also outwardly silvered, form the positive poles, B, B. I pass the current and you all see the dark space distinctly visible. The pressure here is 0·076 m.m., or 100 M.

The next stage, dealing with more rarefied matter, is

* "The thickness of the dark space surrounding the negative pole is the measure of the mean length of the path of the gaseous molecules between successive collisions. The electrified molecules are projected from the negative pole with enormous velocity, varying however with the degree of exhaustion and intensity of the induction current."—(_Phil. Trans._, Part i., 1879, par. 530).

"The extra velocity with which the molecules rebound from the excited negative pole keeps back the more slowly moving molecules which are advancing towards the pole. The conflict occurs at the boundary of the dark space, where the luminous margin bears witness to the energy of the discharge."—(_Phil. Trans._, Part i., 1879, par. 507).

"Here, then, we see the induction spark actually illuminating the lines of molecular pressure caused by the excitement of the negative pole."—(R. I. Lecture, Friday, April 4th, 1879).

"The electrically excited negative pole supplies the _force majeure_ which entirely, or partially, changes into a rectilinear action the irregular vibration in all directions."—(_Proc. Roy. Soc._, 1880, page 472).

"It is also probable that the absolute velocity of the molecules is increased so as to make the mean velocity with which they leave the negative pole greater than that of ordinary gaseous molecules."—(_Phil. Trans._, Part ii., 1881, par. 719).

† "It has been suggested that the extent of the dark space represents the mean free path of the molecules. . . It has been pointed out by others that the extent of the dark space is really considerably greater than the mean free path of the molecules, calculated according to the ordinary way. My measurements make it nearly twenty times as great. This, however, is not in itself a fatal objection; for, as we have seen, the mean free path of an ion may be different from that of a molecule moving among others.—Schuster, _Proc. Roy. Soc._, xlvii., pp. 556-7.

‡ "Physical Memoirs," Part ii., vol. 1, page 244. The paragraph is italicised in the original.

§ _Loc. cit._ p. 242.

* In a valuable paper read before the Royal Society, November 20th, 1890, by Professors Liveing and Dewar, on finely-divided metallic dust thrown off the surface of various electrodes in vacuum tubes, they find not only that dust, however fine, suspended in a gas will not act like gaseous matter in becoming luminous with its characteristic spectrum in an electric discharge, but that it is driven with extraordinary rapidity out of the course of the discharge.

that of phosphorescence. Here is an egg-shaped bulb, shown in Fig. 19, containing some pure yttria, and a few rough rubies. The positive electrode, B, is on the bottom of the tube under the phosphorescent material; the negative, A, is on the upper part of the tube. See how well the rubies and yttria phosphoresce under molecular bombardment, at an internal pressure of 0·00068 m.m., or 0·9 M.

A shadow of an object inside a bulb can also be projected on to the opposite wall of the bulb by means of an outside pole. A mica cross is supported in the middle of the bulb (Fig. 20), and on connecting a small silvered patch, A, on one side of the bulb with the negative pole of the induction coil, and putting the positive pole to another patch of silver, B, at the top, the opposite side of the bulb glows with a phosphorescent light, on which the black shadow of the cross seems sharply cut out. Here the internal pressure is 0·00068 m.m., or 0·9 M.

Passing to the next phenomenon, I proceed to show the production of mechanical energy in a tube without internal poles. It is shown in Fig. 21 (P. = 0·001 m.m., or 1 3 M.). It contains a light wheel of aluminium, carrying vanes of transparent mica, the poles, A, B, being in such a position outside that the molecular focus falls upon the vanes on one side only. The bulb is placed in the lantern, and the image is projected on the screen; if I now pass the current you see the wheel rotates rapidly, reversing in direction as I reverse the current.

Here is an apparatus, Fig. 22, which shows that the residual gaseous molecules when brought to a focus produce heat. It consists of a glass tube with a bulb blown at one end and a small bundle of carbon wool, C, fixed in the centre, and exhausted to pressure of 0·000076 m.m., or 0·1 M. The negative electrode, A, is formed by coating part of the outside of the bulb with silver, and it is in such a position that the focus of rays falls upon the carbon wool. The positive electrode, B, is an outer coating at the other end of the tube. I pass the current, and those who are close may see the bright sparks of carbon raised to incandescence by the impact of the molecular stream.

You thus have seen that all the old "radiant matter" effects can be produced in tubes containing no metallic electrodes to volatilise. It may be suggested that the sides of the tube in contact with the outside poles become electrodes in this case, and that particles of the glass itself may be torn off and projected across and so produce the effects. This is a strong argument, which fortunately can be tested by experiment. In the case of this tube (Fig. 23, P. = 0·00068 m.m., or 0·9 M.) the bulb is made of lead glass phosphorescing blue under molecular bombardment. Inside the bulb, completely covering the part that would form the negative pole, A, I have painted a substantial coat of yttria, so as to interpose a layer of this earth between the glass and the inside of the tube. The negative and positive poles are silver disks on the outside of the bulb, A being the negative and B the positive pole. If, therefore, particles are torn off and projected across the tube to cause phosphorescence, these particles will not be particles of glass but of yttria, and the spot of phosphorescent light, C, on the opposite side of the bulb will not be the dull blue of lead glass, but the golden yellow of yttria. You see there is no such indication; the glass phosphoresces with its usual blue glow, and there is no evidence that a single particle of yttria is striking it.

Witnessing these effects I think you will agree I am justified in adhering to my original theory that the phenomena are caused by the Radiant Matter of the residual gaseous molecules, and certainly not by the torn off particles of the negative electrode.

(To be continued).

ELECTRICITY
IN TRANSITU:
FROM PLENUM TO VACUUM.*

By WILLIAM CROOKES, F.R.S.,
President of the Institution of Electrical Engineers.

(Continued from p. 93).

Phosphorescence in High Vacua.

I HAVE already pointed out that the molecular motions rendered visible in a vacuum tube are not the motions of molecules under ordinary condition, but are compounded of these ordinary or kinetic motions and the extra motion due to the electrical impetus.

Experiments show that in such tubes a few molecules may traverse more than a hundred times the *mean* free path, with a correspondingly increased velocity, until they are arrested by collisions. Indeed, the molecular free path may vary in one and the same tube, and at one and the same degree of exhaustion.

Very many bodies, such as ruby, diamond, emerald, alumina, yttria, samaria, and a large class of earthy oxides and sulphides, phosphoresce in vacuum tubes when placed in the path of the stream of electrified molecules proceeding from the negative pole. The composition of the gaseous residue present does not affect phosphorescence; thus, the earth yttria phosphoresces well in the residual vacua of atmospheric air, of oxygen, nitrogen, carbonic anhydride, hydrogen, iodine, sulphur, and mercury.

With yttria in a vacuum tube the point of maximum phosphorescence, as I have already pointed out, lies on the margin of the dark space. The diagram (Fig. 24) shows approximately the degree of phosphorescence in different parts of a tube at an internal pressure of 0·25 m.m., or 330 M. On the top you see the positive and negative poles, A and B, the latter having the outline of the dark space shown by a dotted line, C. The curve D, E, F, shows the relative intensities of the phosphorescence at different distances from the negative pole, and the position inside the dark space at which phosphorescence does not occur. The height of the curve represents the degree of phosphorescence. The most decisive effects of phosphorescence are reached by making the tube so large that the walls are outside the dark space, whilst the material submitted to experiment is placed just at the edge of the dark space.

Hitherto I have spoken only of phosphorescence of substances placed under the negative pole. But from numerous experiments I find that bodies will phosphoresce in actual contact with the negative pole.

This is only a temporary phenomenon, and ceases entirely when the exhaustion is pushed to a very high point. The experiment is one scarcely possible to exhibit to an audience, so I must content myself with describing it. A U-tube, shown in Fig. 25, has a flat aluminium pole, in the form of a disk, at each end, both coated with a paint of phosphorescent yttria. As the rarefaction approaches about 0·5 m.m. the surface of the negative pole A becomes faintly phosphorescent. On continuing the exhaustion this luminosity rapidly diminishes, not only in intensity, but in extent, contracting more and more from the edge of the disk, until ultimately it is visible only as a bright spot in the centre. This fact does not prop a recent theory, that as the exhaustion gets higher the discharge leaves the centre of the pole, and takes place only between the edge and the walls of the tube.

If the exhaustion is further pushed, then at the point where the surface of the negative pole ceases to be luminous, the material on the positive pole, B, commences to phosphoresce, increasing in intensity until the tube refuses to conduct, its greatest brilliancy being just short of this degree of exhaustion. The probable explanation is that the vagrant molecules I introduce in the next experiment, happening to come within the sphere of influence of the positive pole, rush violently to it and excite phosphorescence in the yttria, whilst losing their negative charge.

Loose and Erratic Molecules.

In the brief time left to me this evening I cannot touch upon the mass of experiments made to render this result clear, so I will at once show you a piece of apparatus that clearly illustrates the cause of phosphorescence at the positive pole. A drawing of this tube is shown at Fig. 26, but let me first explain the effect I expect to obtain, and then endeavour to show the actual experiment.

A C B is a U-shaped tube with terminals, A and B, at each end; D and E are two mica screens covered with a phosphorescent powder, having at F and G other screens with a small slit in front, so as to allow only a narrow beam of charged molecules to pass through. At first the tube is exhausted to a pressure of 0·076 m.m., or 100 M., and you see how sharp and slightly divergent is the luminous image on screen D, whilst not a trace of phosphorescence is to be seen on the screen E in the other limb of the tube. Now I push the exhaustion to the highest point short of non conduction (0·000076 m.m , or 0·1 M.), and the phenomena change. The initial line of light on D becomes wider and unsteady, whilst at the gate, G, a decided phosphorescence is observable entering the second

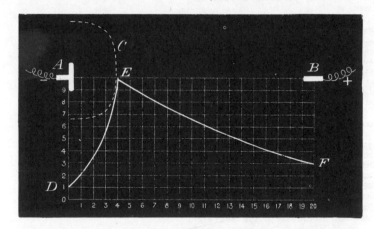

FIG. 24.—P. = 0·25 m.m., or 330 M.

* Inaugural Address delivered January 15th, 1891.

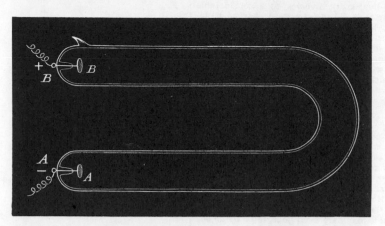

FIG. 25.

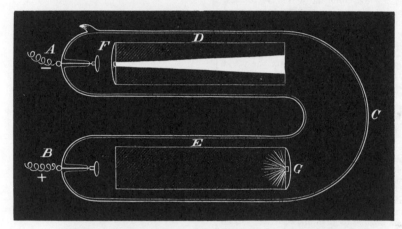

P. = 0·076 m.m.,
or 100 M.

P. = 0·000076 m.m.,
or 0·1 M.

FIG. 26.

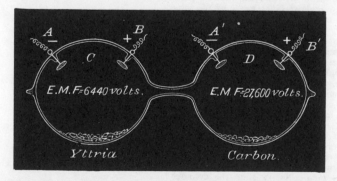

E.M.F. = 6440 volts.

E.M.F. = 27,600 volts.

Yttria

Carbon.

FIG. 27.—P. = 0·02 m.m., or 26 M.

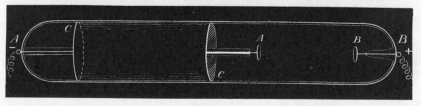

FIG. 28.—P. = 0·00068 m.m., or 0·9 M.

screen, E. This luminosity diverges to a considerable
extent, much more than did the stream of charged molecules observed in the first limb of the tube at the lower
exhaustion. This is only to be expected, inasmuch as
those sparse molecules which have run the gauntlet
of the crowd and have been hunted and buffetted
round the corner will have little or no lateral support, but
wander to the furthest end of the tube, and altogether
behave differently to the orderly procession of molecules
at more moderate exhaustion.

The Resistance of High Vacua.

I must only lightly touch this phenomenon to-night; it
is a subject of deep interest, and one to which I have
lately devoted much time. Shortly I hope to publish a
full account of results recently obtained.

The passage of an induction current at a high vacuum
through a tube depends much upon the material of the
tube or the substance enclosed within it. Given the same
degree of exhaustion, and the same distance between the
terminals inside the tube, the E.M.F. necessary to force
the current through may vary from 3000 volts to 20,000
volts, according to the particular material used.

Here is a very striking example that will serve to illustrate this phenomenon. Fig. 27 is a double tube joined
by a narrow open channel and therefore in the same state
of exhaustion throughout (P. = 0·02 m.m., or 26 M.). Each
tube is furnished with a pair of poles, A, B, and A', B'. One
tube contains the phosphorescent earth yttria, the other
contains finely divided carbon. I first connect the yttria
tube, C, to the coil, and place in parallel circuit with it a
spark gauge. To begin with, this gauge is set to a gap
of 1 m.m. (An E.M.F. of 920 volts is necessary to strike
across 1 m.m. of air; therefore the difference of potential
at the terminals inside the yttria tube is 920 volts, and
this you perceive is not sufficient to force the current
through the tube). I now gradually open the gauge until
the potential of the inner terminals has risen to a point
high enough to allow the spark to pass through the tube,
making the yttria phosphorescent.

The gap is at 7 m.m. equal to an E.M.F. of 6440 volts.
I next attach the coil wires to the tube D containing
the carbon, and the exhaustion in the tubes being the
same as before, I repeat the experiment; you now see
that the gap has to be opened to 30 m.m., equivalent to an
E.M.F. of 27,600 volts, before the current will pass
through the tube. The fact of whether the vacuum tube
contains yttria or carbon makes a difference of 21,160 volts
in the E.M.F. required to cause a discharge between the
terminals.

One other experiment I would like to show in further
illustration of this resistance. The idea suggested itself
that possibly differences in the material or conductivity of
the particular bulbs might influence the results you have
just seen. Here (Fig. 28) is a long cylindrical tube of
phosphorescent Bohemian glass containing a pair of
terminals, A, B; it also contains a shorter cylinder of glass,
C, C, brighly silvered inside. The internal pressure is
0·00068 m.m., or 0·9 M. The silver cylinder is now at one
end of the tube out of the way of the terminals, which
therefore have phosphorescent glass around them. I turn
on the coil and find that the E.M.F. necessary to force
the current through, now that the terminals are in a phosphorescent chamber, is 1380 volts. I slide the cylinder
down to the end of the tube so as to inclose the terminals
in a metallic silver chamber, and you see the E.M.F.
necessary to pass the current rises to 6440 volts. Metallic
silver does not phosphoresce, whilst Bohemian glass
phosphoresces very well. It appears that the greater the
phosphorescing power of the substance surrounding the
poles so much the easier does the induction spark pass.
Surround the poles with Bohemian glass or yttria—two
phosphorescent non-conductors of electricity—and the induction spark passes easily; immediately I surround the
terminals with a non-phosphorescent conductor the current refuses to pass.

(To be continued).

ELECTRICITY
IN TRANSITU :
FROM PLENUM TO VACUUM.*

By WILLIAM CROOKES, F.R.S.,
President of the Institution of Electrical Engineers.

(Concluded from p. 100).

What Occasions Phosphorescence ?

I SHOULD like to interest you in a question that has exercised my mind for some time past. What is it that occasions the phosphorescence of yttria and other bodies in vacuo under molecular bombardment ?

So far I have found this phosphorescence to be an attribute of non-conductors only. We know that in the act of phosphorescence the molecules of yttria are in a state of intense vibration. Each molecule may be viewed as the radiating centre of the entire bundle of rays, which, when decomposed by the prism, displays a discontinuous spectrum. We may also suppose that the residual atoms of gas charged with negative electricity give off their electricity on coming in collision with a phosphorescent body, and on their return take up a fresh charge.

The Electrolysis Hypothesis.

There is a certain amount of evidence in favour of an electrolytic hypothesis of the passage of electricity through rarefied gases. This has been ably advocated by Professor Schuster in the Bakerian Lecture before the Royal Society, March 20th, 1890.†

A molecule of hydrogen gas, for instance, may be made up of one group of atoms of hydrogen having an equivalent of negative electricity inherent in it, and one group of atoms having an equivalent of positive electricity bound up with it. These atoms are also charged with *additional* equivalents of positive or negative electricity, which they carry about as a ship carries its cargo. We are not concerned with the inherent electricity—of which we are ignorant—but with the extra or "cargo" charge. Let us imagine a molecule of hydrogen near the face of the negative pole in a vacuum tube. I turn on the current, and the atoms of the hydrogen molecule are dragged apart. The positive atom is attracted to the negative pole, where the violence of impact, or the discharge of electricity, renders it apparent with evolution of light. The internal luminous layer, which is closely adjacent to the negative electrode, is due, therefore, to the positive atoms rushing to the negative pole, and not, like the glow round the edge of the dark space, to the negative atoms projected from it. The negative atom, on the other hand, is driven violently from the negative pole, in virtue of the mutual repulsion existing between any two bodies similarly electrified, with a velocity varying with the intensity of the electrification and the degree of the vacuum ; the more perfect the vacuum the greater the velocity, the atoms flying outwards in straight lines until they meet with an obstacle. Such an obstacle may be a procession of positively charged atoms from the positive pole ; in this case the two kinds of atoms mutually discharge each other's cargos with a display of light. This phenomenon occurs at the margin of the dark space when the vacuum is only moderate. Or the obstacle may be produced by the vacuum being so high that the atoms of gas present are too few to form a continuous procession. (*Why* a high vacuum should be non-conductive does not clearly appear, but the fact itself is beyond doubt ; it is probably connected with the inability of electrified atoms to leave the poles.) Or again, the obstacle may be a phosphorescent body like yttria. In this case the negatively charged atoms deliver up their charge of electricity to the yttria, which is so constituted— perhaps after the manner of a Hertz resonator—that its atoms charge and discharge, vibrating about 550 billion

times in a second, and producing waves in the ether of the length, approximately, of 5·74 ten-millionths of a millimetre, and occasioning in the eye the effect of citron light.

We are not under the necessity of supposing that this number of hydrogen atoms are driven against the yttria in the second, although even at a high vacuum there are quite enough atoms left in the bulb to keep up such a supply. All that is needed is that a succession of shocks, not necessarily rhythmical, may strike the yttria at frequencies which will set up such a number of vibrations, just as a series of slow impacts on a gong causes it to emit sound waves of much greater frequency.

In a low vacuum only very few atoms can run the gauntlet among the crowd of inrushing atoms, and those few which succeed in reaching the yttria arrive with much reduced velocity, and so faint is the phosphorescence they set up that it is completely obscured* by the brighter phosphorescence of the residual gas. As the vacuum becomes higher, more and more atoms find their way across, and their speed being at the same time accelerated, the phosphorescence becomes intensified.

At a good vacuum most of the atoms hit the yttria, their velocity is increased, and the rhythmical excitation reaches its maximum.

The Dark Space in Mercury Vapour.

In applying the electrolytic hypothesis, I have used for illustration's sake the gaseous residue of hydrogen, which is known as a diatomic gas. I have found, however, the phenomena of the dark space, &c., to occur in the vapour of mercury, which is a monatomic gas. This important result induced me to patiently investigate this subject, and the result of one experiment is before you (Fig. 29). The tube is furnished with aluminium terminals, and is so arranged that the induction spark can be kept passing during exhaustion to drive off occluded gases. When at the highest attainable vacuum, the tube is filled with pure mercury by simply raising a reservoir. On applying heat the entire contents of the tube are boiled away and pass down the fall tubes of the pump, exhaustion going on at the same time. When the whole of the mercury has thus been boiled away *in vacuo*, except a little condensed at the upper part of the tube, the results on passing the spark are as followed :—When the tube is cold the induction current refuses to pass ; on gently heating with a gas-burner the current passes and the dark space is distinctly visible. Continuing the heat so as to volatilise the drops of mercury condensed on the sides, the whole tube becomes filled with a green phosphorescent light, the dark space gets smaller and smaller, and ultimately the negative pole becomes covered with a luminous glow. On allowing the tube to cool the same phenomena ensue in inverse order. The luminous halo expands, showing the dark space between it and the pole, and this dark space gradually grows larger as the tube becomes cooler. The mercury again condenses on the side of the tube, the green phosphorescence grows paler and paler, until at last the induction spark from the large coil refuses to pass.

At first sight this result appears fatal to the electrolytic hypothesis, for if the molecule of mercury contains only one atom how can we talk of its separation into positive and negative atoms by the electric stress ? It must be remembered, however, that we are as yet ignorant of the absolute mass of the atom of any element. All that can be said is that a molecule of free hydrogen becomes halved in combining chemically with certain other elements, whilst a molecule of free mercury does not suffer division on yielding any known compound of mercury : the physical atoms in the one behave as two separate groups, and those in the other as one undivided group. It has been agreed by chemists, for simplicity's sake and for facili-

---

* Inaugural Address delivered January 15th, 1891.
† *Roy. Soc. Proc.*, xlvii , 526.

---

* This faint phosphorescence at a low vacuum can be rendered visible by the electrical phosphoroscope described in my lecture at the Royal Institution in 1887.

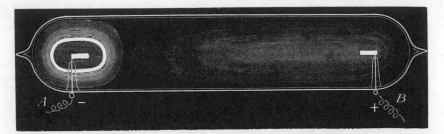

FIG. 29.

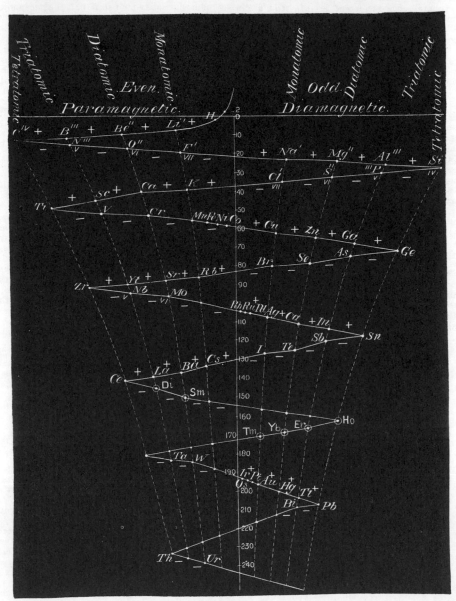

FIG. 30.

tating chemical calculations, to reduce the units to the lowest term, consistent with the avoidance of fractions; we therefore say that the atoms in a molecule of free hydrogen act in chemistry as two separate groups, each of a minimum relative weight of 1, whilst those in a molecule of free mercury act as one undivided group of the relative weight 200. But to what number of atoms the 1 and 200 correspond respectively no chemist knows.

To show how intimately chemistry and electricity interlock, I may here remark that one of the latest theories in chemistry renders such a division of the molecule into groups of electro-positive and electro-negative atoms necessary for a consistent explanation of the genesis of the elements. This is so important that I may be excused for digressing a little into this development of theoretical chemistry.

Genesis of the Elements.

It is now generally acknowledged that there are several ranks in the elemental hierarchy, and that besides the well-defined groups of chemical elements, there are underlying sub-groups. To these sub-groups has been given the name of meta-elements. The original genesis of atoms assumes the action of two forms of energy working in time and space—one operating uniformly in accordance with a continuous fall of temperature, and the other having periodic cycles of ebb and swell, and intimately connected with the energy of electricity (Fig. 30). The centre of this creative force in its journey through space scattered seeds or sub-atoms that ultimately coalesced into the groupings known as chemical elements. At this genetic stage the new born particles vibrating in all directions and with all velocities, the faster moving ones would still overtake the laggards, the slower would obstruct the quicker, and we should have groups formed in different parts of space. The constituents of each group whose form of energy governing atomic weight was not in accord with the mean rate of the bulk of the components of that group, would work to the outside and be thrown off to find other groups with which they were more in harmony. In time a condition of stability would be established, and we should have our present series of chemical elements each with a definite atomic weight—definite on account of its being the average weight of an enormous number of sub-atoms or meta-elements, each very near to the mean. The atomic weight of mercury, for instance, is called 200, but the atom of mercury, as we know it, is assumed to be made up of an enormous number of sub-atoms, each of which may vary slightly round the mean number 200 as a centre.

We are sometimes asked why, if the elements have been evolved, we never see one of them transformed, or in process of transformation, into another? The question is as futile as the cavil that in the organic world we never see a horse metamorphosed into a cow. Before copper, *e.g.*, can be transmuted into gold it would have to be carried back to a simpler and more primitive state of matter, and then, so to speak, shunted on to the track which leads to gold.

This atomic scheme postulates a to and fro motion of a form of energy governing the electrical state of the atom. It is found that those elements generated as they approach the central position are electro-positive, and those on the retreat from this position are electro-negative. Moreover the degree of positiveness or negativeness depends on the distance of the element from the central line; hence calling the atom in the mean position electrically neutral, those sub-atoms which are on one side of the mean will be charged with positive electricity, and those on the other side of the mean position will be charged with negative electricity, the whole atom being neutral.

This is not a mere hypothesis, but may take the rank of a theory. It has been experimentally verified as far as possible with so baffling an enigma. Long-continued research in the laboratory has shown that in matter which has responded to every test of an element, there are minute shades of difference which have admitted of

selection and resolution into meta-elements, having exactly the properties required by theory. The earth yttria, which has been of such value in these electrical researches as a test of negatively excited atoms, is of no less interest in chemistry, having been the first body in which the existence of this sub-group of meta-elements was demonstrated.

Conclusion.

I frankly admit I have by no means exhausted the subject which daily and nightly fills my thoughts. I have ardently sought for facts on which to base my theory. I have struggled with problems which must be conquered before we can arrive at exact conclusions—conclusions which, so far as inorganic Nature is concerned, can only be reached by the harmonious interfusion—not confusion—of our present twin sciences, electricity and chemistry. Of this interfusion I have just endeavoured to give you a foretaste. In elaborating the higher physics, the study of electrical phenomena must take a large, perhaps the largest, share.

We have invaded regions once unknown, but a formidable amount of hard work remains to be completed. As we proceed we may look to electricity not only to aid, as it already does, our sense of hearing, but to sharpen and develop other powers of perception.

Science has emerged from its childish days. It has shed many delusions and impostures. It has discarded magic, alchemy, and astrology. And certain pseudo-applications of electricity, with which the present Institution is little concerned, in their turn will pass into oblivion.

There is no occasion to be disheartened at the apparent slow pace of elemental discovery. The desponding declare that if Roger Bacon could re-visit "the glimpses of the moon" he would shake his head to think we have got no further, that we are still in a haze as to the evolution of atoms. As for myself I hold the firm conviction that unflagging research will be rewarded by an insight into natural mysteries such as now can scarcely be conceived. Difficulties, said a keen old statesman, are things to be overcome, and to my thinking Science should disdain the notion of Finality. There is no stopping half way, and we are resistlessly driven to ceaseless inquiry by the spirit "that impels all thinking things, all objects of all thought, and rolls through all things."

Section III

ISOMERISM AND THE NATURE OF THE ELEMENTS

This section begins with another of Faraday's early lectures, this time deploring the way in which the number of metallic elements kept on increasing. It is one of the ironies of history that Davy, who is the 'chemist of our own nation' alluded to, sought to decompose several elements but in fact added many new ones to the list. The techniques of electrolysis and spectroscopy revealed numerous elements during the century; but the cathode ray tube did in the end put the bases of the metals in our hands.

Liebig was one of the greatest chemists of the generation following Berzelius. With Dumas he proposed the radical theory; his laboratory at Giessen provided the best training available in chemistry; and he applied organic chemistry successfully to physiology and to agriculture. He was also a very good populariser; and his *Familiar Letters on Chemistry* is a very good example of its kind. The first edition, from which this extract is taken, was quite a slim volume; in later editions some material was deleted, but much more was put in and the book is a tribute to Liebig's breadth of interest.

George Wilson was an attractive man, interested in chemistry and its history—he wrote the *Life of Cavendish* which gave Ramsay a clue to his work on argon—as well as in natural theology and in technology, particularly the electric telegraph. The *Edinburgh New Philosophical Journal* was another general journal, valuable to the historian; it grew out of David Brewster's *Edinburgh Philosophical Journal*. Wilson's review is useful and readable; those to whom he refers sound crazy but were mostly respected scientists.

Berzelius' paper of 1846 was published in Richard Taylor's *Scientific Memoirs*, a journal devoted entirely to translations; particularly from the German, which was otherwise a stumbling block to many. It shows how ready Berzelius was to speculate; by this time he was elderly, but still evidently lively; and he was never, as perhaps his demolitions of Thomson might have led one to suppose, a devotee of mere facts.

Edward Frankland was one of the first generation of Englishmen who went to Germany, in his case to Bunsen at Marburg and Liebig at Giessen, to get the chemical education which was not available at home. In 1851 he became first Professor of Chemistry at Owens College, Manchester. His studies of the radicals of organic chemistry, some of which he believed he had isolated in the free state, led him towards the idea that each element has a fixed combining power. The *Proceedings of the Royal Society* grew through the nineteenth century from abstracts of papers in the Society's *Philosophical Transactions* into a journal of major importance in its own right.

Herman Kopp, perhaps best known as an historian of chemistry, was Professor of Chemistry at Giessen and then from 1863 at Heidelberg. He made numerous measurements of physical quantities; his studies of specific heats, reported at length in the *Philosophical Transactions* for 1865 and in brief in the paper reprinted here, were based upon the empirical law of Dulong and Petit; they led him to speculate that the metals were compound bodies.

Henry Roscoe followed Frankland at Manchester in 1857, when he was only twenty-four; be became a Fellow of the Royal Society in 1863, and in 1885 a Member of Parliament. His address to the British Association in 1884 is a useful historical survey of the progress of chemistry during the nineteenth century; stressing the change from the period when the properties of compounds were seen as governed by the nature of their components, to the present when they seemed the result of the number and arrangement of the atoms within the molecule.

FURTHER READING

J. H. Brooke, 'Wöhler's Urea and its Vital Force?—A Verdict from the Chemists', *Ambix*, XV (1968), 84–114.

R. Fox, 'The Background to the Discovery of Dulong and Petit's Law', *British Journal for the History of Science*, IV (1968), 1–22.

S. C. Kapoor, 'Dumas and Organic Classification', *Ambix*, XVI (1969), 1–65.

S. H. Mauskopf, 'The Atomic Structural Theory of Ampère and Gaudin', *Isis*, LX (1969), 61–74.

R. Siegfried, 'Composition, a neglected aspect of the Chemical Revolution', *Annals of Science*, XXIV (1968), 275–93.

I.

In 1818 his lectures still give the best insight into his mind.

At the end of his thirteenth lecture, on gold, silver, mercury, platinum, palladium, rhodium, iridium, osmium, he says :—

'As in their physical properties so in their chemical properties. Their affinities being weaker, (the noble metals) do not present that variety of combinations, belonging to the more common metals, which renders them so extensively useful in the arts; nor are they, in consequence, so necessary and important in the operations of nature. They do not assist in her hands in breaking down rocks and strata into soil, nor do they help man to make that soil productive or to collect for him its products.

'The wise man, however, will avoid partial views of things. He will not, with the miser, look to gold and silver as the only blessings of life; nor will he, with the cynic, snarl at mankind for preferring them to copper and iron. He will contemplate society as the proper state of man, and its artificial but necessary institutions and principles will appear to him the correct and advantageous result of natural causes. That which is convenient is that which is useful, and that which is useful is that which is valuable. It is in the relative position of things one to the other that they are to be considered and estimated; and whilst a man makes use of them no otherwise than wisely to supply his wants and virtuous pleasures, the avaricious trader has no reason to call him a fool of nature, nor the moral philosopher to name him the victim of society.'

His fourteenth and fifteenth lectures were on copper, iron, tin, lead, and zinc.

At the end of the sixteenth lecture, on antimony, arsenic, cobalt, manganese, nickel, bismuth, tungsten, molybdenum, uranium, cerium, tellurium, titanium, columbium, he says :—

'I have now, in the progress of the three last lectures, brought before you by far the most important of the metallic bodies. There yet remain a few, the result of late, even of contemporaneous research, which will form the subject of the next and the concluding lecture of this course. They are the bases of the alkalies and alkaline earths. It is interesting to observe the progress of this branch of chemistry, its relations to the ages through which it has passed, and the continual refinement of the means which have urged it in its career.

'The ancients knew but of seven reduced metals, but they are those which, amongst the extensive range we now possess, are the most important; and in gold, silver, mercury, copper, iron, tin, and lead they found abundant resources for weapons of war and for implements of art, for economical applications and for ornamental uses.

'The metals in use in old times were obtained almost by accident—either pure from the hands of nature or by the rudest and simplest workings. But, excited by the result of their labours, and by a rude perception of the important ends to be obtained, men would, in the course of years, not only have their curiosity, but their interest engaged in the pursuit; and, improved by the experience of past ages, we find the alchemists and their followers, in spite of the self-created obscurity which surrounded them, still frequently successful in withdrawing from the concealed stores of nature new metallic wonders, and giving to mankind at one time amusing, at another useful information.

'As the views of men became clearer, as their growing means continually improved by practice in their hands, new individuals were added to the metallic species, and each addition drew forth applause for the genius of the discoverer, and for his contribution to general chemical science. Stimulated by the due and awarded commendation given to prior merit, all exerted themselves; and the result at this day is, that in place of the seven metals known to the ancients, at least forty have been distinguished from each other and from other bodies, and have had their properties demonstrated.

'At present we begin to feel impatient, and to wish for a new state of chemical elements. For a time the desire was to add to the metals, now we wish to diminish their number. They increase upon us continually, and threaten to enclose within their ranks the bounds of our fair fields of chemical science. The rocks of the mountain and the soil of the plain, the sands of the sea and the salts that are in it, have given way to the powers we have been able to apply to them, but only to be replaced by metals. Even the gas of our atmosphere puts on at times a metallic appearance before us, and seems to indicate a similar base within. But a few combustibles and a few supporters of combustion remain to us of a different nature, and some (men of celebrity too) have found metals even among these.

'To decompose the metals, then, to reform them, to change them from one to another, and to realise the once absurd notion of transmutation, are the problems now given to the chemist for solution. Let none start at the difficult task, and think the means far beyond him; everything may be gained by energy and perseverance. Let us but look to the means which have given us these bodies, and to their gradual development, and we shall then gain confidence to hope for new and effective powers for their removal from the elementary ranks. Observe the first rudiments of metallurgical knowledge in the mere mechanical separation of native gold and silver from the encumbering substances; mark

VOL. I.   Q

the important step made in the reduction of copper, of iron, and of tin from their ores—rude, indeed, in the hands of the ancients, but refinement itself compared with their prior knowledge; consider the improvement when, by a variety of manipulations, the early chemist of the last century separated a small quantity of a metallic substance from five or six other bodies, where it existed in strong combination, and then pass to the perfection of these means as exhibited in the admirable researches of Tennant and Wollaston; lastly, glance but at the new, the extraordinary powers which the chemist of our own nation put in action so successfully for the reduction of the alkalies and earths, and you will then no longer doubt that powers still progressive and advanced may exist and put at some favourable moment the bases of the metals in our hands.'

At the conclusion of his course, at the end of the seventeenth lecture, on alkalies and earths, he says:—

'The substances that have now been described complete the series of metallic bodies which I enumerated to you among the present elements of chemical science. With them we conclude our consideration of simple bodies, and at this point too it is my intention to close the present series of lectures.

'During the seventeen lectures which have formed this course, I have constantly endeavoured rather to enlarge upon the powers and properties of the simple substances, and upon a few of their more proximate combinations, than to describe to you their ultimate compounds, and their applications and subserviency to the purposes of man. I have consequently drawn my

matter from more hidden and obscure sources, of the
character of which it has partaken, and have, to a certain degree, neglected to use that influence over your attention which would have been in my power if I had reverted to effects frequently in common life before you, and to bodies the uses of which are daily in your hands.

'It certainly is not necessary that I should give reasons for the adoption of a particular plan, though it may appear to be imperfect and badly designed; but courtesy claims some attention even in science; and, influenced by this, and in justification of myself (for whatever independence we may assume, no one is inclined to permit a censure, however unjust or however slight, to rest upon him which he can remove), I *shall* give reasons.

'On considering at the commencement of the course what ought to be the nature of these lectures, they appeared to me to admit with propriety of two distinct characters: they might be illustrative of the various processes and applications in the arts which are of a chemical nature; or they might be elementary in their nature, and explanatory of the secret *laws and forces* on which the science, with all its uses, is founded.

'These two modes in which chemistry may be said to exist, you will observe, form what may be called the extremes of our knowledge on the subject; the first, or applicative chemistry, is identical with the experimental knowledge and practice of the artisan and manufacturer, and is also that from which we gain our first perceptions and ideas of this kind; the second, or elementary chemistry, is the result of our researches *in the science*; and though, being the result of long and laborious trains of inductive reasoning and experimental investigation, it is the last production of the mind, yet it is also the basis upon which nature and art have raised practical science.

'To have made these lectures, therefore, illustrative of the arts and manufactures, would have been to repeat what the man of observation had already noticed, i.e. the results of general experience: and as it cannot but be supposed that a knot of men drawn together for mutual improvement must consist of men of observation, it would have been to repeat at least much of what must be known to the members of the Society. On the contrary, to treat of the first laws and principles of the science would be to explain the causes of the effects observed by those attentive bystanders on nature, and to point out to them by what powers and in what manner those effects were produced.

'It is possible, and likely, that to suppose men not observant would be to incur their distaste and enmity, but to suppose them ignorant of elementary chemistry (and by lecturing I have supposed, in form at least, ignorance) could be no cause of offence. The acute man may observe very accurately, but it does not follow that he should also reason perfectly or extensively. To tell a person that a stone falls to the ground would be to insult him, but it would not at all compromise his character for sagacity to be informed of the laws by which the stone descends, or the power which influences it. And though perhaps this illustration does not apply *very strongly* to the case of *applicative* and elementary chemical science, it is still satisfactory, by assuring me that I have not given offence by any imaginary depreciation of the knowledge of others. But,

further, I know of no illustration of the arts and manufactures of the civilised world, which would have been
worth my offering, or your acceptance, not founded on
first principles.   I have endeavoured to give those first
principles, in an account of the inherent powers of
matter, of the forms in which matter exists, and of
simple elementary substances.   It is probable that at a
future time I may make use of this groundwork in
illustrating the chemical arts to you; in the meantime
I shall rest confident that, should I find it convenient,
the sufferance and kindness you *have* extended to me
will still attend my efforts.

'Before I leave you, I may observe that, during the
last year or two of the Society's proceedings, I have
been intruded on your attention to an extraordinary
degree : this has been occasioned partly by my eager-
ness to use the opportunities this Society affords for
improvement ; partly by the necessity we are under,
from the constitution of our laws, to supply every required effort of whatever nature from among ourselves ;
but much more by *your* continued indulgence.

'At present I shall retire awhile from the public
duties of the Society, and that with the greater readiness, as I have but little doubt that the evenings devoted
to lectures will be filled in a manner more worthy of
your attention.   I retire gratified by the considerations
that every lecture has tended to draw closer the ties of
friendship and good feeling between the members of
the Society and myself ; that each one of them has
shown the advantages and uses of the Society ; and still
more, by the consciousness that I have endeavoured to
do, and the belief that I have done, my duty to the
Society, to myself, and to you.'

Professor Johnston then brought forward his Report in reference to Dimorphous bodies. The report, he conceived, would occupy too much time, if read at length ; he would, therefore, give a verbal account of its leading points, of which the following is a summary :—

When a body, chemically the same, is capable of assuming forms which belong to two different systems of crystallization, it is said to be dimorphous. Thus, as carbonate of lime occurs in forms reducible, some to the rhombohedron, and some to the right rhombic prism, it is a dimorphous body, and, *pari ratione*, if a substance were found to occur in shapes referable to three distinct crystallographic systems, it would be trimorphous. Isomorphous bodies, on the other hand, are such as, though chemically different, occur in the same crystalline form, of which the carbonates of lime and lead may be quoted as familiar examples. These two classes, therefore, the dimorphous and isomorphous, are obviously distinguished, in that the former can be separated from each other only by their mechanical or physical properties, whereas the latter differ in chemical constitution, and do not necessarily agree in any character save that of external form, and probably internal structure. Isomorphous substances, however, though chemically different, are so far analogous in their constitution as to admit, generally speaking, of having their composition represented by the same general formula. Thus $CO_2 + RO$ will express the composition of the carbonate of lime, or carbonate of iron, accordingly as R stands for one or other of these metals. To isomorphous and dimorphous bodies, Professor Johnston suggested, that a third class might be added—namely, the isodimorphous, or those whose corresponding dimorphous forms were the same. Thus, carbonate of lime and carbonate of lead are isodimorphous, as they both crystallize, either as rhombohedrons or right rhombic prisms. Isomeric bodies were next compared with those which are dimorphous, and the two classes found to differ, in that the dimorphous are necessarily isomeric, but not *vice versâ*. The proximate cause of dimorphism is altogether unknown ; but the circumstances which would appear most influential in causing the same body to assume two distinct forms, would appear to be variations of temperature. Thus, the crystals of bichromate of potash and of biniodide of mercury, obtained at a high and a low temperature, are different, and it is remarkable that, in the case of these two salts, the forms acquired at high heats are not permanent, but alter rapidly when subjected to the slightest mechanical injury, such as would be caused by friction or percussion.

These remarks were not, we presume, put forward as possessing any novelty, but intended as introductory to the exhibition of a table, in which the physical and mechanical properties, relations to light, electricity, &c. of the two forms of all known dimorphous bodies, were (as far as they have been explained) registered in separate columns. This table will be useful in presenting, in an intelligible and compendious form, the amount of our knowledge and of our ignorance in a most interesting department of physico-chemical research, and stimulating philosophers to fill the many hiatus with which it abounds, a task which must be performed before we can hope to arrive at a just conception of the physical cause or causes of dimorphism, or reduce its phænomena to general laws.

Mr. Whewell inquired whether any column existed in Mr. Johnston's table, showing the angles between the axes of dimorphous crystals. Mr. Whewell stated that he was opposed to the adoption of any rule of philosophizing, but that which is founded on a knowledge of the laws of phænomena, and took exception to certain expressions employed by Professor Johnston—such as, that " the cause of dimorphism is deeper than that of the ordinary laws of crystallization." He also objected to the speaking of atoms as if they had a real existence, and recommended the examination of the relations of crystals to light, as subsidiary to the investigation of the laws of dimorphism and isomorphism.—Sir William Hamilton concurred with Mr. Whewell in objecting to the use of the term " atom," usually made, and conceived, with Newton, that phænomena being referred to the affections of atoms, may be due solely to attractive and repulsive forces. The present atoms of chemistry may not be incapable of decomposition, and are no doubt very different from those which chemists will admit a thousand years hence. As for his part, he rejected the existence of atoms altogether, and would replace them by attractions and repulsions acting upon mathematical points.—Dr. Kane conceived that the abuse of the term " atom," alluded to by Mr. Whewell and Sir William Hamilton, could not, with truth, be charged upon any well-informed chemist. He then observed, as well as we could collect, that arragonite may be brought to the state of ordinary calcareous spar, and that this, as well as analogous cases, are no doubt due to a tendency of the atoms to assume the condition of greatest stability ; and finally stated, that he could not agree with physical philosophers in their propositions to supersede, by the application of polarized light, as an instrument of analysis, the researches of the chemist.—Dr. Faraday also objected to the employment of the term " atom" in chemistry, as he conceived that atoms were not only hypothetical, but that their existence was obviously disproved, even by the report of Professor Johnston. Dr. Faraday emphatically stated, that he was not an atomic chemist.—Dr. Apjohn, we believe, with the view of preventing any misconception, inquired whether the substances in Mr. Johnston's list of isomorphous bodies, were not merely plesiomorphous, to which Professor Miller replied that they undoubtedly were, and this was assented to by Mr. Johnston.

LETTER V.

—•—

My dear sir,

Until very recently it was supposed that the physical qualities of bodies, *i. e.* hardness, colour, density, transparency, &c., and still more their chemical properties, must depend upon the nature of their elements, or upon their composition. It was tacitly received as a principle, that two bodies containing the same elements in the same proportion, must of necessity possess the same properties. We could not imagine an exact identity of composition giving rise to two bodies entirely different in their sensible appearance and chemical relations. The most ingenious philosophers entertained the opinion that chemical combination is an inter-penetration of the particles of different kinds of matter, and that all matter is susceptible of infinite division. This has proved to be altogether a mistake. If matter were infinitely divisible in this sense, its particles must be imponderable, and a million of such molecules could not weigh more than an infinitely small one.

But the particles of that imponderable matter, which, striking upon the retina, gives us the sensation of light, are not in a mathematical sense infinitely small.

Inter-penetration of elements in the production of a chemical compound, supposes two distinct bodies, A and B, to occupy one and the same space at the same time. If this were so, different properties could not consist with an equal and identical composition.

That hypothesis, however, has shared the fate of innumerable imaginative explanations of natural phenomena, in which our predecessors indulged. They have now no advocate. The force of truth, dependent upon observation, is irresistible. A great many substances have been discovered amongst organic bodies, composed of the same elements in the same relative proportions, and yet exhibiting physical and chemical properties perfectly distinct one from another. To such substances the term *Isomeric* (from ισος *equal* and μερος *part*) is applied. A great class of bodies, known as the volatile oils, oil of turpentine, essence of lemons, oil of balsam, of copaiba, oil of rosemary, oil of juniper, and many others, differing widely from each other in their odour, in their medicinal

effects, in their boiling point, in their specific
gravity, &c., are exactly identical in composition,
—they contain the same elements, carbon and
hydrogen, in the same proportions.

How admirably simple does the chemistry of
organic nature present itself to us from this point
of view. An extraordinary variety of compound
bodies produced with equal weights of two elements! and how wide their dissimilarity! The
crystallised part of the oil of roses, the delicious
fragrancy of which is so well known, a solid at
ordinary temperatures, although readily volatile,
is a compound body containing exactly the same
elements, and in the same proportion, as the gas
we employ for lighting our streets; and, in short,
the same elements, in the same relative quantities,
are found in a dozen other compounds, all differing
essentially in their physical and chemical properties.

These remarkable truths, so highly important
in their applications, were not received and admitted as sufficiently established, without abundant
proofs. Many examples have long been known
where the analysis of two different bodies gave
the same composition; but such cases were regarded as doubtful: at any rate, they were isolated

observations, homeless in the realms of science, until, at length, examples were discovered of two or more bodies whose absolute identity of composition, with totally distinct properties, could be demonstrated in a more obvious and conclusive manner than by mere analysis; that is, they can be converted and reconverted into each other without addition and without subtraction.

In cyanuric acid, hydrated cyanic acid, and cyamelide, we have three such isomeric compounds.

Cyanuric acid is crystalline, soluble in water, and capable of forming salts with metallic oxides.

Hydrated cyanic acid is a volatile and highly blistering fluid, which cannot be brought into contact with water without being instantaneously decomposed.

Cyamelide is a white substance very like porcelain, absolutely insoluble in water.

Now if we place the first,—cyanuric acid,—in a vessel hermetically sealed, and apply a high degree of heat, it is converted by its influence into hydrated cyanic acid; and, then, if this is kept for some time at the common temperature, it passes into cyamelide, no other element being present. And, again inversely, cyamelide can be converted into cyanuric acid and hydrated cyanic acid.

E

We have three other bodies which pass through similar changes, in aldehyde, metaldehyde, and eltaldehyde; and, again two, in urea and cyanuret of ammonia. Further, 100 parts of aldehyde hydrated butyric acid and acetic ether contain the same elements in the same proportion. Thus one substance may be converted into another without addition or subtraction, and without the participation of any foreign bodies in the change.

The doctrine that matter is not infinitely divisible, but on the contrary, consists of atoms incapable of further division, alone furnishes us with a satisfactory explanation of these phenomena. In chemical combinations, the ultimate atoms of bodies do not penetrate each other, they are only arranged side by side in a certain order, and the properties of the compound depend entirely upon this order. If they are made to change their place—their mode of arrangement—by an impulse from without, they combine again in a different manner, and another compound is formed with totally different properties. We may suppose that one atom combines with one atom of another element to form a compound atom, while in other bodies two and two, four and four, eight and eight, are united; so that in all such compounds the amount per cent. of the elements is absolutely equal; and yet their physical and chemical properties must be totally different, the constitution of each atom being peculiar, in one body consisting of two, in another of four, in a third of eight, and in a fourth of sixteen simple atoms.

The discovery of these facts immediately led to many most beautiful and interesting results; they furnished us with a satisfactory explanation of observations which were before veiled in mystery, —a key to many of nature's most curious recesses.

Again, solid bodies, whether simple or compound, are capable of existing in two states, which are known by the terms *amorphous* and *crystalline*.

When matter is passing from a gaseous or liquid state slowly into a solid, an incessant motion is observed, as if the molecules were minute magnets; they are seen to repel each other in one direction, and to attract and cohere together in another, and in the end become arranged into a regular form, which under equal circumstances is always the same for any given kind of matter; that is, crystals are formed.

Time and freedom of motion for the particles of bodies are necessary to the formation of crystals.

E 2

If we force a fluid or a gas to become suddenly solid, leaving no time for its particles to arrange themselves, and cohere in that direction in which the cohesive attraction is strongest, no crystals will be formed, but the resulting solid will have a different colour, a different degree of hardness and cohesion, and will refract light differently; in one word, will be *amorphous*. Thus we have cinnabar as a red and a jet-black substance; sulphur a fixed and brittle body, and soft, semitransparent, and ductile; glass as a milk-white opaque substance, so hard that it strikes fire with steel, and in its ordinary and well-known state. These dissimilar states and properties of the same body are occasioned in one case by a regular, in the other by an irregular, arrangement of its atoms; one is crystalline, the other amorphous.

Applying these facts to natural productions, we have reason to believe that clay-slate and many kinds of greywacke are amorphous feldspar, as transition limestone is amorphous marble, basalt and lava mixtures of amorphous zeolite and augite. Anything that influences the cohesion, must also in a certain degree alter the properties of bodies. Carbonate of lime, if crystallised at ordinary temperatures, possesses the crystalline form, hardness, and refracting power of common spar ; if crystallised at a higher temperature, it has the form and properties of arragonite.

Finally, *Isomorphism* or the equality of form of many chemical compounds having a different composition, tends to prove that matter consists of atoms the mere arrangement of which produces all the properties of bodies. But when we find that a *different arrangement* of the *same* elements gives rise to *various* physical and chemical properties, and a *similar arrangement* of *different* elements produces properties very much the *same,* may we not inquire whether some of those bodies which we regard as elements may not be merely modifications of the same substance? whether they are not the same matter in a different state of arrangement? We know in fact the existence of iron in two states, so dissimilar, that in the one, it is to the electric chain like platinum, and in the other it is like zinc ; so that powerful galvanic machines have been constructed of this one metal.

Among the elements are several instances of remarkable similarity of properties. Thus there is a strong resemblance between platinum and iridium ; bromine and iodine ; iron, manganese and magnesium ; cobalt and nickel ; phosphorus and arsenic :

but this resemblance consists mainly in their forming
isomorphous compounds in which these elements
exist in the same relative proportion. These compounds are similar, because the atoms of which
they are composed are arranged in the same manner. The converse of this is also true : nitrate of
strontia becomes quite dissimilar to its common state
if a certain proportion of water is taken into its
composition.

If we suppose selenium to be merely modified
sulphur and phosphorus, modified arsenic, how does
it happen, we must inquire, that sulphuric acid
and selenic acid, phosphoric and arsenic acid, respectively form compounds which it is impossible
to distinguish by their form and solubility? Were
these merely isomeric they ought to exhibit properties quite dissimilar!

We have not, I believe, at present the remotest
ground to suppose that any one of those substances
which chemists regard as elements can be converted
into another. Such a conversion, indeed, would
presuppose that the element was composed of two
or more ingredients, and was in fact not an element;
and until the decomposition of these bodies is accomplished, and their constituents discovered, all
pretensions to such conversions deserve no notice.

Mr. Brown of Edinburgh thought he had converted iron into rhodium, and carbon or paracyanogen into silicon. His paper upon this subject was published in the Transactions of the Royal Society of Edinburgh, and contained internal evidence, without a repetition of his experiments, that he was totally unacquainted with the principles of chemical analysis. But his experiments have been carefully repeated by qualified persons, and they have completely proved his ignorance: *his* rhodium is iron, and *his* silicon an impure incombustible coal.

THE

EDINBURGH NEW

PHILOSOPHICAL JOURNAL.

On Isomeric Transmutation, and the Views recently published concerning the compound nature of Carbon, Silicon, and Nitrogen. By GEORGE WILSON, M.D., Lecturer on Chemistry, Edinburgh. Communicated by the Author.*

I propose, in the following Memoir, to offer some observations on the views recently published by Dr Samuel Brown, Mr Knox, and Mr Rigg, concerning the compound nature of silicon, nitrogen, and carbon. Before entering, however, at any length on the discussion of these, I would consider, very briefly, some points connected with the general question of the simplicity and unity of matter.

The great majority of chemists acknowledge the existence of some 55 simple or elementary substances. These are declared to be simple, not in virtue of any test of simplicity which the chemist has discovered and applied to them, but solely because they resist the decomposing or modifying action of all the forces which are, or at least are known to be, at man's disposal. The chemist, as it were, begins with the globe itself, and breaks it down into some thousand organic and inorganic compounds; these, in their turn, he resolves into some hundred less complex substances; and the latter, last of all, into the 55 bodies which are called simple. Here his analysis, in the meanwhile, has ended; all the forces which are at his command, for the modification of matter, having

---

* This memoir contains nearly verbatim the substance of a lecture delivered on 6th May 1844.

been spent in vain on these refractory substances. The single and combined agencies of heat, light, electricity, magnetism, mechanical pressure, and the like, have been directed in innumerable ways against them. But they have emerged from every trial, except those we are soon to consider, without betraying any sign of non-simplicity, or unfolding, if they are compound, the hidden secret of their true nature.

On the negative evidence of this insusceptibility of decomposition, the residual undecomposed bodies have been termed simple or elementary : they are the visible elements out of which all things are made. It cannot be denied, however, that in the minds of many, the term " *simple*" has passed for something more than the expression of " *hitherto undecomposed*," and has been accepted as fully equivalent to essentially " *indecomponible.*" But it would be doing injustice to the majority of chemists, to affirm that they have not employed the word " *elementary*" in its restricted and negative meaning, and have been willing to acknowledge the possible compositeness of all the so-called simple bodies. I refer to this the more particularly, that, in a curious volume recently published by Professor Low, exception has been taken to the maxim acknowledged among chemists, that a body should be considered simple till it can be shewn to be compound, and the opposite opinion advocated, " that a body is to be regarded as compound, when we are not able to prove it to be simple."[*] Mr Low is at great pains to shew, that the maxim he objects to " is unsound, and is arrived at, not ' by the just logic of Chemical Philosophy,' but by a chemical dogma which ought, long ere now, to have been banished from the science into which it has been introduced."[†] Every chemist, however, will smile at this correction ; for the proposition that all bodies, which cannot be resolved into something less complex than themselves, shall be accepted as simple, is quite accurate, and of much practical value, when taken in the sense in which he uses it. The simplicity of the so-called elementary bodies, is not affirmed to be *intrinsic, essential,* or *absolute,* but only to exist in relation to the decomposing or modifying forces which

---

[*] An Inquiry into the Nature of the Simple Bodies of Chemistry, by David Low, F.R.S.E., &c. p. 9.

[†] Ibid. pp. 11, 12.

chemistry supplies.   It remains competent to whomsoever chooses, to affirm, on the grounds of analogy, probability, direct experiment, or whatever else may seem to warrant it, that any or all of them are not simple substances: all that the chemist contends for is, that, tested by their power to resist the weapons and agents he can direct against them, they preserve their simplicity.

Professor Low would have the chemical elements included among compound substances, because it violates the law of continuity in nature, to suppose some 55 bodies simple, whilst all the rest are compound.   This may or may not be true ; but it could do no service to term the elements *compound*, unless we were prepared immediately to follow up the statement, by shewing of what they are compounded.   Such a proposition is consistent enough on Mr Low's part, since he offers a scheme of their composite constitution, founded on certain hypothetical views ; but it is not competent to the chemist.   All the knowledge he possesses of the composition of bodies, has been obtained by decomposing, or combining them, or by transforming them without decomposition, into each other.   According to the characters they have shewn, when thus treated, they have been named and classified in the order of their complexity, and so as to shew, within certain limits, the nature and number of their several components.   But the elementary bodies being insusceptible of resolution into substances more simple than themselves, cannot be affirmed to be compound in the sense the other bodies the chemist considers, are ; and it is not his office to discuss their complexity on other grounds than those afforded by their behaviour, when subjected to analytic, synthetic, or purely transformative forces.

Whilst, therefore, I fully sympathise with the speculative spirit that has led Mr Low to propose a scheme of elemental constitution, and differ from most of my fellow chemists, in believing that some such scheme will hereafter be realized in our laboratories, I dissent from him in thinking that the chemist has erred, in demanding that every undecomposed body shall be considered simple.   The term residual, or residuary, might indeed be better than simple, as indicating more clearly *undecompounded* as distinguished from *indecomponible ;* but, after all, that, or any other novel phrase is unnecessary, for the explicit sense in which the term *simple* is applied by the chemist, leaves the question of the *bona fide* simplicity of the elementary bodies quite open to discussion.   How fully it does so, may best be gathered from the fact, that Sir H. Davy and Berzelius, two of the warmest advocates of the maxim I have been discussing, were the freest in speculating on the nature of the elementary bodies, and the foremost in endeavouring to decompose them.

Further, I would observe, before leaving the subject, that it is not necessary, or, indeed, desirable, in the discussion of many chemical problems, that the possibility of the elementary bodies being simple should be considered.   The study of most of the properties of the suite of oxides of a metal, or of a series of organic compounds, would not be facilitated, especially to a beginner, by shaking his faith in the stability and unchangeability of their simplest components.   The elementary bodies stand in truth, in relation to all the more complex substances into whose composition they enter, like arithmetical ciphers, possessing in regard to all numbers higher than their own a fixed value, unalterable by any discovery which may be made concerning the lower figures which make up them.   Silicon may be a simple body, as many believe, or a modification of carbon, as Dr Brown supposes, or a compound of carbon, hydrogen, and oxygen, or of carbon and hydrogen, as Mr Low thinks probable.   But whichever of these it be, if any, is as indifferent to the chemist, while ascertaining the proportion in which it combines with oxygen to form silica, and a multitude of its other relations, as it was to the builder of an Egyptian pyramid, whether the bricks he made use of, so long as they possessed the proper shape and weight, and coherence, consisted of clay alone, or of clay mixed with sand, or of clay and sand mingled with straw.

We are free then to speculate to the uttermost on the nature of the elementary bodies ; and if we consider from what direction we are likely to obtain the means of lessening their number, we shall find that the hopes of chemists (*i. e.* of those who hope at all on the matter) are fixed at present on three different quarters, from any, or from all of which, the power to effect the desired reduction may come. One method open to the chemist, is analysis; another synthesis. The experiments I am about to notice illustrate the application of both; for the same researches which seem to Dr Brown to establish synthetically the compound nature of silicon, appear to Mr Knox to demonstrate analytically the compound nature of nitrogen. The third method is not so easily defined; it may be termed that of reduction, by mutual isomeric transmutation.

The application of analysis to the reduction of the elementary bodies is easily understood. Without any addition to the resources, in the way of agent and instrument, we at present possess, it may suffice to produce more remarkable decompositions than we have yet seen it effect. If Mr Rigg's and Mr Knox's experiments are confirmed, it certainly will. Moreover, we may anticipate the discovery of novel agents, or of new powers in those we are familiar with, as we have recently become aware of the presence of marvellous modifying forces in the sunbeam, and in light and heat from other sources, of the existence of which we had scarcely a suspicion ten years ago. We may farther expect greatly to improve our instruments, and thereby to increase enormously our power over matter. Not to speak of what we should effect, could we realize certain improvements which theory indicates as possible in our voltaic batteries, the simple discovery of a substance which would resist the action of very high temperatures as effectually as platina and fireclay do our ordinary furnace heats, would put in the chemist's hands a weapon for conquest of the highest value. Many of the bodies which appear at present, to use the quaint words of old Sir Thomas Browne, " to lie immortal in the arms of fire," might then be found susceptible of resolution into simpler forms of matter. The possibility of all this is so apparent, that it is needless to enlarge on it at greater length. Before passing from it, however, I would observe, that the attempts of chemists to decompose the so-called simple bodies, appear to me to have been hitherto too much directed against the naked elements themselves, and not upon them in a state of combination; and, further, to have too much implied an expectation, that their decomposition was to be brought about, by some successful violent effort to tear or force asunder their constituents. Hence, the uselessly large battery which Davy employed when he effected the decomposition of the alkalis. But the more we learn of molecular forces, the more we seem to become aware of the truth, that the simple reversal or neutralization of the affinities which bind the components of a body together, is all that is necessary to effect its decomposition; and that this may be as fully secured by the invisible action of a sunbeam, or the inappreciable influence of an electric current, as by the most gigantic galvanic battery, or a furnace heated seven times more than is wont.

Meanwhile, it remains to be acknowledged, that analysis hitherto has done nothing to lessen the number of elementary bodies; on the other hand, it has continually been adding to them. The ancients acknowledged but four,—air, earth, fire, and water; a later school had their three,—salt, sulphur, and mercury; and no class of chemists, down to the destruction of the Phlogiston School, acknowledged, so far as I am aware, as many as a dozen. Since the era of Lavoisier, we have been steadily increasing the list, till now we count 55. Sir H. Davy only altered the names of the elements with metallic bases, without abridging the roll by one; and since his death, several new bodies have been ranked among simple substances. The further result of analysis, whether with its present or with additional powers, may be of the same kind. The fifty-five elementary bodies may each be resolved into two, or three, or four, unlike, and for the time, indecomponible substances; so that the list of elements shall be doubled, tripled, or quadrupled. But though this may be the first effect, analogy and probability conspire unequivocally to assure us, that it will not be the ultimate result of a victorious analysis of matter. As we find the prevailing elements of the countless organic bodies we examine, to be the constantly recurring four, carbon, hydrogen, oxygen, and nitrogen; and as Davy found a common element, oxygen, in all the earths, so we may expect, if the so-called elements are really compound, to find the same bodies present in many. We can suppose all the metals proving to be compounds in different proportions of but two : fluorine, chlorine, bromine, and iodine, in the same way reduced to two ; carbon, boron, silicon, and the other groups of simple bodies, in like manner diminished to two. In this way, or in some other, we may resolve all the elementary bodies, as Mr Low thinks we shall, into the two lowest on the atomic scale, carbon and hydrogen ; or, descending further, identify them every one, as Mr Rigg thinks we should, with hydrogen ; or, in the lowest deep, finding a lower still, pass beyond even hydrogen to the long dreamed of ῞Υλη πρωτη, the *materia prima*, or material substratum and essence of all things.

The application of synthesis to the reduction of the list of elementary bodies, is not so obvious as that of analysis, but may, on the whole, be made manifest enough. We can conceive the possibility of its being discovered, that two of the lower metals, such as lead and copper, when fused together formed gold ; and that, nevertheless, the compound should be of such a nature, as to resist the decomposing influence of every agent. In such a case, it would be possible to prove gold not to be a simple substance, by shewing our ability to compound it out of lead and copper, though we might for ever remain unable to establish the same point analytically, by resolving it into these metals. Should such a synthetic demonstration of the compound nature of one of the elements ever be obtained, it would prepare us for attacking the problem of their true nature, by endeavouring to compound them out of each other. There is nothing, however, in the present state of Chemistry to warrant the expectation of such a discovery being made ; and it is not in this shape, but as one of the forms of the method of reduction by isomeric transmutation, that synthesis has been applied to the diminution of the list of elementary bodies.

I turn now, therefore, to the consideration of Isomerism. For a long period after the publication of the Atomic Theory, it was universally believed that the same elements could combine in the same proportion to form only one compound, and that difference in physical properties, such as hardness, solu-

bility, specific gravity, &c., was always occasioned by difference at least in the proportion of ingredients, and in most cases by difference also in their nature ; and this is still acknowledged as true in regard to the majority of substances ; water, *e. g.* is the only body containing oxygen and hydrogen, in the proportion of eight parts by weight of the former to one of the latter ; common salt is the only substance, with thirty-five parts of chlorine to twenty-three of sodium, and so with other compounds. But within the last few years many bodies have been discovered containing the same elements, in the same proportion, and yet differing in every physical and chemical property. A striking example of this may be found in a group of organic substances particularly referred to by Liebig in his Familiar Letters :—
" A great class of bodies," says he, " known as the volatile oils, oil of turpentine, essence of lemons, oil of balsam of copaiba, oil of rosemary, of juniper, and many others, differing widely from each other in their odour, their medicinal effects, their boiling points, their specific gravity, &c. are exactly identical in composition ; they contain carbon and hydrogen in the same proportions ;"* viz. five atoms of carbon to four of hydrogen. Bodies which possess this peculiarity are termed, in relation to each other, *Isomeric* (from ισος, *equal,* and μερος, *part*), which may be Englished equiproportional, and marks their possession of an equal proportion of the same elements. The unexpected discovery of this curious law, while it has shewn chemists that the greatest dissimilarity in the properties possessed by bodies may accompany the most perfect coincidence in proportional composition, has, at the same time, directly led to the conclusion, that the elementary bodies may form a group, or a series of groups, related to each other isomerically, or equiproportionally, as the volatile oils referred to, are. Who first detected the applicability of the law of Isomerism to the possible solution of the problem of the true nature of the chemical elements, I do not know ; nor do I profess to offer any historical sketch of the progress of speculation on this subject. I need only mention, that three of our living chemists have published views on the possible Isomerism

---

* Familiar Letters on Chemistry, by Justus Liebig, pp. 47, 48.

of the elementary bodies,—Professor Johnston* in 1837 ; Dr Samuel Brown,† and Professor Kane in 1841.‡

Dr Brown's theory, which I consider first, as it is a scheme of transmutation by synthesis, is founded upon the existence of a class of isomeric bodies, in which, while the equiproportionality of identical elements occurs, the number of atoms combining to produce this, is different in each member of the group. Thus, there exists a series of compounds of carbon and hydrogen, containing these bodies in the proportion of atom to atom. This is satisfied in the lowest, which is termed *methylene*, by 2 atoms of carbon to 2 of hydrogen ; in the next *olefiant gas*, by 4 to 4 ; in the third *oil gas*, by 8 to 8, and in a fourth *cetene*, by 32 to 32. The volatile oils referred to previously, form so far at least, a similar series ; in them the elements are also carbon and hydrogen, in the proportion of five atoms of the former to four of the latter. In oil of citron this is doubled, or we have $C_{10} H_8$; in oil of cubebs tripled, or $C_{15} H_{12}$; in turpentine quadrupled, or $C_{20} H_{16}$.

Groups of isomeric bodies of this kind are supposed by Dr Brown to be formed by successive duplications or doublings. The lowest member of the series, by combining with itself, produces a first multiple ; this, by uniting with itself a second ; that, by combining with itself a third ; and so on *ad infinitum*. It is not essential, however, to the truth of this view, that a full series of duplicate multiples should be shewn to exist, provided no body is found to manifest an unequivocal departure from the rule : the gaps which occur in the known series may be filled up by future discoveries. Dr Brown thinks he has established the truth of his view by experiment, in regard to the isomeric compounds of carbon and nitrogen, cyanogen and paracyanogen ; the latter of which he believes to be produced by the former combining with itself. In like manner he represents the 55 so-called simple substances as a group, or a series of groups, of isomeric bodies, produced by the element of lowest atomic weight (which may

---

* Report of the Seventh Meeting of the British Association, pp. 163–215.

† Transactions of the Royal Society of Edinburgh, vol. xv. pp. 165–176, and 229–246.

‡ Elements of Chemistry, p. 377.

be the lowest at present known to us *hydrogen*, or a lower and more truly elemental body), forming successive combinations in the way already mentioned, so as to reach from, or through hydrogen, which we call 1, up to gold which is 199 times higher. To prevent any mistake, I quote Dr Brown's own words :*—" This view of isomerism, and the relation of cyanogen to paracyanogen, is farther recommended by the consideration, that it affords a practical foundation for a likely hypothesis of the constitution of the so-called chemical elements, and points out the way in which such a hypothesis may be either established or overthrown by experimental observation. Let it be supposed that several of the elemental groups are so many series of isomeric forms, and it is at once to be inferred, that heat, electrolysis, and reagents, shall all be incapable of decomposing them, as has been found in the actual practice of the laboratories of modern Europe, by innumerable trials. If, to take one instance, sulphur (16 or 2) be an isomeric form of oxygen (8 or 1), which it as much resembles in chemical properties, as it is conformably contrasted with it in mechanical condition, it must be impossible to extract oxygen from it by any analytical force which has yet been discovered ; and the only method in which it shall be possible to prove that such is the mutual relation of these two elements, shall be to have recourse to synthesis, and convert oxygen into sulphur. It *is* within the scope of this hypothesis that the various elements may be all isomeric forms of one truly elementary substance."

Dr Brown's scheme of elemental reduction may be termed one by isomeric synthetic transmutation. You will observe, that he supposes transmutation to take place only by synthesis, and in one direction ; so that an element possessing a certain atomic weight, may form, by uniting with itself, another possessing a higher combining proportion ; but the reverse cannot occur. Oxygen, which is eight, may double itself into sulphur, which is sixteen ; but sulphur cannot halve itself into oxygen ; carbon may quadruple itself into silicon, but silicon cannot quarter itself into carbon. All the other

---

* Trans. Royal Society, vol. xv. p. 176.

elements may be transmuted into gold, which has the highest atomic weight; for, in this respect, Dr Brown's views are strictly in accordance with those of the alchymists; but gold can be changed into none of them, and, if it suffer transmutation, must pass into some unknown new body possessing a higher combining proportion. I shall return immediately to the consideration of those experiments by which Dr Brown thought he had proved the truth of his view, so far as carbon and silicon were concerned. Meanwhile, I proceed, very briefly, to explain in what respect Professors Johnston's and Kane's schemes of elemental isomerism differ from that we have been discussing.

Mr Johnston's views are founded on the existence of a class of isomeric bodies not taken into consideration in Dr Brown's speculations. The members of certain isomeric groups possess not only the same proportion of elements, but likewise the same atomic weight. They are not multiples or submultiples of each other, like those already considered, but owe their difference in properties to the relative grouping of their molecules otherwise than by multiplication, or simple superaddition of the atoms on each other. We have a group of three such bodies in cyanuric acid, hydrated cyanic acid, and cyamelide, compounds of carbon, oxygen, and nitrogen. We have another in aldehyde, metaldehyde, and eltaldehyde; and a well-known pair in urea and cyanate of ammonia.* These isomerics possess the character of mutual convertibility: thus, in a group of three, which we may term A, B, C; A is convertible into B and C; B into A and C; C into A and B; and this without addition or subtraction of any of their constituent elements. Guided by these facts, Professor Johnston observes, that " the speculations of chemists in regard to the probable diminution of the number of received elementary bodies, have hitherto run only in the channel of decomposition. * * * * The idea of a possible *transformation* has hitherto hardly been thought of; and yet the doctrine of isomerism, rich already in its numerous discoveries, has shewn that any number of the received elementary bodies *may* be made up of the same ele-

---

* Liebig's Familiar Letters, pp. 49, 50.

ments united in the same proportion."*   After some incidental remarks, he continues, " It may be, however, that the patient study and pursuit of the kindred classes of phenomena we have been considering, shall, in some brighter moment, shew that substances considered elementary are yet *mutually convertible* without decomposition ;" and, again, " It may be, indeed, that all our *supposed* elementary bodies are in *reality* such, and therefore wholly beyond the resolving energy of electricity, or any other agent ; and yet the study of their changes and reactions in the laboratory, in conformity, perhaps, with new views or modes of investigation, may, at some future period, so enlarge our dominion over the molecules, as shall cause them, at our bidding, to assume this or that arrangement—to appear with the properties of chlorine or iodine, of cobalt or nickel, of rhodium, iridium, or osmium."†   Professor Johnston's view, it will be observed, is a wider one than Dr Brown's, inasmuch as it acknowledges a possible mutual convertibility of the elementary bodies ; and, therefore, implies that transmutation may proceed in both directions of the atomic scale. Sulphur may become oxygen, as readily as oxygen sulphur ; silicon carbon, as readily as carbon silicon ; gold hydrogen, as hydrogen gold.   Any one element, in short, may become any other, whatever be their atomic weights.   This scheme might be termed, in opposition to Dr Brown's, a method of elemental reduction by mutual isomeric transmutation.

Professor Kane's views are too slightly sketched, in his work on Chemistry, to enable us to judge exactly in what way he expects the elements to prove isomeric, and they were certainly formed with a knowledge of what Professor Johnston had written on the subject.   But he has indicated, in a way neither of the other chemists referred to have done, some remarkable relations between the atomic weights of certain of the metals, which would strikingly accord with either of their theories of elemental isomerism.

I do not offer any opinion as to the relative probability of Dr Brown's and Professor Johnston's views ; but it is impos-

---

* Report of British Association, vol. vi. p. 211.   † *Ibid.* p. 212.

sible to help wishing that the latter chemist's scheme of Elemental Isomerism should prove the truer of the two. For Dr Brown supplies us with but a one-edged weapon for conquering nature, while Professor Johnston puts in our hands a two-edged sword, smiting both ways, and increasing twofold our power over matter.

Meanwhile, Dr Brown is the only chemist who has had faith and courage enough to test the reality of Elemental Isomerism, by endeavouring to transmute one of the elements into another. This, he believes, he has succeeded in doing in the case of carbon and silicon. His experiments have been made upon certain compounds of the former body with nitrogen, which he subjected to various modifying processes; one general principle, however, runs through them all, which may be explained in a few words. By a special process, instituted for the purpose, or as a product of a general process for transmutation, he obtained paracyanogen, a body consisting of carbon and nitrogen, in the proportion of twelve parts of the former to fourteen (by weight) of the latter; or of two atoms of carbon to one of nitrogen. The atomic weight and exact constitution of this body are unknown, but Dr Brown, as we have already seen, supposes it to be a duplication of cyanogen, and, therefore, to contain four atoms of carbon to two of nitrogen. When this body is treated in various ways, of which the simplest, and the only one we need consider, is that of heating it out of contact with air, alone, or in contact with substances (such as platina or carbonate of potass) having a strong attraction for silicon, its two atoms of nitrogen, according to Dr Brown, pass away unchanged, and its four atoms of carbon combine together, and form silicon. To some, perhaps, the view intended, and its relation to the isomerism of confessedly compound bodies, will be clearer, if they suppose for the moment that carbon is a compound of two elements, which are united in it in certain proportions, and in the same ratio, but in a multiple four times higher in silicon.

The greater number of chemists refused to acknowledge that silicon was, or could be, produced from paracyanogen; and, joining issue with Dr Brown on this point, offered no opinion on his theory of the origin of the silicon which appeared in his experiments. There was one chemist, however, Mr G. J. Knox, who not only accepted Dr Brown's statements as true, so far, at least, as the appearance of silicon was concerned, but advocated the probability of such an occurrence; on grounds, however, quite opposed to those Dr Brown himself built upon. Mr Knox's views are unfortunately not known to us fully, although it is more than a year since they were laid before the Royal Irish Academy. Owing to a peculiarity in the mode of publishing its transactions followed by that Society, the paper has not yet been printed; and the only shape in which its contents have reached us is that of an imperfect and insufficient abstract in one of our own journals.* Mr Knox conceives that the nitrogen of the paracyanogen, and not its carbon, is the source of the silicon which appeared in Dr Brown's experiments. His own words are the following: after referring to certain experiments of Sir H. Davy, which seem to him to warrant the belief that nitrogen is a compound body, he says, " The latest experiments which bear upon this subject, and from which I received the idea which led me to this investigation, are those of Dr Brown, ' upon the Conversion of Carbon into Silicon,'—an explanation of phenomena which appears to me most unreasonable, and contrary to all chemical analogy: while the supposition of the carbon having reduced the nitrogen, is not only a simple, but an unavoidable conclusion to arrive at, if nitrogen be a compound substance. To determine, by experiment, the correctness or incorrectness of this idea, it were only necessary to reduce nitrogen by some other substance than charcoal; and should silicon result from its decomposition, the problem might be considered to be solved." †

Mr Knox then describes several experiments made with a compound of hydrogen, nitrogen, and potassium, heated in different ways with iron, in two of which silicon appeared, although no carbon was present. The compound Mr Knox employed, he terms the " ammonia-nitruret of potassium," by which I understand him to signify the amidide of potassium (KNH^2) of other authors. He rejects one of the two experiments where silica appeared, as inconclusive as to its anoma-

* Chemical Gazette, September 1843.   † Ibid., p. 574.

lous production; and draws from the whole the following conclusion: " From these experiments, together with those of Sir H. Davy mentioned above, one might infer that nitrogen is either a compound of silicon and hydrogen, or of silicon, hydrogen, and oxygen; to determine which synthetically, a current of dry muriatic acid was passed over siliciuret of potassium," and the resulting gases examined. These were found to contain a variable but marked proportion of nitrogen; so that, so far as can be judged from the imperfect account we possess, Mr Knox seems to consider nitrogen a compound of silicon and hydrogen, and to believe that he formed it by the action of muriatic acid on siliciuret of potassium. He does not suppose, however, as some have imagined, that the nitrogen is *transmuted* into silicon; he believes that the former *yields*, but not that it *forms* the latter, in the way Dr Brown supposes that carbon forms silicon. Silicon, according to Mr Knox, *pre-exists* in nitrogen, along with hydrogen, or with hydrogen and oxygen, by combination with which it makes up the nitrogen. He supposes, accordingly, that, in Dr Brown's experiments, the nitrogen was the source of the silicon, and that the carbon was useful only by combining with and removing the non-siliceous element or elements of the nitrogen, and setting the silicon free; and he endeavours to establish this by shewing, that if the other conditions of Dr Brown's experiments were retained, but the carbon replaced by a metal such as potassium (or rather by potassium and iron), the production of silicon went on as well as if carbon had been there. His view, therefore, has the advantage of explaining Dr Brown's results as well as his own; whereas that gentleman's theory affords no explanation of Mr Knox's experiments.* The latter, moreover,

---

* In so far as Dr Brown refers the silicon which appeared in his experiments to carbon, his explanation will, of course, not apply to researches like those of Mr Knox, where silicon was found, though no carbon was present. He may fall back, however, on his general hypothesis, that the higher elements are isomeric forms of the lowest, and affirm that the hydrogen of the ammonia-nitruret of potassium was transmuted, mediately through carbon, or immediately into silicon. When the text was written, I was not aware that Dr Brown had explained

professes not only to have decomposed nitrogen into silicon and hydrogen, but to have combined silicon and hydrogen into nitrogen; so that he offers both synthetic and analytic proof of the truth of his views. It is impossible, however, to judge of the value of Mr Knox's experiments, till we see them reported in full; and there is a hesitation in his view of the constitution of nitrogen, as to whether it contains oxygen or not, which might, and should have been removed by prolonged experiments, before he published on the question at all. Moreover, he determines nothing as to the quantitative constitution of nitrogen, which should surely have been the chief object of investigation, as soon as he saw reason to believe that nitrogen was not a simple body.

As to the relative probability of the rival theories of the origin of the silicon, which appeared when paracyanogen was subjected to Dr Brown's processes, it is impossible at present to give a decision. I have repeated none of Mr Knox's experiments, and it would be presumptuous in me to criticise his results; but I devoted the greater part of last winter, along with my friend Mr John Crombie Brown, to the repetition of Dr S. Brown's processes for the transmutation of carbon into silicon, and I am free to offer an opinion on their value. Those who wish to know in detail the results my colleague and myself arrived at, will find them in the fifteenth volume of the Transactions of the Royal Society of Edinburgh.* Our general conclusions may be stated in a word.

We were able to confirm Dr Brown's phenomenal results thus far, that we obtained silicon in several of our experiments, in circumstances which seemed, to myself at least, to preclude the possibility of its being derived as an impurity or accidental ingredient, from the vessels or materials, or reagents made use of. The quantity was always much less, than by Dr Brown's hypothesis it should have been, and much less than he himself procured; in many experiments, moreover, no silicon was obtained at all. So far, however, as this scanty and precarious

---

Mr Knox's results in this way. His hypothesis affords no explanation of the latter gentleman's synthetic experiments on the formation of nitrogen from silicon and hydrogen.

* Pp. 547–559.

appearance or production of silicon is concerned, we can authenticate Dr Brown's results, but no further. Some misapprehension, I believe, exists on this subject, and I am anxious it should continue no longer. I took the most public opportunity that was open to me last autumn,* of declaring my confident expectation, that a repetition of Dr Brown's processes would establish the truth of his theory ; and I owe it to myself, still more to those I induced by my representations to advocate his cause, but above all to the interests of science, which must be hindered in its progress by the confusion of doubtful with certain knowledge, to take as public an opportunity of saying, that Dr Brown's processes have not, in my hands, yielded proof of the transmutability of carbon into silicon. I have further come to the conclusion, that they are too imperfect to establish the truth of that proposition in the hands of any one ; and that there exists at present no evidence, in the way of demonstration by experiment, to satisfy a chemist, that carbon or any other element has ever suffered transmutation.

A peculiar difficulty attends the reception of the proposition, that carbon is transmutable into silicon ; a difficulty which to many chemists seems insurmountable, and which has not been provided for by Dr Brown in any of his papers, although he was aware of its existence. It results from the irreconcilability of the atomic weights of carbon and silicon, the former of which is 6, the latter 22·22. According to Dr Brown, an atom of silicon consists of 4 atoms of carbon ; but four times six is 24, not 22·22. If, therefore, transmutation by isomeric synthesis of carbon into silicon occur, it must, according to this view, be accompanied by a destruction of matter equal to the difference between 24 and 22·22 ; or, for every 24 parts by weight of carbon subjected to transmutation, only 22·22 of silicon would be obtained. I did not allow this difficulty to stand in the way of my repetition of the silicon experiments, as I saw a way of overcoming it. I shall mention what this was, without entering into any details in the way of vindication

---

* In a letter to the Lord Provost of Edinburgh on Dr Brown's claims to the Chair of Chemistry, which was printed and widely circulated, but not published.

of its truth or probability.   Let the received atomic weight of silicon, 22·22, be diminished by removal of the decimals, and made the round number 22.   Such an alteration will, not improbably, be made by chemists, apart from all consideration of the question of transmutation.   Then divide the received atom of carbon, 6, by 3, a liberty which would be conceded by many of my brethren, and it becomes 2 ; of which silicon is a multiple by the whole number 11.   11 atoms of carbon might, by synthetic transmutation, become 1 atom = 22 silicon, without any difficulty in the way of atomic weights.*

From all that I have said, it will be manifest that no light task awaits those who propose to labour in the cause of transmutation.   In the particular case of silicon, the question between Mr Knox and Dr Brown is one which can be settled only in the laboratory.   It is possible that both of these gentlemen are right in their views.   Nitrogen may be a compound of silicon and hydrogen, and silicon nevertheless, a compound

---

* I need scarcely say, that such a speculation possesses at present not the slightest value, and was pursued only at a time when I believed that there was full demonstration by experiment of the transmutability of carbon into silicon.

The recent researches of Dumas, Erdmann, and other continental chemists, have shewn, that the atomic weights of several of the elementary bodies (carbon, nitrogen, calcium, barium, strontium) are multiples, in whole numbers of that of hydrogen ; and many, both in this country and abroad, encourage the expectation, that the equivalents of all the elements will prove, according to Dr Prout's hypothesis, multiples of hydrogen in the same way.   I was willing to hope, that the atom of silicon at least, which, owing to the difficulty of procuring that substance, has been fixed on the evidence of comparatively few experiments, might prove to be a multiple of that of hydrogen by 22.   This is a point to be decided solely by experiment.

As for the division of the equivalent of carbon by 3, it is acknowledged on all hands, that the received atomic weights may be multiples, or submultiples of the true ones.   Thus, it is matter of dispute among chemists what are the true equivalents of copper, mercury, arsenic, phosphorus, antimony, and several others ; and any alteration is hypothetically justifiable which does not contradict the law of multiple combination, and for which a sufficient necessity can be shewn.   The justifying necessity in this case would have been the transmutation of carbon into silicon, and the acknowledgment of the atomic weight of the latter as 22.

or form of carbon. To do the subject justice, would require a careful repetition of all Dr Brown's and all Mr Knox's experiments, besides a lengthened series of independent researches, which would occupy at least six months of unremitting labour. The fact of an anomalous production of silicon is not beyond dispute ; and till it is, the practical chemist cannot be expected even to consider the question of transmutation. Should the anomalous production of silicon, however, be fully confirmed, I think there are few who will not agree with me in wishing, that, whatever be the fate of Mr Knox's explanation, Dr Brown's theory should prove true. It seems absurd to wish that a law of nature should prove one thing rather than another; as if the law, when discovered, could be other than of God's making, and the best that can be. But what I mean is this :—Mr Knox's view, whilst it cuts off nitrogen from the list of simple bodies, reveals no general principle applicable to the reduction of the number of remaining elements. But if, with Dr Brown, we could effect the transmutation of one of these, sooner or later we should assuredly succeed in effecting the transmutation of all. If we can find a key, that will unlock in this way the intricacies of one group of elementary bodies, we may fully believe that the same instrument, or one fashioned like it, will open for us the mysteries of the rest.

In conclusion, it will be gathered from the brief and imperfect sketch I have offered, that the doctrine of Elemental Isomerism, and the transmutability of the elements, exists at present only as an unrealized idea, little, if at all, further advanced than it was in 1837, when first explicitly announced by Professor Johnston.

We are flung back, therefore, on the general analogies and probabilities that warrant the entertainment of such a doctrine ; but these, I think, are neither slight in force nor scanty in number. All chemistry seems to me to point steadily and increasingly to the necessity of assuming, and, if possible, realizing such a law ; and many of my brethren, I am certain, would agree with me in this. The scepticism so generally expressed as to the truth of Dr Brown's views, was directed rather against the processes and experiments by which he professed to establish his doctrine, than against the doctrine itself; and so far as this implied a resolution to accept nothing but the most rigidly quantitative experiments, in proof of so revolutionary a proposition as that of transmutation, it was quite justifiable. The instrument *par excellence* of chemistry is the Balance; and every chemist must expect to have his discoveries literally and metaphorically weighed in it, and rejected if found wanting. The Familiar Letters of Liebig, *e.g.*, show that, although he unhesitatingly and too summarily condemns Dr Brown's experiments, he willingly speculates on the light which Isomerism may throw on the true constitution of the elements.

And if chemistry is in favour of the doctrine we are considering, the other physical sciences justify it also. The geologist acknowledges the existence of many phenomena, in the relative distribution of the materials forming the earth's crust, which seem inexplicable by our present chemistry. The naturalist affirms that the whole subject of fossil zoology is plunged in mystery; and anxiously demands if the appearance of substances in fossils, which no one can trace to ordinary sources, does not depend upon a transmutation of some of the pre-existing ingredients of these bodies.* The agriculturist is frequently perplexed, in his endeavours to trace the constituents of the plants he cultivates to the soil they have grown upon. The difficulty is generally got over by the accusation of imperfect analysis; but some have courage enough to refuse this *argumentum ad ignaviam aut ignorantiam,* and one, Mr Rigg, who has been studying the subject for years, declares that his observations have led him to the conclusion, " that of the elements, carbon, hydrogen, oxygen, nitrogen, sodium, potassium, calcium, &c., which constitute the organic and inorganic parts of plants, hydrogen is the only ultimate element, the rest being all com-

---

* I refer particularly to a discussion which took place last winter in the Zoological Society of London, as to the source of the fluoride of calcium which appears in fossil bones. Literary Gazette, 2d December 1843, p. 773. The subject was afterwards referred to by Mr E. Solly, in a lecture at the Royal Institution.

pound bodies ; and to question the compound nature of hydrogen."*

Encouraged by these things, I, for one, will, in faith and patience, abide the issue, ready and willing, should I again see as much encouragement as I did last autumn, to spend another winter, or many winters, in endeavouring to bring about a consummation so devoutly to be wished, as the manifestation of the essential simplicity and unity of matter.

---

* Experimental Researches, &c., shewing Carbon to be a compound body made by Plants.   By Robert Rigg, F.R.S.   p. 264.

ARTICLE XXII.

Opinions relating to the Composition of Organic Substances.
By J. BERZELIUS*.

[From Poggendorff's *Annalen*, vol. lxviii. p. 161.]

NO portion of the science of chemistry has given rise to so much difference of opinion as the composition of organic substances. The greater number of those who have occupied themselves with the subject, have often been led by their experiments to quite unexpected modes of combination, have taken up at the moment some peculiar ideas concerning them, and, without noticing the relationship which these may or may not bear to the whole system of the science, have pursued constantly the same course in search of further proof of their own views. In this manner almost innumerable different opinions have been broached. To these I do not intend to add another; but my object in this essay is simply to call attention to the want of unity thus brought into the science, and to the injury which it sustains from these purely fanciful views, to check and restrain the unscrupulous flight of the imagination in theoretical speculations, and to abide steadfastly, whenever it is possible, by that which can in some measure be proved by practical observation. I am prepared in doing this to dispense with the approbation of the many.

Empirical and rational Composition.—The composition of every organic substance can be determined by analysis. The atomic weight of the greater number can also be ascertained with some degree of certainty, by aid of which the number of simple atoms, as likewise the relative number of atoms of each element, can be satisfactorily established. This constitutes the empirical composition of the substance. For instance, the composition of oxamic acid is represented by the empirical formula $C^4 H^4 N^2 O^5$.

Now arises the question, how are these elementary substances coupled together? Can and must this acid be considered as composed on the one hand of 5 atoms of oxygen, and on the other of a compound radical $C^4 H^4 N^2$? The mode of union of the elements in forming this radical is simple and not improbable; but we know that such is not the composition of the acid, but

* Translated by E. Ronalds, Ph.D.

3 A 2

that it is composed of oxalic acid combined with oxamide as a conjunct.   This is its rational composition, and

$$C^2 O^3 + NH^2, C^2 O^2$$

is its rational formula.

The rational composition of bodies is the highest problem of organic chemistry; but it has only been solved satisfactorily, up to the present time, in the case of a very limited number of organic bodies, notwithstanding it has been the object of all the exertions of those who have occupied themselves with organic chemistry.   In endeavouring to ascertain the rational composition of bodies, the greatest caution in drawing conclusions must be combined with the soundest judgement, and it is impossible to make use of too manifold means of proof.   Science suffers, and will long suffer, from rational formulæ, which, invented by a lively imagination, have no other proof of their correctness than that they do not actually contradict the result of analysis. For when this latter corresponds to a number of rational formulæ, nothing can be proved from it.   What I am about to bring forward in the following pages will clearly show how science is led astray and put into confusion by hasty and unfounded conclusions.

Although the composition of organic bodies, at the first glance, appears to be altogether different from that of the inorganic, yet what we know of the latter is the only unerring guide which we possess to enable us to form an opinion upon the former.   In exploring the unknown, our only safe plan is to support ourselves upon the known.   This must also be the right way here, and what we already know concerning the laws of combination which regulate inorganic nature, must be taken as a guide in judging the modes of combination in organic nature. Every other mode of procedure allows full scope to fancy, which, varying only with the inventive faculties of the individual, is always ready to build new castles in the air.   It is thus that numberless different views are set up, and varied in all kinds of ways, no one following the same rule as the other, they cross and contradict each other in all directions; and this will continue, until we are all agreed by what rule the formation of our judgement should be guided.   I therefore repeat, *that the application of that which is already or can hereafter be known concerning the laws of combination amongst elementary bodies in inorganic nature, is the only guide to our researches concerning their mode of combination in organic nature; that by this means alone we can hope to arrive at a correct and unanimous opinion concerning the constitution of those bodies which occur in nature or which arise from the action of chemical agents upon them.*

These principles, although never positively denied, have also never been generally admitted.

Historical Statement of the Views entertained upon Organic Composition.

The first tolerably satisfactory experiments upon the composition of organic substances were published in 1811 by Thénard and Gay-Lussac. The law of chemical proportions was then only partially developed, and had not attracted the attention of these chemists, who were consequently obliged to determine the relative quantities of the elements in hundredths of the analysed substance. The view to which these experiments led, namely, that those organic substances were neutral, in which, as in starch, sugar, wood, &c., the hydrogen and oxygen were in the same relative proportions as in water, whereas those containing oxygen in excess possessed the properties of acids, and that an excess of hydrogen, classed a substance amongst the resins, oils or spirituous fluids, although in accordance with the results of the analyses, has since been found to be incorrect. Some years afterwards, when I had endeavoured to establish the law of chemical proportions for inorganic substances, I undertook, with a similar object in view, the examination of organic composition. It then appeared that all organic bodies containing oxygen, including those not reckoned amongst the acids, could be made to combine with inorganic oxides in definite and often in multiple proportions, by which means, as in inorganic nature, the atomic weight, and with it a check upon the accuracy of the analytical results, could be obtained. At the same time it was observed that the oxygen of the organic substance was a multiple of the oxygen in the inorganic oxide with which it was combined, and that the amounts of the other constituents corresponded to a certain number of atoms, and could, as in inorganic compounds, be expressed by a formula. This similarity between organic substances containing oxygen and the inorganic oxides, led immediately to the idea that the organic bodies were oxides of compound radicals, whereas the inorganic were oxides of simple elements.

This view I expressed in the second edition of my 'Manual

of Chemistry' (pt. 1. S. 544. Stockholm, 1817), in the following words :—" We find the following to be the difference between organic and inorganic bodies; all oxidized bodies in inorganic nature have a *simple radical*, whereas all organic substances are *oxides of compound radicals*. In vegetable substances these radicals are generally composed of carbon and hydrogen; in animal substances, of carbon, hydrogen and nitrogen. Acids with compound radicals are therefore synonymous with acids of organic origin. In like manner as ammonia is an alkali with a compound radical, *i. e.* of organic origin, being chiefly obtained from the animal kingdom, but nevertheless has the greatest analogy to the alkalies which have a simple radical, and are derived from inorganic nature, so we shall find the same analogy to exist between the acids of organic and those of inorganic origin, and as potash and soda are related to ammonia, so are sulphuric, nitric, and phosphoric acids related to acetic, oxalic and citric acids, &c."

The number of analysed organic substances was confined to those which had been the subject of Thénard and Gay-Lussac's and my own experiments. The view thus put forth was probably considered by the greater number of chemists as premature, for it remained twenty years quite unnoticed.

The notions which in the course of time prevailed were of quite a different nature. Organic bodies were looked upon as binary compounds of elementary substances, or as binary compounds united with an element. Prout endeavoured to show that the greater number of vegetable substances, particularly those used as food by animals, could be considered as combinations of water with carbon in different relative atomic proportions. Thénard and Gay-Lussac had already conceived this idea, although they saw at the same time that it was not tenable. Other chemists began to calculate how organic substances, according to circumstances, might be considered as combinations of two or more binary compounds; carbonic acid, water and carburetted hydrogen, in different proportions. There was no other foundation for these calculations but individual opinion and accordance with the per-centage result of analysis; all the views differed, and now no longer belong to the science, but to its history.

The idea of the union of binary compounds had in the mean time received a strong support from Gay-Lussac's examination of alcohol and æther; he showed in the year 1816 that these bodies comport themselves in such a manner as to render it probable that 2 volumes of olefiant gas were combined in æther with 1 volume of the vapour of water, and in alcohol with 2. The accordance of this view with the specific gravity of these bodies in the gaseous form, and with the consequent theory of the formation of æther from alcohol by the withdrawal of water, was so complete, that it could scarcely fail to awaken a complete conviction in all such as had not previously formed a decided opinion upon the mode of combination of organic bodies. Further evidence in support of this view was obtained, for instance, by Mitscherlich's examination of hydrated benzoic acid, which he considers a combination of 2 atoms of carbonic acid with 1 atom of a carburetted hydrogen obtained by him from this acid, and which he called Benzine $= 2CO^2 + C^{12}H^{12}$.

The discussions upon the composition of the æthers, and more particularly the admirable researches conducted by Dumas himself, or under his guidance, he still retaining the views of Gay-Lussac, afforded me an opportunity, in the annual account of the progress of Chemistry (pp. 189–201) laid before the Royal Academy on the 31st of March 1833, of comparing the two views which were entertained of the composition of organic bodies containing oxygen, namely, that they were oxides of a compound radical or compounds of binary bodies; I there showed that all the combinations of the æthers with acids and salt radicals are equally and perhaps more compatible with the view that æther is the oxide of a compound radical; that this oxide can combine, like an inorganic oxide, with anhydrous acids, both of organic and inorganic origin; that exposed to the action of hydracids its radical is reduced by the hydrogen of the acid, which then, with the separation of water, unites with the salt radical to form a kind of æther, having the same relation to those formed by oxyacids that a halogen salt bears to an oxygen salt. I then remarked that the still missing compounds of this radical with sulphur and selenium would doubtlessly be found, and in reality they were discovered a few years afterwards.

This idea attracted some attention. It was adopted by Liebig, who gave the name of ethyle to the radical, which has been retained, though at first it was disputed by Dumas. The interchange of opinion which took place between these two chemists upon this subject soon induced Dumas to coincide with the view adopted by Liebig, that the organic bodies containing oxygen were oxides of compound radicals. Their common opinion upon this subject was presented to the French Academy, Oct. 23, 1837, in a note by Dumas, in the names of both, ' Upon the actual state of Organic Chemistry.'

In that note Dumas proposes the question, How can we, from the known laws of inorganic chemistry, infer those applicable to organic bodies, which, though differently constituted, nearly all consist of carbon, hydrogen and oxygen, with the addition sometimes of nitrogen.

To preserve the vividness of his description, I will give a literal translation of his answer.

" This presents a great and beautiful problem to natural philosophy, a problem, the solution of which may well give rise to the highest degree of emulation amongst chemists, for, when once solved, the grandest triumphs are secured to science. The mysteries both of vegetable and animal life would be unveiled before our eyes, the key would be given to all the wonderfully rapid, often momentary modifications which take place in animals and plants; and still further, we might possibly discover the way to imitate these in our laboratories.

" And now, we fear not to announce it, nor is it on our part a hasty conclusion, this great and beautiful problem is *already* solved. We have now only to make from it all the deductions of which it is capable.

" With three or four elements Nature produces, in a manner as simple as it is productive, quite as various and perhaps more numerous compounds than occur in the whole of inorganic chemistry: from simple elements she produces compounds having all the properties of elements. It is our firm conviction, that in this the whole mystery of organic nature consists.

" Organic chemistry therefore embraces a number of peculiar elements, some of which play the part of oxygen or chlorine, others on the contrary correspond to the metals. Cyanogen, amidogen, benzoyle, the radicals of ammonia, of the fatty acids, of alcohol and similar bodies, are the elements with which organic chemistry works, and not with the simple elements carbon, hydrogen, nitrogen and oxygen, which only show themselves as such, when every trace of their organic origin has disappeared.

" In our opinion organic chemistry includes all bodies which spring from the reciprocal union of simple elements, and organic chemistry must likewise embrace all such substances as are formed from compounds possessing the properties of simple elements.

" In mineral chemistry the radicals are simple, in organic they are compound—this is the whole difference. The laws upon which combination is based are in both the same.

" According to our view, organic chemistry presents us with radicals, which play the part of metals, and with others which play the part of oxygen, sulphur, chlorine, &c. These radicals combine with each other or with simple elements in accordance with the most simple laws of inorganic chemistry, and thus give rise to all organic compounds."

Amongst the details Dumas commits a palpable historical error when he affirms, that he and Liebig have been employed every day during the last ten years in endeavouring to discover and study these radicals.

Since that period I am not aware that Dumas has made any application of these views; Liebig however has retained them, and his ' Organic Chemistry ' begins with the words, " Organic chemistry is the chemistry of compound radicals."

Both these chemists undertook an experiment, each after his own manner, and at the same time that the declaration of the above views was made, for the purpose of explaining theoretically the loss of water which anhydrous double salts of oxide of antimony and other bases experience when heated to 200° C.; both the explanations however which they have advanced appear to prove, that neither of them had given a thought to the theorems just mentioned, which both had so warmly advocated. Some time before he made the statement concerning organic radicals, Dumas began a closer examination of the fact discovered by Gay-Lussac, that wax exposed to dry chlorine gas gradually converted it into hydrochloric acid gas, without effecting any change of volume; from which it follows, that wax parts with its hydrogen and absorbs chlorine in such proportions, that both in the gaseous form occupy the same volume. Dumas felt the great importance which this fact must have upon the doctrine of organic composition, and commenced an investigation of the laws. He showed at the same time that the greater number of organic substances, when they are treated with chlorine or bromine, exchange hydrogen for the salt radical, generally in equivalent proportions of the elements replacing each other.

This gave rise to a theoretical explanation of the phænomenon which he called the Theory of Substitutions; and as most frequently the elements are exchanged in equivalent quantities, he framed for it the name of Metalepsy. He showed that an organic body, in which one or several equivalents of hydrogen had been exchanged for an equal number of equivalents of chlorine or bromine, retained its original saturating power, as well as several of its chemical properties, and that in many cases it could be proved to possess the same crystalline form. From these facts he came to the conclusion, *that the salt radical plays the same part in the new compound as hydrogen played in the original one*; and having discovered and analysed chloracetic acid (chloroxalic acid), he considered this view so thoroughly established, as to build upon it an entirely new theory of organic composition. In the electro-chemical theory chlorine is one of the most highly electro-negative bodies, hydrogen, on the contrary, electro-positive; and as, according to the view which he had taken, one element could play the same part in a chemical combination as another, Dumas concluded that the electro-chemical views were not sufficiently well-founded to find application in the scientific theory; and with reference to this he settled, that the part which an element plays in organic composition, does not depend upon its original properties, but upon the position which it occupies in the compound; thus then chlorine or any other element taking the position of the hydrogen can play exactly the same part as it does.

This induced him to give up the idea of the organic radicals, which however, having previously so strenuously advocated, he avoided openly to dispute; he compared them now no longer to the simple elements, but to carbonic oxide, sulphurous acid, deutoxide of nitrogen, and to the so-called hyponitric acid, NO^4. He now took quite another view of organic composition.

The elements combine in organic nature two or more with each other to form peculiar *types*, and in these types the atoms are arranged, for each particular type, in a fixed, and for those bodies belonging to the same type, in a similar manner; the characters of the compound depend upon the arrangement amongst the compound atoms, so that it becomes quite immaterial what element it is that occupies a certain position in the compound. In this manner the possibility of substitution is extended to other bodies besides hydrogen and the salt radicals. " The law of substitution," says Dumas, " presupposes that all the elements can supplant and replace each other, that simple can be substituted by compound bodies, so that not only oxygen, but also cyanogen, carbonic oxide, sulphurous acid, deutoxide of nitrogen, hyponitric acid, amidogen, or other compound bodies may occur as fundamental principles, take the place of hydrogen, and give rise to new substances."

Dumas has not succeeded in establishing clearly the fundamental idea of these types. But he assumes two kinds of them, *chemical types* and *mechanical* or *molecular* types.

I shall quote his own words :—

" 1. Experience shows us that a body may lose one of its elements and take up in its place another, equivalent for equivalent. This is the general fact of the theory of substitutions.

" 2. When this occurs, we may admit that its molecule always remains unchanged, forming a group, in which one of the elements simply occupies the space previously filled by another. This constitutes in my opinion a natural family.

" 3. Amongst the bodies which are formed by substitution, the greater portion evidently retains the same chemical character as an acid or a base, and in the same degree as it did before the change by substitution. Such bodies form, in my opinion, a *chemical type*, or speaking in the terms of natural history, belong to the same genus."

The chemical types therefore represent, according to Dumas, a kind of natural historical genus of compounds, which all agree in possessing an equal number of equivalents united in the same manner, and have the same properties.

Here then are three generic characters. The *first*, or the number of equivalents, is quite easily and accurately determined by the atomic weight. The *second*, on the contrary, or the similar mode of union of the elements, is without any foundation, unless it can be discovered by the isomorphism of the compounds. Here then individual opinion has free scope to consider the simple atoms united together in the same manner or not. With regard to the *third*, Dumas appears originally to have considered the chemical characters of those bodies produced by substitution as so little changed (in proof of which he mentions acetic and chloracetic acid), that he may have meant merely a general accordance in the chemical characters ; when however objections were raised to this, he replied, that bodies must be understood to possess the same fundamental properties when they afforded the same products of decomposition, on exposure to the same decomposing agents, and that this was the case with bodies belonging to the same chemical type.

This he illustrated by the following example. If acetate of potash be mixed with hydrate of potash and subjected to *dry distillation,* carburetted hydrogen in minimo CH^2 is generated; and when chloroxalate of potash is *boiled with a strong solution of caustic potash,* perchloride of formyle is formed $C^2 H Cl^3$. The mode of decomposition is however here quite different, for chloroxalate of potash distilled with hydrate of potash gives chloride of potassium, carbon, carbonate of potash and carburetted hydrogen, whilst acetate is decomposed into carbonic acid and carburetted hydrogen, and is not acted upon at all when boiled with caustic potash. Such however was the illustration given by Dumas of his view of fundamental properties. In order now to make carburetted hydrogen in minimo belong to the same chemical type as the perchloride of formyle, he assumed that the atom of the former consisted of $C^2 H^4$, and that in the latter, $C^2 H Cl^3$, the chlorine played the part of hydrogen; or that in the former the 3 atoms of hydrogen occupied a similar position to the 3 atoms of chlorine in the latter. It is hardly necessary to add that this view is entirely fanciful.

To give an idea of the views of atomic composition which arise from this theory, I will quote an example given by Dumas of the chemical type to which the oil of bitter almonds belongs, represented by the chemical formulæ required by this theory, in which the elements that play the same part are placed vertically the one above the other.

Oil of bitter almonds . . . . . . C^{14} H^6 O^2

Binacichloride (chloride) of benzoyle $C^{14} \begin{Bmatrix} H^5 \\ Cl \end{Bmatrix} O^2$

Benzoic acid . . . . $C^{14} \begin{Bmatrix} H^5 \\ O \end{Bmatrix} O^2$

Binacisulphuret (sulphuret) of benzoyle $C^{14} \begin{Bmatrix} H^5 \\ S \end{Bmatrix} O^2$

(Benzoate of amidogen) benzamide . $C^{14} \begin{Bmatrix} H^5 \\ NH^2 \end{Bmatrix} O^2$.

Amongst these very different compounds, the rational composition of which is expressed by the names here employed, it is not only presumed that chlorine, sulphur and oxygen play the same part, but that this is likewise the case with hydrogen;

they are therefore supposed to occupy the same position as one equivalent of hydrogen in the oil of bitter almonds, or in their prototype. The fifth example presents the most remarkable circumstance, that the same space which in oil of bitter almonds is occupied by 1 equivalent of hydrogen suffices not only for 2 equivalents of hydrogen, but in addition for 1 equivalent of nitrogen. The mechanical impossibility is no obstacle to this theory, and the supporters of metalepsy find no difficulty in replacing an equivalent of hydrogen by 4 atoms or equivalents of oxygen and 1 equivalent of nitrogen, or by 2 atoms of oxygen and 1 atom of sulphur.

It is consequently the most easily applicable of all theories, and we need not be surprised that at this moment it is most generally adopted for organic combinations; its author's well-earned celebrity in other ways is no doubt the chief reason for this. His treatises since the publication of this theory have mostly borne the title of 'Memoirs upon the Chemical Types' (*Mémoires sur les Types Chimiques*).

Many more objections could doubtless be raised against this theory, but those which we have noticed are sufficient to give an idea of the manner in which the metaleptic type-theory considers the rational composition of bodies in organic chemistry.

Simultaneously with Dumas, Laurent worked out the same theoretical view, and varied it in an endless manner. The extraordinary number of new compounds which he had the good fortune to discover during his really excellent researches, offered him abundant material to test his views upon, and as these researches were all undertaken for that object, his nomenclature has been framed entirely to suit it; thus he has almost forced chemists to accept his view of the mode of combination, and although during his further progress he has sometimes been obliged to alter his mode of conception, he has nevertheless endeavoured to change the views upon inorganic composition to suit his own upon organic. The fundamental idea upon which his opinions are based is however quite the same as that of Dumas, and consists in types and metalepsy.

The same chemical theme has likewise been varied by Gerhardt in a work upon organic chemistry, in which he allows no predecessor to outdo him in setting up fanciful explanations of the rational composition of bodies.

Persoz starts with the fundamental position, that such bodies as consist of carbon, hydrogen and oxygen, should properly be considered as carburetted hydrogen, in which the equivalents of hydrogen are metaleptically replaced by an equal number of atoms of carbonic oxide, and that such a compound can then combine with carbonic acid. This is evidently the doctrine of the union of binary compounds, corresponding to the theory of substitution. Thus, for instance, alcohol consists empirically of $C^4 H^6 O^2$; according to Persoz, its rational formula is

$$= C^2 \left\{ \begin{array}{l} H^6 \\ C^2 O^2 \end{array} \right\},$$

i. e. in the formula $C^2 H^8$, 2 equivalents of hydrogen are exchanged for 2 equivalents of carbonic oxide. Acetic acid, on the contrary, the empirical formula of which is $= C^4 H^3 O^3$, takes its origin from $C^2 H^4$, in which 1 equivalent of hydrogen is exchanged for one of carbonic oxide, therefore $C^2 \left\{ \begin{array}{l} H^3 \\ CO \end{array} \right\}$, and this body is then combined with 1 atom of carbonic acid $= C^2 \left\{ \begin{array}{l} H^3 \\ CO \end{array} \right\} + \overline{CO}^2$.

Löwig takes a view of organic composition, in which certain fundamental compounds of carbon or water, or both, with oxygen, can unite with carbonic oxide or with carbonic acid, or with both at the same time. Thus, for instance, the rational composition of formic acid according to him is $= CHO + CO^2$, that of tartrylic acid $2CHO + CO + CO^2$, that of citric acid $C^2 H^2 O + CO + CO^2$, and that of malic acid isomeric with the last $C^2 H^2 + 2CO^2$.

Graham assumes that the simple atoms of every fundamental principle combine with each other according to certain types consisting of a definite number of atoms in a fixed and unchangeable order. When fundamental principles unite with each other, then the atoms are exchanged from the types by a sort of double decomposition, and when organic compounds are produced, carbon atoms are exchanged from the carbon type for hydrogen, nitrogen or oxygen atoms from their types, the places of these latter being filled by carbon atoms.

In this manner it is clear that anybody who undertakes to explain the rational composition of organic compounds and does not feel satisfied with what his predecessors have done, may fabricate or invent a new view of his own; nor is there any reason why this should have a limit, until some rule is sought for to guide us in the mode of proceeding which ought here to be followed.

I have already stated that such a rule already exists in the laws for the combinations of elements in inorganic nature. But it frequently happens, that even with the aid of the best guide, we are unable to obtain clear ideas. We must then be satisfied with the empirical composition, and defer establishing the rational, until our knowledge is sufficiently ripe to enable us to do so. When that is the case all will immediately admit and acknowledge its correctness. To increase, in the mean time, the number of imaginative theories is only to retard instead of advance the progress of science.

Which views of the rational composition of organic bodies can be considered as corresponding with the laws of combination observed in inorganic nature? That is the question which I shall now endeavour to answer.

The idea of compound radicals and of their union with oxygen, sulphur, salt radicals, &c., is, as I have previously shown, the best guide for our judgement which we can take from inorganic nature. It is not my intention to develope this idea here. That has already long been done, and I refer for further particulars to the fifth German edition of my 'Manual of Chemistry,' vol. i. S. 672.

But this idea, presupposing even that it is correct, as some experiments tried for the purpose seem to prove, instead of unveiling to our view all the mysteries of organic nature, as Dumas conjectured, rather discovers to us nothing at all concerning them. The number of compound radicals which we have as yet been able to establish with any degree of certainty is also very small.

If in every compound of carbon and hydrogen, or of carbon, nitrogen and hydrogen with oxygen, we were to consider all that is not oxygen as a given radical, we should be led into quite as erroneous an opinion, as when, in sulphate of oxide of formyle (Melsen's acide sulfacétique), we consider all that which is not oxygen as the radical of the acid; for the sulphuric acid contained in it, is united in quite a peculiar manner, or, as we express it, is conjoined with oxide of formyle, which accompanies the acid in all its combinations. In the edition of my 'Manual,' quoted above at page 459, vol. i., I have given the theoretical views concerning this class of compounds. They occur generally, and we have always reason to expect them, whenever the number of oxygen atoms in a compound atom is greater than 7, although it does not follow that the conjugate union may not obtain when the oxygen atoms are in smaller number. As long as one of the bodies constituting a conjugate compound is an inorganic oxide, there is no difficulty in recognising the conjugate compound, or in ascertaining the composition of the conjunct. But when both the chemically active oxide and the other body constituting the conjunct are of organic composition, and contain the same elements, analysis then gives us not the slightest idea of the mode of combination, and its nature must be ascertained by some other means. Such means are however very seldom to be met with, and have only been applicable in a very few cases; we, however, possess sufficient to show that such a mode of combination exists.

When we turn our attention to the great number of conjugate compounds into which sulphuric acid enters, and which are already known, and when we learn that the number of organic bodies which do not produce such compounds with sulphuric acid is very limited; when we find that nitric acid, phosphoric acid, and even the chlorine acids, enter into the composition of conjugate compounds, it then appears that this mode of combination, although but quite recently properly understood, is nevertheless one of general occurrence; and it appears to be much more generally made use of in the œconomy of organic nature than in inorganic chemistry, so that the greater number of organic bodies may consist of conjugate compounds, which we should never be able to discover by the separation of their individual components. We only obtain a knowledge of the conjugate compounds themselves, and must be satisfied to consider these as peculiar definitive bodies.

We have discovered in inorganic chemistry that the affinity between the chemically active oxide and its conjunct is very various; sometimes the compound is decomposed by the weakest agencies, at other times the affinity is so strong that it can only be dissolved by the destruction of one of the constituent bodies. The oxygen of the chemically active oxide can be exchanged for sulphur, chlorine, &c., and the conjunct follows the radical from one combination to the other. The conjugate compound of subchloride of platinum with elayle can be decomposed by hydrate of potash, and the conjunct accompanies the platinum in the suboxide; it can be decomposed by zinc, and the platinum is eliminated still coupled with elayle. When such intimate connexion exists between a compound radical and its conjunct, we can easily conceive how impossible it is to ascertain the relationship.

The grand difficulty in the application of the idea of compound radicals to the critical examination of the rational composition of organic compounds, consists in ascertaining whether the substance under examination is a single organic oxide or a conjugate compound. In the first case it is easy to obtain a notion of the radical, in the last it is impossible, for the active oxide has its radical, as has also the conjunct; and when the conjunct is a compound of two substances, each of these has likewise its radical. It is thus evident that the idea of compound radicals may be quite correct and yet not applicable under such circumstances, until we have obtained some knowledge of the composition both of the active oxide and of the conjunct. When this knowledge is unattainable, which it is in most cases, we must, as I have already noticed, consider the organic substance empirically as a single whole.

The number of compound radicals may perhaps not be very great; for it is possible that the endless multiplicity is produced by changes in the conjugate compounds. We have nearly 100 different acids into the composition of which sulphuric acid enters, only varying according to the different conjuncts which they contain.

I have already mentioned, that when the number of oxygen atoms in an organic atom exceeds seven, we have reason to presume that the compound belongs to the conjugate class of combinations, in which the conjoined amount of oxygen in the conjoined oxides together, may sometimes increase the atomic proportion of that element considerably. It occasionally happens that the amount of oxygen in these compounds may be lessened or increased. When this, however, only takes place with one of the combined oxides, the change never amounts to more than a small fraction of the whole quantity of oxygen contained in the combined oxides, and does not correspond with the multiple proportions which we are in the habit of meeting with in inorganic chemistry; but it is evident that it would correspond to these if we could calculate it for the oxide to which it appertains. Thus, for instance, proteine contains 10 atoms of oxygen, but the amount of oxygen can be raised by 2 or 3 atoms, to bin- and tritoxide of proteine with 12 and 13 atoms of oxygen. This uncommonly small addition of oxygen points clearly to the existence of a conjugate compound, in which the increase of oxygen only pertains to the chemically active oxide.

In the case of oxamic acid, we have seen that to the bodies which may be conjuncts the amides likewise belong. Fehling's experiments upon succinic acid have shown that this acid gives with succinamide a similarly constituted conjugate succinic acid, and Laurent's researches prove the existence of similar conjugate acids with amidogen produced from tartaric, lactic, camphoric and phtalinic acids. From a single one, these examples have rapidly increased to a multitude. It is therefore evident that this mode of combination must be more general than we had previously reason to suppose. But this includes the doctrine of ternary radicals, into the composition of which nitrogen enters, in a manner which it is quite beyond our power to explain. As long as the amide is composed of an acid in a low grade of oxidation combined with NH^2, the kind of relationship is easily discovered, for acids and alkalies then convert the amide into ammonia or oxide of ammonium, by oxidizing, at the expense of water, the oxide with which it is combined. But when amides of a different kind, where such higher oxidation of the oxide cannot take place, and consequently the amidogen cannot be converted into ammonia, become conjuncts to organic oxides, nitrogenized or not nitrogenized, these cannot be discovered by the means just mentioned.

From all that has now been stated it is therefore evident, that although the view relating to compound radicals may be quite correct, there still remains a vast deal to be discovered before we can apply it in a satisfactory manner, and before we can with certainty distinguish between a combination of oxygen with a compound radical and a conjugate oxide.

Our rational view extends only to the fact, that we have succeeded in showing the existence of compound radicals, and that the combinations of these with oxygen, sulphur, salt radicals, &c. have a great and general tendency to the formation of conjugate compounds, n which one of the constituent bodies retains its chemical activity, whilst the other in most cases entirely loses it.

What relation do the phænomena of substitution bear to these views? This question arises spontaneously from what has just been stated.

It is evident that when a conjugate compound is acted upon

by chlorine or bromine, and an exchange takes place, this is not effected simultaneously in the conjunct and in the chemically active oxide. But in this manner quite a different view of the subject is obtained than that arising from the metaleptic theory. It cannot be considered merely as a supposition, when we assume that the conjunct is the first to be changed by the action of the salt radical, for I shall prove by striking examples that a conjunct can be thus changed by a salt radical without ceasing on that account to be still a conjunct to the chemically active oxide, which, if it happen to be an acid, retains its acid properties and forms salts, the characters of which are more or less changed, according to the changed constitution of the conjunct. This it was that gave rise to the fiction of chemical types. It however does not prevent the chemically active oxide from being eventually changed by the action of chlorine; but then quite another kind of compound is produced, in which the properties of the chemically active oxide are no longer to be found.

These substitutions have never been theoretically handled but by supporters of the metaleptic theory, and it must be admitted that almost every one who has made an attempt to investigate the subject has soon become a proselyte to that theoretical view. It will not be out of place here to examine these phænomena from another point of view, and with this motive I shall bring forward a few examples.

1. When oil of bitter almonds, or binoxide of picramyle, $C^{14} H^6 + O^2$, is treated with chlorine, an equivalent of hydrogen is exchanged for an equivalent of chlorine. By this process a body is produced which on its first discovery received the name of chlorbenzoyle. According to the metaleptic theory, from what has been already stated, it would be considered as an unaltered type $= C^{14} \left\{ \begin{matrix} H^5 \\ Cl \end{matrix} \right\} O^2$, and the chlorine would play the same part and take the same position in it as the equivalent of hydrogen which has been eliminated. But when the body is now treated in alcohol with hydrate of potash, chloride of potassium and benzoate of potash are produced. This indicates quite another mode of combination: it contains, namely, the same compound of carbon and hydrogen, which is present in benzoic acid, united with 2 atoms of oxygen and 1 equivalent of chlorine. The counterpart to this may be observed in numerous cases with inorganic radicals, for instance, with sulphur, chromium, tung-

3 B 2

sten, and molybdenum.   But the combinations of these consist of 2 atoms of anhydrous acid with 1 atom of perchloride; such should also be the case here, and chlorbenzoyle-binacichloride should be $= 2C^{14} H^5 O^2 + C^{14} H^5 Cl^3$.

Here consequently no substitution has been effected, but out of 3 atoms of binoxide of picramyle 2 atoms of benzoic acid and 1 atom of perchloride of benzoyle have been formed, which enter into combination with each other. The chlorine has destroyed the radical which was presented to it, and formed another, containing the same number of carbon atoms but an equivalent less of hydrogen.   The relation therefore is clear and indisputable.

2. If chlorine is permitted to act upon a conjugate organic acid in which the conjunct is a carburetted hydrogen containing many atoms of each element, then the substitution must be exercised upon the hydrogen of the conjunct.   What results here from the change cannot be positively proved, but the possible cases may be conceived without the necessity for assuming that any one in particular is actually produced.

a. A conjunct may consist of several atoms of the carburetted hydrogen: one of these alone, on substitution taking place, may part with its hydrogen for chlorine, by which means chloride of carbon and the remainder of the carburetted hydrogen united together to form a conjunct are left.   Let the conjunct consist for instance of 3 atoms $C^2 H^3$.   Now if 1 atom $C^2 H^3$ is converted by chlorine into $C^2 Cl^3$, *i. e.* into sesquichloride of carbon, we obtain by substitution three acids, in which the acid body remains unchanged in each, but is united with three different conjuncts, namely, $C^2 Cl^3 + 2C^2 H^3$, $2C^2 Cl^3 + C^2 H^3$, and lastly $3C^2 Cl^3$.

b. These successive exchanges of hydrogen for chlorine may, however, take place in another manner, with the formation of a differently constituted conjunct, of which mode of exchange I will in addition adduce the following proof.   In the edition of my 'Manual' before noticed (vol. i. S. 709), I mentioned that acetic acid might possibly be a conjugate oxalic acid, in which the conjunct was $C^2 H^3$.   I do not think that it can at present be decided whether this is the case or not, but let us take it as an example.   If acetic acid is $= C^2 O^3 + C^2 H^3$, and if an equivalent of hydrogen is exchanged in it for an equivalent of chlorine, we obtain $C^2 H^2 Cl$; or protochloride of elayle, which belongs to the class of bodies forming conjuncts to acids.   If another equivalent of hydrogen is exchanged for chlorine, we get $C^2 HCl^2$, bichloride of formyle, which is a conjunct to oxalic acid; after another exchange we obtain the perchloride of carbon, $C^2 C^3$, as conjunct to the acid, *i. e.* acetic acid has been converted into ohloroxalic (acide chloracétique, Dumas). The experiments upon the action of chlorine on acetic acid have never been carried to their whole extent, it is therefore unknown whether these intermediate grades exist; should this however be the case, it is quite evident that acetic acid comprising $C^2 H^3$, must be a conjugate oxalic acid. If they do not exist, then the portion of acetic acid upon which the chlorine acts is transformed immediately into chloroxalic acid, which can be separated from the unaltered acetic acid. The chlorine in this case effects no substitution, but a decomposition of the acetic acid, and this would then be a reason for considering it an acid with an independent radical.

c. When the conjunct is an oxide with an independent radical, then, by the exchange of hydrogen for chlorine, an oxychloride of a radical with less hydrogen is produced, or, if all the hydrogen is exchanged, an oxychloride of carbon.

It is remarkable how a body which has once entered as a conjunct into combination can undergo one change after another in its composition and yet retain its properties as a conjunct. The compounds of indigo with sulphuric acid offer the most numerous examples. Indigo-blue as a conjunct to sulphuric acid undergoes the most numerous changes of composition by the action of salt radicals, nitric acid, alkalies, and reducing agencies, without giving up its connexion with sulphuric acid. I will now produce the positive proof of the changes which conjuncts undergo, deduced from inorganic chemistry, of which I have spoken before.

Kolbe, who recently published an examination of the changes produced in sulphuret of carbon by the action of chlorine, has shown, that the body produced by the action of chlorine or *aqua regia* upon sulphuret of carbon consists of perchloride of carbon and sulphurous acid, $C Cl^2 + SO^2$. This is a kind of conjugate combination of perchloride of carbon and sulphurous acid. When acted upon by caustic potash, a portion of the chlorine is exchanged for oxygen from the potash, and from 2 atoms $(CCl^2 + SO^2)$ we obtain 1 atom of chloride of potassium and 1 atom of a potash salt, the acid of which consists of 1 atom of sesquichloride of carbon and 1 atom of dithyonic (hyposulphuric) acid $= C^2Cl^3 + S^2O^5$. This is a strong acid, which can be separated and crystallized; it may be called sesquichloride of carbon-dithyonic acid, and is evidently a conjugate dithyonic acid in which the conjunct is the same as in chloroxalic acid.

If metallic zinc is placed in a solution of this acid, it is dissolved without any evolution of gas, and two salts of zinc are obtained dissolved in the fluid. One atomic equivalent of the acid dissolves 2 atoms of zinc, and the solution contains 1 atom of protochloride of zinc and 1 atom of an oxide of zinc salt, the acid of which consists of $C^2HCl^2 + S^2O^5$. An equivalent of chlorine is taken up by the zinc, and the hydrogen which was eliminated on the formation of an atom of oxide of zinc to saturate the acid has taken its place in the conjunct, which now has been converted from sesquichloride of carbon into dichloride of formyle without leaving its position in the conjunct. This substitution, compared with the previous one, has taken place in an inverse order, and at the same time a compound radical has formed itself in the conjunct which was previously not present. When this or the first-named acid, no matter which, is dissolved in water, mixed with an equivalent of sulphuric acid, and zinc then added, the acid becomes saturated with zinc, without the evolution of hydrogen. According to the nature of the conjugate acid used, 3 or 4 equivalents of zinc are dissolved, and we obtain in solution 1 or 2 equivalents of protochloride of zinc, 1 equivalent of sulphate of zinc, and 1 equivalent of a new salt of zinc, in which only 1 equivalent of chlorine remains, the acid of which is $C^2H^2Cl + S^2O^5$, *i. e.* a conjugate acid composed of protochloride of elayle united to dithyonic acid. If more than 1 equivalent of sulphuric acid is used in the formation of this salt the protochloride of elayle is not decomposed, but on the solution of the zinc in the excess of acid hydrogen is given off.

What however the affinity exerted by an excess of sulphuric acid cannot overcome, is effected by an electrical current. Kolbe has shown that when the potash salt of either of these three conjugate dithyonic acids is exposed in a concentrated solution between amalgamated plates of zinc to the electric current produced by 2 or 3 pairs of Bunsen's carbo-zinc battery, by oxidizement of the zinc basic chloride of zinc is formed, and that no hydrogen is evolved, until the whole amount of chlorine in the acid has been exchanged for hydrogen. Thus we obtain a fourth conjugate dithyonic acid $C^2H^3 + S^2O^5$, in which the conjunct has the composition of methyle. This is the same body which we suppositively adduced as a constituent of acetic acid, should this prove to be a conjugate oxalic acid.

Kolbe has further shown that chloroxalic acid can in the same manner be converted into acetic acid; and has also made mention of intermediate members of the series, the proof of the existence of which, once established with certainty, would be of the greatest importance to the science.

These simple experiments, in which the chemically active inorganic oxide is the same for all the acids, and consequently of such a nature, that its constituents do not interfere with the question concerning the changes effected by substitution in the conjunct, lead to the following simple deductions.

1. A conjunct can, notwithstanding its composition be completely changed, retain its position as a conjunct.

2. When the conjunct is a chloride of carbon, and the chlorine is partially exchanged for hydrogen, new radicals are produced consisting of carbon and hydrogen, the proto- and perchlorides of which still constitute the conjunct; and when the whole of the chlorine is exchanged, there results a carburetted hydrogen, which, as conjunct, takes the place of the chlorine compound. It follows from this, that when the exchange takes place in an inverse order, or when hydrogen is exchanged for chlorine, the process must be judged in the same manner.

3. A possibility is afforded by the agency of an electrical current of exchanging chlorine for hydrogen, and in this case carbon and hydrogen form together a radical, in which hydrogen can occupy the same position as chlorine did, but cannot therefore play the same part. This result of Kolbe's experiments completes the refutation of the metaleptic views and the fanciful theory of chemical types.

It is far from my intention to maintain that what has been here stated affords an explanation of all those cases in which hydrogen is empirically replaced by a salt radical. The internal process in these cases and the products may be of the most various nature. The fault which has always been committed, consists in the endeavour to give an explanation that should apply in every case. There are instances in which even hypochlorous and hypobromous acids appear to be constituents of the products.

I am not so sanguine as to hope that these views upon the difficulties attending the formation of a correct opinion of the rational composition of organic substances will be much listened to by the supporters of metalepsy.   They embrace their theory with the full conviction of its infallibility, and little is to be effected against blind faith by argument and proof.   But I nevertheless do not despair, that what has here been brought forward will find some consideration from the great number of chemists who have not previously had occasion to declare themselves in favour of any particular view of these questions.

4. " On a new Series of Organic Bodies containing Metals." By Dr. E. Frankland, Professor of Chemistry, Owen's College, Manchester. Communicated by B. C. Brodie, Esq., F.R.S. Received May 10, 1852.

The author communicates in this memoir the continuation of his researches, a preliminary announcement of which appeared several years ago, upon a new series of organic compounds closely allied to cacodyl in their composition and properties, and which, like that body, are formed by the union of the alcohol radicals with various metals, and are distinguished for their powerful electro-positive character. These remarkable compounds are procured by the action of heat or light upon their proximate constituents, and are thus distinguished from most other organic compounds of this nature by the manner of their formation. The author describes seven of these compounds.

Stanethylium.—When iodide of ethyl and metallic tin are exposed to the influence of heat or light, which is most conveniently done in sealed glass tubes, the tin gradually dissolves in the ethereal liquid, which finally solidifies to a mass of colourless crystals. A quantity of gas, comparatively very small, is generated at the same time. This gaseous product of the reaction proved, on analysis, to be a mixture of hydride of ethyl and olefiant gas, produced from the decomposition of iodide of ethyl by tin into iodide of tin and ethyl, which last is transformed at the moment of its liberation into the two gases just mentioned. The principal and most important reaction, however, consists in the direct union of tin with iodide of ethyl, giving rise to a crystalline body which is the iodide of a new organic radical, *stanethylium.*

By double decomposition the other compounds of stanethylium can be readily formed ; the author has prepared and described the oxide, chloride, bromide and sulphide of stanethylium ; these bodies exhibit a striking resemblance to the corresponding bi-compounds of tin, but are distinguished from them by a peculiar pungent and irritating odour resembling that of the volatile oil of mustard.

If a slip of zinc be immersed in a solution of chloride of stanethylium, dense oily drops soon form on the surface of the zinc, and finally collect at the bottom of the vessel. This oily liquid is the radical stanethylium, which possesses the following properties :—it exists at ordinary temperatures as a thick heavy oily liquid of a yellow or brownish-yellow colour, and an exceedingly pungent odour resembling that of its compounds, but much more powerful. It is insoluble in water, but soluble in alcohol and ether. At about 150° C. it enters into ebullition, but is simultaneously decomposed, metallic tin being deposited. In contact with air stanethylium rapidly absorbs oxygen, and is converted into oxide of stanethylium.

Chloride, iodide and bromide of stanethylium are instantaneously formed by the action of chlorine, iodine and bromine, or their hydrogen acids respectively, upon stanethylium. The two first are in every respect identical with the salts above mentioned, and the identity of the bromide is further proved by an ultimate analysis. The formula of stanethylium is $C_4 H_5 Sn$; that of the oxide $C_4 H_5 SnO$, and that of the bromide $C_4 H_5 Sn Br$. Stanethylium therefore perfectly resembles cacodyl in its reactions, combining directly with the electro-negative elements, and regenerating the compounds from which it has been derived.

Stanmethylium and stanamylium are formed when the iodides of methyl and amyl respectively are exposed to the action of light in contact with tin; their salts are isomorphous with those of stanethylium, but they have not yet been completely investigated.

Zincmethylium.—This radical is formed in an uncombined state when iodide of methyl and zinc are exposed to a temperature of about 150° C. in a sealed tube; the zinc gradually dissolves with an evolution of gas, whilst a mass of white crystals and a colourless mobile liquid refracting light strongly, occupy, after a few hours, the place of the original materials. In this reaction two distinct decompositions take place, viz. the decomposition of iodide of methyl by zinc with the production of iodide of zinc and liquid zincmethylium, and the decomposition of iodide of methyl by zinc with the formation of iodide of zinc and the gaseous radical methyl. The zincmethylium was obtained pure by distillation in an atmosphere of dry hydrogen. Its formula is $C_2 H_3 Zn$, and it possesses the following properties. It is a colourless, transparent and very mobile liquid, possessing a peculiar penetrating and insupportable odour, and boiling at a low temperature. Zincmethylium combines directly with oxygen, chlorine, iodine, &c., forming somewhat unstable compounds. Its affinity for oxygen is even more intense than that of potassium; in contact with atmospheric air it instantaneously ignites, burning with a beautiful greenish blue flame, and forming white clouds of oxide of zinc; in contact with pure oxygen it burns with explosion, and the presence of a small quantity of its vapour in combustible gases gives them the property of spontaneous inflammability in oxygen. Thrown into water, zincmethylium decomposes that liquid with the evolution of heat and light; when this action is moderated, the sole products of the decomposition are oxide of zinc and hydride of methyl.

The extraordinary affinity of zincmethylium for oxygen, its peculiar composition, and the facility with which it can be procured, cannot fail to cause its employment for a great variety of transformations in organic compounds; by its agency there is every probability that oxygen, chlorine, &c. can be replaced atom for atom by methyl, and thus entirely new series of organic compounds will be produced, and clearer views of the rational constitution of others be obtained.

The gaseous methyl formed simultaneously with zincmethylium is identical in composition and properties with the methyl derived from the electrolysis of acetic acid; it was mixed, however, with hydride of methyl generated by the decomposition of accompanying zincmethylium vapour by the water over which the gas was collected.

Zincethylium and zincamylium are homologous bodies formed by similar processes; their investigation is not yet completed.

Hydrargyromethylium.—The author has only yet studied the iodide of this radical, which is formed by the action of sunlight upon iodide of methyl and metallic mercury. After an exposure of several days to sunlight, white crystals begin to form in the liquid, which finally solidifies to a white crystalline mass; ether dissolves out the new compound and deposits it perfectly pure by spontaneous evaporation.

Iodide of hydrargyromethylium ($C_2 H_3 HgI$) is a white solid, crystallizing in minute nacreous scales, which are insoluble in water, moderately soluble in alcohol, and very soluble in ether and iodide of methyl; it is slightly volatile at ordinary temperatures, and exhales a weak but peculiarly unpleasant odour, which leaves a nauseous taste upon the palate for several days. At 100° C. the volatility is much greater, and the crystals are rapidly dissipated at this temperature when exposed to a current of air. At 143° C. it fuses and sublimes without decomposition, condensing in brilliant and extremely thin crystalline plates. In contact with the fixed alkalies and ammonia it is converted into oxide of hydrargyromethylium, which is dissolved by an excess of all these reagents.

A corresponding compound containing amyl is formed, though with difficulty, under similar circumstances, but the attempts to form one containing ethyl have not yet been successful. Preliminary experiments have also been made with other metals, amongst which arsenic, antimony, chromium, iron, manganese and cadmium promise interesting results.

From a review of the composition and habits of all the organometallic bodies and their compounds at present known, the author is of opinion that the view most generally held respecting the constitution of cacodyl, according to which that radical is a conjugate compound consisting of arsenic conjugated with two atoms of methyl, and which view must, if true, be applied to all the organo-metallic bodies, is no longer tenable; and he contends that the behaviour of these bodies clearly indicates that they are compounds formed upon the type of the oxides of the respective metals, a portion of the oxygen being replaced by the several radicals, methyl, ethyl and amyl; the establishment of this new view of their constitution will remove these bodies from the class of organic radicals, and place them in the most intimate relation with ammonia and the bases of Wurtz, Hofmann and Paul Thenard; indeed the close analogy between stibethine and ammonia first suggested by Gerhardt, has been most satisfactorily demonstrated by the behaviour of stibethine with the haloid compounds of methyl and ethyl. Stibethine furnishes us therefore with a remarkable example of the law of symmetrical combination, and shows that the formation of a five-atom group from one containing three atoms can be effected by the assimilation of two atoms, either of the same or of opposite electro-chemical character: this remarkable circumstance suggests the following question. Is this behaviour common also to the corresponding compounds of arsenic, phosphorus and nitrogen, and can the position of each of the five atoms with which these elements respectively combine be occupied indifferently by an electro-negative or an electro-positive element? This question, so important for the advance of our knowledge of the organic bases and their congeners, cannot now long remain unanswered.

PHYSICAL SCIENCE.

The Specific Heat of Solid Bodies ; Deductions Relative to the Compound Nature of So-called Simple Bodies, by M. H. KOPP.

IF from the atomic heat of various metallic oxides we deduct that of the metals which they contain, or from the atomic heat of oxygenated salts that of all the elements combined with the oxygen, we obtain for the atomic heat of oxygen a value much less than six. The figures obtained by such indirect determinations of oxygen do not correspond so well as might be desired; however, I do not think its atomic heat differs much from four. A comparison between the atomic heats of the carbonates, R_2CO_3 and RCO_3, and the oxides R_3O_3 ($=3RO$) and R_2O_3, shows that that of the carbonates is considerably less. Such comparisons show that the atomic heat of carbon in a state of combination is obviously equal to that of the diamond, $=1\cdot8$ for C. Other similar comparisons lead us to the belief that the atomic heat of other elements is much less than those calculated from the law of Dulong and Petit. Thus the atomic heat of hydrogen equals about $2\cdot3$; that of boron is comprised between 2 and 3 ; that of silicium is about 4 ; whilst that of fluorine seems to be considerably less than $6\cdot4$.

By calculating, by means of the numbers thus obtained for the atomic heat of elements, the atomic and specific heat of compounds, results are generally obtained which very satisfactorily accord with those deduced from direct experiments. Differences are, it is true, sometimes observable, but there are differences of the same kind in the experimental atomic heat of similar combinations, even in those containing, as corresponding elements, bodies which, when free, clearly possess the same atomic heat. In M. Regnault's determinations of specific heat, this difference often amounts to one-tenth of the atomic heat, and is sometimes even greater.

My researches confirm and extend the proposition already enunciated by various experimenters, namely, that among the bodies considered as simple, and taken in the solid state, all do not follow the law of Dulong and Petit. For a certain group of elements the law holds good ; but when it is shown not to be general, and not to apply to some determined elements, its application to other elements may appear doubtful. Sulphur, for example, is a doubtful instance. The specific heat of sulphur, determined by M. Regnault, gives for this body an atomic heat $= 6\cdot5$, which nearly approaches that of metals. But the specific heat of sulphur has been determined between 98 degrees and the ordinary temperature, and 98 degrees is near the fusing point of sulphur. Some determinations which I have made, between 47 degrees and the ordinary temperature, have given results according to which the atomic heat of sulphur would only be $= 5\cdot2$, and this number accords with that indirectly deduced from the atomic heat of sulphides. It is in some instances impossible to decide whether one element, compared with another, does or does not follow the law of Dulong and Petit. Were the application of this law general, important conclusions might be drawn from it concerning the bodies regarded as elements, deciding which ought to be regarded as such ; results not less important are arrived at from the knowledge that all simple bodies do not follow this law.

In comparing the atomic heat of solid bodies it will generally be found that this heat increases with the complication of the composition, and with the number of elementary atoms contained in an atom of the compound. This is especially the case with compounds containing only the elements to which the law of Dulong and Petit applies. Were this law general and applicable to all elements, the following conclusion might be drawn : By leaving it an open question as to whether bodies, not decomposable and considered as elements, are really simple bodies, or simply bodies whose composition is inaccessible to our means of analysis, the equality of the atomic heat of these substances shows that, in the latter case, the art of decomposition has reached its limit in bodies offering the same degree of complication.

In other words, if the bodies which we consider as elements are not simple bodies, they are at least combinations of the same kind, and these combinations, it should be remarked, would show a great divergence of properties, as may be seen in the metals sulphur and iodine.

Such a conclusion would be legitimate, and the atomic heat of a body would furnish a certain criterion by which to decide the question as to whether the body should be numbered among the elements or as a compound. The fact that the same atomic heat has been found to belong to iodine, and indirectly to chlorine, as the law of Dulong and Petit assigns to elements, puts it beyond a doubt that these bodies, if they are compounds, are so in the same degree as the other elements to which this law applies.

Such deductions, which would be of great importance in deciding questions as to the nature of certain elements—for instance, whether chlorine is a simple or compound body (a peroxide)—cease to be legitimate when Dulong and Petit's law is no longer recognised as general, but is only applicable to certain groups of bodies considered as elementary. If, on the one hand, the atomic heat is generally supposed to give the amount of molecular complication, and, on the other hand, it is proved that all bodies supposed to be elementary do not possess the same atomic heat, we arrive at the conclusion, in the first place, that the part of decomposition is arrested by compounds of the same kind (metals, for instance), and in the second by substances with a more simple composition. Hence it is not impossible that a recognised compound body may have the same atomic heat as one reputed simple. Thus the atomic heat of a peroxide, containing an element whose atomic heat equals that of hydrogen (let it be about $2 \cdot 3$), may possess an atomic heat $= 2 \cdot 3 + 4 = 6 \cdot 3$, that is to say, obviously equal to that of the metals, or chlorine, or iodine. Chlorine may be such a peroxide; at least the deductions drawn from specific heat are not contrary to this hypothesis.

It may be found astonishing, or even improbable, that bodies supposed to be simple, which replace each other in compounds like hydrogen and the metals, and even enter into isomorphous combinations, like silicium and tin, may nevertheless possess quite different atomic heats. But this is not more extraordinary than another well-proved fact,—that both simple and compound bodies, such as hydrogen and hyponitric acid, or potassium and ammonium, can replace each other in compounds where the same chemical characteristic remains, or even in isomorphous combinations.

But it will, on the other hand, be easily understood that such differences in the metallic heat of elements—differences appearing in their most simple combinations—become less and less apparent as the complications of the combination increase, and contain, independently of atoms of unequal atomic heat, a greater number of similar atoms possessing the same atomic heat.—*Comptes-Rendus*, 47, lvii., 63.

SECTION B

CHEMICAL SCIENCE

OPENING ADDRESS BY PROF. SIR HENRY ENFIELD ROSCOE, PH.D., LL.D., F.R.S., F.C.S., PRESIDENT OF THE SECTION

WITH the death of Berzelius in 1848 ended a well-marked epoch in the history of our science ; with that of Dumas—and, alas ! that of Wurtz also—in 1884 closes a second. It may not perhaps be unprofitable on the present occasion to glance at some few points in the general progress which chemistry has made during this period, and thus to contrast the position of the science in the "sturm und drang" year of 1848, with that in the present, perhaps quieter, period.

The differences between what may probably be termed the Berzelian era and that with which the name of Dumas will for ever be associated show themselves in many ways, but in none more markedly than by the distinct views entertained as to the nature of a chemical compound.

According to the older notions, the properties of compounds are essentially governed by a qualitative nature of their constituent atoms, which were supposed to be so arranged as to form a binary system. Under the new ideas, on the other hand, it is mainly the number and arrangement of the atoms within the molecule, which regulate the characteristics of the compound which is to be looked on not as built up of two constituent groups of atoms, but as forming one group.

Amongst those who successfully worked to secure this important change of view on a fundamental question of chemical theory, the name of Dumas himself must first be mentioned ; and, following upon him, the great chemical twin-brethren Laurent and Gerhardt, who, using both the arguments of test-tube and of pen in opposition to the prevailing views, gradually succeeded, though scarcely during the lifetime of the first, in convincing chemists that the condition of things could hardly be a healthy one when chemistry was truly defined " as the science of bodies which do not exist." For Berzelius, adhering to his preconceived notions, had been forced by the pressure of new discovery into the adoption of formulæ which gradually became more and more complicated, and led to more and more doubtful hypotheses, until his followers at last could barely succeed in building up the original radical from its numerous supposed component parts. Such a state of things naturally brought about its own cure, and the unitary formulæ of Gerhardt began to be generally adopted.

It was not, however, merely as an expression of the nature of the single chemical compound that this change was beneficial, but, more particularly, because it laid open the general analogies of similarly constituted compounds, and placed fact as the touch-stone by which the constitution of these allied bodies should be ascertained. Indeed, Gerhardt, in 1852, gave evidence of the truth of this in his well-known theory of type, according to which, organic compounds of ascertained constitution can be arranged under the four types of hydrogen, hydrochloric acid, water, and ammonia, and of which it is, perhaps, not too much to say that it has, more than any other of its time, contributed to the clearer understanding of the relations existing amongst chemical compounds.

Another striking difference of view between the chemistry of the Berzelian era and that of what we sometimes term the modern epoch is illustrated by the so-called substitution theory. Dumas,

to whom we owe this theory, showed that chlorine can take the place of hydrogen in many compounds, and that the resulting body possesses characters similar to the original. Berzelius opposed this view, insisting that the essential differences between these two elements rendered the idea of a substitution impossible, and notwithstanding the powerful advocacy of Liebig, and the discovery by Melsens of reverse substitutions (that is, the re-formation of the original compound from its substitution-product), Berzelius remained to the end unconvinced, and that which was in reality a confirmation of his own theory of compound radicals, which, as Liebig says, "illumined many a dark chapter in organic chemistry," was looked upon by him as an error of the deepest dye. This inability of many minds to see in the discoveries of others confirmation of their own views is not uncommon ; thus Dalton, we may remember, could never bring himself to admit the truth of Gay-Lussac's laws of gaseous volume-combination, although, as Berzelius very truly says, if we write *atom* for *volume* and consider the substance in the solid state in place of the state of gas, the discovery of Gay-Lussac is seen to be one of the most powerful arguments in favour of Dalton's hypothesis.

But there is another change of view, dating from the commencement of the Dumas epoch, which has exerted an influence equal, if not superior, to those already named on the progress of our science. The relative weights of the ultimate particles, to use Dalton's own words, which up to this time had been generally adopted by chemists, were the equivalent weights of Dalton and Wollaston, representing, in the case of oxygen and hydrogen, the proportions in which these elements combine, viz. as 8 to 1. The great Swedish chemist at this time stood almost alone in supporting another hypothesis ; for, founding his argument on the simple laws of volume-combination enunciated by Gay-Lussac, he asserted that the true atomic weights are to be represented by the relations existing between equal volumes of the two gases, viz. as 16 to 1. Still these views found no favour in the eyes of chemists until Gerhardt, in 1843, proposed to double the equivalent weights of oxygen, sulphur, and carbon, and then the opposition which this suggestion met with was most intense, Berzelius himself not even deigning to mention it in his annual account of the progress of the science, thus proving the truth of his own words : " That to hold an opinion habitually often leads to such an absolute conviction of its truth that its weak points are unregarded, and all proofs against it ignored." Nor were these views generally adopted by chemists until Cannizaro, in 1858, placed the whole subject on its present firm basis by clearly distinguishing between equivalent and molecular weights, showing how the atomic weights of the constituent elements are derived from the molecular weights of their volatile compounds based upon the law of Avogadro and Ampère, or where, as is the case with many metals, no compounds of known vapour-density exist, how the same result may be ascertained by the help of the specific heat of the element itself. Remarkable as it may appear, it is nevertheless true that it is in the country of their birth that Gerhardt's atomic weights and the consequent atomic nomenclature have met with most opposition, so much so that within a year or two of the present time there was not a single course of lectures delivered in Paris in which these were used.

The theory of organic radicals, developed by Liebig so long ago as 1834, received numerous experimental confirmations in succeeding years. Bunsen's classical research on cacodyl, proving the possibility of the existence of metallo-organic radicals capable of playing the part of a metal, and the isolation of the hydrocarbon ethyl by Frankland in 1849, laid what the supporters of the theory deemed the final stone in the structure.

The fusion of the radical and type theories, chiefly effected by the discovery in 1849 of the compound ammonias by Wurtz, brings us to the dawn of modern chemistry. Henceforward organic compounds were seen to be capable of comparison with simple inorganic bodies, and hydrogen not only capable of replacement by chlorine, or by a metal, but by an organic group or radical.

To this period my memory takes me back. Liebig at Giessen, Wöhler in Gottingen, Bunsen in Marburg, Dumas, Wurtz, and Laurent and Gerhardt in Paris, were the active spirits in Continental chemistry. In our own country, Graham, whose memorable researches on the phosphates had enabled Liebig to found his theory of polybasic acids, was working and lecturing at University College, London ; and Williamson, imbued with the new doctrines and views of the twin French chemists, had just been appointed to the Chair of Practical Chemistry in the same College, vacant by the death of poor Fownes. At the same time, Hofmann, in whom Liebig found a spirit as enthusiastic in the cause of scientific progress as his own, bringing to England a good share of the Giessen fire, founded the most successful school of chemistry which this country has yet seen.

At the Edinburgh meeting of this Association in 1850, Williamson read a paper on "Results of a Research on Ætherification," which included not only a satisfactory solution of an interesting and hitherto unexplained problem, but was destined to exert a most important influence on the development of our theoretical views. For he proved, contrary to the then prevailing ideas, that ether contains twice as much carbon as alcohol, and that it is not formed from the latter by a mere separation of the elements of water, but by an exchange of hydrogen for ethyl, and this fact, being in accordance with Avogadro's law of molecular volumes, could only be represented by regarding the molecule of water as containing two atoms of hydrogen to one of oxygen, one of the former being replaced by one of ethyl to form alcohol, and the two of hydrogen by two of ethyl to form ether. Then Williamson introduced the type of water (subsequently adopted by Gerhardt) into organic chemistry, and extended our views of the analogies between alcohols and acids, by pointing out that these latter are also referable to the water-type, predicting that bodies bearing the same relations to the ordinary acids as the ethers do to the alcohols must exist, a prediction shortly afterwards (1852) verified by Gerhardt's discovery of the anhydrides. Other results followed in rapid succession, all tending to knit together the framework of modern theoretical chemistry. Of these the most important was the adoption of condensed types, of compounds constructed on the type of two and three molecules of water, with which the names of Williamson and Odling are connected, culminating in the researches of Brodie on the higher alcohols, of Berthelot on glycerine, and of Wurtz on the dibasic alcohols or glycols; whilst, in another direction, the researches of Hofmann on the compound amines and amides opened out an entirely new field, showing that either a part or the whole of the hydrogen in ammonia can be replaced by other elements or elementary groups without the type losing its characteristic properties.

Again, in 1852, we note the first germs of a theory which was destined to play an all-important part in the progress of the science, viz., the doctrine of valency or atomicity, and to Frankland it is that we owe this new departure. Singularly enough, whilst considering the symmetry of construction visible amongst the inorganic compounds of nitrogen, phosphorus, arsenic, and antimony, and whilst putting forward the fact that the combining power of the attracting element is always satisfied by the same number of atoms, he does not point out the characteristic tetrad nature of carbon; and it was not until 1858 that Couper initiated, and Kekulé, in the same year, thoroughly established, the doctrine of the linking of the tetrad carbon atoms, a doctrine to which, more than to any other, is due the extraordinary progress which organic chemistry has made during the last twenty years, a progress so vast that it is already found impossible for one individual, even though he devote his whole time and energies to the task, to master all the details, or make himself at home with the increasing mass of new facts which the busy workers in this field are daily bringing forth.

The subject of the valency of the elements is one which, since the year above referred to, has given chemists much food for discussion, as well as opportunity for experimental work. But whether we range ourselves with Kekulé, who supports the unalterable character of the valency of each element, or with Frankland, who insists on its variability, it is now clear to most chemists that the hard and fast lines upon which this theory was supposed to stand cannot be held to be secure. For if the progress of investigation has shown that it is impossible in many instances to affix one valency to an element which forms a large number of different compounds, it is also equally impossible to look on the opposite view as tending towards progress, inasmuch as to ascribe to an element as many valencies as it possesses compounds with some other element, is only expressing by circuitous methods what the old Daltonian law of combination in multiple proportion states in simple terms. Still we may note certain generally-accepted conclusions: in the first place, that of the existence of non-saturated compounds both inorganic and organic, as carbon-monoxide on the one hand, and malic and citraconic acids on the other. Secondly, that the valency of an element is not only dependent upon the nature of the element with which it combines, but that this valency is a periodic function of the atomic weight of the other component. Thus the elements of the chlorine group are always monads when combined with positive elements or radicals; but triad, pentad, and heptad with negative ones. Again, the elements of the sulphur group are dyads in the first case, but tetrad and hexad in the second. The periodicity of this property of the atoms, increasing and again diminishing, is clearly seen in such a series as

$$AgCl_1, CdCl_2, InCl_3, SnCl_4, SbH_3, TeH_2, IH,$$

as well as in the series of oxides. The difficulties which beset this subject may be judged of by the mention of a case or two :— Is vanadium a tetrad because its highest chloride contains four atoms of chlorine? What are we to say is the valency of lead when one atom unites with four methyls to form a volatile product, and yet the vapour-density of the chloride shows that the molecule contains one of metal to two of chlorine? Or, how can our method be said to determine the valency of tungsten when the hexachloride decomposes in the state of vapour, and the pentachloride is the highest volatile stable compound? How again are we to define the point at which a body is volatile without decomposition?—thus sulphur tetrachloride, one of the most unstable of compounds, can be vaporised without decomposition at all temperatures below −22°, whilst water, one of the most stable of known compounds, is dissociated into its elements at the temperature of melting platinum.

But, however many doubts may have been raised in special instances against a thorough application of the law of valency, it cannot be denied that the general relations of the elements which this question of valency has been the means of bringing to light are of the highest importance, and point to the existence of laws of Nature of the widest significance; I allude to the periodic law of the elements first foreshadowed by Newlands, but fully developed by Mendeléeff and Lothar Meyer. Guided by the principle that the chemical properties of the elements are a periodic function of their atomic weights, or that matter becomes endowed with analogous properties when the atomic weight of an element is increased by the same or nearly the same number, we find ourselves for the first time in possession of a key which enables us to arrange the hitherto *disjecta membra* of our chemical household in something like order, and thus gives us means of indicating the family resemblances by which these elements are characterised.

And here we may congratulate ourselves on the fact that, by the recent experiments of Brauner, and of Nilson and Pettersen respectively, tellurium and beryllium, two of the hitherto outstanding members, have been induced to join the ranks, so that at the present time osmium is the only important defaulter amongst the sixty-four elements, and few persons will doubt that a little careful attention to this case will remove the stigma which yet attaches to its name. But this periodic law makes it possible for us to do more; for as the astronomer, by the perturbations of known planets, can predict the existence of hitherto unknown ones, so the chemist, though, of course, with much less satisfactory means, has been able to predict with precision the properties, physical and chemical, of certain missing links amongst the elements, such as ekaluminium and ekaboron, then unborn, but which shortly afterwards became well known to us in the flesh as gallium and scandium. We must, however, take care that success in a few cases does not blind us to the fact that the law of Nature which expresses the relation between the properties of the elements and the value of the atomic weights is as yet unknown; that many of the groupings are not due to any well-ascertained analogy of properties of the elements, and that it is only because the values of their atomic weights exhibit certain regularities that such a grouping is rendered possible. So, to quote Lothar Meyer, we shall do well in this, as indeed in all similar cases in science, to remember the danger pointed out in Bacon's aphorism, that "The mind delights in springing up to the most general axioms, that it may find rest, but after a short stay here it disdains experience," and to bear in mind that it is only the lawful union of hypothesis with experiment which will prove a fruitful one in the establishment of a systematic inorganic chemistry which need not fear comparison with the order which reigns in the organic branch of our science. And here it is well to be reminded that complexity of constitution is not the sole prerogative of the carbon compounds, and that before this systematisation of inorganic chemistry can be effected we shall have to come to terms with many compounds concerning whose constitution we are at present wholly in ignorance. As instances of such I would refer to the finely crystalline phospho-molybdates, containing several hundred atoms in the molecule, lately prepared by Wolcott Gibbs.

Arising out of Kekulé's theory of the tetrad nature of the carbon atom, came the questions which have caused much debate among chemists : (1) Are the four combining units of the carbon atom of equal value or not ? and (2) Is the assumption of a dyad carbon atom in the so-called non-saturated compounds justifiable or not ? The answer to the first of these, a favourite view of Kolbe's, is given in the now well-ascertained laws of isomerism ; and from the year 1862, when Schorlemmer proved the identity of the hydrides of the alcohol radicals with the so-called radicals themselves, this question may be said to have been set at rest ; for Lossen himself admits that the existence of his singular isomeric hydroxylamine derivatives can be explained otherwise than by the assumption of a difference between each of the combining units of nitrogen, and the differences supposed by Schreiner to hold good between the methyl-ethyl carbonic ethers have been shown to have no existence in fact. With respect to the second point the reply is no less definite, and is recorded in the fact, amongst others, that ethylene chlorhydrin yields on oxidation chloracetic acid, a reaction which cannot be explained on the hypothesis of the existence in ethylene of a dyad carbon atom.

Passing from this subject, we arrive, by a process of natural selection, at more complicated cases of chemical orientation—that is, given certain compounds which possess the same composition and molecular formulæ but varying properties, to find the difference in molecular structure by which such variation of properties is determined. Problems of this nature can now be satisfactorily solved, the number of possible isomers foretold, and this prediction confirmed by experiment. The general method adopted in such an experimental inquiry into the molecular arrangement or chemical constitution of a given compound is either to build up the structure from less complicated ones of known constitution, or to resolve it into such component parts. Thus, for example, if we wish to discriminate between several isomeric alcohols, distinguishing the ordinary or primary class from the secondary or tertiary class, the existence of which was predicted by Kolbe in 1862, and of which the first member was prepared by Friedel in 1864, we have to study their products of oxidation. If one yields an acid having the same number of carbon atoms as the alcohol, it belongs to the first class and possesses a definite molecular structure ; if it splits up into two distinct carbon compounds, it is a secondary alcohol ; and if three carbon compounds result from its oxidation, it must be classed in the third category, and to it belongs a definite molecular structure, different from that of the other two.

In a similar way orientation in the much more complicated aromatic hydrocarbons can be effected. This class of bodies forms the nucleus of an enormous number of carbon compounds which, both from a theoretical and a practical point of view, are of the highest interest. For these bodies exhibit characters and possess a constitution totally different from those of the so-called fatty substances, the carbon atoms being linked together more intimately than is the case in the latter-named group of bodies. Amongst them are found all the artificial colouring matters, and some of the most valuable pharmaceutical and therapeutical agents.

The discovery of the aniline colours by Perkin, their elaboration by Hofmann, the synthesis of alizarin by Graebe and Liebermann, being the first vegetable colouring matter which has been artificially obtained, the artificial production of indigo by Baeyer, and lastly the preparation, by Fischer, of kairin, a febrifuge as potent as quinine, are some of the well-known recent triumphs of modern synthetical chemistry. And these triumphs, let us remember, have not been obtained by any such "random haphazarding" as yielded results in Priestley's time. In the virgin soil of a century ago, the ground only required to be scratched and the seed thrown in to yield a fruitful crop ; now the surface soil has long been exhausted, and the successful cultivator can only obtain results by a deep and thorough preparation, and by a systematic and scientific treatment of his material.

In no department of our science has the progress made been more important than in that concerned with the accurate determination of the numerical, physical, and chemical constants upon the exactitude of which every quantitative chemical operation depends. For the foundation of an accurate knowledge of the first of these constants, viz., the atomic weights of the elements, science is indebted to the indefatigable labours of Berzelius. But "humanum est errare," and even Berzelius's accurate hand and delicate conscientiousness did not preserve him from mistakes, since corrected by other workers. In such determinations it it difficult, if not impossible, always to ascertain the limits of error attaching to the number. The errors may be due in the firss place to manipulative faults, in the second to inaccuracy of the methods, or lastly to mistaken views as to the composition of the material operated upon ; and hence the uniformity of any series of similar determinations gives no guarantee of their truth, the only safe guide being the agreement of determinations made by altogether different methods. The work commenced by Berzelius has been worthily continued by many chemists. Stas and Marignac, bringing work of an almost astronomical accuracy into our science, have ascertained the atomic weights of silver and iodine to within one hundred-thousandth of their value, whilst the numbers for chlorine, bromine, potassium, sodium, nitrogen, sulphur, and oxygen may now be considered correct to within a unit in the fourth figure. Few of the elements, however, boast numbers approaching this degree of accuracy, and many may even still be erroneous from half to a whole unit of hydrogen. And, as Lothar Meyer says, until the greater number of the atomic weights are determined to within one or two tenths of the unit, we cannot expect to be able to ascertain the laws which certainly govern these numbers, or to recognise the relations which undoubtedly exist between them and the general chemical and physical properties of the elements. Amongst the most interesting recent additions to our knowledge made in this department we may note the classical experiments, in 1880, of J. W. Mallet on aluminium, and in the same year of J. P. Cooke on antimony, and those, in the present year, of Thorpe on titanium.

Since the date of Berzelius's death to the present day, no discovery in our science has been so far-reaching, or led to such unforeseen and remarkable conclusions, as the foundation of Spectrum Analysis by Bunsen and Kirchhoff in 1860.

Independently altogether of the knowledge which has been gained concerning the distribution of the elementary bodies in terrestrial matter, and of the discovery of half a dozen new elements by its means, and putting aside for a moment the revelation of a chemistry not bounded by this world, but limitless as the heavens, we find that over and above all these results spectrum analysis offers the means, not otherwise open to us, of obtaining knowledge concerning the atomic and molecular condition of matter.

Let me recall some of the more remarkable conclusions to which the researches of Lockyer, Schuster, Liveing and Dewar, Wüllner, and others in this direction have led. In the first place it is well to bear in mind that a difference of a very marked kind, first distinctly pointed out by Alex. Mitscherlich, is to be observed between the spectrum of an element and that of its compounds, the latter only being seen in cases in which the compound is not dissociated at temperatures necessary to give rise to a glowing gas. Secondly, that these compound spectra—as, for instance, those of the halogen compounds of the alkaline-earth metals—exhibit a certain family likeness, and show signs of systematic variation in the position of the lines, corresponding to changes in the molecular weight of the vibrating system. Still this important subject of the relation of the spectra of different elements is far from being placed on a satisfactory basis, and in spite of the researches of Lecoq de Boisbaudran, Ditte, Troost and Hautefeuille, Ciamician, and others, it cannot be said that as yet definite proof has been given in support of the theory that a causal connection is to be found between the emission spectra of the several elements belonging to allied groups and their atomic weights or other chemical or physical properties. In certain of the single elements, however, the connection between the spectra and the molecular constitution can be traced. In the case of sulphur, for example, three distinct spectra are known. The first of these, a continuous one, is exhibited at temperatures below 500°, when, as we know from Dumas' experiments, the density of the vapour is three times the normal, showing that at this temperature the molecule consists of six atoms. The second spectrum is seen when the temperature is raised to above 1000°, when, as Deville and Troost have shown, the vapour reaches its normal density, and the molecule of sulphur, as with most other gases, contains two atoms, and this is a band spectrum, or one characterised by channelled spaces. Together with this band spectrum, and especially round the negative pole, a spectrum of bright lines is observed. This latter is doubtless due to the vibrations of the single atoms of the dissociated molecule, the existence of traces of a band spectrum demonstrating the fact that in some parts of the discharge the tension of dissociation is insufficient to prevent the reunion of the atoms to form the molecule.

To this instance of the light thrown on molecular relations by changes in the spectra, others may be added. Thus the low-temperature spectrum of channelled spaces, mapped by Schuster and myself, in the case of potassium, corresponds to the molecule of two atoms and to the vapour-density of seventy-nine, as observed by Dewar and Dittmar. Again, both oxygen and nitrogen exhibit two, if not three, distinct spectra : of these the line spectrum seen at the highest temperatures corresponds to the atom ; the band spectrum seen at intermediate temperatures represents the molecule of two atoms ; whilst that observed at a still lower point would, as in the case of sulphur, indicate the existence of a more complicated molecule, known to us in one instance as ozone.

That this explanation of the cause of these different spectra of an element is the true one, can be verified in a remarkable way. Contrary to the general rule amongst those elements which can readily be volatilised, and with which, therefore, low-temperature spectra can be studied, mercury exhibits but one spectrum, and that one of bright lines, or, according to the preceding theory, a spectrum of atoms. So that, judging from spectroscopic evidence, we infer that the atoms of mercury do not unite to form a molecule, and we should predict that the vapour-density of mercury is only half its atomic weight. Such we know, from chemical evidence, is really the case, the molecule of mercury being identical in weight with its atom.

The cases of cadmium and iodine require further elucidation. The molecule of gaseous cadmium, like that of mercury, consists of one atom ; probably, therefore, the cadmium spectrum is also distinguished by one set of lines. Again, the molecule of iodine at 1200° separates, as we know from Victor Meyer's researches, into single atoms. Here spectrum analysis may come again to our aid ; but, as Schuster remarks, in his report on the spectra of the non-metallic elements, a more extensive series of experiments than those already made by Ciamician is required before any definite opinion as to the connection of the different iodine spectra with the molecular condition of the gas can be expressed.

It is not to be wondered at that these relations are only exhibited in the case of a few elements. For most of the metals the vapour-density remains, and probably will remain, an unknown quantity, and therefore the connection between any observed changes in the spectra and the molecular weights must also remain unknown. The remarkable changes which the emission spectrum of a single element—iron, for instance—exhibits have been the subject of much discussion, experimental and otherwise. Of these, the phenomenon of long and short lines is one of the most striking, and the explanation that the long lines are those of low temperature appears to meet the fact satisfactorily, although the effect of dilution, that is, a reduction of the quantity of material undergoing volatilisation, is, remarkably enough, the same as that of diminution of temperature. Thus it is possible, by the examination of a spectrum by Lockyer's method, to predict the changes which it will undergo, either on alteration of temperature, or by an increase or decrease of quantity. There appears to be no theoretical difficulty in assuming that the relative intensity of the lines may vary when the temperature is altered, and the molecular theory of gases furnishes us with a plausible explanation of the corresponding change when the relative quantities of the luminous elements in a mixture are altered. Lockyer has proposed a different explanation of the facts. According to him, every change of relative intensity means a corresponding change of molecular complexity, and the lines which we see strong near the poles would bear the same relation to those which are visible throughout the field, as a line spectrum bears to a band spectrum ; but then almost every line must be due to a different molecular grouping, a conclusion which is scarcely capable of being upheld without very cogent proof.

The examination of the absorption-spectra of salts, saline and organic liquids, first by Gladstone, and afterwards by Bunsen, and by Russell, as well as by Hartley for the ultra-violet, and by Abney and Festing for the infra-red region, have led to interesting results in relation to molecular chemistry. Thus Hartley finds that, in some of the more complicated aromatic compounds, definite absorption-bands in the more refrangible region are only produced by substances in which three pairs of carbon atoms are doubly linked, as in the benzene ring, and thus the means of ascertaining this double linkage is given. The most remarkable results obtained by Abney and Festing show that the radical of an organic body is always represented by certain well-marked absorption-bands, differing, however, in position, according as it is linked with hydrogen, a halogen, or with carbon, oxygen, or nitrogen. Indeed, these experimenters go so far as to say that it is highly probable that by this delicate mode of analysis the hypothetical position of any hydrogen which is replaced may be identified, thus pointing out a method of physical orientation of which, if confirmed by other observers, chemists will not be slow to avail themselves. This result, it is interesting to learn, has been rendered more than probable by the recent important researches of Perkin on the connection between the constitution and the optical properties of chemical compound.

One of the noteworthy features of chemical progress is the interest taken by physicists in fundamental questions of our science. We all remember, in the first place, Sir William Thomson's interesting speculations, founded upon physical phenomena, respecting the probable size of the atom, viz. " that if a drop of water were magnified to the size of the earth, the constituent atoms would be larger than small shot, but smaller than cricket balls." Again, Helmholtz, in the Faraday Lecture, delivered in 1881, discusses the relation of electricity and chemical energy, and points out that Faraday's law of electrolysis, and the modern theory of valency, are both expressions of the fact that, when the same quantity of electricity passes through an electrolyte, it always either sets free, or transfers to other combinations, the same number of units of affinity at both electrodes. Helmholtz further argues that, if we accept the Daltonian atomic hypothesis, we cannot avoid the conclusion that electricity, both positive and negative, is divided into elementary portions which behave like atoms of electricity. He also shows that these charges of atomic electricity are enormously large as compared, for example, with the attraction of gravitation between the same atoms ; in the case of oxygen and hydrogen, 71,000 billion times larger.

A further subject of interest to chemists is the theory of the vortex-ring constitution of matter thrown out by Sir William Thomson, and lately worked out from a chemical point of view by J. J. Thomson, of Cambridge. He finds that if one such ring be supposed to constitute the most simple form of matter, say the monad hydrogen atom, then two such rings must, on coming into contact with nearly the same velocity, remain enchained together, constituting what we know as the molecule of free hydrogen. So, in like manner, systems containing two, three, and four such rings constitute the dyad, triad, and tetrad atoms. How far this mathematical expression of chemical theory may prove consistent with fact remains to be seen.

Another branch of our science which has recently attracted much experimental attention is that of thermo-chemistry, a subject upon which in the future the foundation of dynamical chemistry must rest, and one which already proclaims the truth of the great principle of the conservation of energy in all cases of chemical as well as of physical change. But here, although the materials hitherto collected are of very considerable amount and value, the time has not yet arrived for expressing these results in general terms, and we must, therefore, be content to note progress in special lines and wait for the expansion into wider areas. Reference may, however, be properly made to one interesting observation of general significance. It is well known that, while, in most instances, the act of combination is accompanied by evolution of heat—that is, whilst the potential energy of most combining bodies is greater than that of most compounds—cases occur in which the reverse of this is true, and heat is absorbed in combination. In such cases the compound readily undergoes decomposition, frequently suddenly and with explosion. Acety-lene and cyanogen seem to be exceptions to this rule, inasmuch as, whilst their component elements require to have energy added to them in order to enable them to combine, the compounds appear to be very stable bodies. Berthelot has explained this enigma by showing that, just as we may ignite a mass of dynamite without danger, whilst explosion takes place if we agitate the molecules by a detonator, so acetylene and cyanogen burn, as we know, quietly when ignited, but when their molecules are shaken by the detonation of even a minute quantity of fulminate, the constituents fly apart with explosive violence, carbon and hydrogen, or carbon and nitrogen, being set free, and the quantity of heat absorbed in the act of combination being suddenly liberated.

In conclusion, whilst far from proposing even to mention all the important steps by which our science has advanced since the year 1848, I cannot refrain from referring to two more. In the first place, to that discovery, more than foreshadowed by Faraday, of the liquefaction of the so-called permanent gases by Pictet and Cailletet ; and secondly, to that of the laws of dissociation as investigated by Deville. The former, including Andrews's discovery of the critical point, indicates a connection, long unseen, between the liquid and the gaseous states of matter ; the latter has opened out entirely fresh fields for research, and has given us new views concerning the stability of chemical compounds of great importance and interest.

Turning for a moment to another topic, we feel that, although science knows no nationalities, it is impossible for those who, like ourselves, exhibit strong national traits, to avoid asking whether we Anglo-Saxons hold our own, as compared with other nations, in the part we have played and are playing in the development of our science. With regard to the past, the names of Boyle, Cavendish, Priestley, Dalton, Black, Davy, are sufficient guarantees that the English have, to say the least, occupied a position second to none in the early annals of chemistry. How has it been in the era which I have attempted to describe? What is the present position of English chemistry, and what its look-out for the future? In endeavouring to make this estimate, I would take the widest ground, including not only the efforts made to extend the boundaries of our science by new discovery, both in the theoretical and applied branches, but also those which have the no less important aims of spreading the knowledge of the subject amongst the people, and of establishing industries dependent on chemical principles by which the human race is benefited. Taking this wide view, I think we may, without hesitation, affirm that the progress which chemistry has made through the energies of the Anglo-Saxon race is not less than that made by any other nation.

In so far as pure science is concerned, I have already given evidence of the not inconsiderable part which English chemists have played in the progress since 1848. We must, however, acknowledge that the number of original chemical papers now published in our language is much smaller than that appearing in the German tongue, and that the activity and devotion displayed in this direction by the heads of German laboratories may well be laid to heart by some of us in England ; yet, on the other hand, it must be remembered that the circumstances of different countries are so different that it is by no means clear that we should follow the same lines. Indeed our national characteristics forbid us to do so, and it may be that the bent of the Germanic lies in the assiduous collection of facts, whilst their subsequent elaboration and connection is the natural work of our own race.

As regards the publication of so-called original work by students, and speaking now only for myself as the director of an English chemical laboratory, I feel I am doing the best for the young men who, wishing to become either scientific or industrial chemists, are placed under my charge, in giving them as sound and extensive a foundation in the theory and practice of chemical science as their time and abilities will allow, rather than forcing them prematurely into the preparation of a new series of homologous compounds or the investigation of some special reaction, or of some possible new colouring matter, though such work might doubtless lead to publication. My aim has been to prepare a young man, by a careful and fairly complete general training, to fill with intelligence and success a post either as teacher or industrial chemist, rather than to turn out mere specialists, who, placed under other conditions than those to which they have been accustomed, are unable to get out of the narrow groove in which they have been trained. And this seems a reasonable course, for whilst the market for the pure specialist, as the colour chemist for example, may easily be overstocked, the man of all-round intelligence will always find opportunity for the exercise of his powers. Far, however, from underrating the educational advantages of working at original subjects, I consider this sort of training to be of the highest and best kind, but only useful when founded upon a sound and general basis.

The difficulty which the English teacher of chemistry—and in this I may include Canada and the United States—has to contend against is that, whilst in Germany the value of this high and thorough training is generally admitted, in England a belief in its efficacy is as yet not generally entertained. "The Englishman," to quote from the recent Report of the Royal Commission on Technical Instruction, "is accustomed to seek for an immediate return, and has yet to learn that an extended and systematic education, up to and including the methods of original research, is now a necessary preliminary to the fullest development of industry, and it is to the gradual but sure growth of public opinion in this direction that we must look for the means of securing to this country in the future, as in the past, the highest position as an industrial nation."

If, in the second place, we consider the influence which Englishmen have exerted on the teaching of our science, we shall feel reason for satisfaction ; many of our text-books are translated into every European language and largely used abroad ; often to the exclusion of those written by Continental chemists.

Again, science teaching, both practical and theoretical, in our elementary and many secondary schools, is certainly not inferior to that in schools of similar grade abroad, and the interest in and desire for scientific training is rapidly spreading throughout our working population, and is even now as great as, if not greater than, abroad. The universities and higher colleges are also moving to take their share of the work which has hitherto been far less completely done in our country than on the continent of Europe, especially in Germany, where the healthful spirit of competition, fostered by the numerous State-supported institutions, is much more common than with us, and, being of equal value in educational as in professional or commercial matters, has had its due effect.

Turning lastly to the practical applications of our science, in what department does England not excel? and in which has she not made the most important new departures? Even in colour chemistry, concerning which we have heard, with truth, much of German supremacy, we must remember that the industry is originally an English one, as the names of Perkin and of Maule, Simpson and Nicholson, testify ; and if we have hitherto been beaten hollow in the development of this branch, signs are not wanting that this may not always be the case. But take any other branch of applied chemistry, the alkali trade for instance, what names but English, with the two great exceptions of Leblanc and Solvay, do we find in connection with real discoveries? In the application of chemistry to metallurgical processes, too, the names of Darby, Cort, Neilson, and Bell in iron, of Bessemer, Thomas, Gilchrist, and Snelus in steel, of Elkington and Matthey in the noble metals, show that in these branches the discoveries which have revolutionised processes have been made by Englishmen ; whilst Young, the father of paraffin, Spence the alum-maker, and Abel of gun-cotton fame, are some amongst many of our countrymen whose names may be honourably mentioned as having founded new chemical industries.

Hence, whilst there is much to stimulate us to action in the energy and zeal shown by our Continental brethren in the pursuit both of pure and applied chemistry, there is nothing to lead us to think that that the chemistry of the English-speaking nations in the next fifty years will be less worthy than that of the past half-century of standing side by side with that of her friendly rivals elsewhere.

Section IV

THE ARRANGEMENT OF THE ELEMENTS

Lavoisier in his *Elements* proposed the definition of 'element' (a substance which cannot be further analysed) which remained in use until the concept of atomic number was introduced by Moseley in the twentieth century. Lavoisier's classification of elements was based upon purely chemical data; it is interesting to see how he refused to put certain unanalysed substances, notably soda and potash, on the list because by analogy he felt certain that they would be decomposed by some new technique.

Thomas Young was a man of very varied talents; he is probably best known for his revival of the wave theory of light in the first years of the nineteenth century. He made many other contributions to physics; and began the decipherment of the Rosetta Stone. This paper from the *Philosophical Transactions* of the Royal Society, the oldest English scientific journal and certainly at this period the most prestigious, represents one of the last attempts to quantify affinities in the Newtonian manner, instead of by weights.

Ampère was one of the greatest scientists of the early nineteenth century, chiefly remembered for his electrical work; particularly his theory of electro-magnetism propounded in 1820 to account for Oersted's experiment. He appealed for a natural classification of the elements; and in 1817 Thomas Thomson in his annual survey of the progress of chemistry, which forms a useful part of the *Annals of Philosophy*, set out the details of the classification; which, however, never caught on.

J. P. Cooke's tables have already been discussed in the General Introduction; he also wrote on natural theology. His classification was referred to by Sir John Herschel in an address to the British Association in 1858; it represents an attempt to unite atomic weights and chemical properties, but the algebraic series were not to prove the best way to go, and Cannizzaro's atomic weights were needed.

De Chancourtois' paper of 1862 made no stir until it was translated in 1889, when it seemed to contain anticipations of Mendeléeff. But it is hard to take very seriously these Pythagorean numerological speculations; as Crookes remarked, 'no sufficient evidence (exists) that the author disentangled . . . (solid) matter from accompanying speculations'.

Newlands' Law of Octaves must be taken more seriously; and his Tables clearly do have some of the principles upon which Mendeléeff's Table was based. But the gaps he left for undiscovered elements were sporadic, and in his tables some closely related elements are separated, some which are not chemically similar are put in the same family, and certain elements are made to share a place. Newlands was an industrial chemist; he had been a pupil of A. W. Hofmann at the Royal College of Chemistry in London.

Odling was an academic; he had been a delegate at the Karlsruhe Conference of 1860, and had in 1855 translated the *Chemical Method* of the unfortunate Auguste Laurent, who was, with Gerhardt, one of the most important theoretical chemists of the mid-century. Odling's Table is very similar to Mendeléeff's; but it was published in a relatively unimportant journal, Odling seems to have taken little interest in it, and one can only conclude that he did not perceive the importance of what he had done, and lacked Newlands' self-assertiveness.

Mendeléeff's Table was made known through the brief German abstract; and ten years later, when its value had become evident, Mendeléeff's long paper was published in *Chemical News*. In 1866, at the age of thirty-two, he had become Professor of General Chemistry at St Petersburg. His Table gave rise to Proutian and evolutionary speculations, but he insisted upon its empirical basis, notably in his Faraday Lecture of 1889; but near the end of his life he published a strange and speculative work, *A Chemical Conception of the Ether*, reminding us of the evolutionary theory of T. S. Hunt, referred to in section V.

Berthelot was a great chemist, who founded organic chemistry firmly upon syntheses and made thermochemistry really scientific. He was also one of the greatest historians of alchemy, and edited alchemical texts. A summary in English of the passage reprinted from his book on the Origins of Alchemy criticising the Proutians, is provided.

Emerson Reynolds' paper with its zigzag arrangement of the elements influenced Crookes, who referred to it in his paper of 1891, reproduced in section II of this volume. Crookes himself evolved a figure of eight form of the Table, which Ramsay toyed with in his uncertainties over where to put argon in the Table, and what its congeners might be like. Crookes' model of this version of the Table is in the Science Museum, South Kensington. Ramsay was one of the greatest chemists of the day; he had in 1887 become Professor of Chemistry at University College, London. As well as discovering the whole family of inert gases, he speculated on the role of the electron in chemical bonding.

FURTHER READING

M. P. Crosland, *Historical Studies in the Language of Chemistry*, London, 1962.

F. Greenaway, 'A pattern of chemistry: Hundred years of the Periodic Table', *Chemistry in Britain*, V (1969), 97–9.

A. Ihde, *The Development of Modern Chemistry*, London, 1965.

R. P. Multhauf, *The Origins of Chemistry*, London, 1967.

J. W. van Spronsen, *The Periodic System of Chemical Elements: A History of the First Hundred Years*, Amsterdam, 1969.

TABLE OF SIMPLE SUBSTANCES.

Simple fubftances belonging to all the kingdoms of nature, which may be confidered as the elements of bodies.

New Names.	*Correfpondent old Names.*
Light - - -	Light.
Caloric - - -	Heat. Principle or element of heat. Fire. Igneous fluid. Matter of fire and of heat.
Oxygen - - -	Dephlogifticated air. Empyreal air. Vital air, or Bafe of vital air.
Azote - - -	Phlogifticated air or gas. Mephitis, or its bafe.
Hydrogen - -	Inflammable air or gas, or the bafe of inflammable air.

Oxydable and Acidifiable fimple Subftances not Metallic.

New Names.	*Correfpondent old names.*
Sulphur - - -	
Phofphorus - - -	The fame names.
Charcoal - - -	
Muriatic radical -	
Fluoric radical - -	Still unknown.
Boracic radical - -	

Oxydable and Acidifiable fimple Metallic Bodies.

New Names.		*Correfpondent Old Names.*
Antimony -		Antimony.
Arfenic - -		Arfenic.
Bifmuth - -		Bifmuth.
Cobalt - -		Cobalt.
Copper - -		Copper.
Gold - -		Gold.
Iron - - -		Iron.
Lead - - -		Lead.
Manganefe - -	Regulus of	Manganefe.
Mercury - -		Mercury.
Molybdena - -		Molybdena.
Nickel - - -		Nickel.
Platina - -		Platina.
Silver - -		Silver.
Tin - -		Tin.
Tungftein - -		Tungftein.
Zinc - -		Zinc.

Salifiable

Salifiable fimple Earthy Subftances.

New Names.	Correspondent old Names.
Lime	{ Chalk, calcareous earth. { Quicklime.
Magnefia	{ Magnefia, bafe of Epfom falt. { Calcined or cauftic magnefia.
Barytes	Barytes, or heavy earth.
Argill	Clay, earth of alum.
Silex	Siliceous or vitrifiable earth.

SECT. I.—*Obfervations upon the Table of Simple Subftances.*

The principle object of chemical experiments is to decompofe natural bodies, fo as feparately to examine the different fubftances which enter into their compofition.   By confulting chemical fyftems, it will be found that this fcience of chemical analyfis has made rapid progrefs in our own times.   Formerly oil and falt were confidered as elements of bodies, whereas later obfervation and experiment have fhown that all falts, inftead of being fimple, are compofed of an acid united to a bafe.   The bounds of analyfis have been greatly enlarged by modern difcoveries * ; the acids are fhown to be compofed of oxygen, as an acidifying principle common to all, united in each to a particular bafe.   I have proved what Mr Haffenfratz had

before

* See Memoirs of the Academy for 1776, p. 671. and for 1778, p. 535.—A.

before advanced, that thefe radicals of the acids are not all fimple elements, many of them being, like the oily principle, compofed of hydrogen and charcoal.   Even the bafes of neutral falts have been proved by Mr Berthollet to be compounds, as he has fhown that ammoniac is compofed of azote and hydrogen.

Thus, as chemiftry advances towards perfection, by dividing and fubdividing, it is impoffible to fay where it is to end; and thefe things we at prefent fuppofe fimple may foon be found quite otherwife.   All we dare venture to affirm of any fubftance is, that it muft be confidered as fimple in the prefent ftate of our knowledge, and fo far as chemical analyfis has hitherto been able to fhow. We may even prefume that the earths muft foon ceafe to be confidered as fimple bodies; they are the only bodies of the falifiable clafs which have no tendency to unite with oxygen; and I am much inclined to believe that this proceeds from their being already faturated with that element.   If fo, they will fall to be confidered as compounds confifting of fimple fubftances, perhaps metallic, oxydated to a certain degree.   This is only hazarded as a conjecture; and I truft the reader will take care not to confound what I have related as truths, fixed on the firm bafis of obfervation and experiment, with mere hypothetical conjectures.

Z   The

The fixed alkalies, potaſh, and ſoda, are o-
mitted in the foregoing Table, becauſe they
are evidently compound ſubſtances, though we
are ignorant as yet what are the elements they
are compoſed of.

VI. *A numerical Table of elective Attractions; with Remarks on the Sequences of double Decompositions.* *By* Thomas Young, *M. D. For. Sec. R. S.*

Read February 9, 1809.

ATTEMPTS have been made, by several chemists, to obtain a series of numbers, capable of representing the mutual attractive forces of the component parts of different salts; but these attempts have hitherto been confined within narrow limits, and have indeed been so hastily abandoned, that some very important consequences, which necessarily follow from the general principle of a numerical representation, appear to have been entirely overlooked. It is not impossible, that there may be some cases, in which the presence of a fourth substance, besides the two ingredients of the salt, and the medium in which they are dissolved, may influence the precise force of their mutual attraction, either by affecting the solubility of the salt, or by some other unknown means, so that the number, naturally appropriate to the combination, may no longer correspond to its affections; but there is reason to think that such cases are rare; and when they occur, they may easily be noticed as exceptions to the general rules. It appears therefore, that nearly all the phenomena of the mutual actions of a hundred different salts may be correctly represented by a hundred numbers, while, in the usual manner of relating every case as a different experiment, above two thousand separate articles would be required.

Having been engaged in the collection of a few of the principal facts relating to chemistry and pharmacy, I was induced to attempt the investigation of a series of these numbers; and I have succeeded, not without some difficulty, in obtaining such as appear to agree sufficiently well with all the cases of double decompositions which are fully established, the exceptions not exceeding twenty, out of about twelve hundred cases enumerated by FOURCROY. The same numbers agree in general with the order of simple elective attractions, as usually laid down by chemical authors; but it was of so much less importance to accommodate them to these, that I have not been very solicitous to avoid a few inconsistencies in this respect, especially as many of the bases of the calculation remain uncertain, and as the common tables of simple elective attractions are certainly imperfect, if they are considered as indicating the order of the independent attractive forces of the substances concerned. Although it cannot be expected that these numbers should be accurate measures of the forces which they represent, yet they may be supposed to be tolerable approximations to such measures, at least if any two of them are nearly in the true proportion, it is probable that the rest cannot deviate very far from it: thus, if the attractive force of the phosphoric acid for potash is about eight tenths of that of the sulfuric acid for barita, that of the phosphoric acid for barita must be about nine tenths as great; but they are calculated only to agree with a certain number of phenomena, and will probably require many alterations, as well as additions, when all other similar phenomena shall have been accurately investigated.

There is, however, a method of representing the facts, which have served as the bases of the determination, independently of any hypothesis, and without being liable to the contingent necessity of any future alteration, in order to make room for the introduction of the affections of other substances; and this method enables us also to compare, upon general principles, a multitude of scattered phenomena, and to reject many which have been mentioned as probable, though doubtful, with the omission of a very few only which have been stated as ascertained. This arrangement simply depends on the supposition, that the attractive force, which tends to unite any two substances, may always be represented by a certain constant quantity.

From this principle it may be inferred, in the first place, that there must be a sequence in the simple elective attractions. For example, there must be an error in the common tables of elective attractions, in which magnesia stands above ammonia under the sulfuric acid, and below it under the phosphoric, and the phosphoric acid stands above the sulfuric under magnesia, and below it under ammonia, since such an arrangement implies, that the order of the attractive forces is this; phosphate of magnesia, sulfate of magnesia, sulfate of ammonia, phosphate of ammonia, and again phosphate of magnesia; which forms a circle, and not a sequence. We must therefore either place magnesia above ammonia under the phosphoric acid, or the phosphoric acid below the sulfuric under magnesia; or we must abandon the principle of a numerical representation in this particular case.

In the second place, there must be an agreement between the simple and double elective attractions. Thus, if the fluoric acid stands above the nitric under barita, and below it under lime, the fluate of barita cannot decompose the nitrate of lime, since the previous attractions of these two salts are respectively greater, than the divellent attractions of the nitrate of barita and the fluate of lime. Probably, therefore, we ought to place the fluoric acid below the nitric under barita; and we may suppose, that when the fluoric acid has appeared to form a precipitate with the nitrate of barita, there has been some fallacy in the experiment.

The third proposition is somewhat less obvious, but perhaps of greater utility: there must be a continued sequence in the order of double elective attractions; that is, between any two acids, we may place the different bases in such an order, that any two salts, resulting from their union, shall always decompose each other, unless each acid be united to the base nearest to it: for example, sulfuric acid, barita, potass, soda, ammonia, strontia, magnesia, glycina, alumina, zirconia, lime, phosphoric acid. The sulfate of potass decomposes the phosphate of barita, because the difference of the attractions of barita for the sulfuric and phosphoric acids is greater than the difference of the similar attractions of potass; and in the same manner the difference of the attractions of potass is greater than that of the attractions of soda; consequently the difference of the attractions of barita must be much greater than that of the attractions of soda, and the sulfate of soda must decompose the phosphate of barita: and in the same manner it may be shown, that each base must preserve its relations of priority or posteriority to every other in the series. It is also obvious that, for similar reasons, the acids may be arranged in a continued sequence between the different bases; and when all the decompositions of a certain number of salts have been investigated, we may form two corresponding tables, one of the sequences of the bases with the acids, and another of those of the acids with the different bases; and if either or both of the tables are imperfect, their deficiencies may often be supplied, and their errors corrected, by a repeated comparison with each other.

In forming tables of this kind from the cases collected by Fourcroy, I have been obliged to reject some facts, which were evidently contradictory to others, and these I have not thought it necessary to mention; a few, which are positively related, and which are only inconsistent with the principle of numerical representation, I have mentioned in notes; but many others, which have been stated as merely probable, I have omitted without any notice. In the table of simple elective attractions, I have retained the usual order of the different substances; inserting again in parentheses such of them as require to be transposed, in order to avoid inconsequences in the simple attractions: I have attached to each combination marked with an asterisc the number deduced from the double decompositions, as expressive of its attractive force; and where the number is inconsistent with the corrected order of the simple elective attractions, I have also inclosed it in a parenthesis. Such an apparent inconsistency may perhaps in some cases be unavoidable, as it is possible that the different proportions of the masses concerned, in the operations of simple and compound decomposition, may sometimes cause a real difference in the comparative magnitude of the attractive forces. Those numbers, to which no asterisc is affixed, are merely inserted by interpolation, and they can only be so far employed for determining the mutual actions of the salts to which they belong, as the results which they indicate would follow from the comparison of any other numbers, intermediate to the nearest of those, which are more correctly determined. I have not been able to obtain a sufficient number of facts relating to the metallic salts, to enable me to comprehend many of them in the tables.

It has been usual to distinguish the attractions, which produce the double decompositions of salts, into necessary and superfluous attractions; but the distinction is neither very accurate, nor very important: they might be still further divided, accordingly as two, three, or the whole of the four ingredients concerned are capable of simply decomposing the salt in which they are not contained; and if two, accordingly as they are previously united or separate; such divisions would however merely tend to divert the attention from the natural operation of the joint forces concerned.

It appears to be not improbable, that the attractive force of any two substances might, in many cases, be expressed by the quotient of two numbers appropriate to the substances, or rather by the excess of that quotient above unity; thus the attractive force of many of the acids for the three principal alkalies might probably be correctly represented in this manner; and where the order of attractions is different, perhaps the addition of a second, or of a second and third quotient, derived from a different series of numbers, would afford an accurate determination of the relative force of attraction, which would always be the weaker, as the two substances concerned stood nearer to each other in these orders of numbers; so that, by affixing, to each simple substance, two, three, or at most

MDCCCIX.   X

four numbers only, its attractive powers might be expressed in the shortest and most general manner.

I have thought it necessary to make some alterations in the orthography generally adopted by chemists, not from a want of deference to their individual authority, but because it appears to me that there are certain rules of etymology, which no modern author has a right to set aside. According to the orthography universally established throughout the language, without any material exceptions, our mode of writing Greek words is always borrowed from the Romans, whose alphabet we have adopted: thus the Greek vowel Υ, when alone, is always expressed in Latin and in English by Y, and the Greek diphthong ΟΥ by U, the Romans having no such diphthong as OU or OY. The French have sometimes deviated from this rule, and if it were excusable for any, it would be for them, since their *u* and *ou* are pronounced exactly as the Υ and ΟΥ of the Greeks probably were: but we have no such excuse. Thus the French have used the term *acoustique*, which some English authors have converted into " acoustics;" our anatomists, however, speak, much more correctly, of the " acustic" nerve. Instead of glucine, we ought certainly, for a similar reason, to write glycine; or glycina, if the names of the earths are to end in *a*. Barytes, as a single Greek word, means weight, and must be pronounced bárytes; but as the name of a stone, accented on the second syllable, it must be written barites; and the pure earth may properly be called barita. Yttria I have altered to itria, because no Latin word begins with a Y.

Table of the Sequences of the Bases with the different Acids.

In all mixtures of the aqueous solutions of two salts, each acid remains united to the base which stands nearest to it in this table.

SULFURIC ACID.

NITRIC
Barita
Strontia
Lime
(Silver?)
(Mercury?)
Potass
Soda
{ Zinc
{ Iron
{ Copper
Magnesia
Ammonia (1)
Glycina
Alumina
Zirconia

MURIATIC
Barita
Strontia
Lime
Potass
Soda
(Mercury?)
(Iron?)
Magnesia
Glycina
Alumina
Ammonia (2)
Zirconia
(Copper?)
Alumina (2)
Lime

PHOSPHORIC
Barita
Potass
Soda
Ammonia
Strontia
Magnesia (3)
Glycina
Alumina
Zirconia
Lime

FLUORIC
Barita
Potass
Soda
Ammonia
Strontia
Magnesia (4)
Glycina
Alumina
Zirconia
Lime

SULFUROUS
Barita
Potass
Soda
Strontia
Ammonia (5)
Magnesia
Lime
Glycina
Alumina
Zirconia

BORACIC
Barita
Potass
Soda
Strontia
Ammonia (6)
Lime
Magnesia
Glycina
Alumina
Zirconia

CARBONIC
Potass
Soda
Barita
Strontia
Lime
Ammonia
Magnesia?
Glycina
Alumina
Zirconia

(NITROUS)
Barita
Strontia
Lime
Potass
Soda
Magnesia?
Ammonia
Glycina
Alumina
Zirconia
Lead { Zinc { Copper

(PHOSPHOROUS)
Barita
Potass
Soda
Ammonia
Strontia
Magnesia
Glycina
Alumina
Zirconia
Lime?

(ACETIC)
Lead
Mercury
Iron
{ Potass
{ Soda
{ Magnesia
Lead
{ Zinc
{ Copper
X 2

(1) Ammonia stands above magnesia when cold.   (2) A triple salt is formed.   (3) Perhaps magnesia ought to stand lower   (4) A compound salt is formed, and when hot, magnesia stands above ammonia.   (5) FOURCROY says, that sulfate of strontia is decomposed by borate of ammonia.   (6) With heat, ammonia stands below lime and magnesia.

of elective Attractions.

NITRIC ACID.

NITRIC AND MURIATIC ACIDS.

Barita	Potass	Barita	Potass	Barita (10)	Potass
Potass	Soda	Potass	Soda	Potass	Soda
Soda	Ammonia	Soda	Ammonia	Soda	Barita (10)
Strontia	Magnesia	Ammonia	Magnesia	Ammonia	Ammonia (7,11)
Lime	Glycina	Magnesia	Glycina	Magnesia	Magnesia (7)
Magnesia (7)	Alumina	Glycina	Alumina	Glycina	Strontia
Ammonia (7)	Zirconia (8)	Alumina	Zirconia	Alumina	Lime
Glycina	Barita	Zirconia	Barita	Zirconia	Glycina
Alumina	Strontia	Strontia (9)	Strontia	Strontia	Alumina
Zirconia	Lime	Lime	Lime	Lime	Zirconia
MURIATIC	PHOSPHORIC	FLUORIC	SULFUROUS	BORACIC	CARBONIC

(7) A triple salt is formed. (8) FOURCROY says, that the muriate of zirconia decomposes the phosphates of barita and strontia. (9) According to FOURCROY's account, the fluate of strontia decomposes the muriates of ammonia, and of all the bases below it; but he says in another part of the same volume, that the fluate of strontia is an unknown salt. (10) According to FOURCROY's account of these combinations, barita should stand immediately below ammonia in both of these columns. (11) With heat, the carbonate of lime decomposes the muriate of ammonia.

PHOSPHORIC ACID.

Barita	Lime	Barita	Potass	Barita
Lime	Barita	Lime	Soda	Lime
Potass	Potass	Potass	Barita	Potass
Soda	Soda	Soda	Lime (13)	Soda
Strontia	Strontia	Strontia	Strontia	Strontia
Magnesia	Magnesia	Ammonia (12)	Ammonia	Magnesia
Ammonia	Ammonia	Magnesia	Magnesia	Glycina?
Glycina	Glycina	Glycina	Glycina	Alumina
Alumina	Alumina	Alumina	Alumina	Zirconia
Zirconia	Zirconia	Zirconia	Zirconia	
FLUORIC	SULFUROUS	BORACIC	CARBONIC	(PHOSPHOROUS)

(12) According to FOURCROY, the phosphate of ammonia decomposes the borate of magnesia. (13) FOURCROY says, that the carbonate of lime decomposes the phosphates of potass and of soda.

FLUORIC ACID.

Lime	Lime	Potass
Potass	Barita	Soda
Soda	Strontia	Lime
Magnesia	Potass	Barita
Ammonia	Soda	Strontia
Glycina	Ammonia	Ammonia (14)
Alumina	Magnesia	Magnesia
Zirconia	Glycina	Glycina
Strontia	Alumina	Alumina
Barita	Zirconia	Zirconia
SULFUROUS	BORACIC	CARBONIC

(14) According to FOURCROY, the carbonate of ammonia decomposes the fluates of barita and strontia.

SULFUROUS ACID.

Barita	Potass	Lime	Zirconia
Strontia	Soda	Strontia	Alumina
Potass	Barita (15)	Barita	Glycina
Soda	Strontia	Zirconia	Ammonia
Ammonia	Ammonia	Alumina	Magnesia
Magnesia	Lime	Glycina	Strontia
Lime	Magnesia	Magnesia	Soda
Glycina	Glycina	Ammonia	Potass
Alumina	Alumina	Soda	Barita
Zirconia	Zirconia	Potass	Lime
BORACIC	CARBONIC	(NITROUS)	(PHOSPHOROUS?)

BORACIC ACID.

Potass
Soda
Lime
Barita
Strontia
Magnesia
Ammonia
Glycina
Alumina
Zirconia
CARBONIC

(15) FOURCROY says, that the sulfite of barita decomposes the carbonate of ammonia.

Table of the Sequences of the Acids with different Bases.

	BARITA			STRONTIA					LIME				POTASS / SODA	MAGNESIA	
Sulfuric	S	C	S	S	C	S	P	S	C	P	P	P		S	B
Nitric	N	S	P	N	SS	P	S	P	P	F	F	F	MAGN.=AMM.	N	C
Muriatic	M	P	SS	M	F	SS	SS	SS	F	B	B	SS	GLYCINA	M	P
Phosphoric	SS	SS	N	SS	P	F	F	F	B	SS	C	S	ALUMINA	P	F
Sulfurous	P	N	M	C	B	B	B	B	SS	S	SS	B	ZIRCONIA	F	SS
Fluoric	C	M	F	B	S	C	C	N	S	C	S	N	Each with every subsequent base in this order	SS	S
Boracic	B	F	B	F	M	N	N	M	M	N	N	M		B	N
Carbonic	F	B	C	P	N	M	M	C	N	M	M	C		C	M
STRONTIA	LM	PT	MG	LM	PT	MG	AM	GL	PT	MG	AM	GL			AM
		SD	AM		SD				SD			AL			
			GL									ZR			
			AL												
			ZR												

The comparative use of this table may be understood from an example: if we suppose that the nitrate of barita decomposes the borate of ammonia, we must place the boracic acid above the nitric, between barita and ammonia in this table, and consequently barita below ammonia, between the fluoric and boracic in the former: hence the boracic and fluoric acids must also be transposed between barita and strontia, and between barita and potass; or if we place the fluoric still higher than the boracic in the first instance, we must place barita below ammonia between the nitric and fluoric acids, where indeed it is not impossible that it ought to stand.

Numerical Table of elective Attractions.

BARITA.		STRONTIA.		POTASS.	SODA.	LIMB.	
Sulfuric acid	1000*	Sulfuric acid	903*	Sulfuric acid		Oxalic acid	960
Oxalic	950	Phosphoric	827*		894* 885*	Sulfuric	868*
Succinic	930	Oxalic	825	Nitric	812* 804*	Tartaric	867
Fluoric		Tartaric	757	Muriatic	804* 797*	Succinic	866
Phosphoric	906*	*Fluoric*		Phosphoric		Phosphoric	865*
Mucic	900	Nitric	754*		801* 795*	Mucic	860
Nitric	849*	Muriatic	748*	Suberic?	745 740	Nitric	741*
Muriatic	840*	(Succinic)	740	Fluoric	671* 666*	Muriatic	736*
Suberic	800	(Fluoric)	703*	Oxalic	650 645	Suberic	735
Citric		*Succinic*		Tartaric	616 611	Fluoric	734*
Tartaric	760	Citric?	618	Arsenic	614 609	Arsenic	733¾
Arsenic	733½	Lactic	603	Succinic	612 607	*Lactic*	732
(Citric)	730	*Sulfurous*	527*	Citric	610 605	Citric	731
Lactic	729	*Acetic*		Lactic	609 604	Malic	700
(Fluoric)	706*	*Arsenic*	(733¼)	Benzoic	608 603	Benzoic	590
Benzoic	597	Boracic	513*	Sulfurous	488* 484*	*Acetic*	
Acetic	594	(Acetic)	480	Acetic	486 482	Boracic	537*
Boracic	(515)*	Nitrous?	430	Mucic	484 480	Sulfurous	516*
Sulfurous	592*	Carbonic	419*	Boracic	482* 479*	(Acetic)	470
Nitrous	450			Nitrous	440 437	Nitrous	425
Carbonic	420*			Carbonic	306* 304*	*Carbonic*	423*
Prussic	400			Prussic	300 298	Prussic	290

MAGNESIA.		AMMONIA.		GLYCINA?	ALUMINA.	ZIRCONIA?
Oxalic acid	820	Sulfuric acid	808*	Sulfuric acid	718*	709*	700*
Phosphoric		Nitric	731*	Nitric	642*	634*	626*
Sulfuric	810*	Muriatic	729*	Muriatic	639*	632*	625*
(Phosphoric)	736*	Phosphoric	728*	Oxalic	600	594	588
Fluoric		Suberic?	720	Arsenic	580	575	570
Arsenic	733	Fluoric	613*	Suberic?	535	530	525
Mucic	732½	Oxalic	611	Fluoric	534*	529*	524*
Succinic	732¼	Tartaric	609	Tartaric	520	515	510
Nitric	732*	Arsenic	607	Succinic	510	505	500
Muriatic	728*	Succinic	605	Mucic	425	420	415
Suberic?	700	Citric	603	Citric	415	410	405
(Fluoric)	620*	Lactic	601	*Phosphoric*	(648)*	(642)*	(636)*
Tartaric	618	Benzoic	599	Lactic	410	405	400
Citric	615	Sulfurous	433*	Benzoic	400	395	390
Malic?	600?	Acetic	432	Acetic	395	391	387
Lactic	575	Mucic	431	Boracic	388*	385*	382*
Benzoic	560	Boracic	430*	Sulfurous	355*	351*	347*
Acetic		Nitrous	400	Nitrous	340	336	332
Boracic	459*	Carbonic	339*	Carbonic	325*	323*	321*
Sulfurous	439*	Prussic	270	Prussic	260	258	256
(Acetic)	430						
Nitrous	410						
Carbonic	366*						
Prussic	280						

Acids.

SULFURIC.
Barita	1000*
Strontia	903*
Potass	894*
Soda	885*
Lime	868*
Magnesia	810*
Ammonia	808*
Glycina	718*
Itria	712
Alumina	709*
Zirconia	700*

NITRIC.
Barita	849*
Potass	812*
Soda	804*
Strontia	754*
Lime	741*
Magnesia	732*
Ammonia	731*
Glycina	642*
Alumina	634*
Zirconia	626*

MURIATIC.
Barita	840*
Potass	804*
Soda	797*
Strontia	748*
Lime	736*
Ammonia	729*
Magnesia	728*
Glycina	639*
Alumina	632*
Zirconia	625*

PHOSPHORIC.
Barita	906*
Strontia	827*
Lime	(865)*
Potass	801*
Soda	795*
Ammonia	(728)*
Magnesia	736*
Glycina	648*
Alumina	642*
Zirconia	636*

FLUORIC.
Lime	734*
Barita	706*
Strontia	703*
Magnesia	(620)*
Potass	671*
Soda	666*
Ammonia	613*
Glycina	534*
Alumina	529*
Zirconia	524*

OXALIC.   **TARTARIC.**
Lime	960	867
Barita	950	760
Strontia	825	757
Magnesia	820	618
Potass	650	616
Soda	645	611
Ammonia	611	609
Glycina?	600	520
Alumina	594	515
Zirconia?	588	510

ARSENIC.
Lime	733¾
Barita	733½
Strontia	733¼
Magnesia	733
Potass	614
Soda	609
Ammonia	607
Glycina	580
Alumina	575
Zirconia	570

TUNGSTIC.
Lime	
Barita	
Strontia	
Magnesia	
Potass	
Soda	
Ammonia	
Glycina	
Alumina	
Zirconia	

SUCCINIC.
Barita	930
Lime	866
Strontia?	740
(Magnesia)	732¼
Potass	612
Soda	607
Ammonia	605
Magnesia	
Glycina?	510
Alumina	505
Zirconia?	500

SUBERIC.
Barita	800
Potass	745
Soda	740
Lime	735
Ammonia	720
Magnesia	700
Glycina?	535?
Alumina	530
Zirconia?	525?

CAMPHORIC.
Lime	
Potass	
Soda	
Barita	
Ammonia	
Glycina?	
Alumina	
Zirconia?	
Magnesia	

CITRIC.
Lime	731
Barita	730
Strontia	618
Magnesia	615
Potass	610
Soda	605
Ammonia	603
Glycina?	415?
Alumina	410
Zirconia	405

LACTIC.
Barita	729
Potass	609
Soda	604
Strontia	603
Lime	(732)
Ammonia	601
Magnesia	575
Metallic oxids	
Glycina	410
Alumina	405
Zirconia	400

BENZOIC.
White oxid of arsenic	
Potass	608
Soda	603
Ammonia	599
Barita	597
Lime	590
Magnesia	560
Glycina?	400?
Alumina	395
Zirconia?	390?

SULFUROUS.
Barita	592*
Lime	516*
Potass	488*
Soda	484*
Strontia	(527)*
Magnesia	439*
Ammonia	433*
Glycina	355*
Alumina	351*
Zirconia	347*

ACETIC.
Barita	594
Potass	486
Soda	482
Strontia	480
Lime	470
Ammonia	432
Magnesia	430
Metallic oxids	
Glycina	395
Alumina	391
Zirconia	387

of elective Attractions.

MUCIC?		BORACIC.		NITROUS?		PHOSPHOROUS.
Barita	900	Lime	537 *	Barita	450	Lime
Lime	860	Barita	515 *	Potass	440	Barita
Potass	484	Strontia	513 *	Soda	437	Strontia
Soda	480	*Magnesia*	(459)*	Strontia	430	Potass
Ammonia	431	Potass	482 *	Lime	425	Soda
Glycina	425	Soda	479 *	Magnesia	410	Magnesia ?
Alumina	420	Ammonia	430 *	Ammonia	400	Ammonia
Zirconia	415	Glycina	388 *	Glycina	340	Glycina
		Alumina	385 *	Alumina	336	Alumina
		Zirconia	382 *	Zirconia	332	Zirconia

CARBONIC.		PRUSSIC.	
Barita	420 *	Barita	400
Strontia	419 *	Strontia	
Lime	(423)*	Potass	300
Potass ?	306 *	Soda	298
Soda	304 *	Lime	290
Magnesia	(366)*	Magnesia	280
Ammonia	339 *	Ammonia	270
Glycina	325 *	Glycina ?	260
Alumina	323 *	Alumina ?	258
Zirconia	321 *	Zirconia ?	256

XCII. *Essay towards a natural Classification of simple Bodies.* By M. AMPERE*.

WHEN the arbitrary hypotheses which had long led chemists astray were banished from science, and it was ascertained that we were to consider as simple, all the bodies which had not yet been decomposed, the number of these bodies was not two-thirds of what they are now: this number successively increased as the processes of chemical analysis were applied to compounds which had not yet been analysed, or which had been so but imperfectly. Every time that a new simple body was discovered, a further term of comparison was obtained, and new relations were observed: it became necessary sometimes to restrain, and sometimes to generalize, the first views of the fathers of modern chemistry; and the want of arranging simple bodies in an order which renders more sensible their mutual relations, and facilitates the study of their properties, became more and more felt. This order may be purely artificial, like the systematic classifications which were at first resorted to in the other branches of the natural sciences: it may also be deduced from the *ensemble* of the characters of the bodies which we propose to classify; and by constantly uniting those presented by the most numerous and essential analogies, they will be to chemistry what the natural methods are to botany and zoology.

* *Annales de Chimie et de Physique*, tome i. p. 295. March 1816.

Hitherto

Hitherto chemists have confined themselves to ranging simple bodies according to the degree of their affinity for oxygen, and the nature of the combinations which they form with it.   They ought naturally to have adopted this kind of classification, when they thought that the properties which characterized the oxygen belonged to it in a manner so exclusive that no other body could be associated with it.   But nowadays that new facts, and a more accurate interpretation of the facts already known, have rectified whatever was too absolute in the theory established by the celebrated Lavoisier ; and now that other substances have presented similar properties ; it appears to me that we must of necessity banish from chemistry the artificial classifications, and begin by assigning to each simple body the place which it ought to occupy in the natural order, by comparing it successively with all the rest, and uniting it with those which resemble it by a greater number of common characters, and particularly by the importance of those characters.   The first advantage which will result from the employment of such a method, will be to give us a more exact and more complete knowledge of all the properties of simple bodies ; and frequently to refer to general laws a multitude of isolated facts.   Another advantage will arise from this, namely, that after having ascertained among those which we shall have thus united, analogies so multiplied that we cannot refuse to regard them as connected very closely in the natural order, we shall be led to try upon some, experiments similar to those which have been attempted with success on others.   A classification, which should have induced every person from the very origin of modern chemistry to consider all the salifiable bases as belonging to one and the same class of bodies, would have taught chemists to place potash and soda in contact with iron at a high temperature, and potash and soda would have been discovered twenty years sooner.   When it was ascertained that chlore was a simple body, it was at first compared to oxygen ; and it was only when M. Gay-Lussac remarked its analogies with sulphur, that he was led to a discovery, the consequences of which upon the ulterior progress of chemistry can only as yet be guessed at, viz. that of the chloric and iodic acids, and of the perfect analogy of the chlorates and iodates with the nitrates.   A second approximation followed almost instantly by another discovery to which it naturally led, that of cyanogene, and of the true nature of the hydrocyanic acid.   Finally, these very analogies doubtless guided M. Dulong in the work which he communicated to the Institute on the 7th of November 1815, in which we see that the oxalic acid is composed of carbonic acid gas and hydrogen gas combined in the ratio of two to one in volume ; that

E e 4

this

this acid, which he calls in consequence hydro-carbonic acid, conformably to the established nomenclature, is united with the oxides in such a proportion, that the volume of hydrogen which it contains is double that of the oxygen of the oxide, so that when the latter is not very difficult to decompose, water is formed, and the carbonic acid gas remains alone combined with the metal, as happens with cyanogene, sulphur, chlore and iode, in the formation of the cyanures, sulphurets, chlorures and iodures.

A third advantage equally important is, to prepare by the natural classification of simple bodies, that of compound substances, —a work of much more labour, and to which I purpose to devote another paper. I know that the compound bodies have been already classed in a manner much more conformable to their true analogies than their elements have been. Many things have doubtless been done in this respect; but more perhaps remains to be done; and the discovery of the new substances with which the domains of chemistry have been enriched within these few years, cannot fail to lead to a modification and generalization of the principles according to which we now class compound substances, and to determine in a more precise manner the signification of the names which serve to designate the various kinds of combinations, and particularly that of the words acid, alkali, salt, &c.

I shall confine myself in this paper to the simple bodies, and shall divide them into three heads. I shall offer in the first, some general considerations on the order according to which it is proper to arrange bodies, so that this order may be as conformable as possible to their natural analogies; and on the means of avoiding the junction which has been hitherto made of the metals with bodies very different in almost all their other characters, and which have only been brought together because the energy of their affinity for oxygen is nearly the same,—a circumstance certainly remarkable, but to which perhaps too much importance has been attached,—and which certain considerations on the natural order of simple bodies, the principal results of which I shall soon detail, ought to induce us to regard as secondary, when it does not concur with other analogies which embrace the whole of the properties of the body.

Under the second head, I shall unite under natural genera the bodies which present characters of resemblance so multiplied and important that it is impossible to separate them in every classification which shall not be purely artificial; and I shall settle at the same time what places ought to be occupied in the natural order by the simple bodies which seem to form the passage from one genus to another, presenting analogies very striking
with

with substances belonging to two different genera. In this case, they indicate between these two genera an analogy which it will perhaps be difficult to ascertain without their assistance, but which is not the less real, and according to which we ought to place them in succession after each other, so that the body which establishes the link of the chain is at the end of the first, or the beginning of the second, in order to be always between two bodies which they resemble by common characters. Nothing then remains but to determine with which of the two genera it ought to be definitively united, by comparing the properties which it shares with the one, and those which are common to it with the other, in order to decide according to the number and importance of the analogies which result from these properties. The analogies to which these researches will lead us, will fix in an invariable manner the natural order of the simple bodies, conformably to the general idea which I am about to give of them.

The last head of this paper will have for its object to examine once more the various genera into which all these bodies shall have been distributed according to the data laid down in the preceding head, in order to assign to each of them a distinctive character formed by the union of some remarkable properties, chosen in such a way that they cannot be found at once but in a body appertaining to the genus which it is wanted to characterize, and to see at the same time according to what principles of nomenclature we could, if necessary, establish for each genus a denomination common to all the bodies which form part of it.

§ I. *On the impossibility of reconciling the manner in which chemists have hitherto ranged simple bodies, and the distinctions which have been established between them; with a classification deduced from the whole of their properties; and on the order which it is proper to adopt, to unite as much as possible those which present the most characters in common.*

The first source of the artificial classifications hitherto used, seems to me to have arisen from the old distinction of the metals and non-metallic bodies. I must confess, however, that this division leads us to separate but a very small number of bodies which we ought to unite in the natural order; and that it is in general tolerably conformable to the classification which results from the comparison of all the properties of bodies; and that it will even be sufficient, in order that it may embrace none of the genera which I regard as natural, to separate from the metals three substances which are generally united with them, arsenic, tellurium, and silicium: but then it becomes very difficult to assign a character which distinguishes in every case the metals
from

from the non-metallic bodies. Those which originally had served as the basis for this distinction can no longer be used as such, since most of the metals hitherto regarded as of that class are brittle ; and because some have been discovered even lighter than water ; and iode, and even carbon, when its particles are very close, as in animal charcoal, present the metallic lustre and a perfect opacity ; and because chemists have discovered in carbon the property of being a conductor of the electric fluid, &c. A more important character, viz. that of producing salifiable bases on being united with oxygen, cannot be considered as sufficient for characterizing exclusively the metals ; because some metals do not form any, because the boric and the nitrous acids are combined with the sulphuric acid, and because the products of those combinations have all the characters of the acid salts, which they resemble even more by their easy crystallization than several metallic solutions the oxide of which is precipitated in proportion as they are evaporated ; solutions which are only considered as salts, precisely because they are compounded of an acid and of the oxide of a body which we have been accustomed to regard as a metal. This character, which is admitted besides as exclusive, will remove tellurium from arsenic, and particularly from iode ; whereas the far more important property which it possesses of forming with hydrogen a permanent acid gas places it necessarily between those two bodies. The first object to which I shall turn my attention in the following article, will be to examine to what extent we might preserve the distinction of the metals, and of the non-metallic bodies, by subjecting the character which we shall choose for defining it in a precise manner, to the modification required by the necessity of rendering it conform to the natural order of simple bodies. I shall confine myself to remarking, that in the way in which it has been admitted, it has retarded the progress of the true theory of chemistry, by inducing a neglect of the observation of the properties by which certain metals are connected with the other simple bodies, and to which we cannot pay too much attention, when it is required to ascertain the truly natural order which exists between both the one and the other. The character drawn from the various degrees of affinity has still more contributed to establish between bodies, and particularly between the metals, approximations disavowed by nature. I shall confine myself to quoting in this respect, an example which appears to me very striking. Silver and gold form equally with oxygen combinations which an elevated temperature easily decomposes : from that instance those two metals have been regarded as being entitled to be placed very near each other in every methodical arrangement of simple bodies. Nevertheless the degree of affinity
for

for chlore ought not, in the eyes of the chemist who has precise ideas as to the action, equally energetic at least with that of oxygen, which it exercises on the metals, to be regarded as a character less important than the degree of affinity for oxygen; and the chlorure of gold is decomposable by heat; whereas that of silver, kept free from the contact of water and hydrogen, is absolutely unalterable at the highest temperature.

Here therefore we have two motives nearly equal, one for uniting and the other for separating these bodies. In order to resolve this difficulty, it is indispensable to have recourse to other properties; those of their oxides and their salts which they form with the acids, are then exhibited naturally, and decide the question, by showing us that the supposed analogy between those two metals is not confirmed by the resemblance of their principal characters. In fact, the oxide of silver is very alkaline, a little soluble in water, and completely saturates the acids: the oxide of gold presents nothing similar: and this difference, added to some other properties, less important, it is true, which silver presents in the metallic state, places this body with lead near potassium and sodium, and consequently very far from gold.

Since precise notions have been acquired as to the nature of chlore, several chemists have ceased to give to the properties which depend on the affinity of simple bodies for oxygen, an exclusive preponderance; but then one too great has been given to the assimilation which has been made of oxygen and of chlore. It has been attempted to arrange all the simple bodies in two classes, independent of each other, under the names of *combustible bodies*, and *supporters of combustion*. This division has had the same inconvenience with that of the metals and the nonmetallic bodies, by making us mistake the analogies of bodies which it places in different classes. This inconvenience has even been the more injurious to the theory of chemistry, in as much as it has taught us to neglect analogies much more complete and more striking than those which the separation of the metals and other simple bodies had caused to be neglected. Such are the analogies of the chloric and iodic acids with the sulphuric and nitric acids, of the chlorates and the iodates with the nitrates, and particularly that of chlore and iode with sulphur, greater still than the analogy of the same bodies with oxygen. The combination of those four substances with the metals to which they strongly adhere, takes place by occasioning a greater or less extrication of heat, and frequently of light: the compounds which result present a crowd of common properties: so far there is no occasion to approximate more particularly chlore and iode to oxygen than to sulphur; but this last forms like

iode

iode and chlore a permanent acid gas with hydrogen : the hydro-sulphates present the greatest analogy with the hydro-iodates and hydro-chlorates : they are reduced into sulphurets, as the latter are into iodures and chlorures, when they are insoluble, or after having evaporated their solutions, and dried their residues—properties which still more closely approximate iode and chlore to sulphur than to oxygen. It is impossible to separate iode from chlore, from which it differs only because the same characters are manifested at a less degree of energy; and nevertheless, why should iode be a supporter of combustion rather than sulphur, which is united to most of the metals with a greater extrication of light and heat? Shall we say that the sulphur is combined with the oxygen of the atmospheric air when we expose it to a sufficient temperature, and that this is not the case with iode also? But is the latter not also combined with oxygen when we place it in contact with the gaseous oxide of chlore, discovered by the celebrated chemist who was the first to decompose the alkalis, and demonstrate that chlore is a simple body, conformably to the opinion already given by the French chemists as an admissible hypothesis? It is of no use to insist longer on considerations of this kind : what precedes is sufficient to show how easy it is to be led into error, and to give too much importance to certain analogies, when we commence by establishing between the bodies which we purpose to classify, general divisions founded on a single character only. We should then deceive ourselves still more, if we thought to be able to arrange the simple bodies in an order conformable to their natural analogies, by forming of them a series dependent on their various degrees of affinity for one of them, for oxygen for instance. On comparing all the properties which those substances present, we find that they form a system in which every body belongs, on one side or the other, to bodies adjoining them, by analogies so strong that we are not able to establish in any way the complete separation which will be required by the reduction of the system into a single series; so that we must represent it to ourselves as a sort of circle, in which two bodies placed at the two extremities of the chain formed by all the rest, approach and unite mutually by common characters. I have long endeavoured to establish a natural order among simple bodies, by arranging them in a single series, which should commence with those whose properties presented the most complete opposition to those of the bodies which I attempted to place at the end of the series : these attempts have not been attended with any success; and it is by them that I have been led to adopt an order quite different, of which I shall in the first place attempt to give a general idea,
which

which shall be developed and defined in the two following articles. The bodies which have been hitherto considered as non-metallic, all possess the property of forming acids with some among them, the distinctive character of which is to make acid all combinations into which they enter in sufficient quantity. Several metals, and even some of those best entitled to this name, also present this property, and produce acids on being combined with the same acidifying substances. Those metals form two groupes, distinguished besides from each other by numerous differences. Some are eminently fixed and infusible; and it seems to be the same case with their combinations with chlore, at least if we may judge by the chlorure of chrome which presented this property to M. Dulong. It is with oxygen that they produce the acids from which the chemists have drawn the character which distinguishes them. The others are very fusible; the combinations which they form with oxygen present but feeble acid characters; for we ought not to reckon among them arsenic, which, as I have already said, ought to be united to the bodies which are considered as not metallic. And but for the labours of Messrs. Chevreul and Berzelius, the acidity of the peroxides of tin and antimony would have been still unknown; but these metals produce with chlore compounds liquid, or of a butyraceous consistency and volatile, and which possess the most essential properties of the acids. It is easy to see that the non-metallic simple bodies are united on the one hand with the infusible metals acidifiable by carbon and borium, and on the other with tin and antimony by phosphorus and arsenic, the combinations of which with chlore have the greatest analogy with the chloro-stannic acid (*spiritus Libavii*). All the other metals ought therefore to be placed between tin and antimony on the one hand, and the infusible acidifiable metals on the other. Those which add to the greatest affinity for oxygen, the property of forming with it alkaline combinations, occupy in some measure the middle part of this interval, and are connected on the one hand with tin and antimony, and on the other with tungsten, columbium, chrome, and molybdenum, by two series of metals, which present in the one and the other series all the degrees of affinity for oxygen, and the oxides of which also pass gradually through the various degrees of alkalinity and acidity which characterize that kind of compounds; but the bodies of which the two series are composed, present besides sufficient differences to enable us always easily to determine that to which they belong.

Such is the order by which I have been led, not by systematic views, which I disavow; but after having made a great number of attempts, in order to see if we could not adopt another with-

out

out wandering from natural analogies, and after having compared simple bodies under every point of view which can be presented by the properties which they possess.

I shall return to the subject in another paper.

II. NEW CLASSIFICATION OF CHEMICAL BODIES.

The recent discoveries in chemistry have occasioned a very considerable revolution in the theory of that science.   Whoever has paid sufficient attention to these improvements must be sensible that the present arrangement of chemical bodies is in many respects imperfect and inconvenient. The undecomposed bodies at present known amount to about 48, all of which, except eight, are considered as metals. I had occasion to touch upon this subject some months ago in my review of Professor Jameson's Mineralogy (*Annals*, viii. 136). and pointed out one or two alterations which appeared to me necessary. About the same time an elaborate dissertation on this subject by M. Ampere appeared in the Ann. de Chim. et de Phys. (i. 295, 373; ii. 5, 105). He examines the properties of all the simple bodies in detail, with much acuteness and discrimination, and endeavours to form them into a natural system, in which they follow each other according to their properties. I have not room at present to examine this arrangement with the minuteness which would be requisite in order to determine its accuracy, or to point out the reasons which induce me to dissent from some of his conclusions. I shall satisfy myself with giving the following outline of the classification.

The simple substances naturally subdivide themselves into three classes, namely,

1. GAZOLYTES, or substances capable of forming permanent gases with each other.

2. LEUCOLYTES, or metals fusible below 25° Wedgewood, and whose oxides form colourless solutions with the colourless acids.

3. CHROICOLYTES, or metals requiring a higher temperature for fusion than 25°, and whose oxides form coloured solutions in colourless acids.

CLASS I. GAZOLYTES.

Genus 1.   BORIDES. (From *boron*.)

*Bodies forming permanent Acid Gases with Phthore.**

Sp. 1. Silicon.   Sp. 2. Boron.

Genus 2.   ANTHRACIDES. (From ανθραξ.)

Bodies combining with one of the Elements of Air when exposed to it at a sufficient Temperature, and forming permanent Gases with the other Element.

Sp. 1. Carbon.   Sp. 2. Hydrogen.

---

* *Phthore* is the name by which M. Ampere has thought proper to distinguish the hypothetical body called *fluorine* by Sir H. Davy.

Genus 3. THIONIDES. (From θειον.)

Bodies capable of uniting with the preceding Genus, and of forming gaseous or very volatile Compounds.

Sp. 1. Azote.   Sp. 3. Sulphur.
2. Oxygen.

Genus 4. CHLORIDES. (From *chlorine*.)

Bodies unalterable in the Air at all Temperatures, forming with Hydrogen Acid Compounds gaseous or very volatile.

Sp. 1. Chlorine.   Sp. 3. Iodine.
2. Phthorine.

Genus 5. ARSENIDES. (From *arsenic*.)

Bodies oxidated in the Air when exposed to it at a sufficient Temperature, forming solid Compounds with Oxygen, and permanent Gases with Hydrogen.

Sp. 1. Tellurium.   Sp. 3. Arsenic.
2. Phosphorus.

CLASS II. LEUCOLYTES.

Genus 1. CASSITERIDES. (From κασσιτερος.)

Bodies whose Combinations with Oxygen are decomposed by Carbon, but not by Iodine.

Sp. 1. Antimony.   Sp. 3. Zinc.
2. Tin.

Genus 2. ARGYRIDES. (From αργυρος.)

Bodies whose Oxides are decomposed by Ioaine and Hydrogen.

Sp. 1. Bismuth.   Sp. 3. Silver.
2. Mercury.   4. Lead.

Genus 3. TEPHRALIDES. (From τεφρας and αλς.)

Bodies whose Oxides are decomposed by Iodine, and not by Hydrogen.

Sp. 1. Sodium.   Sp. 2. Potassium.

Genus 4. CALCIDES. (From *calcium*.)

Bodies whose Oxides are not decomposed by Carbon or Iodine, but by Chlorine.

Sp. 1. Barium.   Sp. 3. Calcium.
2. Strontium.   4. Magnesium.

Genus 5.   ZIRCONIDES.   (From *zirconium*.)

Bodies whose Oxides are not decomposed by Chlorine, Iodine, or Carbon.

Sp. 1. Yttrium.   Sp. 3. Aluminium.
    2. Glucinium.   4. Zirconium.

CLASS III.   CHROICOLYTES.

Genus 1.   CERIDES.   (From *cerium*.)

Bodies brittle and infusible at the Temperature at which Iron melts.

Sp. 1. Cerium.   Sp. 2. Manganese.

Genus 2.   SIDERIDES.   (From σιδηρος.)

Bodies whose Oxides dissolve in Acids in a State of Purity, and form coloured Solutions only when concentrated, and whose Peroxides have not Acid Properties.

Sp. 1. Uranium.   Sp. 4. Nickel.
    2. Cobalt.   5. Copper.
    3. Iron.

Genus 3.   CHRYSIDES.   (From χρυσος.)

Metals unalterable in the Air at all Temperatures.

Sp. 1. Palladium.   Sp. 4. Iridium.
    2. Platinum.   5. Rhodium.
    3. Gold.

Genus 4.   TITANIDES.   (From *titanium*.)

Infusible Bodies whose pure Oxides do not dissolve in Acids, and do not form with the Alkalies Compounds which can be considered as true Salts.

Sp. 1. Osmium.   Sp. 2. Titanium.

Genus 5.   CHROMIDES.   (From *chromium*.)

Bodies infusible at the Temperature at which Iron melts, acidifiable by Oxygen.

Sp. 1. Tungsten.   Sp. 3. Molybdenum.
    2. Chromium.   4. Columbium.

III. AFFINITY.

1. *Effect of Trituration on Chemical Combination.* — In the *Annals*, vii. 426, I have published a set of experiments by Mr. Link to determine what happens when dry salts, that mutually decompose each other when in solution, are triturated together. He found that when the two salts were destitute of water of crystalliza-

tion, no decomposition took place. But if either of them contained water of crystallization, they in that case mutually decomposed each other. When, after the trituration, a liquid capable of dissolving any of the constituents is poured on, decomposition takes place. It would appear from these experiments that the water of crystallization, though solid. still continues to exert its solvent powers.

2. *Structure of Solid Bodies.*—A very curious and important paper, by Mr. Daniell, has been published in the Journal of the Royal Institution (i. 24), which throws much new light upon the structure of solid bodies. If a lump of alum, or borax, or of nitre, be immersed in a vessel of water, and left at rest for three or four weeks, the solution will be found to have gone unequally on ; the uppermost portion will be found most wasted, and the undermost least ; so that the undissolved part of these salts will have assumed a conical form. The lower part of these bodies, after this treatment, will be found embossed over with numerous crystalline forms. These in alum are octahedrons, or figures formed by different sections of the aluminous octahedron. In borax they are fragments of eight-sided prisms, and so on. Mr. Daniell has shown in a satisfactory way that these embossments are not formed by the crystallization of that portion of the salt which has been dissolved ; but that they are brought into view by the unequal solution of the lump of salt subjected to the action of the water. Hence it follows that all these apparently amorphous masses are in reality composed of crystals, though such a structure cannot be distinguished by the eye previous to this natural dissection of it. The same crystalline structure was developed when carbonate of lime, carbonate of strontian, and carbonate of barytes, were slowly acted on by vinegar. Bismuth, antimony, and nickel, treated with very dilute nitric acid, likewise exhibited a crystallized structure. From these experiments we may infer, with considerable probability, that the structure of most bodies is in reality crystallized, even when they appear amorphous. If this mode of natural dissection could be applied to minerals in general, it would greatly extend the Haüyan method, and remove most of the objections to which it is at present exposed.

Mr. Daniell terminates his paper by an ingenious examination of the structure of crystals, and shows that Dr. Wollaston's hypothesis, that the integrant particles of bodies are spherical or spheroidal, will alone agree accurately with all the phenomena.

Art. XLI.—*The Numerical Relation between the Atomic Weights, with some Thoughts on the Classification of the Chemical Elements;* by Josiah P. Cooke, Jr., A.M., Erving Professor of Chemistry in Harvard University.*

Numerical relations between the atomic weights of the chemical elements have been very frequently noticed by chemists. One of the fullest expositions of these relations was that given by M. Dumas of Paris, before the British Association for the Advancement of Science, at the meeting of 1851. This distinguished chemist at that time pointed out the fact, that many of the elements might be grouped in triads, in which the atomic weight of one was the arithmetical mean of those of the other two. Thus the atomic weight of bromine is the mean between those of chlorine and iodine ; that of selenium is the mean between those of sulphur and tellurium, and that of sodium, the mean between those of lithium and potassium. M. Dumas also spoke of the remarkable analogies between the properties of the members of these triads, comparing them with similar analogies observed in organic chemistry, and drew, as is well known, from these facts arguments to support the hypothesis of the compound nature of many of the now received elements. Similar views to those of Dumas have been advanced by other chemists.

The doctrine of triads is, however, as I hope to be able to show in the present memoir, a partial view of this subject, since these triads are only parts of series similar in all respects to the series of homologues of organic chemistry, in which the differences between the atomic weights of the members is a multiple of some whole number. All the elements may be classified into six series, in each of which this number is different, and may be said to characterize its series. In the first it is nine, in the second eight, in the third six, in the fourth five, in the fifth four, and in the last three. The discovery of this simple numerical relation, which includes all others that have ever been noticed, was the result of a classification of the chemical elements made for the purpose of exhibiting their analogies in the lecture-room. A short notice of this classification will, therefore, make a natural introduction to the subject.

Every teacher of chemistry must have felt the want of some system of classification like those which so greatly facilitate the acquisition of the natural-history sciences. In most elementary text-books on chemistry, the elements are grouped together with little regard to their analogies. Oxygen, hydrogen, and nitrogen are usually placed first, and therefore together, although there are hardly to be found three elements more dissimilar ; again, phos-

* Communicated to the American Academy, Boston, Feb. 28, 1854.

phorus and sulphur, which are not chemically allied, are frequently placed consectively, while arsenic, antimony, and bismuth in spite of their close analogies with phosphorus, are described in a different part of the book.   This confusion, which arises in part from retaining the artificial classification of the elements into metals and metalloids, is a source of great difficulty to the learner, since it obliges him to retain in his memory a large number of apparently disconnected facts.   In order to meet this difficulty, a classification of the elements into six groups, differing but slightly from that given in the table accompanying this memoir, was made. The object of the classification was simply to facilitate the acquisition of chemistry, by bringing together such elements as were allied in their chemical relations considered collectively.   As the classification has been in use for some time in the courses of lectures on chemistry given in Harvard University, I have had an opportunity for observing its value in teaching, and cannot but feel that the object for which it was made has been in a great measure attained.   The series which is headed the Six Series will illustrate the advantage gained from the classification in a course of lectures, the elements which compose it being among those especially dwelt upon in lectures to medical students, and, generally, very widely separated in a text-book on the science. As chemistry is usually taught, the properties of the members of this series, nitrogen, phosphorus, arsenic, and antimony, as well as the composition and properties of their compounds, make up a large body of isolated facts, which, though without any assistance for his memory, the student is expected to retain.   Certainly it cannot be wondered at, that he finds this a difficult task.   The difficulty can, however, be in a great measure removed, if, after he has been taught that nitrogen forms two important acids with oxygen, NO_3 and NO_5, that it unites with sulphur and chlorine to form NS_3 and NCl_3, and also with three equivalents of hydrogen to form NH_3, he is also told, that, if in these symbols of the nitrogen compounds he replaces N by P, As, or Sb, he will obtain symbols of similar compounds of phosphorus, arsenic, and antimony ; for he thus learns, once for all, the mode of combination of all four elements, so that when he comes to study the properties, in turn, of phosphorus, arsenic, and antimony, he has not to learn with each an entirely new set of facts, but finds the same repeated with only a few variations.   Moreover, these very variations he will learn to predict, if he is shown that the elements are arranged in the series according to the strength of their electro-negative properties, or, in other words. that their affinities for oxygen, chlorine, sulphur, etc. increase, while those for hydrogen decrease, as we descend.   He will then readily see why it is that, though nitrogen forms NO_3 and NO_5, it forms only NCl_3 and NS_3, and that this reason is correct he will be pleased to find confirmed when he learns that phosphorus, which is more electro-positive than nitrogen, and has, therefore, a stronger affinity both for chlorine and sulphur, forms not only PCl_3 and PS_3, but also PCl_5 and PS_5. Again, he will not be surprised, after seeing the affinity of the elements for hydrogen growing constantly weaker as he descends in the series, to learn that a compound of bismuth and hydrogen is not certainly known. Should he inquire why, though NH_3 has basic properties, PH_3, AsH_3, and SbH_3 have not, he can be shown that the loss of basic properties in passing from NH_3 to PH_3 corresponds to a decrease in the strength of the affinity between the elements, and that if in PH_3, SbH_3, or AsH_3, atoms of methyle, ethyle, or other organic radicals analo-gous to hydrogen, are substituted for the hydrogen atoms, and more stable compounds thus obtained, strong bases are the result. The other series would afford similar illustrations, and, from my own experience, I am confident that no teacher who has once used a classification of the elements like that here proposed, would ever think of attempting to teach chemistry without its aid.

Classifications of the elements, more or less complete, have been given by many authors; but the fact that no one has been generally received, is sufficient to prove that they are all liable to objections, and would, indeed, also seem to show that a strictly scientific classification is hardly possible in the present state of the science. The difficulty with most of the classifications is, undoubtedly, that they are too one-sided, based upon one set of properties to the exclusion of others, and often on seeming, rather than real resemblances. This is the difficulty with the old classi-fication into metals and metalloids, which separated phosphorus and arsenic, sulphur and selenium, because arsenic and selenium have a metallic lustre, while phosphorus and sulphur have not, though there could hardly be found another point of difference. For a zoölogist to separate the ostrich from the class of birds be-cause it cannot fly, would not be more absurd, than it is for a chemist to separate two essentially allied elements, because one has a metallic lustre and the other has not. Yet it is surprising to see how persistently this classification is retained in every ele-mentary work on the science; and if it is sometimes so far modi-fied as to transfer elements analogous to selenium and arsenic to the class of metalloids, this is only acknowledging the worthless-ness of the principle, without being willing to abandon it. If there were any fundamental property common to all the elements, the law of whose variation was known, this might serve as the basis of a correct classification. Chemistry, however, does not as yet present us with such a property, and we must, therefore, here, as in other sciences, base our classification on general analo-gies. The most fundamental of all chemical properties is, undoubtedly, crystalline form ; but a classification of the elements based solely on the principles of isomorphism is defective in the same way as it is in mineralogy. It brings together, undoubtedly, allied elements, but it also groups with them those which resemble each other only in their crystalline form. The mode of combining seems to be also a fundamental property ; but, like crystalline form, it would bring together in some instances elements differing very widely in their chemical properties. A classification of the elements which shall exhibit their natural affinities, must obviously pay regard to both of these properties. It must at the same time seek to group together isomorphous elements, and those which form analogous compounds. Moreover, in such a classification, other less fundamental properties must not be disregarded. There are many properties both physical and chemical, which, although they cannot be exactly measured, and are oftentimes difficult to define, (such properties as those by which a chemist recognizes a familiar substance, or a mineralogist a familiar mineral,) and which on account of their indefinite character cannot be used as a basis of classification, may, nevertheless, render important aid in tracing out analogies. Judging from such properties as these, chemists are generally agreed in grouping together carbon, boron, and silicon, although they cannot be proved to be isomorphous, and are not generally thought to form similar compounds.

It is, however, much easier to point out what a classification should be, than to make one which shall fulfil the required conditions. Indeed, as has been already said, past experience would seem to show that a perfect scientific classification of the elements is hardly possible in the present state of chemistry. At best, the task is attended with great difficulties, and it cannot be expected that these should be surmounted at once. The classification which is offered in this memoir will, undoubtedly, be found to contain many defects. If, however, it is but one step in advance of those which have preceded it, it will be of value to the science. It was originally made, as has already been said, simply for the purpose of teaching, and never would have been published had it not led to the discovery of the numerical relation between the atomic weights.

On turning to the table which accompanies this memoir, it will be seen that the elements have been grouped into six series. These correspond entirely to the series of homologues of organic chemistry. In the group of volatile acids, homologues of formic acid, for example, we have a series of compounds yielding similar derivatives, and producing similar reactions, and many of whose properties, such as boiling and melting points, specific gravity, etc., vary as we descend in the series according to a determinate law. From formic acid, a highly limpid, volatile, and corrosive fluid, the acids become less and less volatile, less and less fluid, less and less corrosive; first oily, then fat-like, and finally hard, brittle solids, like wax. As is well known, the composition of these acids varies in the same way, and the variation follows a regular law, so that by means of a general symbol we can express the composition of the class. This symbol for the volatile acids may be written $(C_2H)O_3$, $HO + n(C_2H_2)$.

This description of the well-known series of the volatile acids, applies, word for word, *nominibus mutatis*, to each of the six series of chemical elements. The elements of any one series form similar compounds and produce similar reactions; moreover, they resemble each other in another respect in which the members of the organic series do not. Their crystalline forms are the same, or, in other words, they are isomorphous. Although this may be true of the volatile acids, yet it cannot be proved in the present state of our knowledge. Still further, many of their properties vary in a regular manner as we descend in the series. In one case, at least, the law of the variation is known, and can be expressed algebraically, though in most instances it cannot be determined. Finally, as one general symbol will express the composition of a whole organic series, so a simple algebraic formula will express the atomic weight, or, if you may please so to term it, the constitution of a series of elements.

These points may be illustrated with any of the series in the table; with the first, for example, which consists of oxygen, fluorine, cyanogen, chlorine, bromine, and iodine. All these elements form similar compounds, as will be seen by inspecting the symbols of their compounds given at the right hand of the list of names, where the similar or homologous compounds are arranged in upright columns. Moreover, they are all isomorphous, as may be seen by referring to the left-hand side of the list, where the similar compounds in each upright series are isomorphous, the numbers at the heads of the columns indicating the systems of crystallization, as described in the explanation accompanying the table. That the properties of these elements vary as we descend, can be easily shown. Oxygen is a permanent gas, as is also fluorine. Cyanogen is a gas, but may be condensed to a liquid. Chlorine, a gas also, can be condensed more easily than cyanogen. Bromine is a fluid at the ordinary temperature; and, finally, iodine is a solid. Moreover, starting from cyanogen, the solubility of these elements in water decreases as we descend in the series; and, again, the specific gravity of their vapors follows the inverse order of progression, gradually increasing from oxygen down. The atomic weights vary in the same order, and admit of a general expression, which is $8 + n9$, or, in other words, the differences between the atomic weights of these elements are always a multiple of nine. This general formula may be said to represent the constitution of these elements, in the same way that the symbol $(C_2H)O_3$, $HO + n(C_2H_2)$ represents the composition of the volatile acids before mentioned. In the place of $(C_2H)O_3$, HO we have $8=O=$ the weight of one atom of oxygen, and in the place of C_2H_2 we have nine. What it is that weighs nine (for it must be remembered that those numbers are weights) we cannot at present say, but it is not impossible that this will be hereafter discovered. In order to bring the general symbol of the volatile acids into exact comparison with that of the Nine Series, we must reduce the symbols to weights, when the two formulæ become

$$46+n14, \quad \text{where } 46=(C_2H)O_3, HO \quad \text{and } 14=C_2H_2 ;$$
$$\text{and } 8+n \ 9, \quad \text{where } 8=O \qquad\qquad \text{and } 9=x.$$

The numbers 46 and 14 are known to represent the weights of aggregations of atoms. The number 8 represents the weight of one oxygen atom, but we cannot as yet say what the 9 represents. After this comparison, it does not seem bold theorizing to suppose that the atoms of the members of this series are formed of an atom of oxygen as a nucleus, to which have been added one or more groups of atoms, the weight of which equals nine, or perhaps one or more single atoms each weighing nine, to which the corresponding element has not yet been discovered. As it will be convenient to have names to denote the two terms of the formulæ which represent the constitution of the different series, we will call the first term, in accordance with this theory, the nucleus, and the number in the second term multiplied by n the common difference of the series.

From what has been said, it will be seen that the idea of the classification is that of the organic series. It is in this that the classification differs from those which have preceded it. Other authors, in grouping together the elements according to the principles of isomorphism, have obtained groups very similar to those here presented. Indeed, this could not be otherwise, since, as has been already said, the members of each series are isomorphous, while, as a general rule, to which, however, there are many exceptions, no isomorphism can be established between members of different series. These groups, however, have been merely groups of isomorphous elements, and not series of homologues like those in which the elements are here classed.

These general remarks will suffice to indicate the principles upon which the classification has been made, and the character of the numerical relation between the atomic weights which has been established. The details of the classification can be best studied by referring to the table, so that it will be only necessary to speak of those points which are of special interest, or which may require explanation, or in regard to which there may be doubt. The series I have named from their common differences.

The first I have called the Nine Series, the second the Eight Series, &c. Let us examine the doubtful points in each, commencing with the first.

The last five members of the Nine Series are connected by so many analogies, that they have been invariably grouped together in the elementary books. There can be no doubt, therefore, in regard to the propriety of placing them in the same series, on the ground of general analogies. Fluorine, it is true, presents some striking points of difference from the rest. Fluorid of calcium is almost insoluble in water, while the chlorid, bromid, and iodid of calcium are all very soluble. We must, however, remember that we have to do with series, and must not therefore expect to find close resemblances except between adjacent members. If, then, we consider that oxygen is one of the series, and that fluorine stands but one step removed from oxygen, while it is two steps removed from chlorine, the discrepancy in a measure vanishes, for lime CaO is but slightly soluble in water. Nevertheless, the difficulty does not entirely disappear; for CaFl is much less soluble than CaO, although it should be more soluble judging from the law of the series and the fact that CaCl is so much more soluble than CaFl.

The solubility of a series of homologous elements or compounds in water, may be regarded as a function of one or more variables. In the case of the elements there may be but one variable, but it is easy to see that in the case of compounds there must be several. One of these variables is probably the same which determines the common difference of the series to which the elements or compounds belong; (it will be hereafter shown that the atomic weights of the homologous compounds are related in the same way as those of the elements;) the other variables are perhaps the atomic forces which determine the hardness, density, &c. of the solid. We may, therefore, with justice, compare the relative solubilities of a series of homologues to a curve which should be the same function of the same variables, and what mathematics teaches we ought reasonably to expect in the case of this curve, we ought to expect also in the variations of solubility of these substances. Now every mathematician is familiar with the remarkably rapid changes which a curve undergoes that is a function of several variables, and we cannot be surprised that similarly rapid changes should be observed in the solubility of homologous substances in passing from one to the next in the series. In the curve which corresponds to the relative solubility of CaO, CaFl, CaCy, CaCl, CaBr, and CaI, it would seem that at CaFl there is a singular point where the curve, after rising for some distance above the axis, bends down again towards it. Several of the other series of compounds of these elements present similar anomalies; for example, KO, KFl, KCy, KCl, KBr, and KI.

Here the solubility diminishes until we come to KCl, which is less soluble than KCy; then it increases to the last. Here, of course, the singular point is at KCl. With the corresponding compounds of sodium, the solubility diminishes to NaFl, which is the least soluble of the series, and then increases constantly to the end.

These facts at least seem to show that apparent variations from the law of series in properties, which evidently are unknown functions of several variables, should not be allowed to outweigh strong analogies, and certainly the analogies between Fluorine and the other haloids are very marked. Fluorine itself possesses properties such as we should expect to find in a member of the series above chlorine. The strong and active affinities of fluorine might be indeed predicted, after seeing the rapid increase both in the strength and activity of the affinities in passing from iodine to chlorine. In passing from bromine to chlorine, we pass from a liquid to a gas, permanent under any natural conditions; and we should expect, therefore, in rising still higher in the series, to find in fluorine a gas less easily reduced to a liquid than chlorine. Now although, on account of its remarkably active affinities, this fact cannot be demonstrated on the gas itself, it can, nevertheless, be inferred with perfect certainty from its compounds. Finally, the isomorphism of fluorine and the other haloids may be urged as indicating close analogy. From these considerations, I cannot but think that those chemists who have questioned the propriety of classing fluorine with the other haloids will, on reviewing the facts, and regarding the haloids in the light of a series, and not simply as a group of elements possessing certain general properties, be led to change their opinion.

Cyanogen, though a compound radical, has been classed with the other haloids, not only from its atomic weight, but also from its other analogies. Its properties are in most cases those which we should expect from an element occupying its position in the series; but in others it presents remarkable variations, owing probably to the fact that it contains a radical which is easily decomposed. As is well known, it is perfectly isomorphous with chlorine.

The propriety of classing oxygen in this series seems to be placed beyond doubt by the discovery of ozone, which, although it does not seem to possess such energy as we should expect in an element higher in the series than fluorine, may, nevertheless, be found to fulfil all anticipations should it ever be obtained in a perfectly unmixed condition. The isomorphism of oxygen with chlorine, and therefore with the other haloids, seems sufficiently established by the determination both of Proust and Mitscherlich of the tetrahedral form of Cu_2Cl. It must, however, be admitted that oxygen presents as strong analogies with sulphur as it does with chlorine; and since, not only from its analogies, but also from its atomic weight, it appears to be the nucleus in all the first three series, I have placed it at the head of each. It may be mentioned here, that in all cases the fact that the atomic weight of an element is included in the general formula of a series, is an argument for classifying it in that series, if the relation between the atomic weights pointed out in this memoir is admitted to be a law of nature; but as I wish to show that the relation is not that of a mere accidental group of numbers, but is connected with the most fundamental properties of the elements, and has, therefore, the claims of a law, I have endeavored to show the correctness of the classification which conforms to the law, and, indeed, in fact suggested the law on other grounds.

The atomic weights of the numbers of the Nine Series, as determined by experiment, present greater deviations from the numerical law already explained, than are to be found in any of the others. The weights which would exactly conform to the general formula $8 + n9$ are given in the column of the table headed Theoretical, while in the next column at the right are given the weights of experiment. These for the most part (in this as well as in the other series), have been taken from the table of Atomic Weights given in the last volume of Liebig and Kopp, *Jahresbericht* for 1852, which was supposed to give the latest and most accurate results. In the few cases in which the numbers have not been taken from this table, the initial letter of the name of the observer has been annexed. It will be seen, on comparing the two columns that the greatest deviation from the law is in the case of fluorine, if we consider the care which was taken both by Berzelius and Louyet in the determination of the atomic weight of this element. It may, however, be remarked, that, as the processes used by both experimenters were essentially identical, any hidden constant source of error would produce the same effect on both results; so that the atomic weight of fluorine cannot be regarded as yet absolutely fixed. Nevertheless, it is not possible that so great a difference between the true and observed weights as two units could have escaped detection in the numberless analyses which have been made, by the most experienced chemists, of the fluorine compounds. It must, therefore, be admitted, and not only in the case of fluorine, but also in other instances, that there are deviations from the law; but these deviations are not greater than those from similar numerical laws in astronomy and other sciences, and indeed, judging from the analogy of these sciences, they ought to be expected.

Those who are not familiar with the amounts of probable error in the determination of the different atomic weights would judge, on comparing together the columns of theoretical and observed values, that the deviations from the law were much greater than they are in reality.   It should, therefore, be stated, that, in by far the larger number of instances, the deviations are within the limit of possible errors in the determinations, leaving only a few exceptional cases to be accounted for.   It must be remembered that, other things being equal, the amount of probable error is the greater the greater the atomic weight, so that a difference of 1·9 in the case of iodine is not a greater actual deviation from the law than only 0·5 in the case of chlorine.   Indeed, it is very possible that on more accurate determinations the atomic weight of iodine will be found to correspond to the law, which cannot be expected of that of chlorine.   It is well known that many of the larger atomic weights, especially those of the rarer elements, cannot be regarded as fixed within several units.

I have calculated, as well as the data I have would permit, the amount of probable error in the determinations of many of the atomic weights, and by comparing the results from different processes, and by different experimenters, I have endeavored to detect the existence of constant errors, which seem to be the great errors in all these determinations, those accidental errors which are made in the repetitions of the same process by equally careful experimenters being comparatively insignificant.   The results of this investigation will be published in a subsequent memoir.   It is sufficient for the present purpose to state, that, while they show that, in the greater number of cases, the apparent variations from the law are within the limit of probable error, there are yet several instances, where, after allowing for all possible errors of observation, there is a residual difference.   I do not therefore look alone to more accurate observations for a confirmation of the law, but, regarding the variations as ascertained facts, hope that future discovery will reveal the cause.   Whether the variations will be found to be a secondary result of the very cause which has determined the distribution of the atomic weights according to a numerical law, as the perturbations in astronomy are a necessary consequence of the very law they seemed at first to invalidate, or whether they are due to independent causes, can of course, for the present, be only a matter of speculation.   There are, however, facts which seem to indicate that the variations are not matters of chance, but correspond to variations in the properties of the elements.

From the beautiful discovery of Professor Schönbein we have learnt that oxygen has two allotropic modifications, and that besides its ordinary condition, it is capable of assuming another highly active state when its properties resemble those of chlorine.   Cyanogen is known only in a quiescent state.   The other haloids, fluorine, chlorine, bromine, and iodine, are known only in a highly active state.   Now it will be seen on examining the table, that the atomic weights of the highly active elements, as determined by experiment, exceed slightly the theoretical numbers, and that where the affinities are the most intense, in fluorine, the deviation is the greatest. A similar fact may be observed in the atomic weights of the members of the Six Series. Arsenic has been proved to be capable of existing in two allotropic modifications. In its ordinary state, it has a crystalline form belonging to the rhombic system. In the other condition, in which it may be obtained by sublimation at a low temperature, it crystallizes in regular octahedrons. The other members of this series are probably isodimorphs with arsenic. The ordinary condition of phosphorus is its monometric modification, while the rhombic state seems to be the normal condition of arsenic, antimony, and bismuth. Now the atomic weights of the last three are either equal to, or slightly exceed, the theoretical number, while that of the first falls short, perhaps even by a unit. Other facts, which also tend to show that the deviations are not matters of chance, may be found in the affiliations of the series. There are some elements which seem to be most remarkably double-faced, having certain properties which connect them closely with one series, and at the same time others which unite them nearly as closely to another. In such cases we find that the atomic weight either falls naturally into both series, or, not corresponding exactly with the theoretical number of the series to which the element properly belongs, it inclines towards that of the other, and sometimes equals it. Such is the case with chromium, manganese, and gold, as will be seen by referring to the affiliations at the bottom of the Nine Series. These various facts force upon me the conviction, that this relation between the atomic weights is not a matter of chance, but that it was a part of the grand plan of the Framer of the universe, and that in the very deviations from the law, there will, hereafter, be found fresh evidence of the wisdom and forethought of its Divine Author.

The general formulæ for the Eight Series are, $8+n8$ and $4+n8$. The two nuclei correspond to two different sets of elements, or sub-series, one consisting of oxygen, sulphur, selenium, and tellurium, the other of molybdenum, vanadium, tungsten, and tantalum. The atomic weights of the first are all equal to $8+n8$; those of the second to $4+n8$. The sub-series exhibit marked analogies, as well as certain differences. They resemble each other chiefly in that the members of both form analogous acids with oxygen, while they differ in that though the members of the first sub-series form compounds with hydrogen, those of the second do not. The isomorphism of the members of each sub-series among themselves, with the exception of vanadium, is complete; but there seems to be no proof of any isomorphism between the sub-series. Johnston attempted to establish the isomorphism of chromic and molybdic acids from the red variety of molybdate of lead from Rezbanya, which he supposed to be a chromate; but the fact has been disproved by G. Rose, who has shown that the supposed chromate is a molybdate mixed with a small amount only of chromate. There seems, nevertheless, to be some reason for believing that chromic acid may replace molybdic acid to a certain extent. If this is proved, it establishes another link of connection between the members of the two sub-series, since chromic acid is isomorphous with sulphuric acid. For the present, however, we must regard them as sub-series, related, but distinct, the second being in a measure supplementary to the first. They are distinguished in the table by printing the names of the second sub-series a little to the right of those of the first, and the fact that their atomic weights are intermediate to those of the first, I have indicated to the eye by giving to the names also an intermediate position.

The analogies between oxygen and sulphur are so numerous, that, were we to place oxygen in but one series, we should place it in this. HO and HS, HO_2 and HS_2, resemble each other very closely, as do also the oxygen salts the corresponding sulphur salts. Moreover, there can be no doubt in regard to the isomorphism of the two elements, since it has been established upon the authority both of Mitscherlich and Becquerel, and from two different compounds. The only doubtful case in the series was that of vanadium, which in some of its properties resembles arsenic more closely than it does molybdenum. The reasons for giving it the place which it occupies were the facts that its acids correspond to those of molybdenum, and that it forms remarkably highly colored oxyds which are repeated also in molybdenum. It is true that the properties of the element itself are not those we should expect from the position which it occupies in our table; yet, if it were placed in the Six Series, it would fall between phosphorus and arsenic, which on the whole it resembles less than it does molybdenum, for although it is combustible, yet neither it nor its oxyds are volatile. I consider it, therefore, as a member of the Eight Series, but affiliating very closely with the Six. Its atomic weight favors this hypothesis. Vanadate of lead has been considered isomorphous with the phosphate; but as this isomorphism does not rest on any measurement of angles, and as, moreover, the received symbols of the two minerals, vanadanite and pyromorphite, on whose crystalline forms the isomorphism was determined, show a very different constitution, I have not given much weight to this fact.* The observed atomic weights of the members of this series are almost precisely the same as the theoretical members, and with the exception, perhaps, of molybdenum, there appears to be no instance in which the difference is greater than the amount of possible error.

* See G. Rose's Mineral System.

The members of the Six Group form a well characterized family, so that, with the exception of oxygen, there can be no doubt in regard to the justice of classifying them together, and any discrepancies will disappear on considering the group in the light of a series. They form acids containing three and five atoms of oxygen which are completely homologous, and make two series parallel to that of the elements. They form also a remarkable series of compounds with three atoms of hydrogen. The idea which has been advanced by some authors, that NH_3 is the nitrid of hydrogen, while PH_3 is the hydruret of phosphorus, or, in other words, that hydrogen is electro-positive with reference to nitrogen and electro-negative with reference to phosphorus and those lower in the series, does not seem to me correct, since the remarkable bases which may be formed from PH_3, AsH_3, SbH_3, and BiH_3, by replacing the hydrogen atoms by organic radicals, seem to indicate that they have the same type as NH_3, and are therefore homologues of it.

The isomorphism of the four lower members of the series is perfect. It has been shown in the table, both by the crystalline forms of the elements themselves, as well as by those of their compounds. In the other series, wherever it was possible, the same double proof has been given. The doubt expressed by G. Rose in regard to the dimorphism of arsenic, as I hope to be able to show in a paper soon to be published, has been removed. In one state arsenic crystallizes in perfect octahedrons of the regular system, and is therefore isomorphous, not only with antimony and bismuth, but also, in its allotropic state, with phosphorus. Isomorphism, as is well known, is not absolute, except in forms of the regular system. The rhombic angles of the crystals of arsenic, antimony, and bismuth, are respectively, 85° 41′, 87° 35′, 87° 40′, and therefore conform to the general rule. It will be observed that the angle varies constantly in the same way as we descend in the series. Now, although these few instances do not afford sufficient ground for any general conclusion, yet they show that similar variations are possible in the other systems, and therefore that we cannot be expected to establish isomorphism in any case except between merely consecutive members.

The atomic weights of the members of this series, with the exception of phosphorus, do not present any important deviations from the theoretical numbers, taking into account always, of course, the amount of possible error. The deviation in the case of phosphorus has already been noticed. Oxygen, it must be admitted, is not connected with the series from any similarity of properties though the phosphids, arsenids, and antimonids, present certain analogies with the oxyds. As has already been said, oxygen was placed at the head of this, as well as of the last two series, because its atomic weight seemed to be the nucleus of all three.

The Five Series is the shortest of all, consisting of only three members, carbon, boron and silicon. Of these, the last two are as closely allied as are any two members of the other series, silicon having precisely the properties we should expect in a homologue of boron, which was lower in the series; and the same is also true of their compounds. The analogies, however, between these two elements and carbon are by no means so close, for not only carbon cannot be proved to be isomorphous with them, but it does not form similar compounds. Carbonic acid, it is true, presents some points of resemblance to boracic and silicic acid; like them it unites in a large variety of proportions with bases, its alkaline salts give a basic reaction, &c.; but according to the generally received opinion, its symbol is CO_2, while those of boron and silicon are BO_3 and SiO_3. In its uncombined state, however, carbon resembles boron and silicon, not only in its outward properties, but also in its action before the blowpipe. Two of the allotropic states of carbon, graphite and charcoal, are probably repeated in boron, and are known to be in silicon. The principle of exclusion would also seem to place carbon in this series, for it certainly presents no analogies with the members of any other. The correspondence of the atomic weights of the members of this series to the law is remarkably close.

The four series is by far the largest of all, including the greater number of what are generally known as the heavy metals. The members of the series resemble each other in the following respects. First, they are isomorphous; for although each member cannot be directly proved to be isomorphous with every other, yet isomorphism can be established between consecutive members, which, as has before been said, is all that can be expected. Second, the members of this series all form, by uniting with oxygen, either protoxyds or sesquioxyds, or both, which, as a general rule, are strong bases. Third, these oxyds are either insoluble, or nearly insoluble, in water. And finally, the elements of the series have all those physical properties which are known as metallic properties.

This series may be naturally divided into two sub-series. The first contains those elements whose protoxyd bases are their characteristic compounds, and which do not form acids with oxygen. The second contains those elements whose characteristic compounds are their sesquibases. They generally unite with two or more equivalents of oxygen, and form acids. These sub-series are distinguished in the table in the same way as those of the Six Series. Corresponding to these sub-series we have two sets of atomic weights, each having the same common difference, but differing in their starting-point or nucleus. The first set is expressed by the formula $4+n4$, the second by $2+n4$.

The sub-series affiliate with each other in a most remarkable manner. Manganese, for example, not only forms a strong protoxyd base, but also unites with a larger amount of oxygen, forming both a sesquibase and acids. Its atomic weight places it in the first group, and it has therefore been classed there, although by its properties it is equally allied to the second. Cobalt and nickel certainly resemble much more closely the members of the first than of the second sub-series, although their atomic weights place them in the second. With this exception, the subdivision of the series which the atomic weights require does not differ from that suggested by the properties of the elements. The members of this series may of course be still further subdivided into groups according to their special properties, as they are in all works on chemistry. They are placed together here because the atomic weights form but one numerical series.

The isomorphism of the members of this series will be found well established with the limitations before given. In order to establish the isomorphism of cobalt and nickel with iron, the isomorphism of one atom of arsenic with two atoms of sulphur has been assumed. This is generally admitted; but if it is not, no one can doubt in regard to the isomorphism of these three metals, as they constantly replace each other. Glucinum, zirconium, lanthanum, cerium, and thorium, cannot be shown to be isomorphous with the other metals by any of their compounds, but their oxyds are known to replace the analogous oxyds of the other metals. So also is ruthenium known to replace rhodium. There have been doubts expressed in regard to the existence of a monometric form of zinc; but as we have established its isomorphism with the other members of the series, not only by its own crystalline form, but also by those of its compounds, the fact is of no importance to the present question.

The atomic weights of the members of this series, as determined by observation, very nearly correspond with the theoretical numbers, which is the more remarkable, as the limit of error in the determination of the atomic weights of the greater number, especially of the rarer metals, is quite wide.

The Three and last Series is composed of hydrogen and the metals of the alkalies. The analogies between lithium, sodium and potassium, are very close, as is well known, and there can be no doubt in regard to the propriety of classing them together. It may be said however, in regard to hydrogen, that it resembles as closely some of the metals of the four series as it does those of the alkalies. Though this cannot be denied, yet the fact that the atomic weight of hydrogen is the nucleus of the series, and the great solubility of the alkalies in water, may be urged as reasons for placing it at the head of the Three Series.

The isomorphism of lithium, sodium, and potassium, is fully established; but I can find no data which prove hydrogen isomorphous either with them or with the metals of the other group.

The unit of the atomic weights which has been used thus far throughout the table, is the double atom of hydrogen; but the nucleus of the Three Series is the weight of the single atom, so that the unit in this series is one half of the unit of the weights in all the other series. This fact must be kept in mind in comparing the atomic weights of this with those of the other series. All the weights might have been made uniform by doubling them throughout; but as this would not have changed the relation, and would have been departing from the general custom, it was thought best to confine the doubling to the Three Series into which alone hydrogen enters. The general symbol of this series is $1+3n$, where of course the unit is one half of that of the symbols at the head of the other series. The observed atomic weights will be found to correspond very closely with the theoretical numbers; indeed, the two coincide, except in the case of potassium, where the difference is 0·6. This, however, it must be remembered, is 0·6 of the single hydrogen atom. Compared with the double atom, as the weight of potassium is generally given, the difference amounts to but 0·3.

One of the most remarkable points of the classification which has been now explained, is the affiliation of the series. We find in chemistry, as in other sciences, that Nature seems to abhor abrupt transitions, and shades off her bounding lines. Many of the elements, while they manifestly belong to one series, have properties which ally them to another. Several examples of this have already been noticed. In such cases, we find invariably, that there is a similar affiliation of the atomic weight. Of all the elements chromium and manganese are the most protean. Two atoms of these elements unite with seven atoms of oxygen and form acids analogous to perchloric acid, and, as has already been shown, the weight of two atoms of either element falls into the Nine Series. Moreover, one atom of chromium or of manganese, unites with three atoms of oxygen, to form chromic or manganic acid. Chromic acid is a strong oxydizing agent, and resembles closely nitrous acid, and the atomic weight of chromium falls into the Six Series just below that of nitrogen. Manganic acid, on the other hand, resembles sulphuric acid, with which it is isomorphous, and the atomic weight of manganese would place it in the Eight Series. In like manner osmium in many of its properties resembles platinum and the other metals with which it is associated in nature; but, unlike them, it forms a very remarkable volatile acid, whose insupportable and suffocating odor, as well as composition, reminds one of the acids of the Nine Series, and its atomic weight seems to justify the apparent analogy. Gold likewise, though the noblest of metals, yet in some of its chemical relations resembles much more closely the members of the Nine than of the Four Series, and here again its accommodating atomic weight seems to account for its double-sided character. Several other examples of similar affiliations are given in the table, but do not need explanation.

In the description just concluded of the classification of the chemical elements, which is offered in this memoir, I have not entered into details, for to have done so would have been to write a treatise on chemistry. I have confined myself almost exclusively to general points, and referred only to those particulars which I thought might present doubts. I hope that I have been able to show, first, that the chemical elements may be classified in a few series similar to the series of homologues of organic chemistry; second, that in those series the properties of the elements follow a law of progression; and finally, that the atomic weights vary according to a similar law, which may be expressed by a simple algebraic formula. As already intimated, I have endeavored to prove the correctness of the classification on general grounds, in order that it might appear that the simple numerical relation which has been discovered between the atomic weights is not a matter of chance, but is connected with the most fundamental properties of the elements. I might leave the subject at this point, but the existence of the law which I wish to establish will be proved more conclusively if it can be shown, not simply that the general properties of the members of each series vary in a regular manner, but also if in one or more cases the exact law of the variation can be pointed out.

There are but few properties of the elements which are subjects of measurement, and which therefore can be compared numerically. Such are the specific gravity in which the three states of aggregation, the boiling and melting points, the capacity of heat, and a few others. It is easy to see that there are but few of these properties the law of whose variation in the series we could reasonably expect to discover in the present state of science. Most of them evidently depend upon molecular forces with which we are entirely unacquainted. Such in solids is undoubtedly the case with so simple and fundamental a property as specific gravity, and most, if not all, of the other properties of solids belong to the same category. It cannot therefore be expected that we should point out the laws by which these properties vary, although the remarkable investigations of Kopp, Dana, Filhol, Schröder, and others, on the relations between the density of substances and their atomic weights, and those of Kengott on the relation of hardness to atomic volume, give grounds for expecting that even they will before long be discovered. In liquids and gases, however, most of these molecular forces which produce the ap-

parent irregularities in solids have less influence, as we should
naturally expect, probably because the atoms are removed out of
the sphere of their action. We may therefore hope, on comparing together the properties of the liquid or gaseous states of the
elements in any series, to discover some numerical relation between them. Unfortunately, however, we have not sufficient
data for making such a comparison except in the case of one
property, the specific gravity. The boiling point, which would
be a very valuable property for the purpose, is known only in a
few instances.

That the specific gravity of the elements in their gaseous state
varies in each series according to a numerical law, follows necessarily from what is already known. It is a well-known fact, that
the specific gravities of the gaseous states of the elements divided
by their atomic weights give quotients which are either equal, or
which stand in a very simple relation to each other. For any
series, as far as we have data, this quotient is the same for all the
elements with only a few exceptions. That is $\frac{\text{Sp. Gr.}}{\text{At. W.}} = p$. But
we have found that At. W. may be expressed in general by $a + nb$,
and substituting this for for At. W. in the above equation, it be-
comes $\frac{\text{Sp. Gr.}}{a + nb} = p$, or sp. gr. $= pa + npb$; so that $pa + npb$ is a
general expression for the specific gravity of all the elements of
any series, in the same way that $a + nb$ is for the atomic weight.
The value of p will differ according as the specific gravities used
are referred to hydrogen or air. Below will be found tables which
give the calculated and observed specific gravities of the elements
of the Nine and Six Series referred to hydrogen, which has been
taken as the unit instead of air, as we thus in a great measure
avoid fractions. In the Nine Series $p = 1$, so that the numbers
representing the specific gravities are the same as those representing the atomic weights. In the Six Series it equals two, so that
the numbers representing the specific gravities are in this series
twice as large as those representing the atomic weights. When
the specific gravity has not been observed, the calculated number
only is given. The observed numbers are taken from the "Table
of Specific Gravity of Gases and Vapors," in Graham's *Elements
of Chemistry*, which is a very complete collection of all known
data. For the other series, we have only occasional data, so that
no complete tables of their specific gravities are possible.

It is evident, then, that at least one property of the elements
varies in the series according to an ascertained numerical law.
But, it may be said, this proves nothing, for these specific gravities are connected so closely with the atomic weights that what
is true of one must be to the same extent true of the other.
It must be remembered, however, that the specific gravities are a
distinct set of observed facts, and that the probability of a law is

in exact proportion to the number of facts which accord with it. Moreover, the closeness of the connection is unimportant. Whether the value of p be expressed by a single digit, or by a complicated algebraic formula, is evidently a matter of indifference so far as the confirmation of the law is concerned.

THE NINE SERIES. $\dfrac{\text{Sp. Gr.}}{\text{At. W.}} = 1.$ Sp. Gr. $= 8 + n9.$			THE SIX SERIES. $\dfrac{\text{Sp. Gr.}}{\text{At. W.}} = 2.$ Sp. Gr. $= 16 + n12.$		
Names.	SPECIFIC GRAVITIES.		Names.	SPECIFIC GRAVITIES.	
	Theoret.	Observed.		Theoret.	Observed.
Oxygen	8	16	Oxygen	16	16
Fluorine	17		Nitrogen	28	14
Cyanogen	26	26	Phosphorus	64	64
Chlorine	35	35·5	Arsenic	148	150
Bromine	80	78	Antimony	256	
Iodine	125	126	Bismuth	412	

I regret exceedingly that there are not sufficient data in the case of any of the other properties of the elements in the state of gas to allow comparison, as I feel confident that the law which governs their variation in the series might easily be discovered; but I look forward to the time when in the general formula $pa + npb$ the value of p shall be known, not only for the properties of the elements in their gaseous state, but for every property capable of numerical expression.

In this memoir I have confined myself entirely to the elements, but it is evident that the classification here offered, and the numerical law here explained, may be extended to all compounds. The elements of any one series, by combining, give rise to perfectly parallel series of homologous binaries, some of which are given in the table. The binaries of those series which have the greatest common difference are generally acids; and of those which have the smallest, they are generally bases. These acids and bases unite together and form series of homologous salts. As in organic chemistry, many of the series are very incomplete; but they are much more generally perfect than in that newer department of the science, and almost every day fills up some gap.

It will be seen, then, that not merely a plan has been given for classifying the elements, but one which will also embrace all inorganic compounds, and affiliate with the similar classification which has already been established in organic chemistry. We have not attempted to develop such a classification, since to do it would require a volume; nor is it necessary, as any one can develop it for himself.

That the atomic weights of the series of homologous compounds follow the same numerical law as those of the elements is easily shown. Take as an example the series of salts homologous with KO, NO_5, which may be expressed in general by

KO, RO$_5$, where R is any member of the Six Series after oxygen, and whose atomic weight, therefore, equals $8+n6$. The atomic weight of KO, RO$_5$ must be necessarily $39\cdot5+48+(8+n6)$, or $95\cdot5+n6$. As this symbol differs from that of the Six Series only in the nucleus, the atomic weights of the salts which are represented by it must progress by the same differences as those of the corresponding elements.

The properties of these series of homologous compounds will also be found to vary in a regular manner, and the law of the progression of the specific gravities in the gaseous state can be easily expressed algebraically, since in each series the quotient of the specific gravity divided by the atomic weight is a constant quantity. As an illustration, we may take the series of binaries homologues of water given in the Nine Series of our table. It follows from what has been said, that the atomic weights of these compounds equals $9+n9$. With each $\frac{\text{Sp. Gr.}}{9+n9}=\frac{1}{2}$, therefore Sp. Gr. $=4\cdot5+n4\cdot5$. We give below a table of the observed or calculated specific gravities, not only of these compounds, but also of those homologues of NH$_3$ whose specific gravity has been observed.

HOMOLOGUES OF WATER. Sp. Gr./At. W. = ½. Sp. Gr. = 4·5 + n4·5.			HOMOLOGUES OF AMMONIA GAS. Sp. Gr./At. W. = ½. Sp. Gr. = 5·5 + n3.		
Symbols.	SPECIFIC GRAVITIES.		Symbols.	SPECIFIC GRAVITIES.	
	Theoret.	Observed.		Theoret.	Observed.
HO	4·5	9	NH$_3$	·8·5	8·5
HFl	9		PH$_3$	17·25	17·5
HCl	13·5	13·5	AsH$_3$	39	38·5
HBr	40·5	39·5			
HI	63	·63·5			

As the series of compounds give a greater scope for investigating the relations of properties than is presented by those of the elements, we may expect that these relations will be first discovered in the former, and to my conceptions chemistry will then have become a perfect science, when all substances have been classed in series of homologues, and when we can make a table which shall contain, not only every known substance, but also every possible one, and when by means of a few general formulæ we shall be able to express all the properties of matter, so that when the series of a substance and its place in its series are given, we shall be able to calculate, nay, predict, its properties with absolute certainty; and when our chemical treatises shall have been reduced to tables of homologues, and our laws comprised in a few algebraic formulæ, then the dreams of the ancient alchemist will be realized, for the problem of the transmutation of the elements will have been theoretically, if not practically, solved.

EXPLANATION OF THE TABLE.

The formula at the head of each series is a general expression for the atomic weights of that series. The names of the series are derived from the "Common Differences," which are the numbers multiplied by n in the general formulæ. In the columns headed "Theoretical" are given the atomic weights calculated from these formulæ and the values of n given in the last columns at the right of each division of the table. In the columns headed "Observed" will be found the observed values of the same atomic weights. These have been taken from the table of atomic weights given in the last volume of Liebig and Kopp's *Jahresbericht* (for 1852), with the exception of those against which are placed the initials of the observers. The last were taken from Weber's *Atomgewichts Tabellen*. In some cases the atomic weight is taken at twice its received value, but it is then underlined. The compounds in any one column at the right of the names of the elements are homologous. In the same way, those in any one at the left are isomorphous. The numbers at the head of these last columns indicate crystalline systems as follows: 1. Monometric; 2. Dimetric; 3. Trimetric; 4. Monoclinic; 5. Triclinic; 6. Hexagonal. The data from which the table was compiled were drawn from numerous sources, but especially from the following works: Gmelin's *Handbook of Chemistry*, Graham's *Elements of Chemistry*, Phillips's *Mineralogy* by Brooke and Miller, and Gustav Rose's *Krystallo-chemische Mineralsystem*. References have been given only in a few cases, to avoid crowding the tables. For authorities in other cases, the author would refer to the abovementioned works.

A FIRST FORESHADOWING OF THE PERIODIC LAW.

IT is well known that the Newlands-Mendeleeff classification of the elements was preceded by the discoveries of certain numerical relations between the atomic weights of allied elements, due to Döbereiner, Dumas, and others; but what has been almost entirely ignored is the immense advance made by M. A. E. Béguyer, de Chancourtois,[1] a French geologist of note, Professor at the École des Mines, who was the first to publish *a list of all the known elements in the order of their atomic weights.*

M. de Chancourtois embodied his results in two memoirs presented to the French Academy of Sciences in April 1862 and March 1863. These memoirs have never been printed *in extenso*,[2] but extracts from them, and additional notes relating to the subject, were published in the *Comptes rendus* for 1862 (liv. pp. 757, 840, and 967; lv. p. 600), 1863 (lvi. pp. 253 and 479), and 1866 (vol. lxiii. p. 24). The first note bears the date of April 7, 1862, so that there can be no doubt as to de Chancourtois's claim to priority in this important matter.[3]

I have in my possession a thin quarto pamphlet, by de Chancourtois, entitled " Vis Tellurique : Classement naturel des corps simples ou radicaux, obtenu au moyen d'un système de classification hélicoïdal et numérique" (Paris, Mallet-Bachelier,[4] 1863), which contains nearly all the extracts from the *Comptes rendus*, together with some additional matter. It contains, also, what is absolutely essential to the comprehension of de Chancourtois's ideas, the graphic representation of his system, which is not to be found in the *Comptes rendus*.

I propose to give here a translation of the first communication to the Academy, followed by certain explanatory comments and brief extracts from the other papers :—

" Geological studies in the field of research opened up by M. Elie de Beaumont in his note on volcanic and metalliferous intrusions (*émanations*) have led me, for the completion of a lithological memoir on which I am now engaged, to a natural classification of the simple bodies and radicles by a table in the form of a helix, founded on the use of numbers which I call *characteristic numbers* or *numerical characteristics*.

" My numbers, which are immediately deduced from the measure of the equivalents or other physical or chemical capacities of the different bodies, are, in the main, the proportional numbers given by the treatises on chemistry, these being reduced to half in the case of hydrogen, nitrogen, fluorine, chlorine, bromine,

iodine, phosphorus, arsenic, lithium, potassium, sodium, and silver; in other words, I either divide the equivalents of these bodies by two in the system in which oxygen is taken as 100, or multiply by two the equivalents of the other bodies in the system in which hydrogen is taken as unity.

" On a cylinder with a circular base, I trace a helix which cuts the generating lines at an angle of 45°. I take the length of one turn of the helix as my unit of length, and starting from a fixed origin, I mark off on the helix lengths corresponding to the different *characteristic numbers* of the system in which the number for oxygen is taken as unity. The extremities of the lines thus marked off determine points on the cylinder which I call indifferently *characteristic points* or *geometrical characters*, and which I distinguish by the ordinary symbols for the different bodies. The same points will evidently be obtained if we take as the unit of length the $\frac{1}{10}$ of a turn of the helix, and mark off on the curve lengths corresponding to the numbers of the system in which hydrogen is represented by unity.

" The series of points thus determined constitutes the graphic representation of my classification, which may easily be traced on a plane surface by supposing the surface of the cylinder developed; by its aid I am enabled to enounce the fundamental theorem of my system : *The relations between the properties of different bodies are manifested by simple geometrical relations between the positions of their characteristic points.*

" For instance, oxygen, sulphur, selenium, tellurium, bismuth,[1] fall approximately on the same generating line, while magnesium, calcium, iron, strontium, uranium, and barium, fall on the opposite generating line. On either side of the first of these lines we find hydrogen and zinc on the one hand, bromine and iodine, copper and lead on the other; parallel to the second line we find lithium, sodium, potassium, manganese, &c.

" Simple relations of position on a cylindrical surface would be obviously defined by means of helices, of which the generating lines are only a particular case; hence, as a complement to the first theorem, we may add the following : *Each helix drawn through two characteristic points and passing through several other points or only near them, brings out relations of a certain kind between their properties; likenesses and differences being manifested by a certain numerical order in their succession, for example, immediate sequence or alternation at various periods.*

" In order to attain a greater degree of accuracy, it is necessary to discuss the results of different measurements with respect to the same body.

" One question is all-important in this discussion; it is to know if the divergencies which occur may have causes other than the error of experiment. I reply to this question in the affirmative.

" I think that here, as in all determinations of constants which we wish to compare, they must be reduced to the same conditions. This idea seems to me the indispensable complement to the notion of an absolute characteristic number. Once the existence of this absolute number or *numerical characteristic* guaranteed by the possibility of connecting it afresh with observed facts, certain limits of variation being allowed [*literally*, varying within certain limits], we promptly arrive at Prout's law, which presents itself as furnishing a means for reducing experimental observations to a comparable state by a series of trials, without this state being even a completely defined one, but, on the contrary, in order to be able to define it. The combination of this principle with the rules for alignment allow me to give the most striking form to my invention. I am thus led to formulate the table of integral numbers, which, as I must not omit to mention, exhibits under certain aspects the *résumé* of the work of M. Dumas on this subject.

" In the construction of this table I have had recourse to the determinations of specific heats, not only as a means of control, but also to find new numbers unattainable by the methods of chemical investigation. By adopting as the constant product of specific heat by atomic weight, the number which corresponds both to sulphur and to lead, I have deduced from the series of results given by M. Regnault, purely *thermic* quotients or numbers, which take their places on my alignments in the most felicitous way. I will only quote two examples : firstly, the number 44, obtained from the specific heat of the diamond, which finds its place on the generating line of the characteristic, 12, of carbon, by the side of the characteristic, 43, which corresponds to one of the equivalents generally accepted for silicon; and another

---

[1] Wurtz (" The Atomic Theory," p. 170) and Berthelot (" Les Origines de l'Alchimie," p. 302) give a bare mention of de Chancourtois's name. Mendeleeff, in his Faraday Lecture (Journ. Chem. Soc., October 1889), couples his name with those of Newlands and Strecker, and shows greater appreciation of his work.

[2] M. Friedel, the eminent Professor of Organic Chemistry at the Sorbonne, has kindly procured for me the information that the original manuscripts of these memoirs are preserved in the archives of the Institut; I hope to be able to examine them at some future period.

[3] Mr. Newlands' first paper, chiefly devoted to showing that the numerical differences between the atomic weights of allied elements are approximately multiples of 8 was published on February 7, 1863 (*Chemical News*, vol. vii. p. 70); his second paper, in which he arranges the elements in the order of their atomic weights, was published on July 30, 1864 (*Chemical News*, vol. x. p. 39). See J. A. R. Newlands "On the Discovery of the Periodic Law," &c. (Spon, 1884).

[4] Now Gauthier-Villars.

[1] This is probably a misprint, as bismuth does *not* fall on the same generating line in the table.

characteristic, 36, of silicon deduced from an equivalent proposed by M. Regnault, and which is very remarkable, from its coincidence with the characteristic of ammonium.

"By the discussion, which has shown me the advisability of accepting various results hitherto looked on as scarcely reconcilable, I have been led to conceive the possibility of reproducing the *series of natural numbers* in the series formed by the numerical characteristics of the real or supposed simple bodies supplemented by the characteristics of the compound radicles formed from gazolytic [1] elements, such as cyanogen, the ammoniums, &c., and doubtless also by the compound radicles formed from metallic elements, of which the alloys offer us an example. In this natural series, the bodies which are really simple, or at least irreducible by the ordinary means at our disposal, would be represented by the *prime numbers*. It will be at once seen that there are in my table at least twelve bodies, which, like sodium (23), have characteristics which are prime numbers. This is what led me to perceive this law, which, I believe, is destined, when established, to form one of the bases for the discovery of the law of molecular attraction. The predominance of the law of divisibility by 4 in the series of my table, a predominance which is also to be found in the elements of the theory of numbers, has confirmed me in the idea—an idea in itself really simple—that there is a perfect agreement between bodies, the elements of the material order, and numbers, the elements of the abstract order of things (*éléments de la variété matérielle, de la variété abstraite*). This goal once caught sight of, it will seem natural that I should have recourse to the theory of numbers to help me attain it. It will seem not less natural that I should also have recourse to higher geometry; for the series of relations it offers cannot fail to afford resources which may enable one to establish connections between the different numerical characteristics.

"My helicoidal system in this way leads me on towards abstract views of an extremely general nature; and on the other hand it should, it seems to me, find an application in the natural [2] sciences, as a method of classification throughout their whole domain, from the series of simple bodies which forms the prototype, to the opposite extreme of our natural divisions; in it will be found, I believe, the means of bringing into connection simultaneously, and by all their characters, the different terms of those parallel series, orders, families, genera, species, and races, in each natural kingdom, of which men of science have in vain tried to show the affiliation. In geology, as is evident, the application is implicit.

"Whatever may be the import of these considerations, and to return to the principal object of the present memoir, I think that the efficacy of the helicoidal system will be admitted as a means towards hastening the advent of the time when chemical phenomena shall be amenable to mathematical investigations.

"My table, by the distribution of bodies in simple or coupled series, by its indication of the existence of conjugate groups, &c., traces a plan for diverse categories of syntheses and analyses already executed or to be executed; it draws up very definite programmes for the execution of several researches which are exciting attention. Will not my series, for instance, essentially chromatic as they are, be a guide in researches on the spectrum? Will not the relations of the different rays of the spectrum prove to be derived directly from the law of numerical characteristics, or *vice versâ*? This idea, which occurred to me before we were taught the identification of the lines in the spectrum, and the admirable applications of this discovery, seems to me now even more than probable. Finally, looking upon it only as a concise representation of known facts, and reducing it to the points which offer no matter for discussion, the geometrical table of numerical characteristics affords a rapid method for teaching a large number of notions in physics, chemistry, mineralogy, and geology. I hope, therefore, that my natural classification of the simple bodies and radicles being capable of rendering manifold services, will need, like every object in habitual use, a name of easy application; hence, on account of its graphic representation and its origin, I give it the significant name of *telluric helix*."

It will be well to point out immediately that M. de Chancourtois's system assigns to the *numerical characteristics* of the elements a general formula of the form $(n + 16n')$, where n' is necessarily an integer; [3] and his table thus brings out the fact

[1] Metalloid.
[2] The term includes physical science.
[3] *n* is always represented in the author's table as integral, but he expressly states that he looks on this as by no means necessary. "The construction of the telluric helix rests on the use of proportional numbers derived from

that the differences between the atomic weights of allied bodies approximate in many cases to multiples of 16.

Thus we get the parallel series of which our author speaks—

Li		Na		K		Mn
7	...	$7 + 16 = 23$	...	$7 + 2 \cdot 16 = 39$	...	$7 + 3 \cdot 16 = 55$

$$Rb$$
$$7 + 5 \cdot 16 = 87. [2]$$

O		S		Se		Te
16	...	$16 + 16 = 32$	...	$16 + 4 \cdot 16 = 80$	...	$16 + 7 \cdot 16 = 128. [3]$

As we glance at the first two turns of de Chancourtois's helix, we ask ourselves if the discovery of Newlands and Mendeleeff does not lie before us.

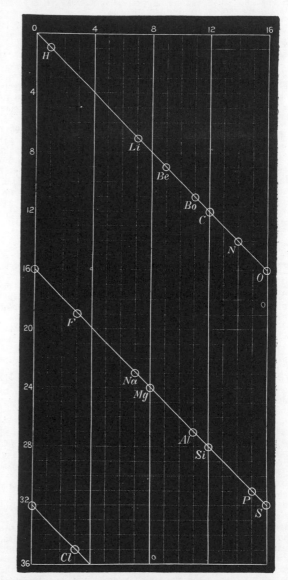

But the discovery of the "octaves" or "periods" cannot be ascribed to our author, although it seems almost impossible that chemists should not have perceived their existence on looking at his table.

experiment. It would remain valid with fractional numbers, and often the helicoidal alignments would be even more easily applicable to these than to integers" (*Comptes rendus*, vol. liv. p. 842).

[1] This fact, now familiar, has again been noticed by your correspondent, Mr. A. M. Stapley, in the issue of November 21, 1889.
[2] The atomic weight of rubidium should be 85. We may notice that the author gives as probable also $Cs = 135 = 7 + 8 \cdot 16$, which is thus placed on the same generating line.
[3] Certainly too high a value; according to Brauner, the exact atomic weight of tellurium remains to be determined.

Three important points, however, do exist in common between de Chancourtois's system and that of Mendeleeff:—

Firstly, all the known elements are arranged in the order of their combining weights.

Secondly, the combining weights chosen as best suited to bring out clearly the numerical relations existing between them are those adopted by Cannizzaro in 1858, a striking fact when we recollect that de Chancourtois wrote only in 1862, at a date long before these numbers had gained anything like general acceptance.

Lastly, an attempt is made to show that simple numerical relations exist, not only between the combining weights, but between all the measurable properties (*toutes les capacités physiques et chimiques*) of allied elements.

The reasons for the neglect of de Chancourtois's researches and the oblivion into which they have fallen are not far to seek. His style was heavy and at times obscure, and, moreover, his ideas were presented in a way most unattractive to chemists.

A geologist by profession, de Chancourtois had been powerfully impressed by the facts of isomorphism in the case of the feldspars and pyroxenes, which form such important constituents of the volcanic rocks he was studying, and he was thus led to seek for a system of classification which should bring out some simple relationship between the elements they contained.

To quote from his paper (*Comptes rendus*, vol. liv. p. 969): "The parallelism of the groups of manganese $(7 + 3 \cdot 16)$ and iron $(8 + 3 \cdot 16)$, of potassium $(7 + 2 \cdot 16)$ and calcium $(8 + 2 \cdot 16)$, of sodium $(7 + 16)$ and magnesium $(8 + 16)$, is the origin of my system"; and again, suggesting the expediency of adopting $55 (= 7 + 3 \cdot 16)$ as a *characteristic* for aluminium, which would bring the element on the sodium and potassium generating line, "this would render perfect the parallelism between the elements of the feldspars and the pyroxenes, the starting-point of my system" (*Comptes rendus*, lvi. p. 1479).

Thus the correct idea of seeking for a relationship between the combining weights of isomorphous elements was marred by a somewhat imperfect comprehension of the facts of isomorphism. No chemist would certainly have tried to show any close relationship between aluminium on the one hand and the group of the alkalies on the other, notwithstanding their union in the feldspars and pyroxenes; and a suggestion of this kind served to cast discredit on de Chancourtois's really important views.

Notwithstanding his frequently eccentric ideas, de Chancourtois had the merit, so rare in an inventor of this stamp, of not considering his system as final. We cannot do better than let him speak for himself; and quote the conclusion of his last paper on the subject (*Comptes rendus*, lvi. p. 481):—"In presence of the rapid increase in the list of elements which engage the attention of chemists and physicists, it has become urgent to unite in one synthesis all the notions of chemical and physical capacities, of which the exposition would otherwise become an impossible task.

"It is, therefore, perhaps not unnecessary to recall the ideas of Pythagoras, or what I may better term the *Biblical truth* which dominates all the sciences, and of which I propose to make practical use by the following concrete example,[1] the first general conclusion of my essay:—

"THE PROPERTIES OF BODIES ARE THE PROPERTIES OF NUMBERS.

"It is easily perceived, that a helicoidal system of some kind or another, which is necessarily a graphic table of divisibility, offers the most convenient means for rendering manifest the relations between the two orders of ideas. It is evident, also, that the particular system which I have adopted brings into relief the relations of the most important and usual of the properties of matter, because the case of divisibility by 4, which is the basis of my plan, is the first which presents itself in arithmetical speculation after the case of divisibility by 2, to which there corresponds directly, as one perceives by a first glance at my table, the existence of the natural couples of elements, with consecutive odd and even characteristics.

"I hope, therefore, that the *telluric helix* will offer, until it is replaced by some more perfect invention, a practical framework, a convenient scale, on which to set down and compare all measurements of capacities, whatever the point of view which may be taken, whatever elasticity or variation, whatever interpretation may be given to the *numerical characteristics*, by which these capacities must always be represented.

[1] The French is *vulgarisation*, literally *popularization*.

"The development in a plane of the cylinder ruled into squares, with the circumference at the base divided into 16 equal parts, seems to me, in a word, to be a *stave* on which men of science, after the fashion of musicians, will note down the results of their experimental or speculative studies, either to co-ordinate their work, or to give a summary of it in the most concise and clear form to their colleagues and the public."

Lothar Meyer has noted down his classification in the form of a helix,[1] and Dr. Johnstone Stoney, F.R.S., has shown that the numerical values of the atomic weights may be expressed geometrically as functions of a series of integral numbers by points all lying approximately on a logarithmic spiral.

But no simple mathematical formula has so far been discovered to express the relationships of the atomic weights accurately—*i.e.* within the limits of experimental error, and de Chancourtois's predictions still remain but incompletely fulfilled.

I need not comment further on the remarkable breadth and originality of our author's views, taken as a whole. Strange to say, it was only a year or two before his death that he heard, through a colleague, of the immense development they had undergone; nor did he ever set up any claims to priority. But when we speak of the greatest generalization of modern chemistry, and recall the names of Newlands and Mendeleeff, it is only just that we should no longer forget their distinguished precursor, de Chancourtois.

P. J. HARTOG.

[1] "Die modernen Theorien der Chemie," iv. Auflage, p. 137; English translation, p. 118.

CORRESPONDENCE.

Boiler Incrustation.

To the Editor of the CHEMICAL NEWS.

SIR,—I feel no jealousy from your correspondent, Mr. Napier, setting forth his prior suggestion as to a mode of preventing the great evil of incrustation in steam boilers. Alkalies have been used for this purpose for at least a quarter of a century; in fact, most of the nostrums, patented or otherwise puffed, have soda or soda-ash for their basis. The small merit that belongs to my suggestion arises from its simplicity, and the minimum of attention it requires from the workmen—a single act daily, and that done in two minutes—insures perfect immunity from incrustation. I have now had a satisfactory experience of more than twelve months; and my only object in thus referring to your correspondent's letter is in order, by the further publicity thus given to the matter, to induce some of your numerous readers who may be subject to the annoyance to give my plan a trial, and they will at once cure the evil.
                          I am, &c.   PETER SPENCE.
Pendleton Alum Works, Newton Heath, Manchester.

Preservation of Stone.

To the Editor of the CHEMICAL NEWS.

SIR,—If none but professional chemists perused the pages of your Journal we should not have troubled you further in the matter of preservation of stone; but knowing full well that your columns stand high in estimation as affording valuable data for the non-professional world, we would beg the privilege of a reply to the very specious explanation of Messrs. Rust on the matter of the silicate of alumina.

Messrs. Rust acknowledge the abandonment of their provisional patent for the adoption of our idea of using in one solution the agent and re-agent, so that the grave objections to the use of two solutions were thus seen, and the discovery appropriated by them forthwith, with this unimportant variation in its application, that they profess to keep their materials in stock enclosed in one jar, until required for use, instead of two; and, further, that being doubtful of finding lime or magnesia enough in a lime or magnesian limestone, they add to their solution this very scarce and essential agent, after an assertion that their solution has no free potash, and implying that ours has a detrimental quantity. They inform the readers of a technical

work on an abstruse science that they "need go no further, for sufficient has probably been said to satisfy an unprejudiced mind," &c., &c.

If all conclusions were thus to be arrived at, and rights thus to be set at nought, it would certainly create an unwelcome revolution in that science whose existence almost depends on the laborious and persevering researches of truthful and right-minded men, whose theories and deductions are sought to be balanced by the infinitesimal portion of a grain.   I am, &c.
                          SANDERS TROTMAN,
                          Pro BARTLETT BROS. AND CO.
Devonshire Wharf, Camden-town, N.W.

Electrical Decomposition.

To the Editor of the CHEMICAL NEWS.

SIR,—I will, with your kind permission, ask a question or two, on which I am anxious for exact information. In books on the subject, 33 parts of zinc, dissolved *in* a battery, are represented as precipitating 31 parts of copper, 108 parts of silver, &c., *out* of the battery.

1. Now, does this law hold good in all circumstances, or for composite solutions in which a minute quantity of some good conductor is diffused, and combined with other substances less conducting? 2. Or, for simple, but very weak, solutions of salts of a good conductor, as chlorides of platinum or silver?

3. In these cases, would not a weak battery be best, so that the current may be sufficient only to carry off the trace or traces of the best conductor first?

4. But would not several cells or pairs of elements be necessary in order to overcome resistance, or push the one-cell electricity through the solution, and thus a consumption of zinc be occasioned merely to generate intensity or pushing force?   I am, &c.   C. R.

The Cavendish Society.

To the Editor of the CHEMICAL NEWS.

SIR,—As the time for the annual meeting of the Cavendish Society is drawing near, I wish to throw out a suggestion to the subscribers and council of that society. It seems to be useless to wait for the completion of Gmelin's Handbook before the society is reconstructed. That book may perhaps be finished some day, but, instead of in two years, as the council expected, we shall, at the present rate of publication, have to wait at least six years. During all this time, the present society will stand in the way of a more useful and energetic association.

I propose, then, that a new society be formed immediately, the present to hand over the balance of funds, which are to be entirely devoted to the completion of Gmelin; that those who have paid their subscriptions to the present time receive the books due, but after these have been issued, the remaining volumes and the index be issued by the new society to those who wish them, at cost price, allowing for balance of funds handed over by the old society.

I do not think there would be much difficulty in starting a new society, as, no doubt, all who subscribe to the old would continue their subscription. All that is needed is a little more activity and enterprise in the management, and some care in the selection of books. There will never again be occasion to enter on such an engagement as the publication of a book on the same scale as Gmelin—for which, after all, the subscribers must be grateful; and, unsaddled by such a weight, there is no reason why a regenerated Cavendish Society should not be active and prosperous.   I am, &c.
              A SUBSCRIBER TO THE CAVENDISH SOCIETY.

On Relations among the Equivalents.

To the Editor of the CHEMICAL NEWS.

SIR,—Many chemists, and M. Dumas in particular, have, on several occasions, pointed out some very interesting relations between the equivalents of bodies belonging to the same natural family or group ; and my present purpose is simply to endeavour to proceed a little further in the same direction. I must, however, premise that many of the observations here collected together are well known already, and are only embodied in my communication for the purpose of rendering it more complete.

Before proceeding any further, I may also remark, that in the difficult task of grouping the elementary bodies, I have been guided more by chemical characteristics than by physical appearances, and have, therefore, taken no notice of the ordinary distinction between metals and non-metallics. The numbers which I have attached to the various groups are merely for the purpose of reference, and have no further significance whatever. For the sake of perspicuity, I have employed the old equivalent numbers, these atomic weights being, with one or two exceptions, taken from the 8th edition of "Fownes' Manual."

The following are among the most striking relations observed on comparing the equivalents of analogous elements. (In order to avoid the frequent repetition of the word "equivalent," I have generally used the names of the different elements as representing their equivalent numbers —thus, when I say that zinc is the mean of magnesium and cadmium, I intend to imply that the equivalent of zinc is the mean of those of magnesium and cadmium, and so on, throughout the paper) :—

Group I. Metals of the alkalies :—Lithium, 7 ; sodium, 23 ; potassium, 39 ; rubidium, 85 ; cæsium, 123 ; thallium, 204.

The relation among the equivalents of this group (see CHEMICAL NEWS, January 10, 1863) may, perhaps, be most simply stated as follows :—

1 of lithium + 1 of potassium = 2 of sodium.
1   „   + 2   „   = 1 of rubidium.
1   „   + 3   „   = 1 of cæsium.
1   „   + 4   „   = 163, the equivalent of a metal not yet discovered.
1   „   + 5   „   = 1 of thallium.

Group II. Metals of the alkaline earths :—Magnesium, 12 ; calcium, 20 ; strontium, 43·8 ; barium, 68·5.

In this group, strontium is the mean of calcium and barium.

Group III. Metals of the earths :—Beryllium, 6·9 ; aluminium, 13·7 ; zirconium, 33·6 ; cerium, 47 ; lanthanium, 47 ; didymium, 48 ; thorium, 59·6.

Aluminium equals two of beryllium, or one-third of the sum of beryllium and zirconium. (Aluminium also is one-half of manganese, which, with iron and chromium, forms sesquioxides, isomorphous, with alumina.)

1 of zirconium + 1 of aluminium = 1 of cerium.
1   „   + 2   „   = 1 of thorium.

Lanthanium and didymium are identical with cerium, or nearly so.

Group IV. Metals whose protoxides are isomorphous with magnesia :—Magnesium, 12 ; chromium, 26·7 ; manganese, 27·6 ; iron, 28 ; cobalt, 29·5 ; nickel, 29·5 ; copper, 31·7 ; zinc, 32·6 ; cadmium, 56.

Between magnesium and cadmium, the extremities of this group, zinc is the mean. Cobalt and nickel are identical. Between cobalt and zinc, copper is the mean. Iron is one-half of cadmium. Between iron and chromium, manganese is the mean.

Group V.—Fluorine, 19 ; chlorine, 35·5 ; bromine, 80 ; iodine, 127.

In this group bromine is the mean between chlorine and iodine.

Group VI.—Oxygen, 8 ; sulphur, 16 ; selenium, 39·5 ; tellurium, 64·2.

In this group selenium is the mean between sulphur and tellurium.

Group VII.—Nitrogen, 14 ; phosphorus, 31 ; arsenic, 75 ; osmium, 99·6 ; antimony, 120·3 ; bismuth, 213.

In this group arsenic is the mean between phosphorus and antimony.

Osmium approaches the mean of arsenic and antimony, and is also almost exactly half the difference between nitrogen and bismuth, the two extremities of this group ; thus, $\frac{213-14}{2}=99\cdot5$.

Bismuth equals 1 of antimony + 3 of phosphorus ; thus, 120·3 + 93 = 213·3.

Group VIII.—Carbon, 6 ; silicon, 14·20 ; titanium, 25 ; tin, 58.

In this group the difference between tin and titanium is nearly three times as great as that between titanium and silicon.

Group IX.—Molybdenum, 46 ; vanadium, 68·6 ; tungsten, 92 ; tantalium, 184.

In this group vanadium is the mean between molybdenum and tungsten.

Tungsten equals 2 of molybdenum, and tantalium equal 4 of molybdenum.

Group X.—Rhodium, 52·2 ; ruthenium, 52·2 ; palladium, 53·3 ; platinum, 98·7 ; iridium, 99.

In this group the first three are identical, or nearly so, and are rather more than half of the other two. (I may mention, by the way, that platinum is rather more than the half of gold ; thus, 98·7 × 2 = 197·4, gold being 197.)

Group XI.—Mercury, 100 ; lead, 103·7 ; silver, 108.

Lead is here the mean of the other two.

If we deduct the member of a group having the lowest equivalent from that immediately above it, we frequently observe that the numbers thus obtained bear a simple relation to each other, as in the following examples :—

Member of group having lowest equivalent.		One immediately above the preceding.		Difference.
Magnesium	. 12	Calcium	. 20	8
Oxygen .	. 8	Sulphur	. 16	8
Carbon .	. 6	Silicon	. 14·2	8·2
Lithium .	. 7	Sodium	. 23	16
Fluorine .	. 19	Chlorine	. 35·5	16·5
Nitrogen .	. 14	Phosphorus .	. 31	17

A similar relation, though not quite so obvious as the above, may be shown by deducting the lowest member of a triad from the highest. The numbers thus obtained in the different triads correspond to a great extent. (By a triad I understand a group of three analogous elements, the equivalent of one of which is the mean of the other two.) Of this relation I append a few examples :—

Lowest term of triad.		Highest term of triad.		Difference.
Lithium .	. 7	Potassium .	. 39	32
Magnesium	. 12	Cadmium .	. 56	44
Molybdenum	. 46	Tungsten .	. 92	46
Sulphur .	. 16	Tellurium .	. 64·2	48·2
Calcium .	. 20	Barium .	. 68·5	48·5
Phosphorus .	. 31	Antimony .	. 120·3	89·3
Chlorine .	. 35·5	Iodine .	. 127	91·5

In the relation previously pointed out, the difference between the lowest member of a group, and the next above it, was either 8, or 8 × 2 = 16 ; and in the first of these triads the difference is 8 × 4 = 32 ; in the next four it approaches 8 + 6 = 48 ; and in the two last triads it is nearly twice as great.

The difference between the highest member of the platinum group, viz., iridium 99, and the lowest, rhodium 52·2, is 46·8, a number which approximates very closely to those obtained in some of the above triads ; and it, therefore, appears possible that the platinum metals are the extremities of a triad, the central term or mean of which is at present unknown.   I am, &c.

J. A. R. N.

P.S. With the view of economising space I have omitted most of the calculations, which, however, are very simple, and can be verified in a moment by the reader. The equivalents thus obtained by calculation will be found to approximate to those procured by experiment, as closely as can be expected in such cases.

I also freely admit that some of the relations above pointed out are more apparent than real; others, I trust, will prove of a more durable and satisfactory description.

Chemical Notices from Foreign Sources.

I. MINERAL CHEMISTRY.

Action of Chloride of Phosphorus on some Metallic Sulphides.—M. Baudrimont sums up the results of his experiments as follows (*Comptes-Rendus*, t. lv., p. 378):—1. Protochloride of phosphorus easily attacks metallic sulphides at a red-heat. With those of barium, calcium, &c., it produces a splendid incandescence, and gives immediately the phosphorous sulphide PS_3, and the corresponding metallic chlorides. With the sulphides of antimony, lead, mercury, &c., it furnishes at first sulpho-phosphides by the union of the sulphide of phosphorus with the metals; these are afterwards destroyed by an excess of PCl_3, and are completely changed into PS_3. 2. Perchloride of phosphorus reacts twice on sulphide of hydrogen, producing at first chloro-sulphide of phosphorus, and afterwards, at a higher temperature, phosphoric sulphide, PCl_5. 3. It behaves in the same way towards earthy and alkaline sulphides. 4. With the sulphides of antimony, tin, lead, mercury, &c., it produces the same results in the end, but forms intermediary sulpho-phosphides corresponding to PS_3 rather than PS_5. 5. The action of perchloride of phosphorus on sulphide of antimony is the basis of an easy method for the preparation of chloro-sulphide of phosphorus—(see CHEMICAL NEWS, vol. v., p. 41). 6. Chloro-sulphide of phosphorus attacks metallic sulphides exactly as it does PCl_5. 7. The sulpho-phosphide of mercury obtained either by the reaction PCl_3 or PCl_5 on cinnabar, has for its formula $PS_3,3$ (HgS), and appears to be the type of a group of sulpho-salts different from those described by Berzelius.

Compounds of Perchloride of Phosphorus with other Chlorides—M. Baudrimont has also devoted a long series of experiments to the study of these compounds (*Comptes-Rendus*, t. lv., p. 361). He finds that PCl_5 combines with the chlorides of selenium ($SeCl_2$), of iodine (ICl), of aluminium (Al_2Cl_3), of iron (Fe_2Cl_3), of tin ($SnCl_2$), of mercury ($HgCl$), and of platinum ($PtCl_2$). All these compounds may be obtained either by acting on the simple bodies with PCl_5, or by combining the latter directly with the chlorides. The excess of PCl_5 can always be driven off by careful and prolonged heating to 160° or 180° C. At a higher temperature the double chloride sublimes. All these compounds are solid, volatile, and in some cases partly decomposed by heat. They fume in the air, are changed by moisture, and decomposed by water. The author describes the compounds formed with the chlorides named above, noticing particularly the platinum compound as being the only volatile salt of that metal known, and the iodine compound as being unequalled in causticity.

II. ORGANIC CHEMISTRY.

An Isomer of Amylic Alcohol.—Würtz (*Comptes-Rendus*, t. lv., p. 370) has compared the compound formed by the direct combination of hydriodic acid with amylene, and the iodide of amyl formed with amylic alcohol, and has been led to regard the two compounds, not as identical, but as isomeric. The bodies act very differently on moist oxide of silver, iodide of amyl having no action at the common temperature, while hydriodate of amylene acts energetically even at 0°, the principal product being an organic hydrate, which Würtz regards as an isomer of amyl alcohol. With iodide of amyl and acetate of silver the author obtained an acetate of amyl, with its characteristic odour; but with the hydriodate of

amylene a body having very nearly the composition of acetate of amyl, but possessed of a very different odour. The author points out the relations which exist between the compounds of amyl and of amylene; and expresses his belief that hydrates, homologous with the hydrate of amylene, and isomeric with ordinary alcohols, may be obtained in a similar way from hydrocarbons near to amylene, such as caproylene, œnanthylene, and caprylene.

Transformation of Urea into Sulpho-cyanide of Ammonium.—M. Fleury effected this change (*Comptes-Rendus*, t. lv., p. 519) by heating urea with an excess of bisulphide of carbon according to the equation $C_2H_4N_2O_2 + CS_2 = CO_2 + NH_4.C_2NS_2$. He has not perfectly succeeded in changing sulpho-cyanide of ammonium into urea.

MISCELLANEOUS.

Simpson and Others v. Wilson and Another.—This cause is set down for trial on Monday, the 9th. If any new feature be introduced in the course of the trial we shall report it.

Royal Institution.—On Monday, February 2, a General Monthly Meeting was held, William Pole, Esq., M.A., F.R.S., Treasurer and Vice-President, in the Chair. The Earl of Clanwilliam, Edward W. Cox, Esq., Sir William Augustus Fraser, Bart., General Charles H. Hamilton, C.B., and Peter Vanderbyl, Esq., were elected Members of the Royal Institution. The Secretary reported that the executors of the late James Walker, Esq., F.R.S., M.R.I., had bequeathed to the Institution a marble bust of Professor Faraday, by Mr. Matthew Noble, M.R.I. The thanks of the members were returned to Professor Tyndall, and to his Eminence Cardinal Wiseman, for their discourses on the evening meetings on Fridays, January 23 and 30. The presents received since the last meeting were laid on the table, and the thanks of the members returned for the same. The following lectures will be delivered:—Tuesday, February 10, at 3 o'clock, Professor Marshall, "On Animal Mechanics." Thursday, Feburary 12, at 3 o'clock, Dr. E. Frankland, "On Chemical Affinity." Friday, February 13, at 8 o'clock, Dr. E. Frankland, "On Artificial Illumination." Saturday, February 14, at 3 o'clock, W. S. Savory, Esq., "On Life and Death."

Carbonic Acid as an Anæsthetic.—M. Ozanam has given a mixture of three parts carbonic acid with one part of atmospheric air with success as an anæsthetic. After breathing it for ten minutes the patient became insensible, and an operation was performed without his evincing any sign of pain.

ANSWERS TO CORRESPONDENTS.

*** All *Editorial Communications* are to be addressed to the EDITOR; and *Advertisements* and *Business Communications* to the PUBLISHER, at the Office, 1, Wine Office Court, Fleet Street, London, E.C.

Vol. VI. of the CHEMICAL NEWS, containing a copious Index, is now ready, price 10s. 8d., by post. 11s. 2d., handsomely bound in cloth, gold-lettered. The cases for binding may be obtained at our Office, price 1s. 6d. Subscribers may have their copies bound for 2s. if sent to our Office, or, if accompanied by a cloth case, for 6d. Vols. I. and II. are out of print. All the others are kept in stock. Vol. VII. commenced on January 3, 1863, and will be complete in 26 numbers.

R. H.—We regret the change as much as our correspondent, but circumstances compelled the alteration.

Juvenis.—Noad's lectures, or the volume of Lardner's Handbook of Natural Philosophy containing "Electricity."

Dialyser.—We do not know where the substance can be procured. It is obtained from a sea-weed.

J. A. M. and Co.—Apply to Mr. Wilson, of Price's Patent Candle Company.

A Subscriber to the Cavendish Society should address a letter to the council or secretary of the society.

multiples of eight," any differences between them must also be divisible by eight.

We have here the symbols and the atomic weights of sixty-one elements, placed in their numerical order, and in the third column is the difference between each atomic weight and the one immediately preceding it :—

H .	1		Ca .	40	1	Ce .	92	2·5	V .	137	0
Li .	7	6	Ti .	50	10	La .	92	0	Ta .	138	1
G .	9	2	Cr .	52·5	2·5	Di .	96	4	W .	184	46
B .	11	2	Mn .	55	2·5	Mo .	96	0	Nb .	195	11
C .	12	1	Fe .	56	1	Ro .	104	8	Au .	196	1
N .	14	2	Co .	58·5	2·5	Ru .	104	0	Pt .	197	1
O .	16	2	Ni .	58·5	0	Pd .	106·5	2·5	Ir .	197	0
Fl .	19	3	Cu .	63·5	5	Ag .	108	1·5	Os .	199	2
Na .	23	4	Y .	64	0·5	Cd .	112	4	Hg .	200	1
Mg .	24	1	Zn .	65	1	Sn .	118	6	Tl .	203	3
Al .	27·5	3·5	As .	75	10	U .	120	2	Pb .	207	4
Si .	28	0·5	Se .	79·5	4·5	Sb .	122	2	Bi .	210	3
P .	31	3	Br .	80	0·5	I .	127	5	Th .	238	28
S .	32	1	Rb .	85	5	Te .	129	2			
Cl .	35·5	3·5	Sr .	87·5	2·5	Cs .	133	4			
K .	39	3·5	Zr .	89·5	2	Ba .	137	4			

Now, it will be observed that in all the above differences the number eight occurs but once, and we never meet with a multiple of eight, whereas if the law of "Studiosus" were true the equivalents of the elements, in whatever order they might be placed, should, when not identically the same, differ either by eight or by some multiple of eight in every case.

While upon the subject of "relations among the equivalents," I may observe that the most important of these may be seen at a glance in the following table :—

		Triad.			
		Lowest term.	Mean.	Highest term.	
I.		Li 7 +17 =Mg 24	Zn 65	Cd 112	
II.		B 11			Au 196
III.		C 12 +16 =Si 28		Sn 118	
IV.		N 14 +17 =P 31	As 75	Sb 122	+88 = Bi 210
V.		O 16 +16 =S 32	Se 79·5	Te 129	+70 = Os 199
VI.		F 19 +16·5=Cl 35·5	Br 80	I 127	
VII.	Li 7 +16 =Na 23	+16 =K 39	Rb 85	Cs 133	+70 = Tl 203
VIII.	Li 7 +17 =Mg 24	+16 =Ca 40	Sr 87·5	Ba 137	+70 = Pb 207
IX.			Mo 96	V 137	W 184
X.			Pd 105·5		Pt 197

This table is by no means so perfect as it might be ; in fact, I have some by me of a more complete character, but as the position to be occupied by the various elements is open to considerable controversy, the above only is given as containing little more than those elementary groups the existence of which is almost universally acknowledged.

I now subjoin a few explanatory remarks on the different groups contained in the above table, the number attached to each group being merely for the purpose of reference.

Group II.—Boron is here classed with gold, both these elements being triatomic, although the latter is sometimes monatomic.

Group III.—Silicon and tin stand to each other as the extremities of a triad. Titanium is usually classed along with them, and occupies a position intermediate between silicon and the central term or mean of the triad, which is at present wanting ; thus,

$$\frac{Si\ 28 + Sn\ 118}{2} = 73, \text{ mean of triad, and}$$

$$\frac{Si\ 28 + \text{Mean of triad } 73}{2} = 50·5, \text{ the eq. of Ti being 50.}$$

Group IV.—The equivalent of antimony is nearly the mean of those of phosphorus and bismuth ; thus,

$$\frac{31 + 210}{2} = 120·5, \text{ the eq. of Sb being 122.}$$

Relations between Equivalents.

To the Editor of the CHEMICAL NEWS.

SIR,—In your impression of the 2nd inst. a correspondent, under the name of "Studiosus," has called attention to the existence of a law to the effect "that the atomic weights of the elementary bodies are, with few exceptions, either exactly or very nearly multiples of eight."

Now, in a letter "On Relations among the Equivalents," which was signed with my initials, and inserted in the CHEMICAL NEWS of February 7, 1863, I called attention to the numerical differences between the equivalents of certain allied elements, and showed that such differences were generally multiples of eight, as in the following examples :—

Member of a Group having Lowest Equivalent.	One immediately above the Preceding.	Difference. H=1	O=1
Magnesium 24	Calcium 40	16	1
Oxygen 16	Sulphur 32	16	1
Lithium 7	Sodium 23	16	1
Carbon 12	Silicon 28	16	1
Fluorine 19	Chlorine 35·5	16·5	1·031
Nitrogen 14	Phosphorus 31	17	1·062

Lowest Term of Triad.	Highest term of Triad.		
Lithium 7	Potassium 39	32	2
Magnesium 24	Cadmium 112	88	5·5
Molybdenum 96	Tungsten 184	88	5·5
Phosphorus 31	Antimony 122	91	5·687
Chlorine 35·5	Iodine 127	91·5	5·718
Potassium 39	Cæsium 133	94	5·875
Sulphur 32	Tellurium 129	97	6·062
Calcium 40	Barium 137	97	6·062

In the last of the above columns the difference is given referred to 16, the equivalent of oxygen, as unity, and it will be seen that, generally speaking, the equivalent of oxygen is the unit of these differences, just as the equivalent of hydrogen, in "Prout's law," is the unit of the atomic weights. Exceptions there are, however, in both cases which render it necessary to take one half or one quarter of the equivalent of oxygen in the one case, and of hydrogen in the other, in order to represent all the numbers obtained as multiples by a whole number of the given standard.

Now, if the law of "Studiosus" had any real existence, the above facts would resolve themselves into particular cases of its application. For if "the atomic weights are

Group VII.—The relations which M. Dumas has pointed out between the members of this group are well known; a slight alteration must be made, owing to the atomic weight of cæsium having been raised. The relations, then, will be thus:—

$$Li + K = 2Na,$$ or in figures, $$7 + 39 = 46$$
$$Li + 2K = Rb,$$ „ „ $$7 + 78 = 85$$
$$2Li + 3K = Cs,$$ „ „ $$14 + 117 = 131$$
$$Li + 5K = Tl,$$ „ „ $$7 + 195 = 202$$
$$3Li + 5K = 2Ag,$$ „ „ $$21 + 195 = 216$$

The equivalent of silver is thus connected with those of the alkali metals. It may also, which amounts to the same thing, be viewed as made up of the equivalents of sodium and rubidium, thus, $23 + 85 = 108$. It is likewise nearly the mean between rubidium and cæsium, thus, $\frac{85 + 133}{2} = 109$.

Group VIII.—If lithium may be considered as connected with this group as well as with the foregoing (and by some chemists its oxide is viewed as a connecting link between the alkalies and the alkaline earths), we may perform the same calculations in this group that M. Dumas has done in the preceding, thus,—

$$Li + Ca = 2Mg,$$ or in figures, $$7 + 40 = 47$$
$$Li + 2Ca = Sr$$ „ „ $$7 + 80 = 87$$
$$2Li + 3Ca = Ba$$ „ „ $$14 + 120 = 134$$
$$Li + 5Ca = Pb$$ „ „ $$7 + 200 = 207$$

Again, there are two triads in the group of alkali metals, one which has been long known—viz., lithium, sodium, and potassium, and the other, which was pointed out by Mr. C. W. Quin, in the CHEMICAL NEWS of November 9, 1861—viz., potassium, rubidium, and cæsium. Potassium is thus the highest term of one triad and the lowest term of another.

In like manner, if we include lithium, we shall have among the metals of the alkaline earths two triads, the first comprising lithium, magnesium, and calcium, and the second calcium, strontium, and barium, calcium standing at the top of one triad and at the bottom of the other.

The element lead occupies a position in relation to the metals of the alkaline earths similar to that filled by thallium in the group of alkali metals. Osmium appears to play a similar part in the sulphur group, and bismuth in the phosphorus group. The analogous term in the chlorine group is not yet known.

Thallium, in its physical properties, bears some resemblance to lead, and it frequently happens that similar terms taken from different groups, such as oxygen and nitrogen, or sulphur and phosphorus, bear more physical resemblance to each other than they do to the members of the groups to which, for chemical reasons, we are compelled to assign them.

It will be observed that the difference between the equivalents of tellurium and osmium, cæsium, and thallium, and barium and lead, respectively, is the same in each case—viz., 70.

Group X.—Palladium and platinum appear to be the extremities of a triad, the mean of which is unknown.

So frequently are relations to be met with among the equivalents of allied elements, that we may almost predict that the next equivalent determined, that of indium, for instance, will be found to bear a simple relation to those of the group to which it will be assigned.

In conclusion, I may mention that the equivalents I have adopted in this letter were taken from the highly-interesting and important paper by Professor Williamson, lately published in the *Journal of the Chemical Society*.

I am, &c.
                    JOHN A. R. NEWLANDS, F.C.S.
Laboratory, 19, Great St. Helens, E.C., July 12.

NOTICES OF PATENTS.

Communicated by Mr. VAUGHAN, PATENT AGENT, 54, Chancery Lane, W.C.

Notices to Proceed.

592. Edward Bishop, Headingly, near Leeds, and William Bailey, Halifax, Yorkshire, "Improved means and apparatus for evaporating the water contained in the fœcal or excrementitious matter."

597. John Thomas Way, Leadenhall Street, London, "Improvements in the manufacture of manure from woollen rags, mixed woollen and cotton rags, shoddy, or waste wool."—Petitions recorded March 9, 1864.

639. Thomas Parkinson and Francis Taylor, Blackburn, and Thomas Burton, Padiham, Lancashire, "Improvements in machinery and apparatus for sizing, dressing, dyeing, and drying."—Petition recorded March 14, 1864.

801. James George Beckton, Whitby, Yorkshire, "Improvements in engines or machinery for forcing, blowing, or exhausting the air and other gaseous fluids."—Petition recorded March 31, 1864.

926. Auguste Andigier, Rue Grignan, Marseilles, France, "Embalming and mummyfying dead bodies."—Petitions recorded April 13, 1864.

MISCELLANEOUS.

Trial of Gun-Cotton at Moor Edge Farm.—Mr. Winship's farm, at the Moor Edge, was on Saturday morning last the scene of an interesting trial of the powerful explosive nature of the gun-cotton prepared according to the Austrian process by Messrs. Thomas Prentice and Co., of Stowmarket, Suffolk. The engineering arrangements were conducted by Herr Reve, C.E., assisted by Dr. Richardson, Mr. Hardcastle, of Newcastle, and Mr. Prentice. At the low end of the field, a military stockade, consisting of six heavy piles of the best Memel wood, about 10 feet long, firmly fixed into the ground to about 4 feet, the openings being backed up with five timbers 9 inches square. In front of the stockade was, in military parlance, a bridge of two timbers 14 inches in diameter, resting on sleepers 1 foot from the ground, on which the explosive shell, of a cylindrical construction, was laid, leaning against the piles. The shell contained 25 lbs. of gun-cotton. About a quarter to one all the preliminaries were completed, when the company, who had been previously engaged in minutely examining the stockade, were requested to go to the top of the field to be out of danger. The "destructable" was then fired by electricity at a distance of 220 yards from a heavy battery placed at this distance from the stockade. Immediately the electric spark reached the machine the explosion occurred; the heavy pieces of timber were soon flying about in all directions, and that which the moment before appeared strong enough to withstand the assault of the enemy was a perfect wreck. Two of the centre piles were literally chopped off to within a foot of the ground; and the other four were removed out of their position to an angle of 170 degrees, one being carried to a distance of 130 yards, leaving a clear space of about 5 feet for the advancing enemy to rush through. The timbers forming the bridge on which the shell was laid were both broken in the middle, and were removed between 30 and 40 yards. The shell itself was broken into atoms, and scattered in all directions.—*Northern Daily Express.*

ANSWERS TO CORRESPONDENTS.

A Reader.—Alcoholate of zinc, $\left.\begin{array}{c}2C_2H_5\\Zn\end{array}\right\} O_2$, is formed by the slow oxidation of zinc ethyl.

P. H.—We think it has been described in the *Philosophical Transactions.*

P. F. P.—Dry at 112°, and preserve it in well dried and stoppered bottles; there is no other way.

R.—Received with thanks.

ON THE PROPORTIONAL NUMBERS OF THE ELEMENTS.

By WILLIAM ODLING, M.B., F.R.S.

UPON arranging the atomic weights or proportional numbers of the sixty or so recognized elements in the order of their several magnitudes, we observe a marked continuity in the resulting arithmetical series, the only exceptions to the very gradual increase in value of the consecutive terms being manifested between the numbers 40 and 50, 65 and 75, 96 and 104, 138 and 184, 184 and 195, and 210 and 231·5, thus :—

H	1	Hydrogen.	Fe	56	Iron.	Cd	112	Cadmium.
L	7	Lithium.	Co	59	Cobalt.	Sn	118	Tin.
G	9	Glucinum.	Ni	59	Nickel.	U	120	Uranium.
B	11	Boron.	Cu	63·5	Copper.	Sb	122	Antimony.
C	12	Carbon.	Yt	64	Yttrium.	I	127	Iodine.
N	14	Nitrogen.	Zn	65	Zinc.	Te	129	Tellurium.
O	16	Oxygen.	As	75	Arsenic.	Cs	133	Cæsium.
F	19	Fluorine.	Se	79·5	Selenium.	Ba	137	Barium.
Na	23	Sodium.	Br	80	Bromine.	V	137	Vanadium.
Mg	24	Magnesium.	Rb	85	Rubidium.	Ta	138	Tantalum.
Al	27·5	Aluminium.	Sr	87·5	Strontium.	W	184	Tungsten.
Si	28	Silicon.	Zr	89·5	Zirconium.	Cb	195	Niobium.
P	31	Phosphorus.	Ce	92	Cerium.	Au	196·5	Gold.
S	32	Sulphur.	La	92	Lanthanum.	Pt	197	Platinum.
Cl	35·5	Chlorine.	Dy	96	Dydymium.	Ir	197	Iridium.
K	39	Potassium.	Mo	96	Molybdenum.	Os	199	Osmium.
Ca	40	Calcium.	Ro	104	Rhodium.	Hg	200	Mercury.
Ti	50	Titanium.	Ru	104	Ruthenium.	Tl	203	Thallium.
Cr	52·5	Chromium.	Pd	106·5	Palladium.	Pb	207	Lead.
Mn	55	Manganese,	Ag	108	Silver.	Bi	210	Bismuth.
						Th	231·5	Thorinum.

With what ease this purely arithmetical seriation may be made to accord with a horizontal arrangement of the elements according to their usually received groupings, is shown in the following table, in the first three columns of which the numerical sequence is perfect, while in the other two the irregularities are but very few and trivial :—

			Ro 104	Pt 197
			Ru 104	Ir 197
			Pd 106·5	Os 199
H 1	"	"	Ag 108	Au 196·5
"	"	Zn 65	Cd 112	Hg 200
L 7	"	"	"	Tl 203
G 9	"	"	"	Pb 207
B 11	Al 27·5	"	U 120	"
C 12	Si 28	"	Sn 118	
N 14	P 31	As 75	Sb 122	Bi 210
O 16	S 32	Se 79·5	Te 129	"
F 19	Cl 35·5	Br 80	I 127	"
Na 23	K 39	Rb 85	Cs 133	"
Mg 24	Ca 40	Sr 87·5	Ba 137	"
	Ti 50	Zr 89·5	Ta 138	Th 231·5
	"	Ce 92	"	
	Cr 52·5	Mo 96	V 137	
	Mn 55		W 184	
	Fe 56			
	Co 59			
	Ni 59			
	Cu 63·5			

If we compare together certain pairs of more or less analogous elements, we find in a considerable number of instances, embracing one-half the entire number of elements, a difference in atomic weight ranging from 84·5 to 97, as shown in the following table :—

I	–	Cl	or	127	–	35·5	=	91·5
Au	–	Ag		296·5	–	108	=	88·5
Ag	–	Na		108	–	23	=	85
Cs	–	K		133	–	39	=	94
Te	–	S		129	–	32	=	97
W	–	Mo		184	–	96	=	88
V	–	Cr		137	–	52·5	=	84·5
Hg	–	Cd		200	–	112	=	88
Cd	–	Mg		112	–	24	=	88
Ba	–	Ca		137	–	40	=	97
Bi	–	Sb		210	–	122	=	88
Sb	–	P		122	–	31	=	91
U	–	Al		120	–	27·5	=	92·5
Pb	–	Sn		207	–	118·5	=	88·5
Sn	–	Si		118·5	–	28	=	90·5
Ta	–	Ti		138	–	50	=	88
Pt	–	Ro		197	–	104	=	93
Os	–	Pd		199	–	106·5	=	92·5

In about one-half of the above instances, the two elements associated with one another, are known to be the first and third terms respectively of certain triplet families; and the discovery of intermediate elements in the case of some or all of the other pairs, is not by any means improbable. Consequent upon the existence of these triplet groups, we have a considerable number of pairs of elements, also including more than one-half the entire number of elements, in which the average difference of atomic weight is about half as great as the average difference between the previously cited pairs, thus :—

I	–	Br	or	127	–	80	=	47	or	48
Br	–	Cl		80	–	35·5	=	44·5		44
Cs	–	Rb		133	–	85	=	48		48
Rb	–	K		85	–	39	=	46		48?
Te	–	Se		129	–	80	=	49		48
Se	–	S		80	–	32	=	48		48
W	–	V		184	–	137	=	48		48
V	–	Mo		137	–	96	=	41		40
Mo	–	Cr		96	–	52·5	=	43·5		44
Cd	–	Zn		112	–	65	=	47		48
Zn	–	Mg		65	–	24	=	41		40
Ba	–	Sr		137	–	87·5	=	49·5		48
Sr	–	Ca		87·5	–	40	=	47·5		48
Sb	–	As		122	–	75	=	47		48
As	–	P		75	–	31	=	44		44
Ta	–	Zr		138	–	89·5	=	48·5		48
Zr	–	Ti		89·5	–	50	=	39·5		40

At present there seems no reason to anticipate the existence of an intermediate term between any one of these pairs of elements.

In ten instances we find that more or less analogous elements have a difference in atomic weight of 16, or something approximating

closely thereto ; and in seven of these instances, the element of
lowest atomic weight is the first member, and the element compared
therewith the second member of the group to which they both
belong, or may be considered to belong, as shown in the following
table, which includes nearly one-third of the entire number of
elements :—

Cl	–	F	or	35·5	–	19	=	16·5
K	–	Na		39	–	23	=	16
Na	–	L		23	–	7	=	16
Mo	–	Se		96	–	80	=	16
S	–	O		32	–	16	=	16
Ca	–	Mg		40	–	24	=	16
Mg	–	G		24	–	9	=	15
P	–	N		31	–	14	=	17
Al	–	B		27·5	–	11	=	16·5
Si	–	C		28	–	12	=	16

In looking over the above tables, we can scarcely help noticing
that those elements whose resemblance to one another is most
pronounced, have a difference of about 48 between their respective
atomic weights, that is to say, the largest difference in atomic weight
known to exist between what are conceived to be proximate elements,
as shown in the following table, which also includes nearly one-third
of the entire number of elements.   For example, the resemblance of
cadmium to zinc, where the difference in atomic weight is 47, is
greater than the resemblance of zinc to magnesium, where the
difference is 41 ; while the resemblance of antimony to arsenic,
where the difference is again 47, is greater than the resemblance of
arsenic to phosphorus, where the difference is 44.   Moreover, the
co-resemblances of cæsium, rubidium, and potassium, and of barium,
strontium, and calcium, with a common difference of about 48 between
the proximate members, are far closer than the co-resemblances of
potassium and sodium, and of calcium and magnesium respectively,
with a difference of 16 in each instance :—

,,			Zn	65	+	47	Cd	112	,,
,,			As	75	+	47	Sb	122	,,
,,			Br	80	+	47	I	127	,,
S	32	+	48	Se	80	+	49	Te	129	,,
K	39	+	46	Rb	85	+	48	Cs	133	,,
Ca	40	+	47·5	Sr	87·5	+	49·5	Ba	137	,,
,,			Zr	39 5	+	48·5	Ta	138	,,
,,			,,				V	137	+	48	W	184

If we consider the analogous elements having a difference of about 48 in their respective atomic weights, to stand upon the same level, we may represent those with a difference of 44 or 40 as standing one or two stages above or below the level, as shown in the next table :—

Monads.		"	Br 80	I 127		
	Dif. 4 F 19	Cl 35·5	"	"		
	Dif. 12	"	"	Ag 108		Au 196·5
	L 7	Na 23	"	"		Tl 203
	Dif. 16	K 39	Rb 85	Cs 133		"
Dyads.	O 16	S 32	Se 80	Te 129	"	
	Dif. 8	"	"	V 137	W 184	
	Dif. 8	"	Mo 96	"	"	
	Dif. 4	Cr 52·5	"	"	"	
	Dif. 8	"	Zn 65	Cd 112		Hg 200
	G 9	Mg 24	"	"		Pb 207
	Dif. 16	Ca 40	Sr 87·5	Ba 137		"
Triads.	Dif. 4	"	As 75	Sb 122		Bi 210
	N 14	P 31	"	"		"
	Dif. 4	"	"	U 120		
	B 11	Al 27·5	"	"		
	Dif. 16	"	Ce 92	"		
Tetrads.	Dif. 8	"	"	Sn 118	"	
	C 12	Si 28	"	"	"	
	Dif. 12	"	Zr 89·5	Ta 138	"	"
	Dif. 8	Ti 50	"	"	Cb 195*	Th. 231·5

By a slight modification of the above table, the occupants of similar positions in different groups, having nearly the same

* The analysis of niobic chloride by H. Rose gives 195, while the determination of its vapour density by Deville and Troost gives 173 for the atomic weight of niobium, the mean being 184.

atomic weights, may be brought into association with one another, thus :—

„		„		Ag	108	Au	196·5
„		Zn	65	Cd	112	Hg	200
Na	23	„		„		Tl	203
Mg	24	„		„		Pb	207
„		„		Sn	118	„	
Al	27·5	„		U	120	„	
„		As	75	Sb	122	Bi	210
Si	28	„		„		„	
P	31	„		„		„	
S	32	Se	79·5	Te	129	„	
		Br	80	I	127	„	
Cl	35·5	„		„		„	
K	39	Rb	85	Cs	133	„	
				V	137	„	
Ca	40	Sr	87·5	Ba	137	„	
„		Zr	89·5	Ta	138	„	
„		Ce	92	„		„	
„		Mo	96	„		„	
Ti	50	„		„		Th	231·5
Cr	52·5	„		„		„	

The parallelism between the monatomic and diatomic alkaline groups, is shown still more strikingly below :—

			Dif. 1.		Dif. 2.		Dif. 4.		Dif. 4.	
			„		X	63	Ag	108	Au	196·5
			„		Zn	65	Cd	112	Hg	200
	Dif. 8									
L	7	Na	23		„		„		Tl	203
G	9	Mg	24		„		„		Pb	207
	Dif. 16									
		K	39	Rb	85	Cs	133	„		
		Ca	40	Sr	87·5	Ba	137	„		

Seeing the large number of instances in which the atomic weights of proximate elements differ from one another by 48, or 44, or 40, or 16, we cannot help looking wistfully at the number 4, as embodying somehow or other the unit of a common difference, especially when we find in addition that several pairs of strictly analogous elements differ in atomic weight by this same number, as shown below :—

Fe	—	Cr	or	56	—	52·5	= 3·5
Co	—	Mn	or	59	—	55	= 4
Cu	—	Ni	or	63·5	—	59	= 4·5
Dy	—	Ce	or	96	—	92	= 4

But on the other hand it must be borne in mind that the differences between the several atomic weights compared with one another, are for the most part not exactly but only approximately multiples of 4; whilst in a few instances, at any rate, the approximate difference in atomic weight between closely allied elements, is not 4 or some multiple of 4, but 2 or some odd multiple of 2, and in other instances even 1 or 0.

Since many of the elements occupying analogous positions in different groups have closely approximating atomic weights, it is evident that the mere determination of the atomic weight of a newly-discovered element assists us but very little in deciding to what group it belongs, but only indicates its position in the group; since among the members of every well-defined group the sequence of properties and sequence of atomic weights are strictly parallel to one another.

Doubtless some of the arithmetical relations exemplified in the foregoing tables and remarks are simply accidental; but taken altogether, they are too numerous and decided not to depend upon some hitherto unrecognized general law.

Ueber die Beziehungen der Eigenschaften zu den Atomgewichten der Elemente. Von D. Mendelejeff. — Ordnet man Elemente nach zunehmenden Atomgewichten in verticale Reihen so, dass die Horizontalreihen analoge Elemente enthalten, wieder nach zunehmendem Atomgewicht geordnet, so erhält man folgende Zusammenstellung, aus der sich einige allgemeinere Folgerungen ableiten lassen.

```
                                       Ti = 50    Zr =  90     ? = 180
                                        V = 51    Nb =  94    Ta = 182
                                       Cr = 52    Mo =  96     W = 186
                                       Mn = 55    Rh = 104,4  Pt = 197,4
                                       Fe = 56    Ru = 104,4  Ir = 198
                           Ni = Co = 59           Pd = 106,6  Os = 199
   H = 1                                Cu = 63,4   Ag = 108  Hg = 200
           Be = 9,4   Mg = 24           Zn = 65,2   Cd = 112
           B = 11     Al = 27,4          ? = 68     Ur = 116  Au = 197?
           C = 12     Si = 28            ? = 70     Sn = 118
           N = 14     P = 31            As = 75     Sb = 122  Bi = 210?
           O = 16     S = 32            Se = 79,4   Te = 128?
           F = 19     Cl = 35,5         Br = 80      J = 127
   Li = 7  Na = 23    K = 39            Rb = 85,4   Cs = 133  Tl = 204
                      Ca = 40           Sr = 87,6   Ba = 137  Pb = 207
                       ? = 45           Ce = 92
                     ?Er = 56           La = 94
                     ?Yt = 60           Di = 95
                     ?In = 75,6         Th = 118?
```

1. Die nach der Grösse des Atomgewichts geordneten Elemente zeigen eine stufenweise Abänderung in den Eigenschaften.

2. Chemisch-analoge Elemente haben entweder übereinstimmende Atomgewichte (Pt, Ir, Os), oder letztere nehmen gleichviel zu (K, Rb, Cs).

3. Das Anordnen nach den Atomgewichten entspricht der *Werthigkeit* der Elemente und bis zu einem gewissen Grade der Verschiedenheit im chemischen Verhalten, z. B. Li, Be, B, C, N, O, F.

4. Die in der Natur verbreitetsten Elemente haben *kleine* Atomgewichte

und alle solche Elemente zeichnen sich durch Schärfe des Verhaltens aus. Es sind also *typische* Elemente und mit Recht wird daher das leichteste Element H als typischer Massstab gewählt.

5. Die *Grösse* des Atomgewichtes bedingt die Eigenschaften des Elementes, weshalb beim Studium von Verbindungen nicht nur auf Anzahl und Eigenschaften der Elemente und deren gegenseitiges Verhalten Rücksicht zu nehmen ist, sondern auf die *Atomgewichte* der Elemente. Daher zeigen bei mancher Analogie die Verbindungen von S und Tl, Cl und J, doch auffallende Verschiedenheiten.

6. Es lässt sich die Entdeckung noch vieler *neuen* Elemente vorhersehen, z. B. Analoge des Si und Al mit Atomgewichten von 65 75.

7. Einige Atomgewichte werden voraussichtlich eine Correction erfahren, z. B. Te kann nicht das Atomgewicht 128 haben, sondern 123—126.

8. Aus obiger Tabelle ergeben sich neue Analogien zwischen Elementen. So erscheint Bo (?) als ein Analoges von Bo und Al, was bekanntlich schon längst experimentell festgesetzt ist.   (Russ. chem. Ges. 1, 60.)

THE CHEMICAL NEWS.

VOL. XL. No. 1042.

THE PERIODIC LAW OF THE CHEMICAL ELEMENTS.*

By D. MENDELEEF.

ALTHOUGH seven years have passed since these thoughts absorbed any attention; although other occupations† have drawn my attention from the problem of the elements, which was always getting nearer solution; in short, although I might wish to put this question otherwise than I did seven years ago, still I keep to the same firm conviction that I formerly had on the importance and value of the theorems on which my memoir is based; and this is why I am so gratified to see my ideas exposed to the appreciation of the learned. Several occurrences have aided to make some of the logical consequences of the periodic law popular.

1st. The law I announced has been considered as a repetition in another form of what has already been said by others. The article in the *Berichte der Deutschen Chemischen Gesellschaft*, 1874, p. 348, treats only of the question of priority. It is now certain that the periodic law offers consequences that the old system had scarcely ventured to foresee. Formerly it was only a grouping, a scheme, a subordination to a given fact; while the periodic law furnishes the facts, and tends to strengthen the philosophical question, which brings to light the mysterious nature of the elements. This tendency is of the same category as Prout's law, with the essential difference that Prout's law is arithmetical, and that the periodic law exhausts itself in connecting the mechanical and philosophical laws which form the character and glory of the exact sciences. It proclaims loudly that the nature of the elements depends above all on their mass, and it considers this function as periodic. The formula of the law might be changed; a greater appreciation of this function will be found; but, I believe that the original idea of the periodic law will remain, because it is opposed to the primitive notion that the nature of elements depends on unknown causes, and not on their mass. In 1872 to 1877 Gustavson and Palitzine showed that even the coefficients of simple substitution depend on the atomic weight of elements, in the sense which is in accordance with the periodic law.

2ndly. The periodic law requires changes in the atomic weights of many elements yet incompletely examined; for example, indium, cerium, yttrium, erbium, didymium, &c. Before the existence of this law there was no reason for doubting the generally accepted atomic weights of

---

* Considerable attention having been drawn to M. Mendeleef's memoir " On the Periodic Law of the Chemical Elements," in consequence of the newly discovered elements gallium and scandium being apparently identical with his two predicted elements ekaluminium and ekaboron, it has been thought desirable to reproduce the entire article in the CHEMICAL NEWS. The present translation is from the *Moniteur Scientifique*, July, 1879, to which periodical it has been communicated by M. Mendeleef, who prefaces the memoir by a letter. This letter forms the subject of the present article; the memoir will be commenced next week.—Ed. C.N.

† Touching the compressibility, the dilation, and the resistance of gases, the temperature of the upper layers of the atmosphere, aërostatics, researches on the origin of petroleum, &c.

---

these elements. At present, since the researches of Rammelsberg, Roscoe, Clève, Hoglund, Hildebrand, and others, have led to the same conclusion as that which is derived from the periodic law, it becomes incontestable that it shows us the truth, which is in fact its support.

3rdly. The periodic law has given the first chance of predicting not only unknown elements, but also of determining the chemical and physical properties of simple bodies still to be discovered, and those of their compounds. The discovery of gallium by M. Lecoq de Boisbaudran, whom I have now the honour of counting as a friend, may be considered as the inauguration of the periodic law, and reckoned amongst the brilliant pages in the annals of science. The properties announced in the *Comptes Rendus*, Nov. 22, 1875, have been confirmed by after study. It will be enough to mention four—the formation of gallium alum, the equivalent of the oxide, the specific gravity of the metal (5·9), and the atomic weight of the element.

It must be admitted that before the periodic law there was no possibility of a similar prediction. It is here to the point to direct attention to the fact that the fusing-point of gallium is so low that it melts at the heat of the hand. This property might be considered as foreseen, but that is not the case; it suffices to look at the following series :—

Mg	Al	Si	P	S	Cl
Zn	Ga	—	As	Se	Br
Cd	In	Sn	Sb	Te	I

It is evident that in the group Mg, Zn, Cd, the most refractory element (Mg) has the lowest atomic weight; but in the groups commencing with S and Cl the most difficultly fusible bodies are the heaviest. In a transitory group, as Al, Ga, In, it is necessary to notice an intermediate fact, the extremes, that is, the heaviest (In) and the lightest (Al) should be less fusible than the middle one, as it is in reality. I would further remark that a property such as the fusion-point is characterised chiefly by the molecular weight and not by the atomic weight. If we had a variation of solid sulphur, not as S^{VI} (or perhaps heavier, as S'') but S^{II}, which form it acquires at 800°, its boiling- and fusion-point would have been much lower. In the same way, ozone, O^{III}, will condense much more easily than oxygen, O^{II}. Experiments made in our laboratory confirm this fact.

The three circumstances mentioned have obliged chemists to direct their attention, distracted by the brilliant material acquisitions which distinguish our epoch, towards the periodic law. I ought now myself to complete what is still wanting on this subject, but for the time being I am occupied with other matters, and I should leave the care of developing this question to the future and to new forces, which will I hope endeavour to bring as the first fruits of the periodic law a new philosophical order, in fixing it by pillars strengthened by new experiments, so as to give greater stability to the edifice already begun. I will add but three brief observations :—

1st. The best way of drawing up the table of elements, so as to show the periodic relations, will, I think, be:—

TYPICAL ELEMENTS.

I.	II.	III.	IV.	V.	VI.	VII.
H						
Li	Be	B	C	N	O	F
Na						

EVEN ELEMENTS.

I.	II.	III.	IV.	V.	VI.	VII.	VIII.			
K	Ca	—	Ti	V	Cr	Mn	Fe	Co	Ni	Cu
Rb	Sr	Yt	Zr	Nb	Mo	—	Ru	Rh	Pd	Ag
Cs	Ba	La	Ce	—	—	—	Os	Ir	Pt	Au
—	—	Er	Di (?)	Ta	W	—	—	—	—	—
—	—	—	Th	—	Ur	—	—	—	—	—

ODD ELEMENTS.

I.	II.	III.	IV.	V.	VI.	VII.
	Mg	Al	Si	P	S	Cl
Cu	Zn	Ga	—	As	Se	Br
Ag	Cd	In	Sn	Sb	Fe	I
Au	Hg	Tl	Pb	Bi	—	—
—	—	—	—	—	—	—

The Roman figures show the groups or the forms of combination.

2nd. With regard to new elements recently discovered, I think it necessary to keep silence. Of late years metals have been born and have died, such as Davyum, Mosandrium, &c.; this shows the necessity of caution. It is only necessary to mention Marignac's ytterbium (*Archives des Sciences Physiques et Naturelles*, 1878, Nov. 15, No. 251), because the name of the investigator is itself sufficient guarantee. But, after announcement of the series of new metals of Delafontaine, he himself asked that someone else should make fresh researches in such an inaccessible region as gadolinite. One can still say the same of oxide of didymium; it should be re-examined. The absorption-bands of solutions and the equivalents of oxides are not sufficiently sure to show the individuality of elements; because, in different degrees of oxidation and in salts of different basicity, the absorption-bands and the equivalents of oxides can be heterogeneous for the same elements, as, for instance, in cerium, uranium, iron, chromium, &c.

3rd. I should like to fix the attention of chemists on the three principles which are demonstrated in the last chapter, and also on the consequences which are adduced from their strict application to organic compounds (above all not saturated): by their aid we do without hypotheses, we explain cases of isomerism, and we obtain new consequences as yet unexplained.

D. MENDELEEF.

(To be continued.)

THE ALLEGED DECOMPOSITION OF CHORINE.

DR. H. ENDEMANN, who has written an article in the *Journal of the American Chemical Society* entitled "A Review of the Latest Investigations on the Dissociation of the Elements at High Temperatures," publishes the following letter from Prof. Meyer:—

"Zurich, d 28 Sept., 1879.

"SEHR GEEHRTER HERR,—In Erwiederung auf Ihr geehrtes Schreiben vom 12 Sept., theile ich Ihnen mit, dass die Mittheilungen der englischen Journale ueber meine Arbeiten welche ueber Gewinnung von Sauerstoff aus Chlor berichten, ohne mein Wissen und gegen meinen Willen veroeffentlicht sind, und dass diese Mittheilungen in wesentlichen Punkten voellig inkorrekt sind.—Hochachtungsvoll,

"Prof. VICTOR MEYER."

[TRANSLATION.]

"Zurich, September 28, 1879.

"DEAR SIR,—In reply to your favour of September 12th, I would state that the communications in the English journals concerning my investigations which report the separation of oxygen from chlorine, have been published without my knowledge and against my wishes, and that the said communications are entirely incorrect in essential particulars.—Respectfully,

"Prof. VICTOR MEYER."

BLOWPIPE ANALYSIS.

THE increasing appreciation of the value of Blowpipe Analysis in England is shown by the recent publication of the translation of another German work on the subject by one of our first scientific publishers, Messrs. Macmillan.

We shall take an early opportunity of reviewing this little work by Herr Landauer, of Brunswick, who lately visited England at the head of a sanitary commission sent here by that State, but meantime have the pleasure to offer our readers extracts from a letter to Colonel Ross, whose contributions on the subject have so often appeared in this journal. The writer of this letter, whose name we are not at present at liberty to communicate, is obviously one well acquainted with the subject, but we may mention that he is a credit to a body of scientific officers who are admitted throughout Europe to be a credit to this country, the Corps of Royal Engineers.

"Point de Galle, Ceylon,
"Sept. 13th, 1879.

"DEAR SIR,—I have to acknowledge your letter, with thanks for the information it contained. I long since saw notices of your new system of Blowpipe Analysis, one stating that your work was the best English authority on the subject. It was only recently that I got out "Griffin's Handicraft," and going over the list of your apparatus I saw that your method must be something very different from the old, so I ordered out a special box both for the blowpipe and separately for the wet way. Your book came to hand afterwards. There is no question that your method is greatly superior to Plattner's, which I think may be considered out of date, and that it is a powerful and logical mode of chemical analysis. . . . I was greatly pleased both with boric and phosphoric acid as vehicles of decomposition, and especially so with the lime borate balls. . . . Your book I found extremely interesting, as one written by an enthusiastic master of his subject can hardly fail to be. From your going to publish an epitome, it is evident that you think with myself that Spon's edition is rather too bulky. . . . The chapter on flame, apparatus, reagents, manipulation can hardly be better, being very intelligible and practical. The apparatus is a great advance on the old kinds. . . . The chapter on colour seems to me to be not in the right place, but I have not as yet made any particular use of analysis by colour. . . . I generally go when in search of information to the next chapter, where the ordinary metals and bases are dealt with. This is very clear throughout, and easily worked from. . . . The last half of your book appeals to three classes of investigators. It might, therefore, form usefully three divisions—one for mineralogists, who would be chiefly geologists and engineers; the next to relate to metals intended for the guidance of chemists and manufacturers; and the third for the use of farmers. This division would be very convenient, because one generally employs the blowpipe according to locality, and oftener for some specific branch than for mere study of the multitudinous specimens, seldom found in any number together out of a cabinet or museum. As to the chapter on farming, it almost seems necessary to follow the order of some textbook: "Church's Laboratory Guide" (Macmillan) is one of the best on agricultural chemistry that I know. What is required would be to show *seriatim* how all that he accomplishes in the wet way can be attained by the blowpipe, including the physical properties of the soil. The best way of insuring this would be to go right through the book, altering it in brief to blowpipe methods. At present one has to do this each time, wading backwards and forwards through descriptions in pyrology intended for the general student till a result of some kind is obtained. . . . A working chapter on the blowpipe applied to agricultural chemistry would supply a great want. . . . I am more familiar with the wet processes, and as Church lays down very complete instructions can depend upon what I would obtain by this method. But there is no reason except this, I think, for considering the wet way preferable; in fact, the blowpipe system now rests upon exactly the same numerical basis the other does, and the apparatus is much more compact, manageable, and cheap.

"There are, however, processes, such as the volumetric analysis of water, or the analysis of water generally, which do not appear to come within the scope of the blowpipe.—Yours very faithfully,

"A. F. F."

THE CHEMICAL NEWS.

VOL. XL. No. 1043.

THE PERIODIC LAW OF THE CHEMICAL ELEMENTS.

By D. MENDELEEF.

(Continued from page 232.)

EVEN as, up to the time of Laurent and Gerhardt, the words "molecule," "atom," and "equivalent" were used one for the other indiscriminately in the same manner, so now the terms "simple body" and "element" are often confounded one with the other. They have, however, each a distinct meaning, which it is necessary to point out, so as to prevent confusion of terms in philosophical chemistry.

A "simple body" is something material, metal or metalloid, endowed with physical properties, and capable of chemical reactions. The idea of a molecule corresponds with the expression of a "simple body;" a molecule is made up of one atom, as in the case of Hg, Cd, and others; or of several atoms, such as S_2, S_6, O_2, H_2, Cl_2, P_4, &c.

A simple body is able to show itself under isomeric and polymeric modifications, and it is only distinguished from a compound body by the homogeneity of its material parts. But in opposition to this, the name of "element" must be reserved for characterising the material particles which form simple and compound bodies, and which determine their behaviour from a chemical and physical point of view. The word "element" calls to mind the idea of an atom; carbon is an element; coal, diamond, and graphite are simple bodies.

The principal end of modern chemistry is to extend our knowledge of the relations between the composition, the reactions, and the qualities of simple and compound bodies, on the one hand; and, on the other hand, the intrinsic qualities of elements which are contained in them; so as to be able to deduce from the known character of an element all the properties of all its compounds.

For example, saying that carbon is a tetratomic element, is making known a fundamental property which appears in all its combinations.

The elements count among their properties, which can be measured exactly, their atomic weight, and the power of showing themselves under the form of different compounds.

Alone amongst their properties, the two above mentioned bring in their train a number of facts. The last has given rise to a special theory on the atomicity (valency) of elements. Amongst the other properties of elements which influence the character of bodies, the physical properties (such as cohesion, capacity for heat, coefficient of refrangibility, spectral phenomena, &c.) have been up to the present time too incompletely studied for us to be able to generalise them in a rigorously philosophical manner. What we know of these properties is still insufficient and defective in comparison with our knowledge on the atomic weights and the atomicity of elements. However, it has already been often noticed that the physical properties depend one on another, that the atomic weights, and principally the molecular weights of compounds, are equally in intimate relation with them. It is principally the fact, that it is easy to measure these properties exactly, that has induced us to make these comparisons. It is by studying them, more than by any other means, that we can conceive the idea of an atom and of a molecule. By this fact alone we are enabled to perceive the great influence that studies carried on in this direction can exercise on the progress of chemistry.

The above-mentioned measurable properties are by no means the only ones possessed by the elements. They have beyond these a series of properties which have not yet been able to be measured, but which still contribute to their recognition. These last have received the name of chemical properties. Certain elements do not combine with hydrogen; they have, according to the recognised term, a basic character, or, in other words, they absorb oxygen and form bases; they form salts when combining with chlorine; other elements (called acidifying elements) do combine with hydrogen; with oxygen they only form acids, and with chlorine only chloranhydrides. Thirdly, there are elements which form the link between the first and second classes; and fourthly, there are elements which in their forms of higher oxidation have an acid character, and when less oxidised a basic character. Science does not as yet possess any process by which these properties can be measured, but still they are counted among the number of qualitative characteristics which distinguish the elements. Further, these lastly-named elements possess properties which determine the greater or less stability of compounds; these, again, are chemical properties. It is in this manner that some elements can unite with all the others in compounds capable of being decomposed with a relative facility, while we cannot obtain analogous decompositions in the corresponding compounds of other elements. Not being susceptible of exact measurement, the above-mentioned chemical properties can hardly serve to generalise chemical knowledge; they alone cannot serve as a basis for theoretical considerations. However, these properties should not be altogether neglected, as they explain a great number of chemical phenomena. It is known that Berzelius and other chemists considered these properties as being among the principal characteristics of elements, and that it is on them that the electro-chemical system was based.

As a general rule when we study the properties of elements, bearing in mind practical conclusions and chemical previsions, it is necessary to give equal attention to the general properties of the other bodies of the group, and to the individual properties of the given element in that group; it is only after such comparative studies, and laying stress on an accurately measurable property, that we can generalise the properties of an element. The atomic weight furnishes us now, and will long continue to furnish us, with a property of this nature; for our conception of the atomic weight has acquired an indestructible solidity, above all latterly, since the use of Avogadro's and Ampère's law: thanks to the efforts of Laurent, Gerhardt, Regnault, Rose, and Cannizzaro, one can even state boldly that the notion of the atomic weight (considered as the smallest part of an element contained in a molecule of its compounds) will remain without change, whatever may be the modifications that the philosophical ideas of chemists may undergo. The expression atomic weight* implies, it is true, the hypothesis of the atomic structure of bodies; but, then, we are not here discussing denomination, but a conventional idea. It seems that the best method of extending our chemical knowledge would be to elaborate the correlations between the proportions of elements and their atomic weights. It is thus that we should obtain the most natural and most fruitful results in the extension of the study of elements. To determine this dependence seems to me to be one of the principal tasks of future chemists; for this problem has the same philosophical importance as the study of the conditions of isomerism has. In the present memoir I shall try and show the already mentioned relation between the atomic weights of elements and their other properties, particularly the faculty of giving different forms of combination.

This last faculty has already been carefully experimented on; a still more precise expression has recently

---

* By replacing the expression of *atomic weight* by that of *elementary weight*, I think we should, in the case of elements, avoid the conception of atoms.

been found for it, in the theory relative to the limits of chemical combinations, to the atomicity of elements, and to the manner of attachment of atoms in the molecules. It is known that Dalton called combinations in multiple proportions the mode of combination of an ideal element, R, with other elements (of the form RX, RX_2, RX_3. &c.); Gerhardt called them types; they are now used for fixing the atomicity of elements.

The incompleteness which exists in the theory now accepted, with regard to the atomicity of elements, arises from the fact that the opinions of chemists do not coincide in respect to elements such as Na, Cl, S, N, P, Ag; some consider the atomicity as an invariable property of atoms, while others affirm the contrary. The uncertainty in the ideas on atomicity come principally from the novelty of their introduction into Science, and from this—that they include the hypothesis of the union of elements by parts of their affinity. It also arises, according to my idea, from the fact that we only study the forms of combination, without comparing these forms with the other properties of the elements. The gaps which I have just pointed out in the theory of combinations—gaps produced by the doctrine actually accepted on the subject of the atomicity of elements—are at their widest, as I shall point out further on, if the study of the principal properties of elements is based on the atomic weights.

Since the year 1868, the year in which the first part of my work "Principles of Chemistry" appeared, in the Russian language, I have been endeavouring to solve this problem. In this paper I take the liberty of making known the results obtained up to the present time in my researches in that direction.* The formation of natural groups, such as the haloids, the metals of the alkalies and alkaline earths, the bodies analogous to sulphur, to nitrogen, &c., furnished me with the first opportunity of comparing the different properties of the elements with their atomic weights. In the beginning we only arranged in groups the elements which resemble one another in several respects, but later on several experimentalists—notably Gladstone, Cooke, Pettenkoffer, Kremers, Dumas, Lenssen, Odling, &c.—observed that the atomic weights of the different members of these groups had a simple and regular relation to each other. The discovery of these relations led to the comparison of the members of different groups with the homologous series, and, later on, to the conception, in a chemico-mechanical manner, of the complex nature of atoms, which has been held as reasonable by the greater number of chemists, but up to the present time it has not received any definite name.

All the relations observed between the atomic weights of elements have not yet led to any logical conclusion or chemical prevision, on account of the gaps in them. This may be the reason why they have not acquired the right of being generally recognised in Science.

First. Nobody that I know of has, up to the present, prepared any comparative table of the natural groups, and the observed relations between the different members of groups have remained without any connection or explanation. Concerning this subject, in 1859, Strecker rightly said,† "It is hardly possible that the relations noticed between the atomic weights of elements which resemble each other in their chemical properties should be purely accidental. However, we must leave to future research the discovery of the 'regular' relations which are betrayed in these numbers."

Secondly. Only small variations in the magnitude of the atomic weights of some analogous elements (Mn, Fe, Co, and Ni,—Pd, Rh, Ru,—Pt, Os, Ir) have been observed. Therefore we were only authorised in saying that the analogy of elements was connected either by approximate agreement or by the increasing amount of their atomic weights.

Thirdly. Nobody has established any theory of mutual comparison between the atomic weights of unlike elements, although it is precisely in connection with these unlike elements that a regular dependence should be pointed out between the properties and the modifications of the atomic weights. The facts published up to now, being too isolated, could not cause any progress in the philosophical development of chemistry; however, they contain the germs of important additions to chemical science, especially as concerns the nature, to us mysterious, of elements.

In the term *periodic law* I designate the reciprocal relations between the properties and the atomic weights of elements. Later on I shall develop the relations which are applicable to all the elements: they are shown in the form of a periodic function.

(To be continued.)

---

* In relation to some historical and polemical observations on this question, see the *Berichte der Deutschen Chemischen Gesellschaft*, 1871, p. 348.
† "Theorieen und Experimente zur Bestimmung der Atom gewichte der Elemente," p. 146.

THE CHEMICAL NEWS.

VOL. XL. No. 1044.

THE PERIODIC LAW OF THE CHEMICAL ELEMENTS.

By D. MENDELEEF.

(Continued from page 244.)

I. PRINCIPLES OF THE PERIODIC LAW.

ANALOGIES have long been noticed between the elements of high atomic weights and those of low ones. Claus observed that Os, Sr, and Pt, whose atomic weights are about 195, have analogous properties with Ru, Rh, and Pd, which have much lower atomic weights, viz., about 105.

Marignac noticed the analogy of $Ta = 182$ and $W = 184$, with $Nb = 94$ and $Mo = 96$.

To Au and Hg correspond the lighter analogues Ag and Cd, as well as the still lighter ones Cu and Zn.

Cæsium and barium are analogues of potassium and calcium, and so on.

Comparisons of this nature lead to the idea of classifying all the elements according to their atomic weights, and in making this list we find reciprocal relations of a remarkable simplicity. As an example we will give all the elements whose atomic weights are between 7 and 36. *These atomic weights are here arranged in arithmetical order according to their value.*

$Li = 7$; $Be = 9.4$; $B = 11$; $C = 12$; $N = 14$; $O = 16$; $Fl = 19$
$Na = 23$; $Mg = 24$; $Al = 27.3$; $Si = 28$; $P = 31$; $S = 32$; $Cl = 35.5$.

It is plainly shown that the character of the elements is subjected to regular modifications, and little by little, step by step, as the atomic weights vary, it is also periodically modified; that is to say, in the same manner in the two series, so that the corresponding members of these series are analogous. Na and Li, Mg and Be, C and Si, O and S, and so on. The corresponding elements in the two series have the same kind of compounds, and they have, as it is generally said, the same atomicity. The most important circumstance is that the forms of compounds of intermediate elements obey the laws which have been discovered in comparing the hydrated and oxygenised compounds of the elements which precede and follow after. This regularity proves that the elements just mentioned form two natural series in the order in which they are placed, and can neither of them have any intermediate members. Thus the last four members alone can combine with hydrogen, forming, if R represents an element :—

$$— \quad — \quad — \quad RH_4 \quad RH_3 \quad RH_2 \quad RH.$$

The constancy or destructibility of these hydrogen compounds under various influences, as well as their acid character or the faculty of changing hydrogen for metals, and other properties, are gradually modified, according to the position of the elements in the series. HCl is a very decided acid of great stability; H_2S is a weak acid, decomposed by heat; in H_3P the acid character has completely disappeared, and the instability has increased; again, it is much more decided in H_4Si.

All the elements of the second series enter into combination with oxygen. It is therefore in these compounds that we can best observe how the properties of elements are gradually modified, according as the atomic weight changes. As examples for comparison of this kind we will take the higher anhydrous oxides, by which I mean those which correspond to water, which can combine with it and form hydrates, and unite amongst themselves to form salts. We will not here take notice of those oxides which by their composition and properties correspond to peroxide of hydrogen (such as Na_2O_2), because there are only a small number of elements capable of forming such compounds.

The seven elements of the second series give the following higher oxides capable of forming salts :—

$$Na_2O \; ; \; Mg_2O_2 \; ; \; Al_2O_3 \; ; \; Si_2O_4 \; ; \; P_2O_5 \; ; \; S_2O_6 \; ; \; Cl_2O_7 \; ;$$
$$\text{or } MgO \; ; \quad \text{or } SiO_2 \; ; \quad \text{or } SO_3$$

Thus, in a determinate order, seven forms of oxidation known to chemists correspond to the seven members of the above series. Although they have long been discovered, their connection with the fundamental properties of elements remains unknown. We see, by examining these seven forms, that to their position corresponds a diminution of the basic properties and an increase of the acid properties. They give the following normal saline derivatives :—

$$NaX, \; MgX_2, \; AlX_3, \; SiX_4, \; PX_5, \; SX_6, \; ClX_7.$$

For example—

$$Na(NO_3) \; ; \; MgCl_2 \; ; \; Al(NO_2)_3 \; ; \; Si(KO)_4 \; ; \; PO(NaO)_5 \; ;$$
$$SO_2(NaO)_2 \; ; \; ClO_3(KO).$$

We can represent X by the following elements :—

$$X = H, \; Cl, \; NO_3, \; OH, \; CH_3, \; K, \; OK, \; \&c.$$

And, further—

$$X_2 = O, \; S, \; SO_4, \; CO_3, \; \&c.$$

To PX_5 correspond :—

$$PCl_3, \; POCl_3, \; P_2O_5, \; PO(OH)_3, \; POH(OH)_2, \; POH_2(OH),$$
$$PH_4I, \; PH_3HI.$$

In the composition of hydrates we again notice a complete regularity—

$$Na(OH) \; ; \; Mg(OH)_2 \; ; \; Al(OH)_3 \; ; \; Si(OH)_4 \; ; \; PO(OH)_3 \; ;$$
$$SO_2(OH)_2 \; ; \; ClO_3(OH).$$

In the hydrates there are only four hydroxyls, $(OH)_4$; and in the same manner in the oxides there are only four oxygen atoms, O_4. The highest forms of combination known with an element R are RO_4, RH_4, and $R(OH)_4$. No stable hydrate, $S(OH)_6$, corresponds with the type SX_6, although corresponding basic salts are known. This hydrate is transformed by elimination from $2H_2O$ into $SO_2(HO)_2$, which contains O_4, resembling in this manner,

$$Si(OH)_4, \; PO(OH)_3, \; ClO_3(OH).$$

It is not only in the forms of compounds of elements placed according to the amount of the atomic weights that we notice a regular connection, but also in their other chemical and physical properties.

At the beginning of the series are the bodies of a decided metallic character; at the end are the representatives of the metalloids. The former possess basic properties, the latter acid ones; while the bodies in the middle have intermediate properties. In Li_2O and Na_2O the basic properties are more clearly defined than in BeO and MgO; in B_2O_5 and Al_2O_3 they are only feebly shown, and these compounds have some acid properties. CO_2 and SiO_2 have only acid properties, although in only a small degree. This character is more pronounced in N_2O_5 and P_2O_5; and again in SO_3 and Cl_2O_7.

To show with what regularity the physical properties are modified in the above-mentioned series we give the modifications of the specific weights and the atomic volumes of the members of the second series :—

	Na.	Mg.	Al.	Si.	P.	S.	Cl.
Density	0·97	1·75	2·67	2·49	1·84	2·06	1·33
Atomic volume	24	14	10	11	16	16	27

	Na_2O.	Mg_2O_2.	Al_2O_3.	Si_2O_4.	P_2O_5.	S_2O_6.	Cl_2O_7.
Density	2·8	3·7	4·0	2·6	2·7	1·9	?
Atomic volume	22	22	25	45	55	82	?

The volatility diminishes in the members of the commencement of the first series from Na to Si; from thence it increases. The same observation applies to the oxides, amongst which the middle ones, MgO, Al$_2$O$_3$, SiO$_2$, are the most refractory.

The compounds of metals at the beginning of the series with other metals are called alloys; they have the look and the properties of metals. Even in the case of phosphorus and sulphur these properties are not completely effaced, for phosphide of copper, sulphide of lead, and other compounds still retain something of a metallic look. On the contrary, oxides and metallic chlorides (such as many compounds of phosphorus and sulphur) resemble salts.

All the other elements can be arranged in analogous series, more or less complete; for example, the silver series.

Atomic weights.

Ag = 108; Cd = 112; In = 113; Sn = 118; Sb = 122; Te = 125 (?); I = 127.

I shall only give the density of these bodies, because their concordance with the preceding series with regard to the other properties needs no explanation.

Density.

Ag = 10·5; Cd = 8·6; In = 7·4; Sn = 7·2; Sb = 6·7; Te = 6·2; I = 4·9.

(To be continued.)

ON THE

SOLUBILITY OF SOLIDS IN GASES.*

By J. B. HANNAY, F.R.S.E., F.C.S., and JAMES HOGARTH.

(PRELIMINARY NOTICE.)

THIS investigation was undertaken in the hope that, by an examination of the conditions of liquid matter up to the " critical " point, sufficient knowledge might be gained to enable us to determine under what particular conditions liquids are dynamically comparable, in order that the microrheometrical method† (which the Royal Society has done one of us the honour of publishing in the *Philosophical Transactions*) might be applied to determine their molecular mass and energy relations. It seemed that as the laws relating to gases and liquids merge at what was called by Baron Cagnaird de la Tour‡ " l'état particulier," and by Dr. Andrews‖ the " critical point," an examination of matter up to the limit of the liquid state would be likely to yield us much information. The time we have to devote to scientific work being very limited, we found that it was quite impossible to make much advance by using the apparatus devised by Dr. Andrews, as the time required to change from one liquid to another was more than we had at our disposal. We therefore devised a new apparatus, which will be described in a more lengthy communication, but which, we may state, can be opened, the liquid changed, and again closed for a new experiment, in about one minute.

The question as to the state of matter immediately beyond the critical point being considered by Dr. Andrews to be at that time incapable of receiving an answer, we imagined that some insight might be gained into its condition by dissolving in the liquid some solid substance whose fusing point was much above the critical point of the liquid, and noticing whether, on the latter passing its critical point and assuming the gaseous condition, the solid was precipitated or remained in solution. We found that the solid was not deposited but remained in solution, or rather in diffusion, in the atmosphere of vapour, even when the temperature was raised 130° above the critical point, and the gas was considerably expanded. When the

* A Paper read before the Royal Society, November 20, 1879.
† 'On the Microrheometer, *Phil. Trans. Roy. Soc.*, 1879.
‡ *Ann. Chim.*, Series 2me, xxi., p. 127; xvii., p. 410.
‖ "Bakerian Lecture," *Phil. Trans. Roy. Soc.*, 1869, p. 588.

side of a tube containing a strong gaseous solution of a solid is approached by a red-hot iron, the part next the source of heat becomes coated with a crystalline deposit, which slowly re-dissolves on allowing the local disturbance of temperature to disappear. Rarefaction seems to be the cause of this deposition, because if the temperature be raised equally and the volume retained at its original value, no deposition takes place. Those experiments have been done with such solvents as alcohol (ethyl and methyl), ether, carbon disulphide and tetrachloride, paraffins, and olefines, and such solids as sulphur, chlorides, bromides, and iodides of the metals, and organic substances such as chlorophyll and the aniline dyes. Some solutions show curious reactions at the critical point. Thus ethyl alcohol, or ether, deposits ferric chloride from solution just below the critical point, but re-dissolves it in the gas, when it has been raised 8° or 10° above that temperature.

It appeared to us to be of some importance to examine the spectroscopic appearances of solutions of solids when their liquid menstrua were passing to the gaseous state; but as all the substances we have yet been able to obtain in the two states give banded spectra with nebulous edges, we are only able to state that the substance does not show any appreciable change at the critical point of its solvent. Such was the case with anhydrous chloride of cobalt in absolute alcohol. It was suggested to us by Prof. Stokes that the substance obtained by the decomposition of the green colouring matter of leaves by acids, and which yields a very fine absorption-spectrum, might be useful for our purpose. We have prepared the substance according to the careful directions kindly furnished us by Prof. Stokes, and find that it shows the phenomenon in a marked manner, whether dissolved in alcohol or ether. The compound is easily decomposed by heat under ordinary circumstances, and yet can be dissolved in gaseous menstrua, and raised to a temperature of 350° without suffering any decomposition, showing the same absorption-spectrum at that elevated temperature as at 15°.

We considered that it would be most interesting to examine by this method a body such as sodium, which, besides being an element, yields in the gaseous state sharp absorption-lines. An opportunity seemed to be afforded by the blue solution of sodium in liquefied ammonia, described by Gore,* but we found that, on raising the ammonia above its critical point, the sodium combined with some constituent of the gas, forming a white solid, and yielding a permanent gas, probably hydrogen.

There seems, in some cases, to be a slight shifting of the absorption-bands towards the red, as the temperature rises, but we have as yet been able to make no accurate measurements.

When the solid is precipitated by suddenly reducing the pressure it is crystalline, and may be brought down as a " snow " in the gas, or on the glass as a " frost," but it is always easily re-dissolved by the gas on increasing the pressure. These phenomena are seen to the best advantage by a solution of potassic iodide in absolute alcohol.

We have, then, the phenomenon of a solid with no measurable gaseous pressure dissolving in a gas, and not being affected by the passage of its menstruum through the critical point to the liquid state, showing it to be a true case of gaseous solution of a solid.

Determination of the Elements of a Vibratory Movement : Measurement of Amplitudes.—M. Mercadier.—The author uses a so-called vibratory micrometer, a scale divided into millimetres, or fractions of a millimetre, intersected by a transverse line, extending to the zero of the scale, and forming with it a small angle. This may be either engraved upon the vibrating body itself, or upon a plate of glass, &c., fixed to the body so that this angular micrometer may share its vibrations, the scale being perpendicular to their direction.—*Comptes Rendus.*

* *Proc. Roy. Soc.*, vol. xxi., p. 145.

THE CHEMICAL NEWS.

VOL. XL. No. 1045.

THE PERIODIC LAW OF THE CHEMICAL ELEMENTS.

By D. MENDELEEF.

(Continued from page 256.)

I SHALL show further on, and it can be seen from the accompanying tables, that relations of this kind can be drawn up for *all the elements*, showing that there is an intimate dependence between their properties and their atomic weights.

We could have foreseen this by means of the atomic theory, because the atomic weight forms one of the variable magnitudes which determine the functions of atoms. A similar consideration led me to discover the above-mentioned dependence, and this is the reason why I mention it here.

From the preceding, as well as from other relations that I have succeeded in finding, it results that all the functions which show how the properties depend on the atomic weight are periodic functions. First, the properties of elements become modified as the atomic weights increase; then they repeat themselves in a new series of elements, *a new period*, with the same regularity as in the preceding series. The periodic law can therefore be formulated in the following manner:—*The properties of simple bodies, the constitution of their compounds, as well as the properties of these last, are periodic functions of the atomic weights of elements.*

We now pass on to the manner of finding out the function which shows us this dependence; to effect this we should begin by finding the length, or, in better words, the number of members in a period. As for expressing this function, it appears of its own accord in some cases (such as the forms of oxidation); in other cases, there are not up to the present time any means of measuring it exactly, but, nevertheless, it keeps its periodic character.

Already the above-mentioned relations have brought to light the existence and the properties of a period of *seven* elements, corresponding to the period Li, Be, B, C, N, O, F. Let us call it the *small period* or *small series.* If H is attributed to the first series, Li, &c., will belong to the second, Na...... to the third, and so on.

However, all the elements known up to the present time do not belong, as might be believed, to the little series, and, what is still more important, there exists between the corresponding members of the odd and even series (the first two excepted, as will be seen further on) a very marked difference, while the members of the odd or even series show a greater analogy between themselves. An example will prove this sufficiently.

4th series	..	K	Ca	—	Ti	V	Cr	Mn
5th series	..	Cu	Zn	—	—	As	Se	Br
6th series	..	Rb	Sr	—	Zr	Nb	Mo	—
7th series	..	Ag	Cd	In	Sn	Sb	Te	I

The members of the fourth and sixth series show a greater analogy amongst themselves than they do with the members of the fifth or seventh series. Among the members of the even series there are no metalloids so marked as in the odd series; the last members of the even series resemble in many respects (as the forms of lower oxidation) the first members of the odd series. It is thus that Cr and Mn resemble Cu and Zn in their basic oxides. On the other hand, there are marked differences between the last members of the odd series (haloids), and the first members (metals of alkalies) of the even series

which follow them. But at the same time, between the last of the even series and the first of the odd are arranged in order according to their properties and atomic weights, *all* the elements which cannot find room in the small periods. It is in this manner that Fe, Co, and Ni in placing themselves between Cr and Mn, on the one side, and Cu and Zn, on the other, form the following series of transition:

$$Cr=52; \quad Mn=55; \quad Fe=56; \quad Co=59; \quad Ni=59; \quad Cu=63; \quad Zn=65.$$

In the same manner as Fe, Co, and Ni follow the fourth series, Ru, Rh, and Pd follow the sixth, and Os, Ir, and Pt the tenth. These two series (one even and one odd) with the intermediate series of elements, which have just been named, form a large period comprising seventeen members. As the intermediate members (such as Fe, Co, Ni) do not correspond to any of the seven groups of the small period, they form an independent group (the eighth); the members of this group—

$$
\begin{array}{lll}
Fe = 56; & Ni = 59; & Co = 59 \\
Ru = 104; & Rh = 104; & Pd = 106 \\
Os = 193?; & Ir = 195?; & Pt = 197
\end{array}
$$

are analogous between themselves, in the same manner as the corresponding members of the even series; such as V, Nb, Ta or Cb, Mo, W, &c., This analogy is derived from the following facts:—

1st. The metals of the eighth group are all of a grey colour, and are difficultly fusible. The fusibility increases from Fe to Co and to Ni; from Ru to Rh and to Pd; from Os to Ir and to Pt.

2nd. These metals possess, even compared with the neighbouring members, very low atomic volumes; for example, the atomic volume of $Cr=7\cdot6$, of $Mn=7\cdot0$, of $Fe=7\cdot2$, of $Co=7\cdot0$, of $Ni=7\cdot0$, of $Cu=7\cdot2$, of $Zn=9\cdot2$. The volume of $Mo=11\cdot2$, whilst the volumes of Ru, Rh, and Pd are approximately 9, that of Ag is $10\cdot3$, that of Cd $13\cdot0$; the volumes of Os, Ir, and Pt are about $9\cdot5$, and of W $10\cdot1$; that of Au is 10; and, lastly, that of Hg is 15. The smallness of volumes, or of the distances between the atomic centres, render the metals of the eighth group difficultly fusible, leaves them only a small amount of chemical energy, and also determines other properties.

3rd. These metals possess in the highest degree the power of condensing and abandoning oxygen, as has been shown for Ni, Pd, Fe, and Pt by Graham and Raoult.

4th. Their highest forms of oxidation are either bases or acids of such a feeble character that they are easily transformed into lower oxides with a more marked basic character.

5th. In this group we only meet with metals whose form of combination is RO_4 or R_2O_8, such as OsO_4 and RuO_4;* (it is for this reason that these metals have been collected together in a special group.) We notice that in each series the highest oxides capable of being formed by the metals from Fe to Cu, from Ru to Ag, from Os to Au, contain less and less oxygen. The highest oxide of iron is FeO_3; of cobalt, CoO_2; of nickel, Ni_2O_3. In the same manner, osmium gives us OsO_4; iridium with difficulty forms IrO_3; platinum will only give PtO_2; and gold, Au_2O_3.

6th. They give stable alkaline double cyanides. Fe, Ru, and Os give combinations analogous to K_4RCy_6; Co, Rh, and Ir form salts whose composition is according to the formula K_3RCy_6; Ni, Pd, and Pt give salts according to the general formula K_2RCy_4.

7th. They form stable ammoniacal compounds, and resemble one another in many respects. For example, Claus has prepared salts of rhodium and of iridium, which correspond to the roseo-salts of cobalt, $RX_3,5NH_3$, such as $RhCl_3,5NH_3$.

8th. Some forms of combination of these metals, particularly the higher forms, are distinguished by their characteristic colours, &c.

---

* Ferric acid would apparently be FeO_4?

TABLE I.

LARGE PERIODS.

K = 39	Rb = 85	Cs = 133	,,	,,
Ca = 40	Sr = 87	Ba = 137	,,	,,
,,	? Yt = 88 ?	? Di = 138 ?	Er = 178 ?	,,
Ti = 48 ?	Zr = 90	Ce = 140 ?	La = 180 ?	Th = 231
V = 51	Nb = 94	,,	Fa = 182	,,
Cr = 52	Mo = 96	,,	W = 184	Ur = 240
Mn = 55	,,	,,	,,	
Fe = 56	Ru = 104	,,	Os = 195 ?	,,
Co = 59	Rh = 104	,,	Ir = 197	,,
Ni = 59	Pd = 106	,,	Pt = 198 ?	,,
Cu = 63	Ag = 108	,,	Au = 199 ?	,,
Zn = 65	Cd = 112	,,	Hg = 200	,,
,,	In = 113	,,	Tl = 204	,,
,,	Sn = 118	,,	Pb = 207	,,
As = 75	Sb = 122	,,	Bi = 208	,,
Se = 78	Te = 125 ?	,,	,,	,,
Br = 80	I = 127	,,	,,	,,

TYPICAL ELEMENTS.

H = 1	Li = 7	Na = 23
	Be = 9·4	Mg = 24
	B = 11	Al = 27·3
	C = 12	Si = 28
	N = 14	P = 31
	O = 16	S = 32
	F = 19	Cl = 35·5

TABLE II.

SERIES.	GROUP I. R_2O	GROUP II. RO	GROUP III. R_2O_3	GROUP IV. RH_4 RO_2	GROUP V. RH_3 R_2O_5	GROUP VI. RH_2 RO_3	GROUP VII. RH R_2O_7	GROUP VIII. RO_4
1	H=1	,,	,,	,,	,,	,,	,,	
2	Li=7	Be=9·4	B=11	C=12	N=14	O=16	Fl=19	
3	Na=23	Mg=24	Al=27·3	Si=28	P=31	S=32	Cl=35·5	
4	K=39	Ca=40	−=44	Ti=48	V=51	Cr=52	Mn=55	{ Fe = 56 ; Co = 59. Ni = 59 ; Cu = 63 .
5	(Cu=63)	Zn=65	−=68	−=72	As=75	Se=78	Br=80	
6	Rb=85	Sr=87	?Yt=88	Zr=90	Nb=94	Mo=96	−=100	{ Ru=104 ; Rh=104 Pd=106 ; Ag=108.
	(Ag=108)	Cd=112	In=113	Sn=118	Sb=122	Te=125	I=127	
	Cs=133	Ba=137	?Di=138	?Ce=140	,,	,,	,,	
	,,	,,	?Er=178	?La=180	Ta=182	W=184	,,	{ Os=195 ; Ir =197 ; Pt =198 ; Au=199 .
	(Au=199)	Hg=200	Tl=204	Pb=207	Bi=208	,,	,,	
12	,,	,,	,,	Th=231	Ur=240	,,	,,	,, ,, ,, ,,

Above are two tables in which will be found supplementary particulars about what has been already said. In the first the elements are placed in *large periods*, with their atomic weights. In the second they are arranged in groups and series, that is to say, in small periods, in such a manner that the differences between the odd and even series become very apparent.

Observations on Table I.—For the sake of brevity the atomic weights have been given in these tables in round numbers, because in the greater number of cases we cannot be sure of the exactitude of the tenths nor of the units. A point of interrogation (?) *before* the symbol of an element means that the incomplete state of the researches on that element do not allow us yet to give it a determinate position in the system. A point of interrogation *after* the atomic weight indicates that in respect to the atomic weight of that element we are still in doubt ; or, in other words, the equivalent of the element does not appear to have been fixed exactly up to the present day. Some atomic weights have been modified in the table according to the periodic law (see chapter 5). Thus the atomic weight of tellurium is put at 125 ? which is in accordance with the periodic law, and not 128, as found by Berzelius and others.

Observations on Table II.—In this table the groups are indicated by Roman figures. The seven first groups correspond to the seven members of each series ; the eighth group has been already characterised (see above). Cu, Ag, and Au have been placed in the eighth group, because of their analogous properties ; they could also have been put in the first group, because of their forms of lower oxidation. The first two series have been separated from the others, for reasons which will be explained later on. They are *typical* series.

(To be continued.)

THE CHEMICAL NEWS.

VOL. XL. No. 1046.

THE PERIODIC LAW OF THE CHEMICAL ELEMENTS.

By D. MENDELEEF.

(Continued from page 268.)

THE metallic character appears more clearly in the members of the even series, while the corresponding members in the odd series rather possess acid properties. Thus there is a marked difference between V, Nb, and Ta of the even series in the fifth group and P, As. Sb, and Bi, although they all give an oxide corresponding to R_2O_5; further, the first-named give acids less energetic than the latter. The members of the even series do not give, as far as is known, any hydrated or volatile metallo-organic compounds like the corresponding members in the odd series. As, among the odd elements, Zn, Cd, As, Sb, Se. Te. Br, I, Sn, Pb, Hg, and Bi can form, in a general way, metallo-organic compounds, we can foresee with certainty that the elements In and Tl, comprised amongst those which have just been cited, will also give metallo-organic compounds, such as $InAe_3$ and $TlAe_3$. Up to the present time there is no member of the even series of the higher groups which has formed a metallo-organic compound. The attempts made by Buckton, Cahours, &c., to prepare $TiAe_4$ by means of $TiCl_4$ have not succeeded, in spite of the great analogy between $TiCl_4$, $SiCl_4$, and $SnCl_4$. If, then, the even elements should give any metallo-organic compounds, these compounds would behave in an altogether different manner to the metallo-organic bodies already known, for the same reason that the hydrogenated compounds of Pd, Cu, Nb have not the same properties as the corresponding compounds of the odd series. It would be difficult to obtain volatile compounds of hydrogen and ethyl with Zr, Nb, Mo, W, and Ur.

It might seem that the establishment of the second series would be in opposition to the general division into odd and even series, because the members of this even series (Li, Be, B, C. N, O, F) possess acid properties, give hydrogenous and metallo-organic compounds (BAe_3, $CAe_4 = C_9H_{20}$, NAe_3, OAe_2, FAe), and some of them are gaseous, that is to say, that they behave like the odd elements. But in any case, it is necessary to remark, relatively to this series: 1st. That it is not connected with the eighth group like the other even series are; 2nd. That the difference between the atomic weights of elements of this series and the corresponding atomic weights of the following series is about 16, while the difference for all the other series is from 24 to 28. The difference between the atomic weights of the successive even series is about 46, whilst the elements of the second and fourth series have only a difference in their atomic weights of from 32 to 36.

	Li	Be	B	C	N	O	F	Na	Mg	Al	Si	P	S	Cl
	K	Ca	„	Ti	V	Cr	Mn	Cu	Zn	„	„	As	Se	Br
Difference	32	31	„	36	37	36	36	40	41	„	„	44	46	45

In this manner the apparent anomalies are explained, and the fundamental idea that we have conceived concerning the dependence between the properties of the elements and their atomic weights is but confirmed by these observations. For, since there exists another difference in the atomic weights, the reciprocal relation between the properties of the elements should also be different.

The elements of the second series should not have any properties coinciding with those of the elements in the fourth series, except when they would have atomic weights less than those which they really possess. These differences are still noticed between Na and Cu, Mg and Zn; but they disappear when we compare P and As, S and Se, Cl and Br, elements whose differences of atomic weights and properties remain within ordinary limits. It is because of these differences noticed amongst the elements of the second series that I distinguish them by the name of *typical elements*. We can add Na and Mg to them, independently of H, for reasons which have been mentioned above.

If we *compare* the reciprocal relations of the other analogues with the relations that the compounds of the homologous series have amongst themselves, we see that the typical elements correspond to the initial members of the homologous series, which, as we know, do not possess all the properties of the higher homologues. In the same manner the initial members (H_2O and CH_4O) of the alcoholic series $C_nH_{2n}+2O$ possess many properties peculiar to themselves, which disappear in the higher members.

From what precedes, appertains the isolated independent position of hydrogen, which has the lowest atomic weight. From the nature of the saline oxide H_2O and the salts HX, it ought to belong to the first group; the nearest analogue of hydrogen is sodium, which should be placed in an odd series of the first group. Cu, Ag, and Au are very distant analogues of H. All the five give compounds corresponding to RO and R_2O_2 (H_2O, Na_2O_2,* Cu_2O_2, Ag_2O_2, and Au_2O_2). If oxide of copper at the minimum of oxidation is represented by CuO, it is also necessary to represent the preceding peroxides by RO, NaO, AgO, AuO. It is true that CuO forms salts corresponding to CuX_2, while the peroxides, such as AgO,† do not give, as far as is known, any salts of this kind. However, Ni, which immediately precedes Cu in the system, and Pd, which immediately precedes Ag, have an analogous difference; Ni gives no salts like NiX_4, while Pd forms salts like PdX_2, although they are not very stable.

Further, we notice, in each even series, from the first group to the eighth, an increase in the quantity of oxygen which can enter into combination with an element; in the series of the eighth group this faculty diminishes as the atomic weight increases (see above), and it attains its minimum for Cu, Ag, and Au; from thence (for Zn, As, for Cd, In, for Hg, Tl) it commences to increase. Therefore Cu, Ag, and Au, as is shown in the second table, occupy two places, one in the first group and the other in the eighth; in the lower forms of combination they correspond to H and Na (first group). This analogy is above all evident in Ag, which needs no other explanation; the analogy of the compounds of suboxide of copper and suboxide of gold with the compounds of oxide of silver is quite as certain; the comparison of the properties of CuCl, AgCl, and AuCl, afford sufficient proof. In spite of a great analogy between the compounds HX and NaX, there exists between the first and the last a series of differences well known to everybody; it calls to mind the differences between carbonic acid and the higher homologues (glycolic acids), which are analogous to it in many respects.

As a noticeable example of the analogy between the compounds of H, Na, and Cu (as suboxides), Ag and Au (also as suboxides), I will give a table of their volumes. (See next column.)

We see that the corresponding members have very nearly equal volumes. Li and K, on the contrary, have in the same form of combination different volumes. It is in this manner that the volume of $LiCl = 21$, of $KCl = 37$,

---

* It is known that K gives a peroxide KO_2. It would be interesting to further the study of peroxide of lithium, to decide whether its formula is LiO, and if it has not basic properties, even in a small degree. AgO should also be studied in this respect.

† We know that I, in reacting on peroxide of sodium, forms Na_2OI_2, which has a composition analogous to that of Cu_2OCl_2.

	Density.	Volume.
HCl	1·27	29
NaCl	2·16	27
CuCl	3·68	27
AgCl	5·55	26
AuCl	9·3 ?	25

	Volume.		Volume.		Volume.
Na_2CO_3 ..	43	H_2O.. ..	20	H_2SO_4 ..	53
Ag_2CO_3 ..	46	Na_2O ..	21	Na_2SO_4 ..	54
				Ag_2SO_4 ..	58
HNO_3 ..	41	NaHO ..	19		
$NaNO_3$ ..	39	Cu_2O ..	25	$NaClO_3$ ..	46
$AgNO_3$ ..	39	Ag_2O ..	28	$AgClO_3$ ..	44

of $LiNO_3 = 29$, of $KNO_3 = 48$, of $Li_2SO_4 = 50$, of $K_2SO_4 = 66$. The compounds of K have greater volumes (and a constantly less density) than the corresponding compounds of Na or Li.

(To be continued.)

THE CHEMICAL NEWS

VOL. XL. No. 1047.

THE PERIODIC LAW OF THE CHEMICAL ELEMENTS.

By D. MENDELEEF.

(Continued from page 280.)

In spite of the similarity of the corresponding compounds of Na, Cu, and Ag, the metals can be very easily distinguished one from another, which agrees with the fact that the volume of the metal alone is not the same as the volume of the metal when combined with another element. The volume of $Na = 24$, of $Cu = 7$, of $Ag = 10$. The atoms of Na are further apart from each other and are more susceptible to reaction than the atoms of Cu and Ag. When Na enters into combination with another body, there is often a considerable contraction takes place. For example, the volume of $Na = 24$, the volume of $NaHo = 19$; 24 volumes of Na give 27 volumes of NaCl, whilst 10 volumes of Ag give 26 volumes of AgCl. This and many other examples show how arbitrary it is to conclude from the known volume of a compound the unknown volumes of its elementary parts.

We see, from the inspection of Table I.,* that the members at the beginning of the large periods (as well as the small periods commencing with Na and Li) are metals of a very strongly pronounced alkaline nature, whilst the members at the end are energetic haloids. Already, in the time of electro-chemists, the above-mentioned elements occupied the same positions of extreme members in the system. It is incontestable that the electro-chemical theory has had a great deal of influence on the later progress of chemistry. However, I am far from wishing to set myself up as a defender of this theory. To show this I will mention that, as can be seen in Table I., the first and last members of the *large periods* are the only ones which have a clearly defined chemical character. In other words, they are the only elements which react easily, and at temperatures which are not too high, with the greater number of other bodies; they alone possess atomic weights of about the same value, and some other analogous properties. The volume of a compound body increases nearly in equal quantities when H is replaced by Cl $(=35.5)$, or by K $(=39)$. When metaleptic hydrogen is replaced by chlorine the volume increases in a nearly constant ratio. For examples:—

Volume	C_5H_{12} $= 110$	C_6H_6 $= 87$	C_7H_8 $= 104$
„	$C_5H_{11}Cl = 117$	$C_6H_5Cl = 97$	$C_7H_7Cl = 110$
Diff. of vol. when H is replaced by Cl	7	10	6

The volume increases in an analogous manner, and in nearly the same quantity, when H is replaced in the acids by K.

Volume	$HCl = 29$	$H_2O = 20$	$HNO_3 = 41$	$H_2SO_4 = 53$
„	$KCl = 37$	$KHO = 28$	$KNO_3 = 48$	$K_2SO_4 = 66$
Diff. of vol. when H is replaced by K	8	8	7	13

We can bring together, according to their atomic weights, the following unlike elements:—

O $= 16$	F $= 19$	Na $= 23$	Mg $= 24$
S $= 32$	Cl $= 35.5$	K $= 39$	Ca $= 40$
Se $= 78$	Br $= 80$	Rb $= 85$	Sr $= 87$
Te $= 125$?	I $= 127$	Cs $= 133$	Ba $= 137$

* See CHEMICAL NEWS, vol. xl., p. 268.

The transition from Cl to K, &c., corresponds also, in many respects, to a certain analogy between these elements, although in the same period there are no elements having such similar atomic weights and such different properties. Because of this last fact, it might be better to separate the series here, in such a manner that the large periods commence with K and terminate with Cl. But in reality the series of elements is uninterrupted, and represents in a certain degree a spiral function.

In consequence of this, the most dissimilar elements are found at the beginning and at the end of large periods; in the middle are the transitory elements, and the most analogous are the nearest together. It is in this manner that after the elements of the alkalies and the alkaline earths come these rare elements (metals of gadolinite and cerite, Ti, V, Cr, Nb, Mo, Ta, W, Ur), which, even from an analytical point of view, are similar amongst themselves. They are not volatile; they are difficultly fusible, and difficult to reduce; they possess, even in their highest oxides, a feeble power of reaction; they are often found together in nature, and rarely in large quantities, &c. These elements are rare, and I even think that that is because of their character and not by chance. We can mention as a comparison that among the hydrocarbides the series C_nH_{2n+2} and C_nH_{2n-6} are often found together in nature, both being formed in many reactions; they possess a pronounced reaction and furnish many derivatives, while the members of the intermediate series, particularly C_nH_{2n} and C_nH_{2n-2} are formed more rarely, and do not give so many independent combinations as the preceding series. Our knowledge about the rare elements that I have just mentioned is unfortunately very defective, and if we had not before us the classic researches of Marignac on the compounds of Zr, Nb, and Ta, the entire group would be a collection of elements without any importance in the system. The works of Blomstrand, of Roscoe, of Delafontaine, and of Bunsen have thrown much light on the characters of the rare elements; but these elements are still the object of a series of questions which have not yet been answered.

After these rare elements (see Table I.) comes first in importance the group of elements called *noble*; in the system they all grouped together. The useful metals are attached to them, and by the help of As, Sb, and Te, they lead on to the metalloids.

In the groups of analogous elements, the elements whose atomic weights are the highest possess more strongly marked basic properties, or form weaker acids. Bunsen has shown that Cs is more electro-positive than K and Rb; the basic properties are more apparent in BaO than in CaO, in ThO_2 than in ZrO_2 or TiO_2; HgO separates MgO or BeO from compounds; Bi_2O_3 is a more energetic base than Sb_2O_3 or As_2O_3; P_2O_3 does not exhibit any basic properties, unless it is in this circumstance, that PH_3O_3 is not a tribasic acid, but a dibasic acid. For the same reason Ta gives a more feeble acid than Nb and V, Te an acid more feeble than Se and S. The acid properties are very feebly marked in PbO_2,* although Frémy obtained a salt, $K_2PbO_3,3H_2O$, corresponding to salts prepared by means of stannic and silicic acids; SnO_2 possesses feeble basic and acid properties; and SiO_2 cannot act except as an acid, although as a weak acid.

Besides this, when the atomic weight increases in the members of a determined group, it is not only that the faculty of being reduced to the state of a simple body increases (compare Te and Se with S, I with Cl, Au with Cu, &c.), but also the power of giving lower forms of oxidation, exhibits itself frequently by a relative stability and by a great aptitude for forming compounds. Bi forms Bi_2O_5 with difficulty—generally the compounds of bismuth correspond to Bi_2O_3, or BiX_3; in the same manner

* A true analogue of SnO_2, PbO_2, should also be able to act as a base. However, the attempts that I have made, above all in that which concerns the action of HF, have led to no results. To all appearance the known form, PbO_2, answers to metastannic and metatitanic acid. We must wait for the discovery of a variety of PbO_2 with basic properties.

Pb does not only form PbO_2, but also the very stable oxide PbO, which Sn and Si are not capable of doing to the same degree; Tl does not only form Tl_2O_3 but also Tl_2O, which has not been observed either in the case of In or of Al.

In the group Mg, Zn, Cd, Hg, we notice, with the augmentation of the atomic weight, a marked increase in the volatility, basic properties of the oxide RO, reducibility to the metallic state, and the power of forming a lower oxide, R_2O.[*] The volatility does not increase with the augmentation of the atomic weight, except in this and the neighbouring series; on the contrary, it diminishes in the last series, as is well shown by Cl, Br, and I.

It is precisely for the motives expressed above that the so-called noble metals occupy the place which has been given to them in the system,—that is, in the middle of the large periods, amongst the members having high atomic weights, which comprises elements which are very easily reducible, and whose reactions are very weak.

All that has preceded shows the nature of the periodic law. No natural law acquires any scientific importance unless it introduces, so to speak, some practical conclusions, or, in other words, unless it admits of logical conclusions capable of explaining what has before remained unexplained, and, above all, unless it raises questions which can be confirmed by experience. In a case of this nature the use of the law is evident, and it is possible to control its correctness.

This law will at least incite research into the new, little-known parts of science. This is why I propose to study attentively some consequences of the periodic law, and to examine how it can be applied—

To the system of elements;

To the determination of the atomic weights of insufficiently studied elements;

To the determination of the properties of elements up to the present time unknown;

To the correction of atomic weights;

And to the enlargement of our knowledge of the forms of chemical combinations.

I shall not form any hypotheses, either here or further on, to explain the nature of the periodic law; for, first of all, the law itself is too simple;[†] and, secondly, this new subject has been too little studied yet, in its diverse parts, for us to be able to form any hypothesis: the most important reason is the third one,—that is, that we cannot put the periodic law and the atomic theory in accord without upsetting the known facts about the exact values of the most carefully found atomic weights. There is, however, I believe, between the series of elements and the homologous series, an analogy which, although small, is not less real, which authorises me to admit that it is so; it is the comparison of the physical properties of a group to which I shall return later on.

(To be continued.)

THE DISSOCIATION OF CHLORINE.

VICTOR MEYER'S ANSWER TO SEELHEIM'S CRITIQUE.[‡]

SINCE I communicated to you Seelheim's experimental critique[§] of V. Meyer's famous experiments this investigator has published an answer to that critique, which (I rejoice to say) amounts to a complete and (I believe) unanswerable refutation.

What V. Meyer says is easily condensed in a few words. It is quite true that at a "yellow" heat platinum-metal can be volatilised in a current of chlorine. V. Meyer himself has often repeated Troost and Hautefeuille's experiments; but, in looking into the matter quantitatively, he found that at $-1570°$ C. a given quantum of metallic platinum, when exposed to a rapid current of dry chlorine for a whole hour, lost only 1 per cent of its weight. In his vapour-density determinations the platinum produced (from the $PtCl_2$) is exposed to a stagnant atmosphere of chlorine for only a few seconds. It is difficult to assume that, under these circumstances, more than perhaps a few hundredths of a per cent of the metal could possibly get gasified. But, what is more, he always took care, after the experiment, to examine the wee bucket which served for the introduction of the platinous chloride, and he invariably found it full of spongy metal, the weight of which (whenever the test was applied) agreed almost absolutely with that calculated from the weight of the chloride used. Of crystalline or sublimed platinum no trace could be discovered.

Moreover, iodine, as V. Meyer stated in his first communication, behaves at high temperatures as chlorine does, although it naturally was used as such, and not in the shape of a platinum compound at all.

It appears, then, that V. Meyer after all is right, and that we must admit that the substance Cl_2 at high temperatures suffers decomposition, perhaps according to equation $2Cl_2 = Cl_2 + Cl + Cl$. This from the first was my theory of the phenomenon, and I mean to retain it until it is disproved. Not that I claim any credit for an idea which is so obvious that it must have suggested itself to any chemist who took the trouble of reasoning on the matter.

Lieben[*] (besides offering this very obvious explanation) suggests that possibly V. Meyer's result may be owing only to the fact that the coefficient of expansion of chlorine increases rapidly at high temperatures, and he does not see that he thereby only re-states the problem. In any gas of moderate tension the molecules must be assumed to be mere points compared with their average distances. If such a thing expands on being heated it can do so only through two causes, viz., first, an increase in the energy of the progressive motion of the molecules; and secondly, a multiplication of molecules, i.e., through dissociation. The expansion owing to the first cause amounts to 1-273rd of the volume at 0° C. per degree Centigrade. To say that a certain gas, within a certain interval of temperature, expands at a greater rate is only another way of stating that the number of molecules in it increases in consequence of the respective elevation of temperature.

W. D.

Anderson's College, Glasgow

---

* It can be foreseen that cadmic oxide will give a basic suboxide, Cd_2O,—certainly very unstable in air,—or else salts CdX, which correspond to it.

† However, I do not ignore that to completely understand a subject we should possess, independently of observations (and experiences) and of laws (as well as systems), the meanings of both one and the other.

‡ From an article in the *Ber. der Deutsch. Chem. Ges.*, No. 18. Communicated by Prof. Dittmar.

§ CHEMICAL NEWS, vol. xl., p. 244.

THE CHEMICAL NEWS.

VOL. XL. No. 1048.

THE PERIODIC LAW OF THE CHEMICAL ELEMENTS.

By D. MENDELEEF.

(Continued from page 292.)

II. ON THE APPLICATION OF THE PERIODIC LAW TO THE SYSTEM OF ELEMENTS.

THE system of elements possesses not only a purely pedagogic importance as a means of learning more easily the different facts which are grouped in a systematic manner and united one to another, but also a scientific importance, inasmuch that it shows us new analogies, and thus opens up new roads for the research for elements. All the systems known up to the present can be placed in two distinct categories.

Into one of these categories (artificial systems) enter the systems which are based on some few characters of the elements. For example, the systems of division of the elements according to their affinity, according to their electro-chemical properties, according to their physical properties (or division into metals and metalloids), according to the manner in which they behave in the presence of oxygen and of hydrogen, according to their atomicity, &c. These systems, in spite of their evident defects, deserve some consideration, for they have the merit of a certain exactness, and each of them has contributed in a special direction to the progressive elaboration of chemical ideas.

The systems of the second category (natural systems) establish, according to members of different and purely chemical properties, groups of analogous elements. The well-known results of these systems have obtained for them the preference over the artificial systems. However, they are also stained with serious faults.

1st. The partition of elements does not rest upon invariable principles; also such elements as Tl, and even Ag and Hg, &c., have been placed in different groups. Na and K, Li and Rb, and often Tl also, have been frequently placed among the alkaline metals, although there exist between K and Na, in spite of the small difference between their atomic weights, more difference of properties than there is between K, Rb, and Cs. Na and all compounds of Na have a greater density than K and its corresponding compounds, although the atoms of sodium are lighter than the atoms of potassium.

NaCl and KCl crystallise in cubes (KCl often crystallises in a mixture of cubes and octahedra). However, we do not meet with isomorphous mixtures in nature, and we have obtained none artificially. According to Stassfurt we often meet with distinctly separate crystals of NaCl and KCl, one by the side of the other.

In the same manner the situation of Mg relative to Ca, or of Pb relative to Ba, Sr, and Ca, or of Tl relative to K, Rb, and Cs, is doubtful. In consequence of this want of fixed principles of separation the results of the natural systems are very uncertain.

2ndly. Some elements are without analogies, such as Au, Al, B, F, Ur, &c.

3rdly. There is no general expression for the reciprocal relations between the different groups; they could not be united in one whole. Thus these systems are always incomplete.

The periodic law possesses, in the forms of oxides and in the atomic weights, invariable numerical values for the distribution of the elements. It permits us, then, to group together the elements which have in reality a great deal of resemblance, and it answers at the same time to the principles which have successively led to artificial systems. It gives us the means to erect an unarbitrary system, as complete as possible. The preceding considerations, as well as those which are to follow, show up its advantages and properties. I will only insist on the manner of employing it to determine the places of some elements which have given rise to different interpretations. I must now commence with some general remarks.

The position of an element, R, in the system is found by the series and the group to which this element belongs, and therefore by the elements X and Y, which are next to it, one in front and the other behind in the same series, and also by two elements in the same group, the element R', whose atomic weight is the next smaller, and the element R'', whose atomic weight is the next larger. The properties of R can be determined from the known properties of X, Y, R', and R''. We find, for example, in the system the following series:—

$$\begin{array}{llll} \text{Series of order} & (n+2) & X'\ R'\ Y' \\ \text{„} & \text{„} & (n) & X\ R\ Y \\ \text{„} & \text{„} & (n+2) & X''R''Y'' \end{array}$$

$$R'' - R = \text{nearly } R - R' = \text{about } 45.$$

Therefore, to find the properties of analogous compounds, we can establish proportions and find average values; the properties of all the elements are, properly speaking, in an intimate and mutual dependence. I call *atomanalogy* of an element the relation which joins R to X and Y, on the one hand, and to R' and R'' on the other hand. Thus As and Br on the one hand, and S and Te on the other, are the atomic analogues of Se, whose atomic weight is the medium one, viz.,—

$$78 = \left(\frac{75+80+32+125}{4}\right);$$

and, for example, the properties of SeH_2 answer to this atomic weight, because SeH_2 occupies the middle place between $AsH_3 - BrH$ and $SH_2 - TeH_2$, &c. It is only in the series and in the ends of groups that the atomanalogies are not quite valid, although we can there also see clearly the reciprocal relations, which can be shown conditionally by the arithmetical proportions, for example—

$$X' : X = R' : R = Y' : Y \text{ or } X' : R' = X : R = X'' : R''.$$

These reciprocal relations, to which the system founded on the periodic law leads us, furnish the possibility of explaining many isolated and doubted facts.

Different opinions have been given on the position of beryllium in the system since the researches of Awdeeff. He gives to the oxide the formula of magnesia; besides, glucina is very like magnesia in its properties. The periodic law gives the following proofs in favour of the formula BeO. If we adapted the formula Be_2O_3, the atomic weight of beryllium would be $\frac{3}{2}9 \cdot 4 = 14 \cdot 1$. In this case this element could not enter the system, because it would be by the side of nitrogen, and would have to possess very acid properties, and also to form higher oxides, having as formulæ Be_2O_5 and BeO_3, of which it is not capable. If, on the contrary, we adopt BeO as the formula for its oxide, and $9 \cdot 4$ for the atomic weight of the metal, this element, conforming to the formula of oxides and to all its properties, would be between $L = 7$ and $B = 11$. The following proportions will suffice to show what I mean:—

1st. $Be : Li = B : Be.$

In effect the basic properties are more feebly shown in BeO than in Li_2O, and much more feebly in Be_2O_3 than in BeO. Chloride of beryllium is more volatile than chloride of lithium, and much more volatile than chloride of boron.

2ndly. $Be : Mg = Li : Na = B : Al.$

Oxide of beryllium is a less energetic base than MgO. Again, Li_2O reacts more feebly than Na_2O, and B_2O_3 more feebly than Al_2O_3. This is why glucina dissolves in KHO.

The salts of BeO and of MgO do not present an absolute isomorphism; occasionally they show considerable differences in their crystalline forms. This fact is without importance, because the salts of Li and of Na and the compounds of B and of Al are none of them completely isomorphous.

Fluoride of beryllium is soluble in water, fluoride of magnesium is not; fluoride of boron is soluble, and fluoride of aluminium is not.

3rdly.   Be : Al = Li : Mg = B : Si.

If, in spite of the difference of the formulæ of the oxides, oxide of beryllium corresponds in many respects to aluminium, Li_2O and MgO will also possess analogous properties; the same is true for B_2O_3 and SiO_2. If, still further, the volume of an equivalent of BeO equal to—

$$\frac{25\cdot4}{3\cdot05} = 8\cdot3,$$

approaches in volume an equivalent of aluminium equal to—

$$\tfrac{1}{3}Al_2O_3 = \frac{102\cdot6}{4\cdot0} = 8\cdot5,$$

this concordance is repeated for the corresponding compounds of Li and $\tfrac{1}{2}$Mg, as well as for $\tfrac{1}{3}B_2O_3$ compared with $\tfrac{1}{2}SiO_2$.

The volume of LiCl = 21; therefore the volume of $Li_2Cl_2 = 42$, which corresponds to $MgCl_2 = 44$; the volume of $BCl_3 = 87$; therefore that of $\tfrac{2}{3}BCl_3 = 58$, which corresponds to $\tfrac{1}{2}SiCl_4 = 56$ ($SiCl_4 = 112$); the volume of $B_2O_3 = 39$; therefore the volume of $\tfrac{1}{3}B_2O_3 = 13$; whilst the volume of SiO_2 (amorphous) = 27, and, therefore, $\tfrac{1}{2}SiO_2 = 13\cdot5$.

Again, does not Rose's observation, relative to the proximity which the volumes of equivalents of glucinum and of aluminium present, show the concordance which exists between the formulæ of the oxides. If BeO crystallises in the characteristic forms for Al_2O_3, ZrO_2 crystallises in the same manner.

Therefore, the position of Be in the system would be elucidated on all points, so that we could form an opposition group to Li, Na, K, Rb, Co, viz., the group Be, Mg, Ca, Sr, Ba, parallel in all respects.

In the same way the position of B relatively to C, Si, Al is indicated by what precedes. So I think it is sufficient to make the following proportions:—B : Al = Be : Mg; B : C = Be : B;   B : Si = Be : Al.   We can still add B : P = C : S, as it appears, if we examine the tables, and if we consider that B furnishes the compounds BCl_3, B_2O_3, BH_3O_3, which correspond to PCl_3, P_2O_3, PO_3H_3, &c., these circumstances coincide with those that we observe relative to CO_2, C_2, C_2H_2, CH_2O_2, and to SO_2, S_2H_2, SH_2O_2.

(To be continued.)

THE PERIODIC LAW OF THE CHEMICAL ELEMENTS.

By D. MENDELEEF.

(Continued from vol. xl., page 304.)

THE remarkable researches of Roscoe, on vanadium, have shown the quantitative, and also in many cases the qualitative (such as in the oxychloride), analogy of this element with phosphorus. The position that vanadium holds in the system is expressed by the following proportions which, I think, do not need further explanation:—

(1.) $V : P = Nb : As$, or $V : As = Nb : Sb = Ta : Bi$;

(2.) $V : P = Ti : Si = Cr : S$;

(3.) $V : Cr : Ti = Nb : Mo : Zr$.

It is seen from these proportions that the true atomatologues of V are, on the one hand, Ti and Cr, and on the other hand, Nb and Ta, with which it shows more analogy than with P : in the same way Cr is more nearly related to Mo and to W than it is to S. The corresponding compounds of Ti, Cr, and V resemble one another in their reactions, their properties, and even in their aspect. The yellow colour of the acid salts of chromium are very similar to many of the acid salts of vanadium : the green colour of the salts of sesquioxide of vanadium reminds us of the green colour of the salts of sesquioxide of chromium. The distribution of V in nature corresponds with that of Ti and of Cr. The analogy which exists between the oxychlorides $VOCl_3$ and $POCl_3$ has for a companion the analogy between CrO_2Cl_2 and SO_2Cl_2. Further, this circumstance corresponds with the interesting case of isomorphism that Marignac has noticed between R_2TiF_6, R_2NbOF_5, and $R_2WO_2F_4$.[*]

We know that the most different opinions have been expressed relative to the position that *thallium* holds in the system until its properties were fully known. According to the periodic law this element is placed in the eleventh series, between $Au = 197$ and $Hg = 200$ on the one hand, and $Pb = 207$ and $Bi = 208$ on the other. The nature of its highest oxide, Tl_2O_3, gives it at the same time a place in the third group, near $Al = 27$ (and $In = 113$, as we shall see in the following chapter). This place corresponds to all the properties of Tl, although at first sight it gives rise to doubts, the oxide of thallium, Tl_2O_3, showing very little analogy with alumina. However, the proportion—

$$Tl : Al = Hg : Mg = Pb : Si,$$

carries sufficient conviction that such is the natural position of thallium.

It is only the highest oxide of Hg—that is, HgO—which shows any analogy with MgO ; PbO_2, the highest oxide of Pb, only shows any analogy with SiO_2. In the same manner, Tl only shows any analogy with Al in its highest oxide, Tl_2O_3. HgO and MgO are bases which give salts, such as RX_2 ; Tl_2O_3 and Al_2O_3 are less energetic bases which give neutral and basic salts such as RX_3 ; PbO_2 and SiO_2, on the contrary, are but feebly acid oxides. Oxide of thallium is relatively to acids a more energetic base than Al_2O_3, exactly as HgO takes the place of MgO. If Tl gave, independently of Tl_2O_3, a highly basic suboxide, Tl_2O, of which Al is not capable, Hg would give Hg_2O independently of HgO, whilst for Mg we only know of one form of oxide. Pb, differently to Si, gives independently of PbO_2 a highly basic oxide, PbO.

The higher oxides HgO, Tl_2O_3, PbO_2, Bi_2O_5 are peroxides, relatively to the lower forms, Hg_2O, Tl_2O, PbO, Bi_2O_3. If in Bi_2O_3 the basic properties are more feebly expressed than in PbO, Tl_2O, that is explained by the more strongly acid properties of Bi_2O_5, relatively to Tl_2O_3 and to HgO. If Tl_2O in the salts corresponds in part to

---

[*] We can consider $R_2MnO_3F_2$ and $R_2OsO_4F_2$ as the continuation of this series, although it is rather doubtful. (See Marignac *Bibliothèque Universelle de Genève*, t. xxiii. ; *Annales de Chimie et de Physique* [3], t. lxix.)

X_2O, we can also find an analogy between PbO, in the salts PbX_2, and CaO, as well as between Bi_2O_3, in the salts BiX_3, and the elements of the third group, which give salts such as RX_3. The highest oxides, HgO, Tl_2O_3, PbO_2, and Bi_2O_5, are coloured powders; they can be reduced to lower oxides, and when calcined they give off oxygen; mercury forms a stable perchloride, $HgCl_2$, corresponding to the maximum of oxidation; perchloride of thallium has been prepared ($TlCl_3$, or apparently Tl_2Cl_6), but it is easily decomposed by heat with the evolution of chlorine, and the lower form of combination, $TlCl$, is formed. This explains why lead and bismuth do not furnish us with compounds, as $PbCl_4$ and $BiCl_5$, but give stable bodies, $PbCl_2$ and $BiCl_3$. $BiCl_3$ is easily decomposed by water; $PbCl_2$ is only decomposed by steam when red hot; $TlCl$ is stable. Thallium is less volatile than mercury, and more so than Bi or Pb. All the properties that I have mentioned are shown by the following proportions :—

    (1.) $Tl : Hg : Pb = Al : Mg : Si$;
    (2.) $Tl : K = Pb = Ca$;
    (3.) $Tl : Hg = Pb : Tl = Bi : Pb$.'

It should not be forgotten that these relations are not arbitrary; that they not only express the relation of the properties, but also that of the atomic weights. Thus the proportion—

$$Tl : Hg : Pb = Al : Mg : Si.$$

is represented by the members—

$$204 : 200 : 208 = 27 : 24 : 28.$$

If, instead of the sign (:) of geometric proportion, we put the sign (−) used in arithmetic proportion, we get in reality—

$$204 - 200 - 207 = 27 - 24 - 28.$$

However, we must remember, first, that we do not know exactly the real value of atomic weights; and, secondly, it is necessary to observe that we are unable to wait for completely exact relationships, because the true function, expressing how the properties depend on the atomic weights, is unknown to us.

Further, it is certain that the elements possess, independent of the primary properties which determine the analogy and the position of elements in the system, certain individual properties. When the above-mentioned function has to be found exactly, these properties will be explained without doubt by the irregularities in the variations of the atomic weights, irregularities which now appear to be inexplicable.

We can also foresee in the atomic law a kind of perturbation, which cannot cause us to suspect the exactness of the law. The transition of HgO to Hg_2O, and of Tl_2O_3 to Tl_2O, is in the example quoted. A disturbance of this kind for the higher oxides would keep the suboxides of unequal composition subordinate. These suboxides are, among themselves, in the same relation as Cu_2O or Ag_2O is to K_2O or to Na_2O. The suboxide of mercury is, in many respects, analogous to oxide of silver; as suboxide of thallium is analogous to oxide of potassium. However, the salts HgX and TlX concord more between themselves and with AgX than AgX and KX do. These phenomena are already very complicated, and if I mention them here it is only to call attention to the problems raised by the application of the periodic law to the system of elements. The periodic law, I think, should not be neglected when chemistry becomes more elaborated, although it has, I well know, need of development.

I will add still another remark : it is that the use of the periodic law facilitates the learning of chemical facts by beginners. I have come to this conclusion during the courses of lectures that I have given for two years, and during the preparation of my " Traité de Chemie Inorganic," now published (in Russian), which treatise is based on the periodic law.

            (To be continued.)

THE CHEMICAL NEWS.

VOL. XLI. No. 1051.

THE PERIODIC LAW OF THE CHEMICAL ELEMENTS.

By D. MENDELEEF.

(Continued from page 2.)

III. USE OF THE PERIODIC LAW FOR THE DETERMINATION OF THE ATOMIC WEIGHTS OF THE LITTLE KNOWN ELEMENTS.

WHEN an element has very little power of combining with oxygen, and when only a small number of its other compounds are known, it is necessary, for finding the atomic weight of this element, to learn its equivalent (relatively to hydrogen), to determine the physical properties of the body in a simple state (such as its capacity for heat), or those of its compounds (such as vapour density and capacity for heat), or to find a case of isomerism. As some of these determinations have many practical difficulties, and because in a great number of cases the methods of working are defective, the atomic weights of many elements have been determined according to very uncertain data. Therefore, because we are not certain of a sufficient number of properties, we often give the formula RO to strongly basic bodies (this is the reason why the oxides of Ce, Yt, Di, and La, particularly, have been given this formula), the formula R_2O_3 to weak bases (as oxide of uranium), and RO_2, R_2O_5, and RO_3 to acid oxides. This method has been admitted as incorrect, since the formula of oxide of beryllium has been found to be BeO, with feeble basic properties, and the formulæ of the highly basic oxides of Th and Zr to be RO_2.

Lastly, it will be sufficient to remember that the highest compounds of oxygen, such as OsO_4 and RuO_4, have, only weak acid properties.

It is absolutely certain that the basic and acid properties of oxides are determined not only by the number of atoms of oxygen contained therein, but also by the properties of the element. The proof of it has been given in the preceding developments of the nature of the periodic law. The element which comes after Th (twelfth series) and whose atomic weight is about 235, ought to have basic properties, even under the form of oxide, R_2O_5. It is for the same reason that many of the formulæ of oxides, above all the oxides of the rare elements, which have not yet been thoroughly studied, such as, In, Ur, Ce, La, Di, Yt, and Er, are not not legitimate. Hardly any clear cases of isomerism have been observed in the oxides, and some of the isolated facts of this kind, such as the isomorphism of oxide of zirconium, and oxide of beryllium with alumina, cannot lead us to any certain conclusions.

In general, it is necessary to observe that isomorphism gives only a feeble support to conclusions on the atomic composition of bodies; the case of isomorphism presented by bodies of different composition, and the phenomena observed by Marignac, which have been already mentioned, are already sufficient proof. Without pushing any further into this vast and obscure region, I will simply remark that even the capacity for heat, as much in simple as in compound bodies, does not always give us very clear results. Thus we have left only two kinds of permanent criteria for finding the atomic weights; the determination of the vapour density of a large number of compounds of the given element, and a means of controlling, of a purely chemical nature, or conclusions drawn from the composition of the different forms of oxides and from the discovery of analogies with sufficiently well-known ele-

ments, &c. The capacity for heat and isomorphism can only be looked upon as auxiliary means. These methods are almost completely wanting for the above-mentioned rare elements; because, except a few exceptions, they only give volatile compounds, which have been too little studied to authorise any conclusions on the analogies.

The periodic law comes to our aid in cases of this kind, by offering us a new regular relation between the chemical properties and the atomic weights. This law being admitted, we can determine the atomic weight of an element of which the equivalent and some properties are known. If we multiply the equivalent E of an element deduced from the highest oxide (its composition being E_2O, and that of the chloride ECl) by 1, 2, 3, 4, 5, 6, 7, we obtain the values of all possible atomic weights of this element, and one of these numbers expresses the true atomic weights; this number E_n corresponds to a place as yet unoccupied in the system, and at the same time to the atomanalogy of the element. For, to judge of it according to what we know at present, a particular place in the system can only receive one element, and the atomic analogies are of a very simple nature.

Let us suppose that an element gives a basic oxide, not very energetic, and whose equivalent is 38. (It must not be forgotten that this number has a slight inevitable error.) We want to know the atomic weight of the element or the formula of its oxide. If we adopt the formula R_2O for the oxide, R=38 and the element belongs to the first group; but the place is already occupied by K=39; and, further, the atomic analogy indicates a soluble and energetic base for this place. If we adopt for the oxide of the formula RO the atomic weight=76, the atomic weight will not allow it to be in the second group, because Zn=65, and Sr=87, all the places for elements of low atomic weights are filled in this group. If we adopt for the oxide the formula R_2O_3, then the atomic weight, R=144, and the element goes into the third group, where there is in fact between Cd=112 and Sn=118 a place for an element whose atomic weight is near 114. The oxide of the given element should, according to its atomanalogy with Al_2O_3 and Tl_2O_3, as well as with CdO and SnO_2, possess feebly basic properties. Therefore, this element should be placed in the third group. If we adopt for the oxide the formula RO_2, the atomic weight=152, the element in question should not be placed in the fourth group, because only an element having an atomic weight 162, and feebly acid properties, could adapt itself to the empty place in this group (to form the link between PbO_2 and SnO_2). There is still in the eighth group a place for an element whose atomic weight would be 152; but this element forming the transition from Pd to Pt, should have such properties like them, which could not remain hidden. If then the given element does not possess these properties, the above-mentioned atomic weight does not do for it, any more than the place in the eighth group. If we adopt for the oxide the formula R_2O_5, then R=190, which will not do for the fifth group, because Ta=182 and Bi=208. Further, Ta and Bi possess, in the form of R_2O_5, acid properties.

The forms of oxides RO_3 and R_2O_7 do not correspond any better with our element. The only atomic weight which does for it is, therefore, 114, and the formula of its oxide is R_2O_3. Such an element is *Indium*. Its equivalent, according to Winkler, is 37·8; therefore we ought to accept the number 113 for its atomic weight, and the formula In_2O_3 for the composition of its oxide. Up to the present we have made R=75, and we have adopted InO for the formula of its oxide. Al and Tl are indium atomanalogues of the third group, Cd and Sn atomanalogues of the seventh series.

Let us compare the apparent properties of indium, deduced from atomanalogues, with the properties really observed.

The atomanalogues of indium, Cd, and Sn being easily reducible (they can even be reduced from their solutions by zinc), indium should also be reducible. As Ag (seventh series, first group) is more difficultly fusible than Cd, and Sb more difficultly fusible than Sn, it is necessary, according to the atomanalogy of Ag, Cd, In, Sn, that indium should be more easily fusible than cadmium. In fact, indium melts at 176°. Ag, Cd, and Sn are of a greyish-white colour, and so is indium. The density of Cd is lower than the density of Ag, the density of Sb is lower than that of Sn, the density of indium should then be a little lower than the average between Cd and Sn. In reality it is so. The density of $Cd = 8·6$, that of $Sn = 7·2$; therefore the density of In ought to be less than $7·9$. it has been found to be $7·42$. As Cd and Sn oxidise at red heat, and do not rust in the air, these properties should also be found in In, but more feebly than in Cd and Sn, because Ag and Sb oxidise still more difficultly.

All that has just been said has been verified by experience. We arrive at the same conclusions in comparing In with Al and Tl; thus the density of $Al = 2·67$, that of $Tl = 11·8$, the average density would be $7·2$.

We now pass to the properties of the oxide, and to the reactions of the salts. Indium and its atomanalogues are found in the odd series; this is why the higher oxides cannot be strongly basic; the basic character should be more feeble in In_2O_3 than in CdO and Tl_2O_3; it should also be stronger than in Al_2O_3 and SnO_2. These conclusions are confirmed by the following facts: The oxides of Al and Sn dissolve in alkalies, forming definite compounds, while the oxides of Cd and Tl are insoluble in alkalies; In_2O_3 is soluble in alkalies, but without forming any definite compound. The oxides of Cd, Sn, Al, and Tl are difficultly fusible powders, like In_2O_3. Hydrate of In_2O_3 forms, as could be foreseen, a colourless gelatinous mass. The oxides Al_2O_3 and SnO_2 are easily precipitated from saline solutions by carbonate of barium; it is the same with In_2O_3. Sulphuretted hydrogen precipitates Cd and Sn from their acid solutions; In is also precipitated. All these reactions have been confirmed experimentally.

Here are a few more facts deduced from atomic analogies, although they have not been observed, or have not been sufficiently studied: Indium ought to furnish indium-ethyl $InAe_3$, prepared in the ordinary way, because $CdAe_2$ and $SnAe_4$* exist; this compound, according to the atomic analogies, ought to boil at about 150°. As two elements, Sn and Tl, atomically analogues to indium, each give (independently of the higher oxide) a lower more basic oxide, indium would apparently also give a suboxide, InO or In_2O, turning in the air into In_2O_3. Apparently chloride of indium, $InCl_3$ or In_2Cl_6, cannot evolve chlorine when heated, because SnO_4 does not possess that property. The sulphate of cadmium forming with sulphate of potassium a double salt of the form $K_2Cd(SO_4)_2$, and aluminium giving an alum $KAl(SO_4)_2$, indium would no doubt give an analogous double salt; in any case, it is impossible to foresee if this salt will be isomorphous with alum.† To control the exactness of the modifications to be introduced into the atomic weight of indium, and into the formula of its oxide, conformably to the developments given above, I determined the specific heat of this body, and I found it to be $0·055$, which is in accord with the periodic law.‡ A little time before Bunsen employed his new and elegant calorimetric process for determining the specific heat of indium. The two numbers agree. Bunsen found $0·057$. We can then, without the slightest doubt, apply the periodic law for correcting the atomic weights of elements which are little known. Because of this certainty I will offer some considerations relative to other elements. I will only remark now that the periodic law has assigned a determinate position to all the elements more or less known, at the present time, and that this fact, far from being of small value, corroborates the law emphatically. Before leaving indium I would notice this: In is an analogue of Zn and Cd up to a certain point; it accompanies them in nature, and it is near them in the system; it is nearly the same which takes place for Nb relatively to Ti and Zr.

---

* Cadmium-ethyl has not been studied much; still, for many reasons, it merits research. Atomically analogous to Zn and Hg it should boil at 130°. The study of indium-ethyl and of thallium-ethyl would throw a new light on aluminium-ethyl, unhappily little known.

† Only having a small quantity of indium, I have, up to the present, only been able to make one incomplete experiment; an acid solution of sulphate of indium, with an equal quantity of K_2SO_4, left under sulphuric acid, after the addition of alcohol, give crystals formed of microscopic cubes. I recognised it, not only by its form, but also by the absence of double refraction of light.

‡ Bulletin de l'Académie des Sciences de Saint Petersbourg, 1870, p. 445.

THE CHEMICAL NEWS.

VOL. XLI. No. 1052.

THE PERIODIC LAW OF THE CHEMICAL ELEMENTS.

By D. MENDELEEF.

(Continued from page 28.)

THE constitution of uranium compounds again gives rise to numerous doubts, although the classic work of Péligot has cleared up the most important points of the history of this element. The formula of the oxide, such as was announced by him, and as is generally accepted, is Ur_2O_3, and its atomic weight is 120. In admitting these as true we cannot give uranium any determinate place in the system, either according to its atomic weight (for from $Ag = 108$ to $I = 127$ there is no empty place in the seventh series), or according to its properties; all the properties of uranium are even in opposition to the generally accepted hypothesis on the subject of its atomic weight. Although slightly resembling, in some of its properties, the metals of the iron group (particularly by its property of forming oxides, UrO, Ur_2O_3, Ur_3O_4), uranium is distinct from them on several points; its density is 18·4; it gives a volatile chloride, $UrCl_2$; its oxide only gives salts corresponding to the composition $UrOX$, and not UrX_3; having a higher atomic weight than iron it is more difficultly reducible than this metal; the basic properties of the oxide are more feeble than those of oxide of iron. These considerations lead us to imagine that uranium has a different atomic weight to that which was mentioned above. In starting, as I explained fully in the case of indium, from the known equivalent (40 for Ur, which nearly corresponds with 38 for In), we find that Ur ought to be in the sixth group, and that, therefore, the oxide ought to have the formula UrO_3, and the element should have for atomic weight 240. If so, its place is in the twelfth group, where thallium is. The greatest objection that we find here is that oxide of uranium is as rich in oxygen as we have only been accustomed to find the acid oxides.

1. The atomanalogues of Ur (in the even series) are— $Th = 230$, with a strongly basic oxide, ThO_2; and Cr, Mo, and W, which, like Ur, give oxides like RO_3. In these oxides, as in those of the other groups (see Chapter I.), the basic character increases and the acid character diminishes as the atomic weight increases. Thus the acid properties are more sharply defined in CrO_3 than in MoO_3 or WO_3. We can see already that MoO_3 and WO_3 have less acid properties, inasmuch that these oxides are capable of forming different salts with several acids, and with phosphoric, silicic, or sulphuric acid, they give bodies analogous to salts.*

We also notice more acid properties, less marked, it is true, in oxide of uranium. The following facts give us the proof of it:—

2. The precipitates formed by the caustic alkalies in solutions of salts of oxide of uranium are not hydrated oxides, but definite compounds of oxide of uranium with the alkalies. From this is derived the name *uranic acid*, by which name oxide of uranium is sometimes called. The composition of precipitates containing alkalies, $R_2Ur_2O_7$ ($Ur = 240$), corresponds completely with the composition of the bichromates, and of corresponding salts of molybdic and tungstic acids. The alkaline salts of oxide of uranium do not separate from the alkali, either under the influence of water or of acids.

3. The acid character of oxide of uranium betrays itself already in the ordinary composition UrO_2X_2 of all the salts that this body gives. X means an acid residue, $Ur = 240$. These salts, according to their composition, ought to be considered as basic, although they possess strictly acid properties; they recall at the same time the acid chloranhydrides corresponding to—

$$WO_2Cl_2, \quad WO_2F_2, \quad MoO_2Cl_2.$$

UrO_2Cl_2 is decomposed by water; and so is the double salt, which crystallises easily, $R_2UrO_2Cl_4,2H_2O$. The double salt just mentioned will only crystallise in an excess of HCl; therefore it corresponds to TiK_2Cl_6 and to other analogous compounds of the acid chloranhydrides.

4. Oxide of uranium gives perfectly crystallisable double salts, such as—

$$K_4UrO_2(CO_3)_3 \; ; \; KUrO_2(C_2H_3O_2)_3 \; ; \; K_2UrO_2(SO_4)_2,2H_2O, \; \&c.$$

The salt of oxide of uranium in these double salts apparently plays the rôle of an acid element (as in—

$$2K_2CO_3 + UrO_2CO_3).$$

5. The acid character of uranium is corroborated by the volatility of chloride of uranium, $UrCl_4$ ($Ur = 240$). Another circumstance which confirms this character is that this latter, like the acid chloranhydrides, is decomposed by water, so also are $MoCl_4$, $TiCl_4$, $SnCl_4$, and $ThCl_4$.

We can further offer the following considerations to point out the necessity and the advantages of a modification of the atomic weight of uranium.

The atomic volume of $Cr = 7·6$, of $Mo = 11$, of $W = 10$,

$$\text{of } Ur = \frac{240}{18\cdot4} = 13 \; ;$$

the atomic volume therefore increases absolutely the same as in the other groups, according as the atomic weight increases. It is thus that the atomic volume increases rapidly in the group K, Rb, Cs; less in Ti, Zr, Th; still less in Fe, Ru, Os, &c.; until at length the atomic volumes of Cl, Br, and I are nearly equal.

We can establish the following proportion for uranium (see Chapter I.):—

$$Ur : Th = Pb : Hg = Te : Sn ;$$

for, although these elements belong to different groups, that is to say, they form oxides of different formulæ, the chloric compounds of each couple have similar composition and properties. $UrCl_4$ is analogous to $ThCl_4$, so is $PbCl_2$ to $HgCl_2$, and so is $TeCl_4$ to $SnCl_4$.

The salts of suboxide of uranium are not isomorphous with the salts of magnesia, as we might think they were if we adopted the generally-accepted atomic weight, and according to which uranium would be analogous with iron. This difference is seen distinctly by suboxide of uranium forming a double salt, $UrK_2(SO_4)_3,H_2O$, with sulphate of potassium. In this formula $Ur = 240$. We see that for 2K there is $3SO_4$ and not $2SO_4$, as in the salts of the magnesia group.

When chloride of uranium ($UrCl_4$) is calcined in hydrogen it gives a body containing more than its half of chlorine. This phenomena is explained if we admit the possibility of the formation of $UrCl_3$ or Ur_2Cl_3, and it remains unexplained if we keep the accepted atomic weight (120) for Ur, because it is difficult to believe that Ur_2Cl_3 can be formed from $UrCl_2$.

Regnault found that the specific heat of suboxide of uranium (which was thought to be the metal) is 0·062. If we admit the formula for the suboxide to be UrO ($Ur = 120$), the molecular heat of this body is—

$$136 \times 0\cdot062 = 8\cdot4,$$

that is to say, it is less than that of ZnO (10·1) or of HgO (11·2), although the atomic weight of uranium is comprised, according to the above-mentioned hypothesis, between the atomic weights of Hg and Zn.*

If, on the contrary, we adopt the doubled atomic

---

* Phosphoric acid has nearly the same relation with oxide of uranium as it has with molybdic and tungstic acids, which leads us to compare these three last oxygenated compounds with the first. For further information on this subject, see my " Principles of Chemistry" (in the Russian language), vol. ii., p. 281 to 285.

* See my observations on the subject of specific heat in the " Principles of Chemistry," Second Part, Chapter III. Also my memoir published in the *Journal de la Société Chimique Russe*, 1870; and an abstract in the *Zeitschrift für Chemie*, 1870, p. 200.

weight, and if we give the formula UrO_2 to the suboxide the molecular heat = 16·9; it is, then, as we might expect, greater than the molecular heat of MnO_2 (13·8), and of SnO_2 (14·0), and less than that of $PbCl_2$ (18·5). The determination made by Regnault is then more favourable to the formula RO_2 than to the formula RO (for the suboxide of uranium).

The proposed modification of the atomic weight of uranium makes us regard nature and her works from a new point of view, and it invites new researches on the degree of analogy with Cr, Mo*, and W.

The following experiments seem to me to be particularly interesting :—

1. To determine the specific heat of metallic uranium. I have long been wishing to make this determination, but H. Bauer and I, who worked together, have not been able to prepare cast-uranium ; we always obtained it in the state of powder. Not being sufficiently convinced of the purity of this product I did not think it good enough for this determination.

2. Find the vapour-density of volatile chloride of uranium.

3. Observe exactly how it behaves when red-hot in the presence of hydrogen.

4. Examine the salts of suboxide of uranium, and see if they are not isomorphous with the corresponding salts of ThO_2, SnO_2, ZrO_2, TeO_2.

5. Compare the crystalline forms of the corresponding compounds of UrO_3, MoO_3, and WO_3. It would be particularly interesting to pursue the study of $R_2UrO_2F_4$ (Carrington Bolton), and to make it the object of as complete a work as Marignac did on $R_2WO_2F_4$.

6. We can foresee the existence of a soluble oxide of uranium (metoxide) corresponding to the modifications of molybdic and tungstic acids.

7. It would be very interesting to submit the physical properties of corresponding compounds of Cr, Mo, W, and Ur to a comparative study, because we notice a great analogy in the composition, and even in the colour of corresponding compounds, above all of Cr and of Ur.†

The interest presented by a new study on uranium becomes greater if we modify the atomic weight in the manner indicated, because in this case the atom of uranium would become the heaviest of all known elements. I have arranged below the formulæ of some of the compounds of uranium according to the present atomic weight, and the values that they would have if the atomic weight were doubled :—

Ur = 120.	Ur = 240.
Compounds of protoxide—	
Ur_2Cl_3 ; Ur_4O.	$UrCl_3(Ur_2Cl_6$?) ; Ur_2O_3.
Compounds of suboxide—	
$UrCl_2$; UrO.	$UrCl_4$; UrO_2.
$UrSO_4,2H_2O$.	$Ur(SO_4)_2,4H_2O$.
$K_2Ur_2(SO_4)_3,H_2O$.	$K_2Ur(SO_4)_3H_2O$.
Green oxide—	
$Ur_3O_4 = UrOUr_2O_3$.	$Ur_3O_8 = UrO_2,2UrO_3$.
Analogous to—	Analogous to the blue oxide
$FeOFe_2O_3$.	of molybdenum.
	$MoO_2,2MoO_3$.
Compounds of the oxide—	
Ur_2O_3 ; $UrOCl$.	UrO_3 ; UrO_2Cl_2.
$UrO(NO_3),3H_2O$.	$UrO_2(NO_3)_2,6H_2O$.
$2UrONO_3,3H_2O$.	$UrO_2(NO_3)_2,3H_2O$.
$Ur_2O_2(NH_4)_4(CO_3)_3$.	$UrO_2(NH_4)_4(CO_3)_3$.
$UrO(C_2H_3O_2)H_2O$.	$UrO_2(C_2H_3O_2)_2,2H_2O$.
$Ur_2O_2(C_2O_4),3H_2O$.	$UrO_2(C_2O_4),3H_2O$.

(To be continued.)

---

* In pointing out the experiments, based on the periodic law, which I think would be interesting to undertake, I do not claim the right of priority for doing them.

† This fact and some other also (the analogy of Pt and Pd, of Nb and Ta, &c.) suggest the idea that independent of the large and small periods there are still quadruple periods, formed of two large periods. If so, the elements of the sixth group ought in reality to pass from the eighth series to between Cr and Ur. If so, we can explain in a more satisfactory manner, some differences between MoO_3 and CrO_3, as well as the analogy of MoO_3 with WO_3.

ON CHEMICAL REPULSION.*

By EDMUND J. MILLS, D.Sc., F.R.S.

WHILE engaged in some researches on the propagation of chemical change, I have incidentally encountered a new order of phenomena, which the title "chemical repulsion" may serve provisionally to designate. A brief outline of the experiments is given in the following paragraphs.

Upon a glass plate, laid in a horizontal position, is poured enough solution of baric chloride to cover it completely to a considerable depth. On this solution is placed another glass plate, provided with a small centre perforation ; when the two plates are firmly pressed together with the hands, most of the solution is extruded, and only a very thin layer of it left between the plates. All excess of the solution having been removed from the outer surfaces of the plates as well as from the perforation, some dilute hydric sulphate is now introduced into the perforation. This reagent attacks the baric chloride, throwing down a white precipitate of sulphate ; and, proceeding partly by diffusion, partly by flow, does not cease to widen in every direction its figure of advance, until the edges of the plates are attained. If the perforation is circular, the figure of advance is circular ; in other words, the chemical development of a circle is a circle.

Let us now suppose the two plates to be square and equal, and let the upper one have two circular perforations, equidistant from the centre of the square, and situated upon its diagonal. Let also two circular developments of baric sulphate be caused to proceed, as before, from the two perforations simultaneously. At first, nothing remarkable is observed, but in a short time, the two growing circles begin to exercise a visible retardation on each other's progress ; so that the figure of advance is no longer circular, but oval. [This retardation is of course observed only between the perforations, and not outside them, where the motion is entirely free.] As the development of the figures continues, so also does the retardation at their neighbouring edges increase ; the final result being (however long the experiment may be prolonged), that the other diagonal of the square is completely and permanently traced out in a line of no chemical action.

The above experiments are of fundamental importance, and they obviously admit of endless variety. Of this, a few illustrations may suffice.

If the upper plate have three perforations, situated on the points of a central equilateral triangle, there are three repulsion lines ; these end at the centre of the triangle, where they form a trilocular point, and traverse its sides midway at right angles.

When the upper plate has four perforations, situated on the points of a central square, there are four repulsion lines ; these end at the centre of the square, where they form a quadrilocular point, and traverse its sides midway at right angles.

A very beautiful modification of the preceding experiment consists in simultaneously developing a circle from a (fifth) central perforation. This last circle has no means of escape from the surrounding four. The result is that it eventually forms a square figure bounded by repulsion lines, and having four symmetrically situated repulsion lines at its corners.

It is easy to demonstrate that the chemical repulsion in these experiments does not depend upon flow. Two superimposed triangular plates for instance, in neither of which is any perforation, give three repulsion lines on immersion in dilute hydric sulphate. From each corner a line proceeds midway (if the triangle be equilateral) to the centre. In this effect, diffusion is alone concerned.

In addition to hydric sulphate and baric chloride, other pairs of reagents may be used with success ; and I anticipate no difficulty in obtaining results in which precipitation is not concerned. A beginning has also been made with experiments in tridimensional development.

---

* Abstract of a paper read before the Royal Society, Jan. 15, 1880.

THE CHEMICAL NEWS.

VOL. XLI. No. 1053.

THE PERIODIC LAW OF THE CHEMICAL ELEMENTS.

By D. MENDELEEF

(Continued from page 40.)

THE three elements which are found in cerite, viz., cerium, lanthanum, and didymium, have very nearly the same equivalents (about 45)* and similar properties; it is therefore difficult to determine the atomic weights of these elements. Involuntarily we think that they are analogous to those of the iron group, particularly because cerium has some of the properties of manganese. However, we can only admit the analogy if we have simply a superficial knowledge of the metals from cerite. If we admit the formula RO for their ordinary oxides, their atomic weights will be approximatively 92, which will not agree with the eighth group. Even if we give their oxides the formula R_2O_3, these elements will not enter the eighth group, because their atomic weights would then be about 138; and the reason of this exclusion is not only that the elements of group VIII., series 8, should have an atomic weight averaging between Pd and Pl (about—

$$150 = \frac{105 + 195}{2}),$$

but it is above all because the known properties of these elements do not correspond with the properties of the missing members of the eighth group. The metals of cerite are more difficultly reducible—they only give a small number of oxides; the ordinary oxides possess very strongly basic properties; as far as we know they do not give such characteristical ammonic and cyanic compounds; in a word, they are not the atomanalogues of the metals of the eighth group (see Chap. I.).

Let us give still one more proof:—

According to Wœhler, the density of cerium is 5·5. If we accept CeO as the formula of the ordinary suboxide, Ce=92, the atomic volume=17: but if we admit that the suboxide is represented by Ce_2O_3, then Ce=138, and the atomic volume = 25. These two volumes do not correspond with the members of the eighth group, which have much smaller atomic volumes. If we take RO as the formula of the ordinary oxides of the metals of cerite, we cannot place them either in the second or third group. It only remains for us to try if other forms of oxides are applicable, as we did before when determining the atomic weight of indium. For this purpose we will firstly consider cerium only, then afterwards we will examine lanthanum and didymium, superficially and simultaneously, because cerium has been better studied than the other two metals, and it gives at least two basic forms of oxidation. This circumstance facilitates the determination of the atomic weight very much.

If we admit, according to the general opinion, that the atomic weight of cerium is 92, the ordinary strongly basic oxide, which is the lower or suboxide, is represented by CeO, and the highest oxide, which give salts with the acids, is $Ce_3O_4 = CeO,Ce_2O_3$. By attributing this composition to the highest oxide, we still admit an oxide of the formula Ce_2O_3,† although this oxide has never yet been

obtained and none of its salts are known. Rammelsberg, Hermann, Holzmann, Zschiesche, &c., have obtained not only simple salts of the oxide Ce_3X—example, the yellow salt $Ce_3(SO_4)_4,8H_2O$—but also different salts of suboxide of cerium. Their composition is expressed by the formula $Ce_3X_8,nCeX_2$; however, we do not know of any salts which contain more acid residues than what corresponds to the form Ce_3X_8. Therefore we should only admit there to be two actual forms of oxidation of Ce; these are—

CeO and Ce_3O_4 (Ce=92).

Its tenacity for oxygen is in the relation 3 : 4; therefore I think it would be more natural to give the formula Ce_2O_3 to the first of these oxides, and the formula Ce_2O_4 or CeO_2 to the second.* This is as good as giving Ce the atomic weight 138 (46×3), and giving it the corresponding place in the fourth group; this also agrees with the new formula of the highest oxide. The element Ce takes its place in the eighth series; it has as atomanalogues Ti, Zr, and Th in the fourth group, and Cs = 135 and Ba = 137† in the eighth series. The transition from CeO_2 to Ce_2O_3 corresponds to the transition from TiO_2 to Ti_2O_3, and, better still, from PbO_2 to Pb_2O_3 and to PbO.‡ The higher oxide of cerium exhibits the properties of a peroxide, and its reactions resemble those of PbO_2, Tl_2O_3, and TeO_3.

We can explain the feeble basic properties of CeO_2 by the fact that it is placed in the fourth group after TeO_2, whose basic properties are very uncertain, and also after ZrO_2, whose basic character is very doubtful. In the case of ThO_2, which comes after CeO_2, these properties are more doubtful still, its atomic weight being relatively higher than that of Ce. The strongly basic properties of Ce_2O_3 find their explanation in this circumstance, that PbO_2, Tl_2O_3, TiO_2, and CuO also give more basic bodies when reduced. The fundamental properties of the oxides of cerium are explained by reason of the position of Ce in relation to the system: in an even series after Cs and Ba, whose properties are strongly basic. To the atomanalogy of Ce with Ti, Zr, Th, corresponds their association in a great number of minerals—rare, it is true.

To show more completely that the place given to cerium in the system, if we adopt Ce=138 or 140, agrees in reality with its properties, I am going to show, comparatively, the known volumes and densities of the element and of the oxide. If, in an even series, we start from the first group, we notice an increase of density, such as the density of Rb=1·5, that of Sr=2·5, that of Zr=4·2. The density of Cs is unknown; however, according to conclusions which I shall communicate later on, it ought to be about 2·5; the density of Ba=3·6; of Ce=5·5. The volume of Ti=9·3, of Zr=22, of Ce=25, of Th=30; consequently the volume increases here, as in the other groups, according as the atomic weight increases. The volume of $TiO_2=20$, of $ZrO_2=22$, of $CeO_2=24$, of $ThO_2=29$, that is to say, that in this series as in the parallel one, viz., $SiO_2=22$, $SnO_2=22$, $PbO_2=26$, the volumes of the higher oxides increase regularly, although only slightly, as the atomic weight becomes greater.

If, as is thus shown, the new position of cerium in the system corresponds to its properties, both chemical and physical, there are two motives why we should search for a new means of confirming the exactitude of the hypothesis given above, relative to a modification of the atomic weight of cerium. Firstly, CeO_2 is easily reduced to the

---

* According to the determinations of Haeringer, Marignac, Hermann, Bunsen, and Rammelsberg, the equivalent of Ce=46; according to Wolff it is 45·66. For lanthanum, Rammelsberg found the figure 44·4; Zschiesche found 45·1; Holzmann found 46·3; Czudnowicz 46·8; Marignac from 46 to 47; and Erk 45·1. For didymium Zschiesche found from 46·6 to 48·1; Hermann, 46·7; Marignac, 48; and Erk, from 47·4 to 47·8.

† Popp and Hermann have mentioned other oxides, such as Ce_5O_5 and CeO_4, but these assertions have not been confirmed by any other chemists, and Elk has contradicted them.

* Up to the present I do not see how it is possible to give the oxides the formulæ CeO_3 and CeO_4, thus making Ce=276.

† There is good reason for the supposition that, in future experiments, the atomic weight of Ce will be found to be much greater, for at present it is very near that of Ba, and we can better afford to admit errors in the number found for Ce than in the carefully determined atomic weight of Ba. Slight impurities, such as La and Di, the difficulties of analysis, to which Marignac has called attention, as well as the difficulty in obtaining the compounds of suboxide of cerium free from compounds of oxide of cerium, are quite sufficient to account for the low atomic weight admitted by the periodic law. This is the reason I have provisionally given cerium the atomic weight 140.

‡ We can imagine that the ordinary oxides of Zr and Th will give basic suboxides Th_2O_3 and Zr_2O_3. As far as the latter is concerned I purpose doing the necessary experiments, for I have recently received a considerable quantity of zircons from the Ilmen mountains (Siberia.)

state of Ce_2O_3. The highest atomic analogue of cerium, namely, Th, has not, as far as we know, a similar property,* although, from what has been previously said (see Chap. I.), the highest members of a group are the most easily reducible. Secondly, there are other elements which possess the same equivalent as Ce (46 in the suboxide, 34·5 in the oxide), or a very similar one (such as La and Di as suboxides, and Yt as oxide), which ought to be able to occupy a different position in the system. Therefore it seems to me that a more profound knowledge of the compounds of cerium is necessary before we can be certain of the exactitude of the modified atomic weight. Up to the present I have only made one determination of the specific heat of metallic cerium.† This metal was obtained from H. Schuchardt, of Goerlitz. The number 0·5—found from small half-melted lumps which had been dried in hydrogen —shows that the specific heat of cerium is less than that of tin, and equal to that of metallic barium found at the same time. The number found coincides better with the modified atomic weight than with the old one, for $0·05 \times 138 = 6·9$, whilst $0·05 \times 96 = 4·8$. However, it is desirable to repeat this experiment.

(To be continued.)

THE CHEMICAL NEWS.

VOL. XLI. No. 1054.

THE PERIODIC LAW OF THE CHEMICAL ELEMENTS.

By D. MENDELEEF.

(Continued from page 50.)

TURNING to lanthanum and didymium, we find that only the latter gives, independently of its ordinary oxide, a brown-chocolate coloured peroxide, decomposed by heat, not answering to any determined formula, and not furnishing any salts. For this reason we cannot settle the atomic weights of lanthanum and didymium, except from observations on the salts of the ordinary oxides, whose equivalents are very nearly alike (about 46). Again, we cannot be sure of the correctness of these last numbers, above all in the case of didymium, because we have no means of distinguishing a mixture of lanthanum, yttrium, and perhaps even thallium and erbium, from the salts of didymium. According to their equivalents, which are very close together, lanthanum and didymium can occupy one of three places in the system. First, in group III., series 8, between Ba = 137, at Ce = 140? In this case the calculated equivalent is—

$$44\cdot5 = \frac{138}{3}$$

which corresponds with the observed figures. It is very easy, however, to admit an error in the determinations, as the purity of the preparations is only guaranteed by repeated crystallisations, which we well know do not always free us from isomorphous impurities.

The place, "Group III., Series 8," corresponds, according to the atomic analogy, with Cs, Ba, and Ce, to an element which has very basic properties, and which forms very difficultly volatile chlorides; lanthanum and didymium have certainly properties of this kind.

The second place, in which these properties of the oxide would put La and Di, is "Group IV., Series 10," before Ta = 182, in atomic analogy with Ce = 140 (?) and Th = 231. The element which would fill this place should have an atomic weight of about 180, an equivalent 45, corresponding to the oxide RO_2, and nearly identical with the equivalent of La and Di (above all of La).

A third place is free "Group V., Series 12," between Th = 231 and Ur = 240, for an element which gives an oxide, R_2O_5, and which has an atomic weight of about 235, and an equivalent of about 49. It would be difficult for one of these two metals to occupy this place; the equivalent is too high and the oxide should possess feebly basic properties, less pronounced than in ThO_2, and more so than in UrO_3; the oxides of La and Di present, on the contrary, very strongly marked basic properties, and they are soluble in dilute acids, even after calcination. The oxides and the metals which would suit this place ought, like their atomanalogues, to have a great density, whilst the densities of the oxides of La and Di are about 6·5. I therefore think that I am authorised in admitting that La and Di should occupy the two first-named places, that is to say that the oxide of one of these metals possesses the composition R_2O_3, and corresponds to the lower oxide of Ce, and that the oxide of the other is of the form RO_2, and corresponds to the higher oxide of cerium.

The following facts bear witness to the difference of composition of the oxides of La and Di; according to information received personally, Marignac has not noticed one single case of isomorphism between the corresponding salts of La and Di. In one solution Watts obtained different crystals of sulphates of La and of Di. Marignac found that the sulphates had different compositions. If we admit the formula RO for the two oxides, sulphate of didymium is $3DiSO_4,8H_2O$, and sulphate of lanthanum is $LaSO_4,3H_2O$. We have already noticed that the formula R_2O_3 corresponds better to oxide of didymium than it does to oxide of lanthanum, because it changes the unusual formula, $3DiSO_4,8H_2O$, of sulphate of didymium into $Di_2(SO_4)_3$. Further, if we represent oxide of lanthanum by LaO_2, the composition of sulphate of lanthanum will be $La(SO_4)_2,6H_2O$. If we consider that we often meet in nature with oxides of different compositions together, such as Nb_2O_3 and TiO_2 (Marignac, Hermann), WO_3 and Nb_2O_5 (Wöhler), V_2O_5 and CrO_3, then the simultaneous appearance of Ce_2O_3, Di_2O_3, and LaO_2, should not surprise us, as it is by no means an isolated fact; it might be compared to the concomitance of ZnO, CdO, and In_2O_3.

For these reasons we can admit that the atomic weight of Di = 138, that the metal belongs to the third group, and that it forms an oxide, Di_2O_3; and, on the other hand, that lanthanum belongs to the fourth group, and possesses the atomic weight 180[*], the formula of the oxide being LaO_2. However, we cannot be certain of the exactitude of these hypotheses except after new researches in this direction. Up to the present we only know a small number of facts, and we have even interpreted these facts, admitting the formula of the oxides to be RO. The study of the compounds of these metals would present a new interest, because of the modifications that would be introduced in the formulæ of the oxides.

To show how these modifications that I propose to introduce will appear, I extract from Marignac's work the formulæ of several salts of didymium, and I put opposite them the formulæ modified by the new atomic weight; we shall see that these latter are more reasonable.

	According to the Ordinary Atomic Weight.	According to the New Atomic Weight.
The Oxide ..	DiO ;	Di_2O_3.
,, Hydrate ..	$Di(OH)_2$;	$Di(OH_3)$.
,, Oxysulphide	$DiS,2DiO$;	Di_2O_2S.
,, Chloride ..	$DiCl_2,4H_2O$;	$DiCl_3,6H_2O(Di_2Cl_6)$.
,, Oxychloride	$DiCl_2,2DiO,3H_2O$;	$Di_2Cl_2O_2,3H_2O$.
,, Phosphate	$Di_3(PO_4)_2,2H_2O$;	$DiPO_4H_2O$.
,, Sulphate	$3DiSO_4,8H_2O$;	$Di_2(SO_4)_3,8H_2O$.
,, Sulphate	$DiSO_4,2H_2O$;	$Di_3(SO_4)_3,6H_2O$.
,, Basic salt	$DiSO_4,2DiO$;	$Di_2O_2(SO_4)$.
Double salts	$3(DiSO_4)Am_2SO_4,8H_2O$;	$DiAm(SO_4)_2,4H_2O$.
	$3(DiSO_4)Na_2SO_4$;	$DiNa(SO_4)_2$.
	$3(DiSO_4)K_2SO_4,2H_2O$;	$DiK(SO_4)_2H_2O$.

The last three salts are of the same composition as alum, except the small quantity of water they contain. Oxide of didymium, as well as alumina, has the property of being precipitated, although slowly, from its solutions by means of carbonate of barium. The analogy between Di_2O_3 (modified formula) and Al_2O_3 was shown in the last vertical column. If the basic properties of Di_2O_3 are more strongly marked than those of Al_2O_3, we can explain it according to the periodic law by this circumstance, that didymium belongs to one of the even and higher series, whilst aluminium belongs to an odd and lower series.

The composition of the salts of LaO_2 is, according to all that we know, more simple than the composition of the salts of didymium, and it corresponds to the proposed modification of the atomic weight; to $LaCl_4$, for example (or, perhaps, La_2Cl_8, as it is not volatile), corresponds an oxychloride, $La_2O_3Cl_2$. If the formulæ which we give to CeO_2 and LaO_2 are exact, then the salts which correspond to them will turn out to be isomorphous, and will also be

---

[*] It is now possible to obtain lanthanum free from didymium if we guide ourselves by the absorption spectra of the didymium salts, which spectrum was first observed by Gladstone. The numbers found by Zschiesche (45·09), Rammelsberg (44·38), and Erk (45·1) correspond very well to La, because, according to them, La ought to equal 180·36, 177·52, or 180·4.

analogous to the salts of Th and Zr. The study of the double salts of ZrO_2, CeO_2, LaO_2, and ThO_2 would offer a particular interest, as these oxides often form well-crystallised double salts. If the proposed formulæ are correct we might venture to predict the formation of salts as characteristic as the alums. I have already commenced several experiments in this direction.

(To be continued.)

THE CHEMICAL NEWS.

VOL. XLI. No. 1055.

THE PERIODIC LAW OF THE CHEMICAL ELEMENTS.

By D. MENDELEEF.

(Continued from page 62.)

It still remains for me to mention a possible modification in the atomic weights of *yttrium* and of *erbium* (perhaps also of *terbium*, if terbium is really an element existing in nature). However, the history of these interesting elements is still enveloped in darkness, according to the contradictory debates between Mosander and Delafontaine on the one hand, and Bunsen and Bahr on the other hand; they therefore require fresh researches. If we keep to the results obtained by these two last investigators, and if we admit with them that the equivalent of $Yt = 30.85$ (Berzelius found 32.1 and 35.0, Bopp found 34, and Delafontaine found 32), and that the equivalent of $Er = 56.3$,* except from any error resulting from them having been insufficiently studied, these elements can be placed in Group III., which receives the highest forms of oxides, such as R_2O_3. Yttrium, with the atomic weight 88, would come in the sixth series, immediately after $Rb = 85$ and $Sr = 87$, and consequently before $Zr = 90$ and $Nb = 94$; therefore the equivalent of Yt would be $= 29.3 = \dfrac{88}{3}$.

As it is very doubtful whether anyone has yet succeeded in separating Yt from Ce, La, Di, Er (perhaps also from Tb and Th), we can admit that the determinations of Bunsen and Bahr coincide sufficiently with the theoretical number.

This place for yttrium is justified, as the results of applying the periodic law to the preceding elements show, viz., by the strongly basic properties of the oxide, by the non-volatility of $YtCl_3$, by the composition of the sulphate, $(Yt_2SO_4)_3,8H_2O$, similar to the composition of the salt which didymium forms analogous to Yt, by the insolubility of the fluoride, &c. The erbium of Bunsen and Bahr, if we take their results as bordering on the truth, apparently belongs to Group III., Series 10, before $La = 180$, $Ta = 182$, and $W = 184$, and it should possess the atomic weight 178; the form of the oxide is therefore R_2O_3, and its equivalent $= 59.3$. We can explain the low number (56.3) found by Bunsen by supposing that his erbium was not free from yttrium.

To convince us of the exactness of these modifications of the atomic weights of Yt and of Er, we do not know any fact, as we do in the case of the metals of cerite; nobody has determined either the density of their oxides,† their behaviour in the presence of oxidising and reducing agents, or the composition and the form of their double salts. Here, then, more than in any other part of the system, are researches, guided by the periodic law, necessary.

To give another example of the direction in which researches on elements (based on the periodic law) should be made, I propose to pass now to the determination of the properties of elements which are still undiscovered. Without the periodic law, it would be absolutely impossible to foresee any of the properties of unknown elements; and, further, we could not form any idea of what gaps

* Erbium, according to Delafontaine, has an equivalent lower than 39·68. Bunsen thinks that Delafontaine was working with an impure sample.

† The number 48·4 given by Eckeberg for the oxide of yttrium might be thought to relate to an impure substance. However, it corresponds to the periodic law, for the volume of $2SrO = Sr_2O_2 = 49$, the volume of $Yt_2O_3 = 47$, the volume of $Zr_2O_4 = 45$,

were existing in the series of elements. The discovery of new elements was uniquely the results of observations, and was made either by chance or by the extra powers of thought and perception of the investigator. These discoveries did not offer any special philosophical interest; this is the reason we have seen such a small number of investigators venture on the study of the elements, like an exploring party in a new and unknown land, not knowing which way to turn; but now they will find that this vast and important domain of chemistry is not entirely devoid of landmarks, but that the periodic law will serve as a guide, and will facilitate future discoveries.

IV. The Use of the Periodic Law for Determining the Properties of as yet Undiscovered Elements.

The preceding developments show us that the periodic law renders it possible for us to bring to light the unknown properties of elements whose atomanalogues are known to us. Further, we can see, by referring to Tables I. and II., in which the periodic relations are shown, that many elements are missing, and we can confidently predict their discovery. I am therefore going to describe the properties of several as yet undiscovered elements: by this means I hope to show, in a new and perfectly clear manner, the exactitude of the law, although the confirmation of these proofs is reserved for the future. Let us add that the previous determination of the properties of unknown elements will facilitate the discovery of these elements, because knowing them we can foretell the reactions of their compounds.

So as to avoid introducing new denominations for the unknown elements into science I shall designate them by the name of the nearest lower analogue of the odd or even elements in the same group, and placing in front of this word one of the Sanscrit numbers (eka, dui, tri, tschatour, &c.). The unknown elements of the first group will be called ekacæsium ($Ec = 175$), duicæsium ($Dc = 220$), &c. If niobium, for example, were not known we could call it ekavanadium. The denominations will show the analogies very clearly; the names, however, of the fourth series have not this advantage, because they ought to be derived from those of the elements of the second series, and we know from Chapter I. that this typical series is not in complete atomic analogy with the fourth.

Besides, in this series there is only one missing element; it is in the third group, and is called ekaboron, Eb. As it follows $K = 39$, and $Ca = 40$, whilst it precedes $Ti = 48$, and $V = 51$, its atomic weight should be about $Eb = 44$; its oxide should be Eb_2O_3, but it should not have very characteristic properties; it will form in all respects the transition from CaO to TiO_2. In its maximum salts, EbX_3, the equivalent of the metal will be about $15 = \dfrac{44}{3}$; it will not, therefore, be lower than the equivalents of bases which are already known; it will be intermediate between the equivalent of $Mg = 12$, and that of $Ca = 20$.

We have seen in the foregoing chapter that $Yt = 88$ (?), $Di = 138$ (?), and $Er = 178$ (?) belong to this same group. However, the position of this latter is still very uncertain, and the elements have not been sufficiently studied; therefore the properties of Eb can only be determined by means of its atomanalogy with the elements Ca and Ti of the fourth series. The case of this element is therefore more complicated than those of other unknown elements. In consideration of Ca and Ti only giving one stable oxide in the air, we may admit that Eb will only give one stable basic oxide, Eb_2O_3. The oxide in its properties ought to be to Al_2O_3 what CaO is to MgO, or what TiO_2 is to SiO_2; consequently it should be a more energetic base than aluminium, and at the same time it ought to agree with Al not only in its forms of corresponding compounds, but also, in many cases, in its properties. Thus, the sulphate $Eb_2(SO_4)_3$ will not be such an easily soluble body as $Al_2(SO_4)_3$, because sulphate of calcium is more difficultly soluble than sulphate of magnesium. The base formed by Ca being more feeble than that formed by Na, and, at the same time, TiO_2 being a less energetic base than Al_2O_3, Eb_2O_3 will be weaker than MgO. Eb_2O_3 will therefore occupy, in many respects, the place between Al_2O_3 and MgO: numbers of its reactions will be explained by this circumstance.

Duiboron, or yttrium, gives an oxide, in every respect more energetic, as Sr gives a more marked base than Ca; however, even as there exists a great concordance between Ca and Sr, Ti and Zr, so shall we find many analogies between ekaboron and duiboron (yttrium?). If, then, yttrium is veritably duiboron, and ought to be accompanied by ekaboron, the separation of ekaboron from yttrium will be very difficult; in fact, it will only be possible by means of excessively delicate differences; for example, in the solubility of the salts, or in the energy of the basic oxides, &c.

Oxide of ekaboron will naturally be insoluble in alkalies, but we are uncertain as to whether it will drive off the ammonia from chlorhydrate of ammonia. The carbonate of ekaboron will be insoluble in water, and it will be precipitated as a basic salt, if we can judge from what we know of the salts of MgO and Al_2O_3. Its salts will be colourless, and will give gelatinous precipitates with KHO, K_2CO_3, HNa_2PO_4, &c. Sulphate of potassium will give a double salt with $Eb_2(SO_4)_3$, a sort of alum, which will probably not be isomorphous with alum proper. A few salts only of ekaboron will crystallise well, and they will belong to the double salts. The degree of volatility of $EbCl_3$ will depend on the molecular formula of this compound; $EbCl_3$ will be volatile, but Eb_2Cl_6 and the higher molecules will not volatilise. Chloride of ekaboron will, in all probability, be more difficultly volatilised than chloride of aluminium, because $TiCl_4$ boils at a higher temperature than $SiCl_4$; because also $CaCl_2$ is more difficultly volatilised than $MgCl_2$. But we may be permitted to presume that the salts of ekaboron will not be sufficiently volatile to be discovered by means of spectrum analysis. The chloride of ekaboron will naturally be a solid body: its anhydride will be decomposed by water more easily than $MgCl_2$ with the evolution of HCl.

As the volume of $CaCl_2 = 49$, and the volume of $TiCl_4 = 109$, the volume of $EbCl_3$ will be about 78, and its density will be 2.

Oxide of ekaboron will be an infusible powder, which will, after calcination, dissolve in acids, although with difficulty; it will easily give to water its alkaline reactions, but it will not completely saturate acid solutions of litmus so as to make the colour disappear. The density of the oxide will be about 3·5; the volume about 30, because the volume of $K_2O = 35$, of $Cu_2O_2 = 36$, of $Ti_2O_3 = 40$, of $Cr_2O_6 = 72$.[*]

Ekaboron, the simple body, will be a light, non-volatile, difficultly fusible metal. It will only decompose water under the influence of heat, and even then incompletely; it will dissolve in acids, with the evolution of hydrogen. Its density will be about 3·0 (apparently higher), because its volume is about 15; for the volumes of metals in the even series diminish constantly from the first group. Thus the volume of K = 50, of Ca = 25, of Ti and V = about 9, and of Cr, Mn, and Fe = about 7.

(To be continued.)

Variations of the Electromotive Force of the Batteries of Grove, Bunsen, and Daniell, with the Concentration of the Liquids.—C. Fromme.—Grove's battery is most efficacious when the nitric acid contains 40 per cent of water. In the Bunsen battery the strength of the current diminishes with the concentration of the acid. In Daniell's battery the electromotive force increases with the concentration of the solution of sulphate of copper, but with a less rapidity.—*Les Mondes.*

---

[*] Oxide of duiboron (yttria?) should, from an analogical point of view, have for its density the number 4·8 (see above.)

THE CHEMICAL NEWS.

VOL. XLI. No. 1056.

THE PERIODIC LAW OF THE CHEMICAL ELEMENTS.

By D. MENDELEEF.

(Continued from page 72.)

THE two missing elements in the fifth series (third and fourth groups) should have very distinct properties. They will be found in this series between $Zn = 65$ and $As = 75$, and being atomanalogous with aluminium and silicium, they will be called, the one *eka-aluminium*, and the other *eka-silicium*. As they belong to an odd series they will give volatile metallo-organic compounds, as well as volatile anhydrous chlorides, but they will have much more acid properties than their analogues Eb and Ti of the fourth series have. The metals should be very easily obtained by reduction with carbon or sodium. Their sulphides will be insoluble in water, and Ea_2S_3 will be precipitated by sulphide of ammonium, whilst EsS_2 will be apparently soluble in it. The atomic weight of eka-aluminium will be about 68, and that of eka-silicium about 72. Their densities will be about 6·0 for Ea, and 5·5 for Es, and their volumes will be about 11·5 for Ea, and 13 for Es, as the volume of $Zn = 9$, of $As = 14$, and of $Se = 18$. We obtain the same numbers when comparing (for Ea) the volumes of Al, In, and Tl; and (for Es) the volumes of Si, Sn, and Pb, because the first-mentioned elements are the atomic analogues of Ea, and the last those of Es. The volume of $Si = 11$, that of $Sn = 16$; accordingly that of $Es = 13$. In short, in the case now before us, we can find atomanalogues on all sides, and therefore it is possible to determine the properties more exactly than we could when examining Eb. These determinations result from the following atomanalogues :— Eka-aluminium, giving the oxide Ea_2O_3, occupies the place between Ea and As on the one hand, and Al and In on the other hand; eka-silicium, giving the oxide EsO_2, occupies the place between Ea and As on the one hand, and Si and Sn on the other hand. All these circumstances indicating a necessary analogy between Ea and Eb, as well as between Es and Ti, I am going only to submit Es to a comparative examination with Ti, so as to get a clear idea of the properties of these elements.

Eka-silicium will be extracted from EsO_2, or from EsK_2F_6, by the action of sodium; it will only decompose steam with great difficulty; it will hardly react on the acids, but will easily attack the alkalies. It will be a dirty grey metal, difficulty fusible, and will be turned by calcination into a finely-divided oxide, EsO_2, which also will be very difficult to fuse. The density of the oxide will be about 4·7,* corresponding to the volume, which, judging approximatively, according to the volumes of SiO_2 and SnO_2, will be about 22. This oxide in its looks, and apparently in its crystalline forms, in its properties, and in its reactions, will resemble TiO_2. The acid properties being feeble, although decided, as much in TiO_2 as in SnO_2, Es will have the same character, and will form a more decided acid than TiO_2. In this, as in similar cases, we have recourse to proportions. According to the proportion—

$$Es : Ti = Zn : Ca = As : V,$$

the basic properties should be more feebly shown in

---

* Kokscharoff has described, under the name of *ilmeno-utile*, a body analogous to rutile, but which has the density 4·8, whilst ordinary rutile has the density 4·2; 4·8 corresponds to the density of oxide of eka-silicium.

ilmenium, although Marignac* showed later on that the colombites, which had been suspected of containing ilmenic acid, in reality only contained a mixture of titanic, niobic, and tantalic acids. To confirm Marignac's assertion—which was that Hermann's ilmenic acid was only a mixture of titanic and niobic acids—we can say that the equivalent of ilmenium found by Hermann occupies the place between the equivalents of—

$$\text{Ti}\left(12=\frac{48}{4}\right) \text{ and of Nb } \left(19=\frac{94}{5}\right).$$

But Marignac himself has not succeeded in finding an exact method for the separation of titanic from niobic acid : he himself admits that his process (based on the difference of solubility of the double fluorides of potassium) is insufficiently complete. We have, therefore, a perfect right to be sceptical of the exactness of his results, however reasonable they may seem, and I further hold that a repetition of his work would not be superfluous, more especially as Ti and Nb can be separated in the form of chlorides, compounds whose boiling-points show considerable differences.†

The examples that I have given show sufficiently in what manner we can, by means of the periodic law, foresee the properties of unknown elements ; therefore I will not pursue the subject of describing the elements which are wanting in the system. The discoveries which would offer the most interest would be those of the following elements :—Eka-cæsium, Ec=175 : dui-cæsium, Dc=220 ; eka-niobium, En=146 ; eka-tantalum, Et=235 ; and the analogues of manganese, for example :—Eka-manganese, Em=100 ; and tri-manganese, Tm=190.‡ For it is difficult to consider the absence of the whole of a series (the ninth), and nearly all a large period (from Ce=140) as an accidental circumstance : the reason of this gap is very probably caused by the nature of elements.

When we reflect on the problems which form the subject of this memoir another question presents itself :—Is the number of elements limited or unlimited ? If we consider that the system of the known elements is limited, and, so to speak, closed ; that the aërolites, the sun, and the stars contain only the same elements as the earth ; that the acid properties disappear little by little as the atomic weight increases ; and that the greater number of the elements with high atomic weights constitute the heavy and difficultly oxidised metals ; if, I say, we consider all this to be case, we must admit that the number of accessible elements is very small ; and we may presume that if several more heavy metals are found in the interior of the earth, their number and their quantity will be very limited.

A New Voltaic Condenser.—M. d'Arsonval.—The author, seeing the energetic, though brief effects produced by secondary batteries, and in particular by that of M. Gaston Planté, sought for a condenser capable of storing up a much larger quantity of electricity. He forms a couple of a plate of zinc and a plate of coke, surrounded by lead in a very fine state of granulation immersed in a concentrated solution of sulphate of zinc. He also substitutes for the zinc a layer of mercury.—*Comptes Rendus*, tome xc., No. 4.

---

* *Archives des Sciences Naturelles*, January, 1866, and August, 1867.
† I would again drawn attention to the fact that, judging from its atomic analogy with As and Sn, Es might be met with in the minerals which contain these last elements. Wœhler mentions (*Mineralanalyse*, par. 96) that when tungsten is treated with *aqua regia* and ammonia, an insoluble residue of niobic (up to 2 per cent) and silicic acids is left. The further study of niobic acid would be very interesting, as Wœhler's work on that subject was published at a time when this mineral was very little known.
‡ Perhaps Ru and Os occupy these places, if their highest oxides are like R_2O_7 and not RO_4. However, if such were the case, the true analogues of Fe would be missing. A comparative study of OsO_4 and Mn_2O_7 is very desirable.

THE CHEMICAL NEWS.

VOL. XLI. No. 1057.

THE PERIODIC LAW OF THE CHEMICAL ELEMENTS.

By D. MENDELEEF.

(Continued from page 84.)

V. ON THE USE OF THE PERIODIC LAW FOR THE CORRECTION OF ATOMIC WEIGHTS.

EVERYBODY knows the fate of Prout's hypothesis. Careful observations having proved that the numbers of atomic weights can contain fractions—Stas having shown that definite fractions cannot be admitted—we can no longer entertain any doubts, in spite of Marignac's brilliant opposition, but that Prout's hypothesis overstepped the facts. I cannot even see how any reasonable argument can support it. Let us admit that the matter of all the elements is completely homogeneous : there is no reason why we should also admit, that n ponderable particles, or n atoms of an element, in changing into one atom of a second element, will give the same n ponderable particles, or that the atom of the second element will be n times as heavy as the atom of the first. We can consider the law of the invariability of weights as a special case of the law of the invariability of force or of motion. The cause of weight is evidently a particular kind of movement of matter, and there is no reason for denying the possibility of transforming these movements into chemical energy, or into some other form of movement. Two phenomena which we can now notice in the elements, viz., the invariability and the indestructibility of the atomic weights, are, up to the present, in intimate and even historical relation. If, then, a known element was to decompose itself, or if it was to change into another one, these phenomena might be accompanied by a diminution or an increase in weight. In this manner can be explained, up to a certain point, the difference of the chemical energy in different elements. In expressing this idea here I wish to say *only* that there is some possibility of the opinions of chemists on the composed nature of elements becoming uniform without adopting Prout's hypothesis.

From a practical point of view, Prout's hypothesis is defective, in so far as it concerns directly, small numbers. Our usual determinations of atomic weights do not often agree to a fiftieth part with the whole number, often even to nearly five or six units, whilst in Prout's hypothesis it is only a question of tenths. Thus, as we have already seen, the different determinations of the atomic weight of titanium vary from 57 to 48.

But from a theoretical as well as from a practical point of view, it is very important to have a principle which will lead to the discovery of the great errors which are sometimes made in the determination of atomic weights, and to have an idea of greater numbers. If in these last we discover a true regularity, it is certain that eventually we shall discover such a regularity that will enable to determine theoretically, and with great precision, the magnitude of atomic weights. The periodic law, even in its present state, enables us to discover moderate errors in the determinations of atomic weights, as we have seen in the case of titanium. We are now going to give examples of a more rigorous nature.

To correct atomic weights exactly it is necessary to submit the individual properties of elements to an exact and comparative examination, for it is these properties which cause the perturbations in the regular variations of the amount of the atomic weights. Already, in the preceding pages, I have noticed the evident existence of such

properties. The absence of an exact regularity in the variations of atomic weights arises from two comparisons.

It arises, first, from the fact that the differences in the magnitudes of atomic weights of corresponding members of different series and different groups are far from agreeing, even if we take into account the admissible errors of determination. Thus, the difference between the atomic weights of Na and Li, or, for the sake of brevity, $Na - Li = 16$, as also $K - Na = 16$, but $Mg - Be = 14 \cdot 6$; on the other hand, $Ti - Si = 20$, $V - P = 20$, $Pt - Pd = 91$, $Au - Ag = 89$, $Hg - Cd = 88$, $Pb - Sn = 89$, $Bi - Sb = 86$. It is difficult to admit that the gradual diminution of the differences is only a matter of chance. We ought rather to think that the relation between Pt and Pd is not altogether the same as that between Bi and Sb, and therefore the differences cannot be the same. We might notice that the properties of Bi and of Sb differ by a certain number of individual characters, whose origin arises in the slight individual separations of their atomic weights.

A second proof of the fact that periodic differences are in reality irregular, is found in the precise determinations of the atomic weights of the alkaline metals and the haloids, which were done by Stas.

$$
\begin{array}{lll}
\text{If H} = 1 \text{ and O} = 15 \cdot 96 & & \\
\text{Li} = 7 \cdot 004 \} & \text{difference} & 15 \cdot 976 \\
\text{Na} = 22 \cdot 980 \} & & \\
\text{K} = 39 \cdot 040 \} & \text{,,} & 16 \cdot 06 \\
\text{Cl} = 35 \cdot 368 \} & \text{,,} & 44 \cdot 382 \\
\text{Br} = 79 \cdot 750 \} & & \\
\text{I} = 126 \cdot 533 \} & \text{,,} & 46 \cdot 783 \\
\end{array}
$$

We must see that there is a strict comparison in the differences between the values of the atomic weights of elements that correspond in the system, but we also notice by the side of these comparisons individual separations. Therefore the elements possess general properties, dependent periodically on their atomic weights (for example, the property of given determinate forms of oxidations, which are modified spontaneously and gradually), and individual properties which are determined by the separations mentioned above.

We do not really know anything of the just-mentioned relation, except the periodic property ; and even this one is not properly understood. It is therefore impossible to determine exactly the amount of the separations, any more than we can correct positively the magnitude of atomic weights. We can only determine limits, certainly very near together, between which the amount of the atomic weight of a given element should be.

According to the periodic law, the atomic weight of tellurium should be greater than that of $Sb = 122$, and less than that of $I = 127$; that is to say that the atomic weight of tellurium ought to be about 125, because from every point of view, atomic analogies assign it a place between Sb and I. Considering that $Ag - Cu = 45$, that $Cd - Zn = 47$, that $Sb - As = 47$, that $I - Br = 47$, that $Cs - Rb = 48$, that $Ba - Sr = 50$, we are led to presume that in the same manner $Te - Se = 47$ (nearly), for Te is found between Sb and I, as Se is between As and Br; and as Se, whose atomic weight is 78, has been more completely studied, and can be more easily purified than Te, the number found for Se should be accepted before the atomic weight found for Te ; this is why this latter should be about $78 + 47 = 125$. This result does not agree with the numbers last given by Berzélius (who determined the quantity of TeO_2 obtained by the oxidation of Te by means of nitric acid).

$$
\begin{array}{ll}
\text{O} = 15 \cdot 96, & \text{O} = 16 \cdot 0 \\
\text{Te} = 128 \cdot 13, & \text{Te} = 128 \cdot 45 \\
= 127 \cdot 97, & = 128 \cdot 28 \\
= 127 \cdot 96, & = 128 \cdot 28 \\
\end{array}
$$

Hauer (who estimated the bromine of K_2TeBr_6 as AgBr) obtained corresponding numbers ; still, Te is generally taken to be $= 128$. Further, as the first determinations, made by Berzélius (1812), gave the number 116, as the later experiments have given the number 129, the exactitude of the actually recognised atomic weight gives rise to certain doubts, which can only be eliminated by fresh experiments. It is difficult to purify the compounds of tellurium, and even to be certain when they are pure. This may perhaps explain in a measure the errors of the numbers which have been found. It is difficult to admit that the distinctive individual characteristics of tellurium could determine a gap relatively so great (128 to 125) compared with the number of its atomic weight, as it is deduced from the periodic law. Fresh experiments are therefore necessary to show us to what degree the periodic law can be relied upon in the correction of atomic weights.

Os, Ir, and, perhaps, also, Pt and Au, furnish us with a second example of gaps of this nature. These elements are found in the same series of the system (see Chapter I.); they are preceded by $W = 184$, and followed by $Hg = 200$. Their atomic weights are effectively less than that of Hg, and greater than that of W, but the series does not correspond. For, considering that Os, Ru, and Fe are analogous, but that the atomic weights of Ru and Fe are less than the atomic weights of Pd and Ni, we can foresee that the atomic weight of Os will be less than that of Pt, so that Ir, which is between Pt and Os, will be intermediate. According to Berzélius and Frémy, Os = from 199 to 200, $Ir = 197$, $Pt = 198$. The predicted series is the following:—$Pt = 198$,* Ir = about 197, Os = from 196 to 195, whilst Au will be equal to 199.

Besides, it is allowable to suspect errors in the determinations of the atomic weights of the metals of the platinum group, not only because it is difficult to separate them one from another, but also because the compounds which are used for these determinations are very unstable; $IrCl_4$ and $OsCl_4$, for example, easily lose a part of their chlorine.

(To be continued.)

sition. The ordinary organic analysis method was used, but the diamond crystals were laid on a thin piece of platinum-foil, and this was ignited by an electric current, and the combustion conducted in pure oxygen. The result obtained was, that the sample (14 m.grms.) contained 97·85 per cent of carbon—a very close approximation, considering the small quantity at my disposal. The apparatus and all analyses will be fully described in a future paper.

(Received February 25, 1880.)

Extract from a letter from Mr. Hannay, dated Feb. 23 :— " I forgot, in the preliminary notice, to mention that the specific gravity of the diamond I have obtained ranges as high as 3·5 ; this being determined by flotation, using a mixture of bromide and fluoride of arsenic."

ON THE ARTIFICIAL FORMATION OF THE DIAMOND.*

By J. B. HANNAY, F.R.S.E., F.C.S.
PRELIMINARY NOTICE.

WHILE pursuing my researches into the solubility of solids in gases, I noticed that many bodies, such as silica, alumina, and oxide of zinc, which are insoluble in water at ordinary temperatures, dissolve to a very considerable extent when treated with water-gas at a very high pressure. It occurred to me that a solvent might be found for carbon ; and as gaseous solution nearly always yields crystalline solid on withdrawing the solvent or lowering its solvent power, it seemed probable that the carbon might be deposited in the crystalline state. After a large number of experiments it was found that ordinary carbon, such as charcoal, lampblack, or graphite, were not affected by the most probable solvents I could think of, chemical action taking the place of solution.

A curious reaction, however, was noticed, which seemed likely to yield carbon in the nascent state, and so allow of its being easily dissolved. When a gas containing carbon and hydrogen is heated under pressure in presence of certain metals, its hydrogen is attracted by the metal, and its carbon left free. This, as Prof. Stokes has suggested to me, may be explained by the discovery of Profs. Liveing and Dewar, that hydrogen has at very high temperatures a very strong affinity for certain metals, notably magnesium, forming extremely stable compounds therewith.

When the carbon is set free from the hydrocarbon in presence of a stable compound containing nitrogen, the whole being near a red-heat and under a very high pressure, the carbon is so acted upon by the nitrogen compound that it is obtained in the clear, transparent form of the diamond. The great difficulty lies in the construction of an inclosing vessel strong enough to withstand the enormous pressure and high temperature, tubes constructed on the gun-barrel principle (with a wrought-iron coil), of only half an inch bore and 4 inches external diameter, being torn open in nine cases out of ten.

The carbon obtained in the successful experiments is as hard as natural diamond, scratching all other crystals, and it does not affect polarised light. I have obtained crystals with curved faces belonging to the octahedral form, and diamond is the only substance crystallising in this manner. The crystals burn easily on thin platinum-foil over a good blowpipe, and leave no residue, and after two days' immersion in hydrofluoric acid they show no sign of dissolving even when boiled. On heating a splinter in the electric arc it turned black—a very characteristic reaction of diamond.

Lastly, a little apparatus was constructed for effecting a combustion of the crystals and determining their compo-

* A Paper read before the Royal Society, February 26, 1880.

THE PERIODIC LAW OF THE CHEMICAL ELEMENTS.

By D. MENDELEEF.
(Continued from page 94.)

VI. ON THE USE OF THE PERIODIC LAW FOR THE COMPLETION OF OUR KNOWLEDGE RELATIVE TO THE FORMS OF CHEMICAL COMPOUNDS.

AT the time when the discovery of metaleptic phenomena created a change in the ideas of chemists, all the chemical reactions were considered as cases of substitution, and the theory of types was drawn up under several forms. Gerhardt was even inclined to think that the reaction by which C_2H_4 combines with Cl_2 was an act of substitution of H by Cl, and a formation of $C_2H_3Cl + HCl$. Everything which did not correspond with these ideas was declared to be molecular addition. $C_2H_3Cl + HCl$ in the preceding example should, according to this, form a whole body like $CNa_2O_3 + 10H_2O$. Although the interpretation of chemical phenomena has, thanks to the theory of types, acquired a simplicity and harmony which was before impossible, this theory nevertheless shared in a measure the fate of the doctrines which preceded it, and that was because the notions on which it was based were expressed in a too absolute manner, and because it neglected a large class of phenomena and considered them to be molecular additions. It was not until attention was turned towards the study of metallo-organic compounds (towards 1855), in which the phenomena of substitution and addition* are represented both in the same degree, that the phenomena of addition commenced to receive the attention they deserved : Frankland, Cahours, &c., have introduced to chemistry the idea of there being a limit to chemical compounds. These ideas have since been applied to the carbon compounds.†

The ideas which have been entertained about *limits*, about *saturated* compounds, about the aptitude of *non-saturated* compounds to show phenomena of addition, ideas connected with the opinions relating to substitutions, correspond, in my opinion, to the most complicated problems that chemical investigation can propose. However, the theory of limits, in its primitive form and free from everything tending towards hypotheses, has not settled itself in the science of chemistry ; for it is in the destiny of science that the most important discoveries of an era lead first of all to extreme hypotheses. I consider as such the theory of atomicity at present in vogue.

Couper, later on Kékulé, and after them others, in comparing the composition of saturated compounds of an element with chlorine or hydrogen, have thought that each

* I have treated in detail, in the *Bulletin de l'Académie des Sciences de St. Pétersbourg*, 1858, the differences between the phenomena of addition and substitution in the sense of the unitary theory, and bearing in mind the modifications arisen from the physical properties.
† Notably in my memoir, "Essai d'une Théorie sur les Limites (*Ibid.* vol. v., 1861.)

element was capable only of saturating a certain number of equivalents of all the other elements. But not stopping there, these investigators, and afterwards their successors, thought that it was possible to deduce the preceding hypothesis from determined conditions relative to the fixation of elements in the molecule. Everybody is aware of the excellent results which have been arrived at in starting from this hypothesis in the study of isomerics : but all the same it must not be forgotten that other hypotheses, now neglected, have had these periods of glory and popularity. The discoveries due to adepts of the electro-chemical theory have been shown up in a new manner by the theory of types. But the discoveries concerning isomerics are completely explained without it being necessary to have recourse to the hypothesis of the combination of elements in the molecule by parts of their affinity. We have a very simple manner of showing this circumstance, which was discovered and interpreted exactly from cases of isomerism in the aromatic and other compounds, even without using the preceding theory, and uniquely by a clear idea of substitutions. Below are given some considerations directed against the different forms of the hypothesis of valence or of atomicity.

1st. The corner-stone of this hypothesis is the division of all chemical compounds into atomic and molecular compounds; but the hypothesis only touches on the compounds of the first mentioned kind; molecular compounds are neither considered nor generalised. The above-mentioned division—a division which has long existed—lacks any solidity; it is so artificial and so arbitrary that even chloride of ammonium and perchloride of phosphorus are by some people considered as molecular compounds. In reality, there is not one single character which sharply defines the limit of the category of the above-mentioned compounds. That they contain complete independent molecules, one by the side of the other; that they are incapable of being turned into vapours; that there is only a feeble modification in the chemical reactions when combination takes place;—all these are insufficiently distinct characters; if not, $PtCl_4 = PtCl_2 + Cl_2$ and $KClO_4$, &c., might be considered as molecular compounds. But if it is impossible to establish any limit between the molecular and atomic compounds, the possibility of fixing the equivalence or the atomicity of element, above all when based on the compounds with hydrogen and chlorine, disappears. $PtCl_4$ is not a limited compound, for it can still combine with $2HCl + 6H_2O$ or with $8H_2O$, and other molecules can be substituted for these last-named. The formula of hydro-fluosilicic acid, SiH_2F_6, which is reproduced for many elements, shows, so to speak, the hexatomicity of silicium, whilst this element, compared with hydrogen and chlorine, is tetratomic. To prove that this form of combination is not determined by the presence and by the particular properties of fluorine, it suffices to mention PtH_2Cl_3 and the corresponding salts, as well as the double cyanides.

2nd. To appreciate the atomicity of elements, we start with combinations with hydrogen, although out of the sixty-three elements which are known up to the present seventeen only have been combined with it. Further, hydrogen only gives one compound with one atom of an element, whilst other elements, and even chlorine, can furnish several compounds. Therefore, it is at least of not much practical value to judge compounds of an element with other elements according to its compounds with hydrogen. Carbon itself only forms with H one compound, CH_4, and not even CH_2, although it gives with oxygen CO_2 as well as CO. The existence of molecules $SnCl_2$ and $SnCl_4$, HgCl and $HgCl_2$, PCl_3 and PCl_5 has always embarrassed the partisans of the hypothesis of the atomicity of elements, precisely because we started from compounds with hydrogen, and because hydrogen only gives one form of combination.

3rd. All the elements except fluorine can enter into combination with oxygen. These compounds are the most widely spread of any in Nature. Their chemical character is very distinctly shown; many elements give several compounds with oxygen. However, the hypothesis of atomicity does not furnish us with any law relative to the number of atoms of oxygen, because, according to this hypothesis, oxygen can enter into any complete molecule in its character of a biatomic element. A given body, RM, can receive an indefinite number of molecules of oxygen; it then forms what the hypothesis in question calls a chain—

$$R—O—O—O^n—M.$$

The best example which can be given is the following :—

     KCl,
     K—O—Cl,
     K—O—O—O—Cl,
and   K—O—O—O—O—Cl.

In other cases this hypothesis is not considered admissible; thus there is no chain in CO_2. The instability of the cateniform compounds of oxygen cannot serve as a distinctive character, for CO_2, H_2O, Cl_2O, are all one as easily decomposed as the other, although at different temperatures. PtO_2, again, is more easy to decompose than $KClO_3$, in spite of the admitted difference relatively to the mode of fixation of the atoms. Further, we cannot imagine why KClO is more unstable than $KClO_3$, nor why the longest chain, $KClO_4$, is more stable than the shortest chains. But what there is worthy of note is that the forms of combination of oxygen, like hydrogen and other elements, are held between narrow and analogous limits. This may be noticed at first sight if we take care to put next to each other those forms of combination which contain the most oxygen, OsO_4, $KClO_4$, K_2SO_4, K_3PO_4, K_4SiO_4. The periodic law, in fixing the limit of oxygenated compounds, fills up an important gap in the atomic theory.

4th. The most logical partisans of the above-mentioned hypothesis are those who see in atomicity a fundamental or invariable property of atoms. As a rule, they further hold that free affinities cannot exist in molecules. They consider carbon to be invariably tetratomic, and nitrogen always triatomic, &c. However, it is necessary to give in before facts. Thus, at present, the greater number of partisans of atomicity consider this property to be variable. In plainer words, to admit that the atomicity of an element is variable is almost to renounce the hypothesis in question; this hypothesis is no longer anything but that of the number of equivalents in an atom; atomicity varies like the equivalents of elements, according to the law of multiple proportions. If atomicity is a variable property, if we admit that in the greater number of cases a part of the bonds of affinity is latent, we must give up all idea of determining the atomicity. Sulphur was long considered to be a diatomic element, and this diatomicity was deduced from the compounds SH_2, SHK, SK_2, SCl_2; and SO_2 and SO_3 were declared to be compounds in the form of a chain,—

$$S{<}{\overset{O}{\underset{O}{|}}} , \quad \text{and} \quad S{-}{\overset{O}{\underset{O}{{-}}}}{>}O.$$

But in this manner the well-known and distinct analogy between SO_2 and CO_2, in which different structures were admitted, remained inexplicable. Further, the analogues of sulphur, Te and Mo, give $TeCl_4$ and $MoCl_2$. The compound SAe_3I was also discovered, so that sulphur was looked upon as being also tetratomic. Let us add that the existence of WCl_6 and SO_3 caused sulphur to be considered as hexatomic; perhaps Cl may sometimes be heptatomic. What is it that proves to us the hydrogen and oxygen cannot be polyatomic? The first severe blow was dealt to this theory when variable atomicities, free and latent bonds of affinity were admitted; the fundamental principles on which the hypothesis is built were shaken to the foundations.

5th. After having lost, by the adoption of variable atomicities, the value of a rigorously scientific system, and after being changed into a tabulated form of equivalents, the above-mentioned hypothesis could still hold its position in science as a means of representing the forms of combination, on condition of not including the *junction* of elementary atoms in the molecule by means of their bonds of affinity. This part of the hypothesis cannot be put in harmony with the much more certain notions which are now entertained on the construction of matter and on the laws of attraction; we must therefore reject it absolutely. We ought to represent the atoms in a molecule as being in a certain state of unstable equilibrium of reciprocal reactions. The entire system is maintained by forces which belong to each individual particle; for we can only imagine two parts of a whole to be uniquely under the influence of a third part, without giving rise to reciprocal influence, particularly if all that we know concerning these two parts shows that they exercise a distinct and constant chemical action. If we admit that in CH_4 the $4H$ are only held by one atom of carbon, this idea is not, so to speak, unreasonable in itself; but if one H is replaced by Cl it is difficult to admit that the four atoms (H_3Cl) are held only by the carbon—that the reciprocal powers of attraction of Cl and H remain dormant during or after the reaction. This proposition is not admissible except on the ground that Cl is nearly three times as heavy as carbon. If the term " junction " should only be a conditional expression of the distribution of the elements in space; if it signifies that in the just-mentioned example Cl occupies the *same position* as H—that is to say, that it fulfils the same conditions with respect to carbon and other elements—then this would only be a repetition of what has been stated since the discovery of metalepsy. I have often heard the polyatomic elements, and the elements combined with them, compared to the sun and planets; but this comparison is not tenable, because if one planet were replaced by another, whose mass would be even greater than that of the sun itself, the planetary system would not for that reason cease to exist. The partisans of atomicity may suppose that chemical attraction, differently to universal gravitation, does not depend on the mass, but only on the atomicity of the elements. The periodic law in recognising a strict dependence between the atomicity and the mass of atoms completely annuls such an hypothesis, and authorises us to suppose that chemical attraction, like any other attraction, depends upon the masses.

(To be continued.)

PROCEEDINGS OF SOCIETIES.

PHYSICAL SOCIETY.

Ordinary Meeting, February 28, 1880.

Prof. W. G. ADAMS in the Chair.

A PAPER was read by Mr. RIDOUT on " *Some Effects of Vibratory Motion in Fluids.*" It was found by Savart and Tyndall that jets of water were sensitive to notes, or air vibrations, like flames, and the author conceived the idea of vibrating the jet of water from within. To do this he caused an electro-magnetic arrangement to pinch the tube conveying the water 400 to 500 times per second so as to communicate a vibratory motion to the stream of fluid. The issuing jet spread out in two streams, beautifully broken into drops, and representing the fundamental note. When the pinching lever vibrated irregularly, harmonics were observed. When the water was thrown into vibration in two different planes the resulting jet rotated in the tube. Froude's deduction that a liquid moving in a tortuous tube has a tendency to straighten the tube was illustrated by oscillating a pipette with its nozzle in a vessel of water, and filling a coloured liquid into it, which is seen to flow from the nozzle through the water in a tortuous line. By giving the pipette also a motion round its axis the line becomes a spiral; a sounding body produces no disturbance in the stream. The author also showed that the cardboard experiment of M. Clement Desormes can be extended to water. In this experiment a card is attracted to another card by blowing a jet of air through the latter upon the surface of the former. Mr. Ridout allows a jet of water to flow out of a glass tube with a cup-shaped mouth upon the surface of a glass ball, and when the ball is within a certain distance of the mouth it is attracted towards the latter and sticks in the mouth. In explanation of this fact it was shown that the ball and cup remained in such a position that the outflow of water was greater than if the globe had been entirely absent.

Prof. PERRY explained this action by the hydro-dynamical fact that the pressure is less at the centre of the mouth of the cup than at the edges.

Prof. GUTHRIE said that he had tried a similar experiment with a funnel-shaped mouth and a glass cone, but failed. He surmised that perhaps the cohesion of the water for itself as it formed a shell round the ball might help to cause the success of the ball method.

Prof. ADAMS pointed out that with the cup and ball there was less difference of head of water between the centre of the mouth and the edge where the water escaped than with the funnel.

Dr. STONE stated that he had been able recently to imitate many physiological sounds, such as the murmur of the heart, by means of constrictions in tubes through which water and air were flowing. His demonstrations were made before the Royal College of Physicians.

Dr. C. W. WRIGHT then read an important paper on a " *Determination of Chemical Affinity in Terms of Electromotive Force.*" After giving a history of the subject, he described his original experiments. These consisted in performing electrolysis of sulphuric acid and measuring the heat evolved in the process, and by recombustion of the materials. A voltameter with spade-shaped platinum electrodes, soldered to stout copper wires, and sealed by a large plug of gutta-percha, was employed for the electrolysis. An ordinary water calorimeter was used to measure the heat given off, as Bunsen's was found to contain sources of loss of heat. The strength of the current employed was varied from 6 webers to $\frac{6}{10}$ weber. The volume of gas produced was measured by Joule's plan. Radiation loss was corrected for by three methods. From an average of eighteen experiments the value of e, the electromotive force, was found to be $1·5038$ C.G.S. or volts. Taking the formula—

$$J = \frac{e}{(H+h)x},$$

where J is Joule's equivalent, H is the heat actively evolved, h the heat evolved by recomposition, and x a constant to which Kohlrausch gives the value of $0·000105$. Dr. Wright finds that Joule's equivalent should be $4·196 \times 10^7$ instead of $4·20 \times 10^7$ as given, to answer the formula. The author thinks that Joule's water-friction experiments gave the truest value of J, and that his electric heating experiments gave a result about $\frac{1}{4}$ per cent too low, owing to the B.A. unit of resistance being about 2 per cent too high and other causes.

THE CHEMICAL NEWS.

VOL. XLI. No. 1059.

THE PERIODIC LAW OF THE CHEMICAL ELEMENTS.

By D. MENDELEEF.
(Continued from page 106.)

I WOULD add that the theory of atomicity arose from the study of the organic compounds of carbon, to which it is easily applied. The legitimacy of this application is shown by the two following considerations :—

1. *Carbon takes an equal number of equivalents of hydrogen and of oxygen*, and at its limit gives the compounds CH_4, CO_2, CCl_4. It is otherwise with the elements of other groups (see Table II.).

The elements of the fifth group give RH_3 and R_2H_5, that is to say, that they are triatomic compared with hydrogen, and pentatomic compared with oxygen; the elements of the sixth group are biatomic compared with hydrogen, and hexatomic compared with oxygen, &c.

2. As far as we know carbon does not give any so-called molecular compounds, like so many of the other elements do. The combination of oxalic acid and water,—

$$C_2H_2O_4,2H_2O,$$

belongs to the type of hydrates, because it corresponds to the limit compound C_2H_3, in which all the H are replaced by (OH); $C_2(OH_6) = C_2H_2O_4,2H_2O$; in separating by a comma the water of crystallisation, we only want to express this well-known fact that an atom of carbon in a compound retains only a slight residue of water. The incapacity of carbon to form molecular compounds is noticed particularly when we compare it with silicium, analogous in its other properties, and giving like carbon limiting compounds, SiH_4, SiO_2, and $SiCl_4$. If in organic compounds we replace a quantity of carbon by silicium, bodies analogous in many respect to the veritable carbon compounds are produced, as the splendid researches of Friedel and Ladenburg have shown; nevertheless, there exists a great difference between the two elements, and it consists principally in the faculty possessed by Si of forming molecular compounds. Perchloride of silicium combines, like $TiCl_4$, BCl_3, PCl_5, SCl_2, &c., with other chlorides, and forms moderately stable bodies, whilst chloride of carbon, as far as we know, is not capable of so doing.* The aptitude of Si for molecular combinations does not arise only from the existence of SiH_2F_6, and hydrates of silicic acid (we do not know of any hydrates of CO_2), but also from numerous properties of silica. This last point ought to be cleared up. When we compare the compounds of carbon with those of silicium, we constantly notice that *the compounds of silicium boil at lower temperatures than the corresponding compounds of carbon* (or sometimes, but very rarely, at the same temperature), *and that they have a greater* (or equal) *molecular volume.* There is an intimate relationship between the two phenomena :—

	Molecular Volume.	Boiling-points.
CCl_4	94	76°
$SiCl_4$	112	57
$CAe_4 = C_9H_{20}$	,,	about 120
$SiAe_4 = SiC_8H_{20}$	,,	150
$CaCO_3$	37	,,
$CaSiO_3$ (wollastonite) ..	41	,,
$CHCl_3$	78	60
$SiHCl_3$	84	34
$C(OC_2H_5)_4$†	186	158
$Si(OC_2H_5)_4$	201	160

* This question should be very carefully studied.
† Basset's ether.

We can admit that, in other cases also, analogous phenomena will spring up; however, we have contrary evidence in comparing CO_2 with SiO_2. The boiling-point of silicic acid is at a temperature we cannot attain, and in accordance with this the volume is less than that of carbonic anhydride, liquid or solid. The volume of CO_2 is about 44, the volume of amorphous $SiO_2 = 27$. This contradiction can be explained only by the hypothesis of a case of polymerism of silica (Si_nO_n), because the phenomena of polymerisation are always accompanied by a diminution of volume and a rise in the boiling-point.*

But as the substances liable to other additions are the only one which become polymeric (CH_2, C_2H_2, C_5H_8, for example), because the polymerisation is an act of combination of homogeneous molecules, the differences between silica and carbonic acid betray the property which silicium has of forming complicated compounds of this kind, compounds which are unknown in the history of carbon.

Carbon by its principal properties is clearly distinguished from the other elements; it is for this reason that a number of conclusions, completely exact when they apply to carbon compounds, are not true when we wish to apply them to other elements. We have a proof of this in the ideas on constant atomicity, which ideas, confirmed by the study of organic compounds, do not apply to the compounds of other elements. If, from the doctrine relative to the chemical structure of compounds founded on the atomicity of elements, we exclude all that is in contradiction to the mechanical theory of the structure of bodies, and to the ideas which have only been confirmed specially by carbon, there remains that which has already supplied the theories of substitution, and of the limits of chemical combination. The hypothesis of atomicity appears to me to be unsteady, perhaps because it has led to no general law, and is not built on solid foundations. If to what remains of the atomic theory we add what is given by the periodic law, we arrive at the following general idea *of the forms of chemical compounds.*

All the ideas on the structure of chemical compounds rest on three principal propositions. I will try and make them clear.

1. *The Principle of Substitution.*—When a molecule divides into two parts, these parts are equal (they can replace one another). This principle is comparable to the principle of mechanical conservation, and the equality of forces, in virtue of which, action and reaction are equal. It is for this reason that H_2 and O, HO and H, CH_3 and H, CH_2 and H_2, C and H_4, Cl and H, K and Cl, K and H, NH_2 and H, NH_4 and Cl or H or K, &c., are equivalent. The phenomena of substitution, homology, &c., are thus more generalised.

2. *The Principle of Units.*—When a molecule decomposes, one at least of the molecules of decomposition can combine with a quantity of elements equivalent to the second molecule formed simultaneously. C_2H_4 formed from C_2H_6O by eliminating one molecule of H_2O, can combine with Cl_2 (equivalent to H_2O), with HCl, &c. This principle comprises the phenomena of addition, the production of limited forms, as well as of body far from the limit, &c.

3. *The Periodic Principle.*—The highest forms of combination of an element with hydrogen and with oxygen, and consequently with equivalent elements, are determined by the atomic weight of this element, of which they are a periodic function. This principle keeps down the number of possible forms, but it still requires elaboration.

I will now explain the most important consequences of these three principles.

All elements can combine with hydrogen and produce one of the four following forms: RH, RH_2, RH_3, RH_4. In each form one hydrogen can be replaced by an equivalent residue of another form (up to a certain limit deter-

* In adopting polymerisation, we can also explain other properties of silica. See " Principles of Chemistry " (in Russian), 1st edition, chap. xvii., vol. ii.

mined by the nature of the element). For example, from CH_4 we can obtain CH_3NH_2, N acting here as an element which gives NH_3. Another substitution is possible in the body formed, viz., CH_2ClNH_2. The order in which substitution takes place has not any influence on the properties of the bodies; it depends entirely on the form under which the substitution is effected. Thus—

$$CH_3NH_2 = NH_2CH_3,$$

but CH_2ClNH_2 is simply isomeric with CH_3NHCl. In this manner homology and isomerism are explained. When in a compound of an element with hydrogen one hydrogen is replaced by equivalents of hydrogen compounds of the same element, homologues are formed.

The following homologues correspond to CH_4:—

$$C_nH_{2n+2} \begin{cases} CH_3CH_3 \\ CH_3CH_2CH_3 \\ CH_3CH_2CH_2CH_3 \end{cases}$$

or—

$$CH_3CH \begin{cases} CH_3 \\ CH_3 \end{cases}$$

&c.

The following homologues correspond to phosphuretted hydrogen :—

$$P_nP_{n+2} \begin{cases} PH_3 \text{ (gaseous phosphuretted hydrogen)} \\ PH_2PH_2 \text{ (liquid phosphuretted hydrogen)} \\ PH_2PHPH_2, \\ PH_2PHPHPH_2 \text{ and } PH_2P(PH_2)_2, \&c. \end{cases}$$

To sulphuretted hydrogen, S_nH_2 corresponds; SH_2; SHSH; SHSSH.

The number expressed by n can have different values, according to the nature of the element. For C, as far as we know, this number is unlimited; for N it is 1; for P it is not greater than 4; for S it equals 6. The last case seems to show a dependence between the value of n and the number of atoms forming a molecule, because the molecule of sulphur is S_6; we can thus foresee that the molecule of ozone O_3 will give O_3H_2, as O_2 gives O_2H_2; but O_3H_2 will be still more unstable than H_2O_2, because O_3 is more unstable than O_2.

Further, the limited compounds of which we have spoken can form, with elimination of molecules of hydrogen, hydrogen compounds far from the limits. It is thus that P_4H_6 is changed in P_4H_2 (solid phosphide of hydrogen) but losing $2H_2$. It is thus that by means of C_nH_{2n+2}, we obtain the series C_nH_{2n}, C_nH_{2n-2}, &c.

Up to this point our manner of looking only differs in appearance with that adopted by the partisans of the atomicity of elements; they are really similar up to here. But we have only spoken of hydrogen and its compounds. Further on we shall meet with essential differences; but we must notice that there are only some few elements capable of giving hydrogenated compounds, and above all giving homologous bodies. Carbon only, as far as we know, gives the bodies in any great number.

(To be continued.)

THE CHEMICAL NEWS.

VOL. XLI. No. 1060.

THE PERIODIC LAW OF THE CHEMICAL ELEMENTS.

By D. MENDELEEF.

(Concluded from page 114.)

OXYGEN combines with the atoms of each element R, in one or more of the following manners:—

$$R_2O, RO, R_2O_3, RO_2, R_2O_5, RO_3, R_2O_7, \text{ and } RO_4.$$

We often obtain other lower forms of combination, such as R_4O, although in the cases of all the known elements higher forms have been observed. H_2 and O, judging from the constitution of water, are equal between themselves; if so, the first four lower compounds of oxygen are also equivalent to the forms of combination of hydrogen,

$$RH, RH_2, RH_3, RH_4.$$

As the number of equivalents of hydrogen and oxygen which can be fixed by one elementary atom does not surpass eight in all the elements which give RO_4 do not form compounds with hydrogen; the elements which give R_2O_7 give also RH, and those which give RO_3 give also RH_2; those which give R_2O_5 give RH_3, and those which give RO_2 give RH_4. The elements corresponding to the highest form R_2O_3 have not as yet given any compounds with hydrogen, because such a form of combination RH_5 as it would require, does not exist. Therefore the aptitude of elements to combine with hydrogen diminishes when their aptitude for combining with oxygen increases, and *vice versa*; the highest forms or limits of these compounds are RO_4 and RH_4, RH_4 corresponding according to the equivalent of water to the oxygen compound, RO_2.

The formation of complex saline compounds with oxygen is determined by the forms of the simple oxygen compounds; for example, it must be admitted that in the hydrates O is represented by equivalent quantities of $(HO)_2$ or H_2. Therefore, from SO_2 is formed $SO_2(OH)_2$, $SO_2H(OH)$, and SO_2H_2. From CO_2 is derived $CO(OH)_2$, $COH(OH)$, and COH_2. The other phenomena of substitution of this kind would take us too far. Some explain themselves, and I hope to discuss the others, later on, in a memoir developed on molecular combinations. The lower forms of combination are capable of being transformed into the highest forms, either by direct or indirect means. If the highest form that an element can give is RX_n, a given form RX_{n-m} can change into the limit form RX_n by the absorption of X_m, or of an equivalent quantity of other substances; that is easily understood without any further explanation.

The limit compounds of *chlorine*, or in general the *haloid* compounds, correspond to the compounds of oxygen

$$(RO_2 - RCl_4; R_2O_3 - RCl_3; RO - RCl_2);$$

they are often met with in forms far removed from the limit, and then they often correspond to the hydrogen compounds. They have never higher forms than the oxygen compounds. For example, in the case of Te, $TeCl_6$ corresponding to TeO_3 does not exist, but $TeCl_4$ corresponding to TeO_2 does. In the case of I, there is not ICl_7 corresponding to I_2O_7, but there is ICl_5, corresponding to I_2O_5. Again, taking As, there is no $AsCl_5$ corresponding to As_2O_5, but $AsCl_3$, corresponding to As_2O_3. In general, chlorine, like *oxygen*, often gives, besides the higher forms corresponding to the oxygen compounds, some lower forms.

The aptitude of several (but not all) elements, to form different forms of combination, is not expressed completely, either in the hydrogen compounds or in the highest forms of oxygen compounds, above all when a compound is composed of more than two elements. Entire molecules can combine one with the other and produce higher forms; both polymeric and so-called molecular compounds. As far as concerns Si, the form SiX_4 (corresponding to SiH_4, to SiO_2, to $SiCl_4$) does not show the limit to the compounds which this body can form. For Si does not only give SiO_2, but also SiO_{2n}, SiO_2, $SiF_4 2HF$, &c. Pt does not only give PtO_2, $PtCl_4$, or generally PtX_4 and PtX_2, but also $PtX_4 nA$; A here means an entire molecule, and n, in most cases, is a whole number. For example:—

$$PtCl_4 2RCl, PtCl_4 8H_2O, PtCl_4 2HCl 6H_2O, PtX_2 2NH_3,$$
$$PtX_2 4NH_3, PtCy_2 2HCy_5 H_2O, PtCy_2 MgCy_2, 7H_2O, \&c.$$

Some of these forms of compounds are very stable, lend themselves to double decompositions, and are met with in many elements; such is the form RH_2X_6 for the elements which give RO_2 (X means a haloid or a haloid residue). For example:—

$$SiH_2F_6, PtK_2Cl_6, TeK_2Cl_6, ZrK_2(SO_3)_3.$$

These forms of combination serve to characterise the elements and their forms (for example, the form of alum, the form of several salts, $RSO_4 7H_2O$, $RK_2(SO_4)_2 6H_2O$, &c.). They therefore merit a comparative examination as exact as all the other forms with which they have no essential difference.

I do not pretend to erect, from the considerations that I have here set forth, a definite system. I know that many improvements and additions are still necessary, but I believe that the direction indicated and followed in this memoir will lead to the end desired by chemists more easily than any other will. Bold hypotheses have a peculiar attraction for us; they often cause a momentary progress; but more often they lead to inexact conclusions, and they themselves fall into disuse, above all if they are not built upon philosophical laws, for it is to the discovery of such laws that all scientific efforts should tend. I have endeavoured in the preceding developments to find support on the laws of substitution, on those of limits, as well as on the periodic law, and I believe that these laws should be taken as the basis of all generalisations on the forms of combination of the elements.

Independently of all that I have just said, I will add another observation and an example which clear my ideas on this subject in its essential parts. Two facts are necessary to actually identify an element; they are found by observation, by experience, and by comparisons. These are the atomic weight and the atomicity. The periodic law, in bringing to light the mutual dependence of these two values, at the same time affords us the means of determining one by the other; therefore, if the hypothesis of the atomicity of the elements determines the forms of chemical compounds, the periodic law does the same; but the latter goes still further, for it determines at the same time such forms of oxidation that the above-mentioned hypothesis neglected.

Boron and *Aluminium*, judging by their atomic weights, are analogues, and ought to give analogous compounds. Such is really the case, for they give oxides B_2O_3 and Al_2O_3: these oxides are the only ones formed by boron and aluminium. The hydrogenated compounds BH_5 and AlH_5 possible for these elements do not exist, for the general formula for the compounds of B and Al corresponds to RX_3. The form RX_5 is not represented in the isolated condition even in the hydrates; these last correspond to $R(OH)_3$. However, in consideration of the aptitude for more uncommon compounds, bodies composed according to the formula RX_5 are possible. Such bodies do in reality exist; they have been considered up to the present as being molecular compounds; for example, BHF_4. We can say that ordinary octahedric borax, $Na_2B_4O_7 5H_2O$ or $NaB_2H_5O_6$, answers by its composition to the form B_2O_3; that is to say, that this borax would be $B_2(OH)_5(ONa)$;

here again, then, the form BX_3 is preserved. As to the
faculty which borax possesses of absorbing more water of
crystallisation and being transformed into prismatic borax,
$Na_2B_4O_7 10H_2O$, it is explained not only by the property
which is peculiar to boron (of forming compounds like
BX_5), but also by the property possessed by Na, of forming
$Na(OH)3H_2O = NaH_7O_4$ independently of $Na(OH)$. The
aptitude of the form BX_3 to change into BX_5 is again
shown by the reunion of BCl_3 with other chloranhydrides.
H. Gustavson prepared not long ago, in our laboratory, a
compound BCl_3POCl_3, perfectly crystallised. We know
that aluminium possesses the same property, for it not
only gives $AlCl_3NaCl$ and $AlCl_3POCl_3$, but also $AlF_3,3NaF$.
I see in this property a reason for admitting the existence
of combinations of several molecules, either of $AlCl_3$ or of
polymers. The composition—

$$\left.\begin{array}{c} AlCl_3 \\ AlCl_3 \end{array}\right\} \text{ corresponds to } \left.\begin{array}{c} AlCl_3 \\ POCl_3 \end{array}\right\}, \text{ or to } \left.\begin{array}{c} BF \\ HF \end{array}\right\}.$$

We cannot determine from the forms BCl_3 and Al_2Cl_3, corresponding chlorides, the different atomicities of the two
elements. The partisans of atomicity compare Al_2Cl_6
with C_2Cl_6, and from it they conclude that aluminium is
a tetrad : however, we are quite as much authorised to
consider Al as a pentad, for AlH_4 and AlH_2 do not exist.
There is no reason to admit the existence of these hypothetical bodies : the existence of $AlCl_3$ corresponding to
$AlAe_3$, BCl, and BAe_3 is much more natural. A polymeric
transformation of $AlCl_3$ into Al_2Cl_6 would correspond
better with the polymers S_2 and S_6, CH_2 and C_2H_4, &c.
Alumina is apparently a polymeric modification ($Al_{2n}O_{3n}$)
in the same way as $AsCl_3$ does not correspond with As_2O_3,
but with As_4O_6. $AsCl_3$ combines with the chloranhydrides
as As_2O_3 does with different anhydrides : therefore, two
molecules of As_2O_3 can unite in one body—

$$\left.\begin{array}{c} As_2O_3 \\ As_2O_3 \end{array}\right\},$$

analogous to the known compound,—

$$\left.\begin{array}{c} As_2O_3 \\ SO_3 \end{array}\right\}.$$

§ 4. — L'Unité de la matière. — Les multiples de l'hydrogène et les éléments polymères.

Assurément, cette notion de l'existence définitive et immuable de soixante-six éléments distincts, tels que nous les admettons aujourd'hui, ne serait jamais venue à l'idée d'un philosophe ancien ; ou bien il l'eût rejetée aussitôt comme ridicule : il a fallu qu'elle s'imposât à nous, par la force inéluctable de la méthode expérimentale. Est-ce à dire cependant que telle soit la limite définitive de nos conceptions et de nos espérances ? Non, sans doute : en réalité, cette limite n'a jamais été acceptée par les chimistes que comme un fait actuel, qu'ils ont toujours conservé l'espoir de dépasser.

De longs travaux ont été entrepris à cet égard, soit pour ramener tous les équivalents des corps simples à une même série de valeurs numériques, dont ils seraient les multiples ; soit pour les grouper en familles naturelles ; soit pour les distribuer dans celles-ci, suivant des progressions arithmétiques.

Aujourd'hui même, les uns, s'attachant à la conception atomique, regardent nos corps prétendus simples comme formés par l'association d'un certain nombre d'éléments analogues ; peut-être comme engendrés par la condensation d'un seul d'entre eux, l'hydrogène par exemple, celui dont le poids atomique est le plus petit de tous.

On sait en effet que les corps simples sont caractérisés chacun par un nombre fondamental, que l'on appelle son *équivalent* ou son *poids atomique*. Ce nombre représente la masse chimique de l'élément, le poids invariable sous lequel il entre en combinaison et s'associe aux autres éléments, parfois d'après des proportions multiples. C'est ce poids constant qui passe de composé en composé, dans les substitutions, décompositions et réactions diverses, sans éprouver jamais la plus

19

petite variation. La combinaison ne s'opère donc pas suivant une progression continue, mais suivant des rapports entiers, multiples les uns des autres, et qui varient par sauts brusques. De là, pour chaque élément, l'idée d'une molécule déterminée, caractérisée par son poids, et peut-être aussi par sa forme géométrique. Cette molécule demeurant indestructible, au moins dans toutes les expériences accomplies jusqu'ici, elle a pu être regardée comme identique avec l'atome de Démocrite et d'Epicure. Telle est la base de la théorie atomique de notre temps.

Ainsi chaque corps simple serait constitué par un atome spécial, par une certaine particule matérielle insécable. Les forces physiques, aussi bien que les forces chimiques, ne sauraient faire éprouver à cet atome que des mouvements d'ensemble, sans possibilité de vibrations internes; celles-ci ne pouvant exister que dans un système formé de plusieurs parties. Il en résulte encore qu'il ne peut y avoir dans l'intérieur d'un atome indivisible aucune réserve d'énergie immanente.

Telles sont les conséquences rigoureuses de la théorie atomique. Je me borne à les exposer et je n'ai pas à discuter ici si ces conséquences ne dépassent pas les prémisses, les faits positifs qui leur servent de base; c'est-à-dire si les faits autorisent à conclure non seulement à l'existence de certaines masses moléculaires déterminées, caractéristiques des corps simples, et que tous les chimistes admettent; mais aussi à attribuer à ces molécules le nom et les propriétés des atomes absolus, comme le font un certain nombre de savants.

Ces réserves sont d'autant plus opportunes que les

partisans modernes de la théorie atomique l'ont pres-
que aussitôt répudiée dans les interprétations qu'ils
ont données de la constitution des corps simples :
interprétations aussi hypothétiques d'ailleurs que
l'existence même des atomes absolus, mais qui attestent
l'effort continu de l'esprit humain pour aller au delà de
toute explication démontrée des phénomènes, aussitôt
qu'une semblable explication a été atteinte, et pour
s'élancer plus loin vers des imaginations nouvelles.

Retraçons cette histoire : s'il ne s'agit plus d'une
doctrine positive, cependant l'exposé que nous allons
faire offre l'intérêt qui s'attache aux conceptions
par lesquelles l'intelligence essaie de représenter le
système général de la nature. Nous retrouvons ici
des vues analogues à celles des Pythagoriciens, alors
qu'ils prétendaient enchaîner dans un même système
les propriétés réelles des êtres et les propriétés mysté-
rieuses des nombres.

Le premier et principal effort qui ait été tenté dans
cette voie, consiste à ramener les équivalents ou poids
atomiques de tous les éléments à une même unité fondamentale. C'est là une conception *a priori*, qui a donné
lieu à une multitude d'expériences, destinées à la
vérifier. Si le fruit théorique à ce point de vue en a été
minime, sinon même négatif ; en pratique, du moins,
ces travaux ont eu un résultat scientifique très utile :
ils ont fixé avec une extrême précision les équivalents
réels de nos éléments ; c'est-à-dire, je le répète, les
poids exacts suivant lesquels les éléments entrent en
combinaison et se substituent les uns aux autres.

Prout, chimiste anglais, avait proposé tout d'abord

de prendre le poids même de l'un de nos éléments, celui de l'hydrogène, comme unité ; dans la supposition que les poids atomiques de tous les autres corps simples en étaient des multiples. Cette hypothèse, embrassée et soutenue pendant quelque temps par M. Dumas, réduit toute la théorie à une extrême simplicité. En effet, tous les corps simples seraient dès lors constitués par les arrangements divers de l'atome du plus léger d'entre eux. Malheureusement, elle n'a pas résisté au contrôle expérimental, c'est-à-dire à la détermination exacte, par analyse et par synthèse, des poids atomiques vrais de nos corps simples. Cette détermination a fourni, à côté de quelques poids atomiques à peu près identiques avec les multiples de l'hydrogène, une multitude d'autres nombres intermédiaires.

Mais dans les conceptions théoriques, pas plus que dans la vie pratique, l'homme ne renonce pas facilement à ses espérances. Pour soutenir la supposition de Prout, ses partisans ont essayé d'abord de réduire à moitié, puis au quart, l'unité fondamentale.

Or, à ce terme, une objection se présente : c'est que les vérifications concluantes deviennent impossibles. En effet, nos expériences n'ont pas, quoi que nous fassions, une précision absolue ; et il est clair que toute conjecture numérique serait acceptable, si l'on plaçait l'unité commune des poids atomiques au delà de la limite des erreurs que nous ne pouvons éviter.

Ce n'est pas tout d'ailleurs ; le fond même du système est atteint par cette supposition. La réduction du nombre fondamental, au-dessous d'une unité égale au poids atomique de l'hydrogène, enlève à la théorie ce caractère précis et séduisant, en vertu duquel tous les éléments étaient regardés comme formés en définitive par de l'hydrogène plus ou moins condensé. Il faudrait reculer dans l'inconnu jusqu'à un élément nouveau, quatre fois plus léger, élément inconnu qui formerait par sa condensation l'hydrogène lui-même.

Encore cela ne suffit-il pas pour représenter rigoureusement les expériences. En effet, M. Stas, par des études d'une exactitude incomparable, a montré que le système réduit à ces termes, c'est-à-dire réduit à prendre comme unité un sous-multiple peu élevé du poids de l'hydrogène, le système, dis-je, ne peut être défendu. Les observations extrêmement précises qu'il a exécutées ont prouvé sans réplique que les poids atomiques des éléments ne sont pas exprimés par des nombres simples, c'est-à-dire liés entre eux par des rapports entiers rigoureusement définis. La théorie des multiples de l'hydrogène n'est donc pas soutenable, dans son sens strict et rigoureux.

Gardons-nous cependant d'une négation trop absolue. Si l'hypothèse qui admet les équivalents des éléments multiples les uns des autres ne peut pas être affirmée d'une façon absolue, cependant cette hypothèse a pour elle des observations singulières et qui réclament, en tout état de cause, une interprétation. A cet égard les faits que je vais citer donnent à réfléchir.

ENGLISH SUMMARY OF BERTHELOT'S PAPER

Berthelot was an advocate of the study of chemical energetics, and as such was, like Wilhelm Ostwald, an opponent of atomic theories. He was also opposed to *a priori* system-making; and because he saw Proutian speculation as both atomic and *a priori*, and the Periodic Table as the acme of Proutian speculation, he believed that it should be opposed and ridiculed. It is odd that one so tough-minded should have become an historian of alchemy; there must be an attraction of opposites.

Berthelot begins here by remarking that no ancient philosopher would have believed in 66 immutable elements, but would have laughed at such a notion. We are only driven to such a belief by the power of experimental method; but, in fact, all chemists hope that they will be able to pass this barrier. This is demonstrated by the manner in which they try to show that atomic weights are multiples of some basic element; to group elements in natural families; and to arrange them in arithmetical series.

Atomists regard the elements as composed of groups of atoms of one or more real elements. That each element is characterised by a fixed equivalent or atomic weight gave rise to the idea that they are composed of atoms like those of Democritus; but such atoms could not possess internal energy, and to introduce atomic hypotheses is a mistake. The quasi-Pythagorean views of Prout and Dumas, based upon the notion that the properties of real things depend upon the mysterious properties of numbers, had a positive effect only insofar as they led to more accurate determinations of atomic weights and, as modified by chemists like Marignac, such views as well as being *apriori* have become unverifiable.

THE CHEMICAL NEWS.

VOLUME LIV.

EDITED BY WILLIAM CROOKES, F.R.S., &c.

No. 1388.—JULY 2, 1886.

NOTE ON A METHOD OF ILLUSTRATING THE PERIODIC LAW.

By J. EMERSON REYNOLDS, M.D., F.R.S.
Professor of Chemistry, University of Dublin.

THE annexed woodcut is a reproduction of a diagram which has been used in my lecture-room for some years in order to illustrate the periodic character of the relation between the atomic weights and properties of the chemical elements. It was not intended for publication, but some scientific friends considered it likely to prove useful and suggestive outside the limits of a University theatre: hence this note.

The curve represented is intended to give a special picture of the general relations of the elements, and its foundation is Mendeleeff's well-known tabular classification in which he applies and greatly extends the important periodic principle recognised by Newland.

Every chemist is familiar with the annexed table, but it is given to save the trouble of reference elsewhere.

The details of this classification are so well known through the numerous and important papers of Dr. Carnelley and others that it is unnecessary to discuss them here. It will suffice to point out that the chief features are—

1st. The arrangement of the elements according to atomic weight in twelve "series," or "periods" of seven (or in some cases ten) members each. These are found on the same horizontal line.

2nd. The division of the elements which occupy similar positions in the respective periods into "groups" of analogous bodies. These are found in the same vertical column, and there are eight such groups, as the exceptional members of Mendeleeff's "long" or ten-member periods form the eighth group.

3rd. The distinction of the periods into two classes, "odd" and "even," since the corresponding members of *alternate* periods are found to resemble each other more closely than those they immediately succeed or precede.

When we carefully consider the relations of the members of the "periods" it is found that the general properties of the elements vary from one to another with tolerable regularity until the seventh member is reached, which latter is in more or less direct *contrast* with the first element of the same period, and with that of the next. Thus chlorine—which is the seventh member of Mendeleeff's third period—is in chemical contrast with sodium, the first member of the same series, and also with potassium, which is the first member of the next series. On the other hand, sodium and potassium are analogous forms of matter. Now the six elements whose atomic weights lie between those of sodium and potassium vary in properties from one member to another until the sodium contrast—chlorine—is reached ; but from chlorine to the sodium analogue—potassium—the change in properties appears to take place *per saltum*. Similar alternations of gradual and abrupt changes in properties are observed as the atomic weights increase.

The real distinction just pointed out seems to have attracted singularly little attention in the construction of the curves or spirals which are intended to illustrate the relations of the elements in accordance with the periodic principle, yet its recognition is essential if we are to form an adequate mental picture of the relations in question : the particular direction it gives to our conceptions will presently appear.

If the relations of the elements within each period are now considered with a view to graphic translation, it is evident that the recognition of more or less contrast in properties between the first and last members of each series implies the existence of a position of mean variation or transition within each system.

On further comparing the members of the periods, it is found that the *fourth* element of each series in general possesses the properties which a transition member might be expected to exhibit. Thus silicon may be represented at the apex of a tolerably symmetrical curve which should represent for the particular period the direction in which the properties of the series of elements vary with rising atomic weight.

A physical analogy will help to make the meaning of this clear :—

Let the line A B (Fig. 1.) represent part of a string in tension, and *a*, *b*, *c*, *d*, *e*, *f*, and *g* seven knots upon it. The string is now thrown into a number of vibrating segments: *o* and *o'* represent two nodal points, between which one segmental vibration takes place. The several knots oscillate rapidly to and fro in the direction of the dotted lines, *a* moving from the position of rest to *a'* and back again, when it swings to the same extent on the opposite side of the line A B, returns, and starts afresh. Each knot performs similar journeys, but the lengths of the paths vary: thus the length increases from *a* to *b*, from *b* to *c*, and from *c* to *d*, while it diminishes from *d* to *e*, *e* to *f*, and *f* to *g*. The knot *d* is therefore exceptional, in that it suffers the maximum displacement from the mean position ; the knots *c* and *e* perform journeys of comparable lengths, but they are otherwise in more or less direct contrast ; similarly *b* and *f*, *e* and *g*, form contrasted pairs.

Let the knots in the string represent the atomic groupings we call elements, arranged in the order of

Series or Period.	Group I. R_2O	Group II. RO	Group III. R_2O_3	Group IV. RH_4 RO_2	Group V. RH_3 R_2O_5	Group VI. RH_2 RO_3	Group VII. RH R_2O_7	Group VIII. RO_4
1	H=1							
2	Li=7	Be=9	B=11	C=12	N=14	O=16	F=19	
3	Na=23	Mg=24	Al=27	Si=28	P=31	S=32	Cl=35.5	
4	K=39	Ca=40	Sc=44	Ti=48	V=51	Cr=52	Mn=55	Fe=56, Ni=58.5, Co=59
5	Cu=63	Zn=65	Ga=69	—=72	As=75	Se=79	Br=80	
6	Rb=85	Sr=87	Y=89	Zr=90	Nb=94	Mo=96	—=100	Rh=104, Ru=104.5, Pd=106
7	Ag=108	Cd=112	In=113	Sn=118	Sb=121	Te=125	I=127	
8	Cs=133	Ba=137	La—139	Ce=141	Di=144			
9	(—)				Er=166			
10					Ta=182	W=184		
11	Au=196	Hg=200	Tl=204	Pb=207	Bi=208			Ir=192.5, Os=193, Pt=194.5
12				Th=233		U=240		

atomic weight* rising from a to g, and we have a picture of a "period" of seven elements regarded as a vibrating system. Unlike most analogies, this one bears tolerably close examination.

Thus, if we take any well-defined periods, such as the two following,—

2.—Li=7, Be=9, B=11, C=12, N=14, O=16, F=19
3.—Na=23, Mg=24, Al=27, Si=28, P=31, S=32, Cl=35.5

we find that the *fourth* member naturally divides the particular period in two parts. In the period 2 carbon is the middle term, and in 3 silicon. Corresponding elements to these can be found in nearly all the succeeding periods,

* Of course this arrangement is not necessarily symmetrical, as supposed in the case of the knotted string.

and their properties in relation to those of the members of the same period which immediately precede and follow them are on the whole exceptional. They afford strongly marked dioxides whose hydrates exhibit very feeble acid properties, and in all the best known periods they appear as transition elements between two sub-groups of three each. I propose to distinguish such bodies as *meso-elements* to avoid needless repetition. Now the meso-element of a period is analogous to the knot d when at its maximum displacement.

If we examine a particular period—for instance, that one whose meso-element is silicon—we note:—*First*, that the three elements of lower atomic weight than silicon, viz., sodium, magnesium, and aluminium, are distinctly *electro-positive* in character, while those of higher atomic weight, viz., phosphorus, sulphur, and chlorine, are as distinctly *electro-negative*. Throughout the best-known periods this remarkable subdivision is observable, although, as might be anticipated, the differences become less strongly marked as the atomic weights increase. *Secondly*, that the members above and below the meso-element fall into pairs of elements, which, while exhibiting certain analogies, are generally in more or less direct chemical contrast. Thus, in the silicon period we have—

$$\text{Si iv}$$
$$+\text{Al}''' \qquad \text{P}''' -$$
$$+\text{Mg}'' \qquad \text{S}'' -$$
$$+\text{Na}' \qquad \text{Cl}' -$$

This division also happens, in many cases, to coincide with some characteristic valence* of the contrasted elements. It is noteworthy, however, that the members on the electro-negative side exhibit the most marked tendency to variation in atom-fixing power, so that valence alone is an untrustworthy guide to the probable position of an element in a period.

The pairs of more or less contrasted elements may be likened to the pairs of knots on the string whose paths of vibration are of approximately equal length; but it is convenient for the purpose of graphic illustration to assume that the paths of each pair are of the same length, or that the displacements are in the ratio of $1:2:3:4$,—that of d', Fig. 1, which is the longest. On reference to the diagram, Fig. 2, the nature of this arrangement will be evident, and the portions when connected as shown are seen to form an expanding curve such as would be afforded by a string or chain† whose parts are in unequal tension. With the aid of the scale A B there is no difficulty in picturing the elements in the positions of the knots on the string, and so regarding them as members of a vibrating system. All the elements whose constants are well known find places on the curve.‡

Thus the physical analogy helps us to form some conception of the relations of the members of the periods, and of the latter to one another. Moreover, the admission of the periodic principle at all seems to require the recognition of similar relations to those indicated.

It will now be interesting to note some of the points suggested by the foregoing considerations:—

1. If we picture the elements as members of some such vibrating system as supposed, difficulties of the kind noted in the case of the relations of chlorine to potassium disappear, since the potassium knot is in a complementary position to that of chlorine, and is in a certain sense in direct contrast.

2. On Mendeleeff's system hydrogen is the solitary representative of his first period, and it seems to have

* This term is here used in the wide sense in which it was employed by Wurtz.
† My friend Prof. Fitzgerald, F.R.S., suggests that a vibrating metallic chain, suspended from the ceiling and attached to the floor, would afford a more complete picture, as the regular and considerable changes of tension, due to the increasing weight, would lead to the production of regularly expanding loops.
‡ The length of the page of the CHEMICAL NEWS prevents the representation of thorium (233) and uranium (240), but they fall naturally into position on an extension of the curve.

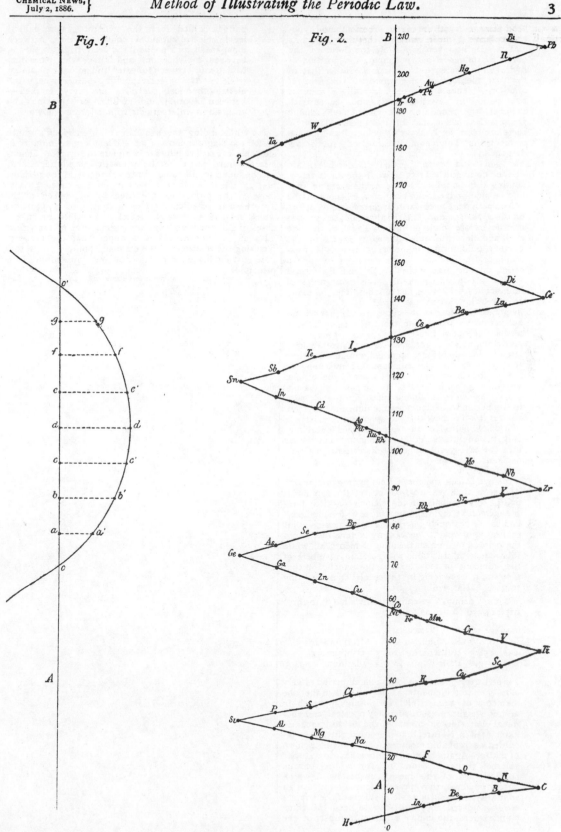

Fig.1.

Fig.2.

been assumed that the other six members (hypothetical) must have atomic weights between 1 and 7. If, however, we allow the form of the curve to influence our judgment, the position of hydrogen in reference to lithium is rather that of the *last* member of one period to the first of another. Hence, if hydrogen is really a member of a period of seven elements, and the seventh member, the remaining six must have atomic weights less than that of hydrogen. There is therefore room for Lockyer's "helium," and for the element that Dumas' modification of Prout's law demands.

3. The "odd" and "even" periods of Mendeleeff are at once distinguished: thus, while the knots representing the carbon period are at the extreme limit on one side of the mean position A B, those of the silicon period are necessarily at corresponding points on the *opposite* side. The strictly analogous periods are therefore found on the *same* side of the line A B, and elements found at *similar points* on the curve should form a homologous series. For example: Mg, Zn, and Cd are members of such a series on one side, while Ca, Sr, and Ba form a similar group on the other; the two groups of homologues exhibiting, at the same time, the many cross relations we should expect.

4. The general contour of the curve is such that we are not permitted to assume the existence of Mendeleeff's *ninth* period (see table *supra*). This is remarkable, as six out of these seven elements postulated by Mendeleeff are admittedly unknown, while the single known element supposed to belong to the period is erbium. Now, even if the atomic weight of erbium be 166, as supposed, it is obviously possible to find a place for the element in the period which includes tantalum and tungsten, so that it is not necessary to assume the existence of a whole period in order to find a place for erbium. But it is doubtful whether the constants of the *pure* element are yet sufficiently known to justify a decision between the two points that can be selected for it.

5. As the eye traces the curve from hydrogen upwards it is seen that near to *three* out of the ten nodal points the symbols of the elements which form Mendeleeff's "eighth group" are to be found. These bodies are obviously *interperiodic*, in the sense that their atomic weights necessarily exclude them from the periods into which the other elements fall, while their chemical relations with certain members of the adjacent periods lead to the conclusion that they are interperiodic in the special sense of being transitional as well.

This "group," so-called, is divisible into three triplets, which are as follow:—

First.	Second.	Third.
Iron = 56.	{ Rhodium = 104.	{ Iridium = 192·5.
{ Nickel = 58·5	{ Ruthenium = 104·5.	{ Osmium = 193.
{ Cobalt = 59.	Palladium = 106.	Platinum = 194·5.

Each triplet seems to form a small period of very closely related elements. It is well known that the members of each triplet so closely approach in atomic weight, present so many physical and chemical characters in common, and are so frequently associated in Nature, that they have often been regarded as probable modifications of a single form of matter. Again, all nine members have low atomic volumes (between 6 and 10), resembling in this respect most of the meso-elements, with which latter, however, they present few other points of similarity. When the scale positions of the triplets are considered they are seen to occur at three points where the known adjacent members of the periods which form the general system exhibit feeble electro chemical contrasts—thus, between manganese and copper, the iron triplet appears, between molybdenum and silver the palladium triplet, and between tungsten and gold the platinum triplet. At the corresponding points where the electro-chemical contrasts are very marked—between fluorine and sodium or chlorine and potassium—interperiodic triplets are unknown.

Notwithstanding the exclusion of Mendeleeff's ninth period the diagram shows that a considerable number of elements are still required to complete the system. Doubtless the rare earth metals, when thoroughly identified, will be found to fill some of the vacancies in the periods, just as the new element, germanium, seems to possess most of the properties, predicted by Mendeleeff, of the hypothetical ekasilicon. The discovery of a principle which renders successful prediction possible, as in the cases of gallium, scandium, and germanium, marks a great advance in science, and it is hoped that the mode of picturing the relations of the elements pointed out in this note may facilitate the recognition and application of the principle.

LXIII.—*The Periodic Law of the Chemical Elements.*

By Professor MENDELÉEFF.

(FARADAY LECTURE delivered before the Fellows of the Chemical Society in the
Theatre of the Royal Institution, on Tuesday, June 4th, 1889.)

THE high honour bestowed by the Chemical Society in inviting me
to pay a tribute to the world-famed name of Faraday by delivering
this lecture has induced me to take for its subject the Periodic Law
of the Elements—this being a generalisation in chemistry which has
of late attracted much attention.

While science is pursuing a steady onward movement, it is convenient from time to time to cast a glance back on the route already
traversed, and especially to consider the new conceptions which aim
at discovering the general meaning of the stock of facts accumulated
from day to day in our laboratories. Owing to the possession of
laboratories, modern science now bears a new character, quite unknown
not only to antiquity but even to the preceding century. Bacon's
and Descartes' idea of submitting the mechanism of science simultaneously to experiment and reasoning has been fully realised in the
case of chemistry, it having become not only possible but always
customary to experiment. Under the all-penetrating control of
experiment, a new theory, even if crude, is quickly strengthened,
provided it be founded on a sufficient basis ; the asperities are
removed, it is amended by degrees, and soon loses the phantom
light of a shadowy form or of one founded on mere prejudice ;
it is able to lead to logical conclusions and to submit to experimental proof. Willingly or not, in science we all must submit not

to what seems to us attractive from one point of view or from another, but to what represents an agreement between theory and experiment; in other words, to demonstrated generalisation and to the approved experiment. Is it long since many refused to accept the generalisations involved in the law of Avogadro and Ampère, so widely extended by Gerhardt? We still may hear the voices of its opponents; they enjoy perfect freedom, but vainly will their voices rise so long as they do not use the language of demonstrated facts. The striking observations with the spectroscope which have permitted us to analyse the chemical constitution of distant worlds, seemed, at first, applicable to the task of determining the nature of the atoms themselves; but the working out of the idea in the laboratory soon demonstrated that the characters of spectra are determined—not directly by the atoms, but by the molecules into which the atoms are packed; and so it became evident that more verified facts must be collected before it will be possible to formulate new generalisations capable of taking their place beside those ordinary ones based upon the conception of simple bodies and atoms. But as the shade of the leaves and roots of living plants, together with the relics of a decayed vegetation, favour the growth of the seedling and serve to promote its luxurious development, in like manner sound generalisations— together with the relics of those which have proved to be untenable —promote scientific productivity, and ensure the luxurious growth of science under the influence of rays emanating from the centres of scientific energy. Such centres are scientific associations and societies. Before one of the oldest and most powerful of these I am about to take the liberty of passing in review the 20 years' life of a generalisation which is known under the name of the Periodic Law. It was in March, 1869, that I ventured to lay before the then youthful Russian Chemical Society the ideas upon the same subject, which I had expressed in my just written "Principles of Chemistry."

Without entering into details, I will give the conclusions I then arrived at, in the very words I used:—

"1. The elements, if arranged according to their atomic weights, exhibit an evident *periodicity* of properties.

"2. Elements which are similar as regards their chemical properties have atomic weights which are either of nearly the same value (*e.g.*, platinum, iridium, osmium) or which increase regularly (*e.g.*, potassium, rubidium, cæsium).

"3. The arrangement of the elements, or of groups of elements in the order of their atomic weights corresponds to their so-called *valencies* as well as, to some extent, to their distinctive chemical properties—as is apparent among other series in that of lithium, beryllium, barium, carbon, nitrogen, oxygen and iron.

2 Y 2

"4. The elements which are the most widely diffused have *small* atomic weights.

"5. The *magnitude* of the atomic weight determines the character of the element just as the magnitude of the molecule determines the character of a compound body.

"6. We must expect the discovery of many yet *unknown* elements, for example, elements analogous to aluminium and silicon, whose atomic weight would be between 65 and 75.

"7. The atomic weight of an element may sometimes be amended by a knowledge of those of the contiguous elements. Thus, the atomic weight of tellurium must lie between 123 and 126, and cannot be 128.

"8. Certain characteristic properties of the elements can be foretold from their atomic weights.

"The aim of this communication will be fully attained if I succeed in drawing the attention of investigators to those relations which exist between the atomic weights of dissimilar elements, which, as far as I know, have hitherto been almost completely neglected. I believe that the solution of some of the most important problems of our science lies in researches of this kind."

To-day, 20 years after the above conclusions were formulated, they may still be considered as expressing the essence of the now well-known periodic law.

Reverting to the epoch terminating with the sixties, it is proper to indicate three series of data without the knowledge of which the periodic law could not have been discovered, and which rendered its appearance natural and intelligible.

In the first place, it was at that time that the numerical value of atomic weights became definitely known. Ten years earlier such knowledge did not exist, as may be gathered from the fact that in 1860 chemists from all parts of the world met at Karlsruhe in order to come to some agreement, if not with respect to views relating to atoms, at any rate as regards their definite representation. Many of those present probably remember how vain were the hopes of coming to an understanding, and how much ground was gained at that Congress by the followers of the unitary theory so brilliantly represented by Cannizzaro. I vividly remember the impression produced by his speeches, which admitted of no compromise, and seemed to advocate truth itself, based on the conceptions of Avogadro, Gerhardt and Regnault, which at that time were far from being generally recognised. And though no understanding could be arrived at, yet the objects of the meeting were attained, for the ideas of Cannizzaro proved, after a few years, to be the only ones which could stand criticism, and which represented an atom as—"the smallest portion of an element which enters into a molecule of its compound." Only such real atomic weights—not conventional ones —could afford a basis for generalisation. It is sufficient, by way of example, to indicate the following cases in which the relation is seen at once and is perfectly clear :—

$$K = 39 \qquad Rb = 85 \qquad Cs = 133$$
$$Ca = 40 \qquad Sr = 87 \qquad Ba = 137$$

whereas with the equivalents then in use—

$$K = 39 \qquad Rb = 85 \qquad Cs = 133$$
$$Ca = 20 \qquad Sr = 43{\cdot}5 \qquad Ba = 68{\cdot}5$$

the consecutiveness of change in atomic weight, which with the true values is so evident, completely disappears.

Secondly, it had become evident during the period 1860–70, and even during the preceding decade, that the relations between the atomic weights of analogous elements were governed by some general and simple laws. Cooke, Cremers, Gladstone, Gmelin, Lenssen, Pettenkofer, and especially Dumas, had already established many facts bearing on that view. Thus Dumas compared the following groups of analogous elements with organic radicles—

Diff.		Diff.		Diff.		Diff.
	Mg = 12		P = 31		O = 8	
Li = 7 ⎱16	Ca = 20 ⎰8	As = 75 ⎱44	S = 16 ⎰8
Na = 23 ⎰	Sr = 44 ⎰3 × 8	Sb = 119 ⎰44	Se = 40 ⎰3 × 8
K = 39 ⎰16	Ba = 68 ⎰3 × 8	Bi = 207 ⎰2 × 44	Te = 64 ⎰3 × 8

and pointed out some really striking relationships, such as the following :—

$$F = 19.$$
$$Cl = 35{\cdot}5 = 19 + 16{\cdot}5.$$
$$Br = 80 = 19 + 2 \times 16{\cdot}5 + 28.$$
$$I = 127 = 2 \times 19 + 2 \times 16{\cdot}5 + 2 \times 28.$$

A. Strecker, in his work "Theorien und Experimente zur Bestimmung der Atomgewichte der Elemente" (Braunschweig, 1859), after summarising the data relating to the subject, and pointing out the remarkable series of equivalents—

$$Cr = 26{\cdot}2 \qquad Mn = 27{\cdot}6 \qquad Fe = 28 \qquad Ni = 29 \qquad Co = 30$$
$$Cu = 31{\cdot}7 \qquad Zn = 32{\cdot}5$$

remarks that : "It is hardly probable that all the above-mentioned relations between the atomic weights (or equivalents) of chemically analogous elements are merely accidental. We must, however, leave to the future the discovery of the *law* of the relations which appears in these figures." *

In such attempts at arrangement and in such views are to be recognised the real forerunners of the periodic law; the ground was prepared for it between 1860 and 1870, and that it was not expressed in a determinate form before the end of the decade, may, I suppose, be ascribed to the fact that only analogous elements had been compared. The idea of seeking for a relation between the atomic weights of all the elements was foreign to the ideas then current, so that neither the *vis tellurique* of De Chancourtois, nor the *law of octaves* of Newlands, could secure anybody's attention. And yet both De Chancourtois and Newlands, like Dumas and Strecker, more than Lenssen and Pettenkofer, had made an approach to the periodic law and had discovered its germs. The solution of the problem advanced but slowly, because the facts, and not the law, stood foremost in all attempts; and the law could not awaken a general interest so long as elements, having no apparent connection with each other, were included in the same octave, as for example :—

1st octave of Newlands..	H	F	Cl	Co & Ni	Br	Pd	I	Pt & Ir
7th Ditto....	O	S	Fe	Se	Rh & Ru	Te	Au	Os or Th

Analogies of the above order seemed quite accidental, and the more so as the octave contained occasionally 10 elements instead of eight, and when two such elements as Ba and V, Co and Ni, or Rh and Ru, occupied one place in the octave.† Nevertheless, the fruit was ripening, and I now see clearly that Strecker, De Chancourtois and Newlands stood foremost in the way towards the discovery of the periodic law, and that they merely wanted the boldness necessary to place the whole question at such a height that its reflection on the facts could be clearly seen.

A third circumstance which revealed the periodicity of chemical elements was the accumulation, by the end of the sixties, of new information respecting the rare elements, disclosing their many-sided relations to the other elements and to each other. The

* "Es ist wohl kaum anzunehmen, dass alle im Vorhergehenden hervorgehobenen Beziehungen zwischen den Atomgewichten (oder Aequivalenten) in chemischen Verhältnissen einander ähnliche Elemente bloss zufällig sind. Die Auffindung der in diesen Zahlen *gesetzlichen* Beziehungen müssen wir jedoch der Zukunft überlassen."

† To judge from J. A. R. Newlands' work, *On the Discovery of the Periodic Law*, London, 1884, p. 149; "On the Law of Octaves" (from the *Chemical News*, 12, 83, August 18, 1865).

researches of Marignac on niobium, and those of Roscoe on vanadium were of special moment. The striking analogies between vanadium and phosphorus on the one hand, and between vanadium and chromium on the other, which became so apparent in the investigations connected with that element, naturally induced the comparison of $V = 51$ with $Cr = 52$, $Nb = 94$ with $Mo = 96$, and $Ta = 192$ with $W = 194$; while, on the other hand, $P = 31$ could be compared with $S = 32$, $As = 75$ with $Se = 79$, and $Sb = 120$ with $Te = 125$. From such approximations there remained but one step to the discovery of the law of periodicity.

The law of periodicity was thus a direct outcome of the stock of generalisations and established facts which had accumulated by the end of the decade 1860—1870: it is an embodiment of those data in a more or less systematic expression. Where, then, lies the secret of the special importance which has since been attached to the periodic law, and has raised it to the position of a generalisation which has already given to chemistry unexpected aid, and which promises to be far more fruitful in the future and to impress upon several branches of chemical research a peculiar and original stamp? The remaining part of my communication will be an attempt to answer this question.

In the first place we have the circumstance that, as soon as the law, made its appearance, it demanded a revision of many facts which were considered by chemists as fully established by existing experience. I shall return, later on, briefly to this subject, but I wish now to remind you that the periodic law, by insisting on the necessity for a revision of supposed facts, exposed itself at once to destruction in its very origin. Its first requirements, however, have been almost entirely satisfied during the last 20 years; the supposed facts have yielded to the law, thus proving that the law itself was a legitimate induction from the verified facts. But our inductions from data have often to do with such details of a science so rich in facts, that only generalisations which cover a wide range of important phenomena can attract general attention. What were the regions touched on by the periodic law? This is what we shall now consider.

The most important point to notice is, that periodic functions, used for the purpose of expressing changes which are dependent on variations of time and space, have been long known. They are familiar to the mind when we have to deal with motion in closed cycles, or with any kind of deviation from a stable position, such as occurs in pendulum-oscillations. A like periodic function became evident in the case of the elements, depending on the mass of the atom. The primary conception of the masses of bodies or of the masses of atoms belongs to a category which the present state of science forbids us to discuss, because as yet we have no means of dissecting or analysing the conception. All that was known of functions dependent on masses derived its origin from Galileo and Newton, and indicated that such functions either decrease or increase with the increase of mass, like the attraction of celestial bodies. The numerical expression of the phenomena was always found to be proportional to the mass, and in no case was an increase of mass followed by a recurrence of properties such as is disclosed by the periodic law of the elements. This constituted such a novelty in the study of the phenomena of nature that, although it did not lift the veil which conceals the true conception of mass, it nevertheless indicated that the explanation of that conception must be searched for in the masses of the atoms; the more so, as all masses are nothing but aggregations, or additions, of chemical atoms which would be best described as chemical individuals. Let me remark by the way that though the Latin word "individual" is merely a translation of the Greek word "atom," nevertheless history and custom have drawn so sharp a distinction between the two words, and the present chemical conception of atoms is nearer to that defined by the Latin word than by the Greek, although this latter also has acquired a special meaning which was unknown to the classics. The periodic law has shown that our chemical individuals display a harmonic periodicity of properties, dependent on their masses. Now, natural science has long been accustomed to deal with periodicities observed in nature, to seize them with the vice of mathematical analysis, to submit them to the rasp of experiment. And these instruments of scientific thought would surely, long since, have mastered the problem connected with the chemical elements, were it not for a new feature which was brought to light by the periodic law and which gave a peculiar and original character to the periodic function.

If we mark on an axis of abscissæ a series of lengths proportional to angles, and trace ordinates which are proportional to sines or other trigonometrical functions, we get periodic curves of a harmonic character. So it might seem, at first sight, that with the increase of atomic weights the function of the properties of the elements should also vary in the same harmonious way. But in this case there is no such continuous change as in the curves just referred to, because the periods do not contain the infinite number of points constituting a curve, but a *finite* number only of such points. An example will better illustrate this view. The atomic weights—

$$Ag = 108 \qquad Cd = 112 \qquad In = 113 \qquad Sn = 118$$
$$Sb = 120 \qquad Te = 125 \qquad I = 127$$

steadily increase, and their increase is accompanied by a modification of many properties which constitutes the essence of the periodic law.

Thus, for example, the densities of the above elements decrease steadily, being respectively—

| 10·5 | 8·6 | 7·4 | 7·2 | 6·7 | 6·4 | 4·9 |

while their oxides contain an increasing quantity of oxygen :—

| Ag_2O | Cd_2O_2 | In_2O_3 | Sn_2O_4 | Sb_2O_5 | Te_2O_6 | I_2O_7 |

But to connect by a curve the summits of the ordinates expressing any of these properties would involve the rejection of Dalton's law of multiple proportions.   Not only are there no intermediate elements between silver, which gives AgCl, and cadmium, which gives $CdCl_2$, but, according to the very essence of the periodic law there can be none; in fact a uniform curve would be inapplicable in such a case, as it would lead us to expect elements possessed of special properties at any point of the curve.   The periods of the elements have thus a character very different from those which are so simply represented by geometers.   They correspond to points, to numbers, to sudden changes of the masses, and not to a continuous evolution.   In these sudden changes destitute of intermediate steps or positions, in the absence of elements intermediate between, say, silver and cadmium, or aluminium and silicon, we must recognise a problem to which no direct application of the analysis of the infinitely small can be made. Therefore, neither the trigonometrical functions proposed by Ridberg and Flavitzky, nor the pendulum-oscillations suggested by Crookes, nor the cubical curves of the Rev. Mr. Haughton, which have been proposed for expressing the periodic law, from the nature of the case, can represent the periods of the chemical elements.   If geometrical analysis is to be applied to this subject it will require to be modified in a special manner.   It must find the means of representing in a special way not only such long periods as that comprising,

K Ca Sc Ti V Cr Mn Fe Co Ni Cu Zn Ga G
As Se Br,

but short periods like the following :—

Na   Mg   Al   Si   P   S   Cl.

In the theory of numbers only do we find problems analogous to ours, and two attempts at expressing the atomic weights of the elements by algebraic formulæ seem to be deserving of attention, although neither of them can be considered as a complete theory, nor as promising finally to solve the problem of the periodic law.   The attempt of E. J. Mills (1886) does not even aspire to attain this end. He considers that all atomic weights can be expressed by a logarithmic function,

$$15(n - 0·9375^t),$$

in which the variables n and t are *whole numbers*. Thus, for oxygen, $n = 2$, and $t = 1$, whence its atomic weight is $= 15\cdot94$; in the case of chlorine, bromine, and iodine, n has respective values of 3, 6, and 9, while $t = 7$, 6, and 9; in the case of potassium, rubidium, and cæsium, $n = 4$, 6, and 9, and $t = 14$, 18, and 20.

Another attempt was made in 1888 by B. N. Tchitchérin. Its author places the problem of the periodic law in the first rank, but as yet he has investigated the alkaline metals only. Tchitchérin first noticed the simple relations existing between the atomic volumes of all alkaline metals; they can be expressed, according to his views, by the formula

$$A(2 - 0\cdot00535An),$$

where A is the atomic weight, and n is equal to 8 for lithium and sodium, to 4 for potassium, to 3 for rubidium, and to 2 for cæsium. If n remained equal to 8, during the increase of A, then the volume would become zero at $A = 46\frac{2}{3}$, and it would reach its maximum at $A = 23\frac{1}{3}$. The close approximation of the number $46\frac{2}{3}$ to the differences between the atomic weights of analogous elements (such as Cs − Rb, I − Br, and so on); the close correspondence of the number $23\frac{1}{3}$ to the atomic weight of sodium; the fact of n being necessarily a whole number, and several other aspects of the question, induce Tchitchérin to believe that they afford a clue to the understanding of the nature of the elements; we must, however, await the full development of his theory before pronouncing judgment on it. What we can at present only be certain of is this: that attempts like the two above named must be repeated and multiplied, because the periodic law has clearly shown that the masses of the atoms increase abruptly, by steps, which are clearly connected in some way with Dalton's law of multiple proportions; and because the periodicity of the elements finds expression in the transition from RX to RX_2, RX_3, RX_4, and so on till RX_8, at which point the energy of the combining forces being exhausted, the series begins anew from RX to RX_2, and so on.

While connecting by new bonds the theory of the chemical elements with Dalton's theory of multiple proportions, or atomic structure of bodies, the periodic law opened for natural philosophy a new and wide field for speculation. Kant said that there are in the world "two things which never cease to call for the admiration and reverence of man: the moral law within ourselves, and the stellar sky above us." But when we turn our thoughts towards the nature of the elements and the periodic law, we must add a third subject, namely, "the nature of the elementary individuals which we discover everywhere around us." Without them the stellar sky itself is inconceivable; and in the atoms we see at once their peculiar individualities, the infinite multiplicity of the individuals, and the submission of their seeming freedom to the general harmony of Nature.

Having thus indicated a new mystery of Nature, which does not yet yield to rational conception, the periodic law, together with the revelations of spectrum analysis, have contributed to again revive an old but remarkably long-lived hope—that of discovering, if not by experiment, at least, by a mental effort, the *primary matter*—which had its genesis in the minds of the Grecian philosophers, and has been transmitted, together with many other ideas of the classic period, to the heirs of their civilisation. Having grown, during the times of the alchemists up to the period when experimental proof was required, the idea has rendered good service; it induced those careful observations and experiments which later on called into being the works of Scheele, Lavoisier, Priestley and Cavendish. It then slumbered awhile, but was soon awakened by the attempts either to confirm or to refute the ideas of Prout as to the multiple proportion relationship of the atomic weights of all the elements. And once again the inductive or experimental method of studying Nature gained a direct advantage from the old Pythagorean idea: because atomic weights were determined with an accuracy formerly unknown. But again the idea could not stand the ordeal of experimental test, yet the prejudice remains and has not been uprooted, even by Stas; nay, it has gained a new vigour, for we see that all which is imperfectly worked out, new and unexplained, from the still scarcely studied rare metals to the hardly perceptible nebulæ, have been used to justify it. As soon as spectrum analysis appears as a new and powerful weapon of chemistry, the idea of a primary matter is immediately attached to it. From all sides we see attempts to constitute the imaginary substance *helium** the so much longed for primary matter. No attention is paid to the circumstance that the helium line is only seen in the spectrum of the solar protuberances, so that its universality in Nature remains as problematic as the primary matter itself; nor to the fact that the helium line is wanting amongst the Fraunhofer lines of the solar spectrum, and thus does not answer to the brilliant fundamental conception which gives its real force to spectrum analysis.

And finally, no notice is even taken of the indubitable fact that the brilliancies of the spectral lines of the simple bodies vary under different temperatures and pressures; so that all probabilities are in favour of the helium line simply belonging to some long since known element placed under such conditions of temperature, pressure, and gravity as have not yet been realised in our experiments. Again, the idea that the excellent investigations of Lockyer of the spectrum of

* That is, a body having a wave-length equal to 0·0005875 millimetre.

iron can be interpreted in favour of the compound nature of that element, evidently must have arisen from some misunderstanding. The spectrum of a compound body certainly does not appear as a sum of the spectra of its components; and therefore the observations of Lockyer can be considered precisely as a proof that iron undergoes no other changes at the temperature of the sun but those which it experiences in the voltaic arc—provided the spectrum of iron is preserved.   As to the shifting of some of the lines of the spectrum of iron while the other lines maintain their positions, it can be explained, as shown by M. Kleiber (*Journal of the Russian Chemical and Physical Society*, 1885, 147), by the relative motion of the various strata of the sun's atmosphere, and by Zöllner's laws of the relative brilliancies of different lines of the spectrum.   Moreover, it ought not to be forgotten that if iron were really proved to consist of two or more unknown elements, we simply should have an increase of the number of our elements—not a reduction, and still less a reduction of all of them to one single primary matter.

Feeling that spectrum analysis will not yield a support to the Pythagorean conception, its modern promoters are so bent upon its being confirmed by the periodic law, that the illustrious Berthelot, in his work *Les origines de l'Alchimie*, 1885, 313, has simply mixed up the fundamental idea of the law of periodicity with the ideas of Prout, the alchemists, and Democritus about primary matter.*   But the periodic law, based as it is on the solid and wholesome ground of experimental research, has been evolved independently of any conception as to the nature of the elements; it does not in the least originate in the idea of an unique matter; and it has no historical connection with that relic of the torments of classical thought, and therefore it affords no more indication of the unity of matter or of the compound character of our elements, than the law of Avogadro, or the law of specific heats, or even the conclusions of spectrum analysis.   None of the advocates of an unique matter have ever tried to explain the law from the standpoint of ideas taken from a remote antiquity when it was found convenient to admit the existence of many gods—and of an unique matter.

When we try to explain the origin of the idea of an unique primary matter, we easily trace that in the absence of inductions from experiment it derives its origin from the scientifically philosophical attempt at discovering some kind of unity in the immense diversity of individualities which we see around.   In classical times

* He maintains (on p. 309) that the periodic law requires two new analogous elements, having atomic weights of 48 and 64, occupying positions between sulphur and selenium, although nothing of the kind results from any of the different readings of the law.

such a tendency could only be satisfied by conceptions about the immaterial world. As to the material world, our ancestors were compelled to resort to some hypothesis, and they adopted the idea of unity in the formative material, because they were not able to evolve the conception of any other possible unity in order to connect the multifarious relations of matter. Responding to the same legitimate scientific tendency, natural science has discovered throughout the universe a unity of plan, a unity of forces, and a unity of matter, and the convincing conclusions of modern science compel everyone to admit these kinds of unity. But while we admit unity in many things, we none the less must also explain the individuality and the apparent diversity which we cannot fail to trace everywhere. It has been said of old, "Give a fulcrum, and it will become easy to displace the earth." So also we must say, "Give anything that is individualised, and the apparent diversity will be easily understood." Otherwise, how could unity result in a multitude?

After a long and painstaking research, natural science has discovered the individualities of the chemical elements, and therefore it is now capable not only of analysing, but also of synthesising; it can understand and grasp the general and unity, as well as the individualised and the multitudinous. Unity and the general, like time and space, like force and motion, vary uniformly; the uniform admit of interpolations, revealing every intermediate phase. But the multitudinous, the individualised—like ourselves, like the chemical elements, like the members of a peculiar periodic function of elements, like Dalton's multiple proportions—is characterised in another way: we see in it—side by side with a connecting general principle—leaps, breaks of continuity, points which escape from the analysis of the infinitely small—a complete absence of intermediate links. Chemistry has found an answer to the question as to the causes of multitudes; and while retaining the conception of many elements, all submitted to the discipline of a general law, it offers an escape from the Indian Nirvana—the absorption in the universal, replacing it by the individualised. However, the place for individuality is so limited by the all-grasping, all-powerful universal, that it is merely a fulcrum for the understanding of multitude in unity.

Having touched upon the metaphysical bases of the conception of an unique matter which is supposed to enter into the composition of all bodies, I think it necessary to dwell upon another theory, akin to the above conception,—the theory of the compound character of the elements now admitted by some,—and especially upon one particular circumstance which being related to the periodic law is considered to be an argument in favour of that hypothesis.

Dr. Pelopidas, in 1883, made a communication to the Russian Chemical and Physical Society on the periodicity of the hydrocarbon radicles, pointing out the remarkable parallelism which was to be noticed in the change of properties of hydrocarbon radicles and elements when classed in groups. Professor Carnelley, in 1886, developed a similar parallelism. The idea of M. Pelopidas will be easily understood if we consider the series of hydrocarbon radicles which contain, say, 6 atoms of carbon :—

I	II	III	IV	V	VI	VII	VIII
C_6H_{13}	C_6H_{12}	C_6H_{11}	C_6H_{10}	C_6H_9	C_6H_8	C_6H_7	C_6H_6

The first of these radicles, like the elements of the Ist group, combines with Cl, OH, and so on, and gives the derivatives of hexyl alcohol, $C_6H_{13}(OH)$; but, in proportion as the number of hydrogen atoms decreases, the capacity of the radicles of combining with, say, the halogens increases. C_6H_{12} already combines with 2 atoms of chlorine; C_6H_{11} with 3 atoms, and so on. The last members of the series comprise the radicles of acids; thus C_6H_8, which belongs to the VIth group, gives, like sulphur, a bibasic acid, $C_6H_8O_2(OH)_2$, which is homologous with oxalic acid. The parallelism can be traced still further—because C_6H_5 appears as a monovalent radicle of benzene—and with it begins a new series of aromatic derivatives, so analogous to the derivatives of the fat series. Let me also mention another example from among those which have been given by M. Pelopidas. Starting from the alkaline radicle of monomethylammonium, $N(CH_3)H_3$, or NCH_6, which presents many analogies with the alkaline metals of the Ist group, he arrives, by successively diminishing the number of the atoms of hydrogen, at a seventh group which contains cyanogen, CN, which has long since been compared to the halogens of the VIIth group.

The most important consequence which, in my opinion, can be drawn from the above comparison is, that the periodic law, so apparent in the elements, has a wider application than might appear at first sight; it opens up a new vista of chemical evolutions. But, while admitting the fullest parallelism between the periodicity of the elements and that of the compound radicles, we must not forget that in the periods of the hydrocarbon radicles we have a *decrease* of mass as we pass from the representatives of the first group to the next; while in the periods of the elements the mass *increases* during the progression. It thus becomes evident that we cannot speak of an identity of periodicity in both cases, unless we put aside the ideas of mass and attraction, which are the real corner-stones of the whole of natural science and even enter into those very conceptions of simple bodies which came to light a full hundred years later than the immortal principles of Newton.*

From the foregoing, as well as from the failures of so many attempts at finding in experiment and speculation a proof of the compound character of the elements and of the existence of primordial matter, it is evident, in my opinion, that this theory must be classed amongst mere utopias. But utopias can only be combatted by freedom of opinion, by experiment, and by new utopias. In the republic of scientific theories freedom of opinions is guaranteed. It is precisely that freedom which permits me to criticise openly the widely diffused idea as to the unity of matter in the elements. Experiments and attempts at confirming that idea have been so numerous that it really would be instructive to have them all collected together, if only to serve as a warning against the repetition of old failures. And, now, as to new utopias which may be helpful in the struggle against the old ones, I do not think it quite useless to mention a *phantasy* of one of my students who imagined that the weight of bodies does not depend upon their mass, but upon the character of the motion of their atoms. The atoms, according to this new utopian, may all be homogeneous or heterogeneous, we know not which; we know them in motion only, and that motion they maintain with the same persistence as the stellar bodies maintain theirs. The weights of atoms differ only in consequence of their various modes and quantity of motion; the heaviest atoms may be much simpler than the lighter ones; thus an atom of mercury may be simpler than an atom of hydrogen—the manner in which it moves causes it to be heavier. My interlocutor even suggested that the view which attributes the greater complexity to the lighter elements finds confirmation in the fact that the hydrocarbon radicles mentioned by Pelopidas, while becoming lighter as they lose hydrogen, change their properties periodically in the same manner as the elements change theirs according as the atoms grow heavier.

The French proverb, *La critique est facile mais l'art est difficile*, however, may well be reversed in the case of all such ideal views, as it is much easier to formulate than to criticise them. Arising from the virgin soil of newly established facts, the knowledge relating to the elements, to their masses, and to the periodic changes of their properties, has given a motive for the formation of utopian hypotheses, probably because they could not be foreseen by the aid of any of the

---

* It is noteworthy that the year in which Lavoisier was born (1743)—the author of the idea of elements and of the indestructibility of matter—is later by exactly one century than the year in which the author of the theory of gravitation and mass was born (1643 N.S.). The affiliation of the ideas of Lavoisier and those of Newton is beyond doubt.

various metaphysical systems, and exist, like the idea of gravitation, as an independent outcome of natural science, requiring the acknowledgement of general laws, when these have been established with the same degree of persistency as is indispensable for the acceptance of a thoroughly established fact. Two centuries have elapsed since the theory of gravitation was enunciated, and although we do not understand its cause, we still must regard gravitation as a fundamental conception of natural philosophy, a conception which has enabled us to perceive much more than the metaphysicians did or could with their seeming omniscience. A hundred years later the conception of the elements arose; it made chemistry what it now is; and yet we have advanced as little in our comprehension of simple bodies since the times of Lavoisier and Dalton as we have in our understanding of gravitation. The periodic law of the elements is only 20 years old: it is not surprising therefore that, knowing nothing about the causes of gravitation and mass, or about the nature of the elements, we do not comprehend the *rationale* of the periodic law. It is only by collecting established laws, that is by working at the acquirement of truth, that we can hope gradually to lift the veil which conceals from us the causes of the mysteries of Nature and to discover their mutual dependency. Like the telescope and the microscope, laws founded on the basis of experiment are the instruments and means of enlarging our mental horizon.

In the remaining part of my communication I shall endeavour to show, and as briefly as possible, in how far the periodic law contributes to enlarge our range of vision. Before the promulgation of this law the chemical elements were mere fragmentary, incidental facts in Nature; there was no special reason to expect the discovery of new elements, and the new ones which were discovered from time to time appeared to be possessed of quite novel properties. The law of periodicity first enabled us to perceive undiscovered elements at a distance which formerly was inaccessible to chemical vision; and long ere they were discovered new elements appeared before our eyes possessed of a number of well-defined properties. We now know three cases of elements whose existence and properties were foreseen by the instrumentality of the periodic law. I need but mention the brilliant discovery of *gallium*, which proved to correspond to eka-aluminium of the periodic law, by Lecoq de Boisbaudran; of *scandium*, corresponding to eka-boron, by Nilson; and of *germanium*, which proved to correspond in all respects to eka-silicium, by Winckler. When, in 1871, I described to the Russian Chemical Society the properties, clearly defined by the periodic law, which such elements ought to possess, I never hoped that I should live to mention their discovery to the Chemical Society of Great Britain as a confirmation of the exactitude and the generality of the periodic law. Now, that I have had the happiness of doing so, I unhesitatingly say that although greatly enlarging our vision, even now the periodic law needs further improvements in order that it may become a trustworthy instrument in further discoveries.*

I will venture to allude to some other matters which chemistry has discerned by means of its new instrument, and which it could not have made out without a knowledge of the law of periodicity, and I will confine myself to simple bodies and to oxides.

Before the periodic law was formulated the atomic weights of the elements were purely empirical numbers, so that the magnitude of the equivalent, and the atomicity or the value in substitution possessed by an atom, could only be tested by critically examining the methods of determination, but never directly by considering the numerical values themselves; in short, we were compelled to move in the dark, to submit to the facts, instead of being masters of them. I need not recount the methods which permitted the periodic law at last to master the facts relating to atomic weights, and I would merely call to mind that it compelled us to modify the valencies of *indium* and *cerium*, and to assign to their compounds a different molecular composition. Determinations of the specific heats of these two metals fully confirmed the change. The trivalency of *yttrium*, which makes us now represent its oxide as Y_2O_3 instead of as YO, was also foreseen (in 1870) by the periodic law, and it now has become so probable that Cleve, and all other subsequent investigators of the rare metals, have not only adopted it but have also applied it without any new demonstration to bodies so imperfectly known as those of the cerite and gadolinite group, especially since Hildebrand determined the specific heats of lanthanum and didymium and confirmed the expectations suggested by the periodic law. But here, especially in the case of didymium, we meet with a series of difficulties long since foreseen through the periodic law, but only now becoming

* I foresee some more new elements, but not with the same certitude as before. I shall give one example, and yet I do not see it quite distinctly. In the series which contains Hg = 204, Pb = 206, and Bi = 208, we can guess the existence (at the place VI—11) of an element analogous to tellurium, which we can describe as dvi-tellurium, Dt having an atomic weight of 212, and the property of forming the oxide DtO_3. If this element really exists, it ought in the free state to be an easily fusible, crystalline, non-volatile metal of a grey colour, having a density of about 9·3, capable of giving a dioxide, DtO_2, equally endowed with feeble acid and basic properties. This dioxide must give on active oxidation an unstable higher oxide, DtO_3, which should resemble in its properties PbO_2 and Bi_2O_5. Dvi-tellurium hydride, if it be found to exist, will be a less stable compound than even H_2Te. The compounds of dvi-tellurium will be easily reduced, and it will form characteristic definite alloys with other metals.

evident, and chiefly arising from the relative rarity and insufficient knowledge of the elements which usually accompany didymium.

Passing to the results obtained in the case of the rare elements *beryllium*, *scandium* and *thorium*, it is found that these have many points of contact with periodic law. Although Avdéeff long since proposed the magnesia formula to represent beryllium oxide, yet there was so much to be said in favour of the alumina formula, on account of the specific heat of the metals and the isomorphism of the two oxides, that it became generally adopted and seemed to be well established. The periodic law, however, as Brauner repeatedly insisted (*Berichte*, 1878, 872; 1881, 53) was against the formula Be_2O_3; it required the magnesium formula BeO, that is, an atomic weight of 9, because there was no place in the system for an element like beryllium having an atomic weight of 13·5. This divergence of opinion lasted for years, and I often heard that the question as to the atomic weight of beryllium threatened to disturb the generality of the periodic law, or, at any rate, to require some important modifications of it. Many forces were operating in the controversy regarding beryllium, evidently because a much more important question was at issue than merely that involved in the discussion of the atomic weight of a relatively rare element; and during the controversy the periodic law became better understood, and the mutual relations of the elements became more apparent than ever before. It is most remarkable that the victory of the periodic law was won by the researches of the very observers who previously had discovered a number of facts in support of the trivalency of beryllium. Applying the higher law of Avogadro, Nilson and Petterson have finally shown that the density of the vapour of the beryllium chloride, $BeCl_2$, obliges us to regard beryllium as bivalent in conformity with the periodic law.* I consider the confirmation of Avdéeff's and Brauner's view as important in the

---

* Let me mention another proof of the bivalency of beryllium which may have passed unnoticed, as it was published in the Russian chemical literature. Having remarked (in 1884) that the density of such solutions of chlorides of metals, MCl_n, as contain 200 mols. of water (or a large and constant amount of water) regularly increases as the molecular weight of the dissolved salt increases, I proposed to one of our young chemists, M. Burdakoff, that he should investigate the beryllium chloride. If its molecule be $BeCl_2$ its weight must be = 80; and in such a case it must be heavier than the molecule of KCl = 74·5, and lighter than that of MgCl = 93. On the contrary, if beryllium chloride is a trichloride, $BCl_3 = 120$, its molecule must be heavier than that of $CaCl_2 = 111$, and lighter than that of $MnCl_2 = 126$. Experiment has shown the correctness of the former formula, the solution $BeCl_2 + 200H_2O$ having (at 15°/4°) a density of 1·0138, this being a higher density than that of the solution $KCl + 200H_2O$ (= 1·0121), and lower than that of $MgCl_2 + 200H_2O$ (= 1·0203). The bivalency of beryllium was thus confirmed in the case both of the dissolved and the vaporised chloride.

history of the periodic law as the discovery of scandium, which, in Nilson's hands, confirmed the existence of the eka-boron.

The circumstance that *thorium* proved to be quadrivalent, and Th = 232, in accordance with the views of Chydenius and the requirements of the periodic law, passed almost unnoticed, and was accepted without opposition, and yet both thorium and uranium are of great importance in the periodic system, as they are its last members, and have the highest atomic weights of all the highest elements.

The alteration of the atomic weight of *uranium* from U = 120 into U = 240 attracted more attention, the change having been made on account of the periodic law, and for no other reason. Now that Roscoe, Rammelsberg, Zimmermann, and several others have admitted the various claims of the periodic law in the case of uranium, its high atomic weight is received without objection, and it endows that element with a special interest.

While thus demonstrating the necessity of modifying the atomic weights of several insufficiently known elements, the periodic law enabled us also to detect errors in the determination of the atomic weights of several elements whose valencies and true position among other elements were already well known. Three such cases are especially noteworthy: those of tellurium, titanium and platinum. Berzelius had determined the atomic weight of *tellurium* to be 128, while the periodic law claimed for it an atomic weight below that of iodine, which had been fixed by Stas at 126·5, and which was certainly not higher than 127. Brauner then undertook the investigation, and he has shown that the true atomic weight of tellurium is lower than that of iodine, being near to 125. For *titanium* the extensive researches of Thorpe have confirmed the atomic weight of Ti = 48, indicated by the law, and already foreseen by Rose, but contradicted by the analyses of Pierre and several other chemists. An equally brilliant confirmation of the expectations based on the periodic law has been given in the case of the series osmium, iridium, platinum, and gold. At the time of the promulgation of the periodic law the determinations of Berzelius, Rose, and many others gave the following figures :—

$$Os = 200 \; ; \; Ir = 197 \; ; \; Pt = 198 \; ; \; Au = 196.$$

The expectations of the periodic law* have been confirmed, first, by new determinations of the atomic weight of *platinum* (by Seubert, Dittmar and M'Arthur), which proved to be near to 196 (taking O = 16, as proposed by Marignac, Brauner, and others); secondly,

* I pointed them out in the *Liebig's Annalen*, Supplement Band viii, 1871, p. 211.

2 z 2

by Seubert having proved that the atomic weight of *osmium* is really lower than that of platinum, and that it is near to 191; and thirdly, by the investigations of Krüss, and Thorpe and Laurie proving that the atomic weight of *gold* exceeds that of platinum, and approximates to 197. The atomic weights which were thus found to require correction were precisely those which the periodic law had indicated as affected with errors; and it has been proved therefore that the periodic law affords a means of testing experimental results. If we succeed in discovering the exact character of the periodical relationships between the increments in atomic weights of allied elements discussed by Ridberg in 1885, and again by Bazaroff in 1887, we may expect that our instrument will give us the means of still more closely controlling the experimental data relating to atomic weights.

Let me next call to mind that, while disclosing the variation of chemical properties,* the periodic law has also enabled us to systematically discuss many of the physical properties of elementary bodies, and to show that these properties are also subject to the law of periodicity. At the Moscow Congress of Russian Naturalists in August, 1869, I dwelt upon the relations which existed between density and the atomic weight of the elements. The following year Professor Lothar Meyer, in his well-known paper,† studied the same subject in more detail, and thus contributed to spread information about the periodic law. Later on, Carnelley, Laurie, L. Meyer, Roberts-Austen, and several others applied the periodic system to represent the order in the changes of the magnetic properties of the elements, their melting points, the heats of formation of their haloid compounds, and even of such mechanical properties as the coefficient of elasticity, the breaking stress, &c., &c. These deductions, which have received further support in the discovery of new elements endowed not only with chemical but even with physical properties which were foreseen by the law of periodicity, are well known; so I need not dwell upon the subject, and may pass to the consideration of oxides.‡

---

* Thus, in the typical small period of

Li, Be, B, C, N, O, F,

we see at once the progression from the alkaline metals to the acid non-metals, such as are the halogens.

† *Liebig's Annalen*, Erz. Bd. vii, 1870.

‡ A distinct periodicity can also be discovered in the spectra of the elements. Thus the researches of Hartley, Ciamician, and others have disclosed, first, the homology of the spectra of analogous elements; secondly, that the alkaline metals have simpler spectra than the metals of the following groups; and thirdly, that there is a certain likeness between the complicated spectra of manganese and iron on the one hand, and the no less complicated spectra of chlorine and bromine on

In indicating that the gradual increase of the power of elements of combining with oxygen is accompanied by a corresponding decrease in their power of combining with hydrogen, the periodic law has shown that there is a limit of oxidation, just as there is a well-known limit to the capacity of elements for combining with hydrogen. A single atom of an element combines with at most four atoms of either hydrogen or oxygen: and while CH_4 and SiH_4 represent the highest hydrides, so RuO_4 and OsO_4 are the highest oxides. We are thus led to recognise types of oxides, just as we have had to recognise types of hydrides.*

The periodic law has demonstrated that the maximum extent to which different non-metals enter into combination with oxygen is determined by the extent to which they combine with hydrogen, and that the sum of the number of equivalents of both must be equal to 8. Thus chlorine, which combines with 1 atom, or 1 equivalent of hydrogen, cannot fix more than 7 equivalents of oxygen, giving Cl_2O_7: while sulphur, which fixes 2 equivalents of hydrogen, cannot combine with more than 6 equivalents or 3 atoms of oxygen. It thus becomes evident that we cannot recognise as a fundamental property of the elements the atomic valencies deduced from their hydrides; and that we must modify, to a certain extent, the theory of atomicity if we desire to raise it to the dignity of a general principle capable of affording an insight into the constitution of all compound molecules. In other words, it is only to carbon, which is quadrivalent with regard both to oxygen and hydrogen, that we can apply the theory of constant valency and of bond, by means of which so many still endeavour to explain the structure of compound molecules. But I should go too far if I ventured to explain in detail the conclusions which can be drawn from the above considerations. Still, I think it necessary to dwell upon one particular fact which must be explained from the point of view of the periodic law in order to clear the way to its extension in that particular direction.

the other hand, and their likeness corresponds to the degree of analogy between those elements which is indicated by the periodic law.

* Formerly it was supposed that, being a bivalent element, oxygen can enter into any grouping of the atoms, and there was no limit foreseen as to extent to which it could further enter into combination. We could not explain why bivalent sulphur, which forms compounds such as

$$S\diagup\raisebox{0.5ex}{O}_{\raisebox{-0.5ex}{O}}\diagdown \quad \text{and} \quad S\diagup\raisebox{0.5ex}{O}_{\raisebox{-0.5ex}{O}}\diagdown O,$$

could not also form oxides such as—

$$S\diagup\raisebox{0.5ex}{O-O}_{\raisebox{-0.5ex}{O-O}}\diagdown \quad \text{or} \quad S\diagup\raisebox{0.5ex}{O-O}_{\raisebox{-0.5ex}{O-O}}\diagdown O,$$

while other elements, as, for instance, chlorine, form compounds such as—

$$Cl-O-O-O-O-K.$$

The higher oxides yielding salts the formation of which was foreseen by the periodic system—for instance, in the short series beginning with sodium—

$$Na_2O, \ MgO, \ Al_2O_3, \ SiO_2, \ P_2O_5, \ SO_3, \ Cl_2O_7,$$

must be clearly distinguished from the higher degrees of oxidation which correspond to hydrogen peroxide and bear the true character of peroxides. Peroxides such as Na_2O_2, BaO_2, and the like have long been known. Similar peroxides have also recently become known in the case of chromium, sulphur, titanium, and many other elements, and I have sometimes heard it said that discoveries of this kind weaken the conclusions of the periodic law in so far as it concerns the oxides. I do not think so in the least, and I may remark, in the first place, that all these peroxides are endowed with certain properties—obviously common to all of them, which distinguish them from the actual, higher, salt-forming oxides, especially their easy decomposition by means of simple contact agencies; their incapacity of forming salts of the common type; and their capacity of combining with other peroxides (like the faculty which hydrogen peroxide possesses of combining with barium peroxide, discovered by Schoene). Again, we remark that some groups are especially characterised by their capacity of generating peroxides. Such is, for instance, the case in the VIth group, where we find the well-known peroxides of sulphur, chromium, and uranium; so that further investigation of peroxides will probably establish a new periodic function, foreshadowing that molybdenum and wolfram will assume peroxide forms with comparative readiness. To appreciate the constitution of such peroxides, it is enough to notice that the peroxide form of sulphur (so-called persulphuric acid) stands in the same relation to sulphuric acid as hydrogen peroxide stands to water :—

$$H(OH), \text{ or } H_2O, \text{ responds to } (OH)(OH), \text{ or } H_2O_2,$$

and so also—

$$H(HSO_4), \text{ or } H_2SO_4 \text{ responds to } (HSO_4)(HSO_4), \text{ or } H_2S_2O_8.$$

Similar relations are seen everywhere, and they correspond to the principle of substitutions which I long since endeavoured to represent as one of the chemical generalisations called into life by the periodic law. So also sulphuric acid, if considered with reference to hydroxyl, and represented as follows—

$$HO(SO_2OH),$$

has its corresponding compound in dithionic acid—

$$(SO_2OH)(SO_2OH), \text{ or } H_2S_2O_6.$$

Therefore, also, phosphoric acid, $HO(POH_2O_2)$, has, in the same sense, its corresponding compound in the subphosphoric acid of Saltzer :—

$$(POH_2O_2)(POH_2O_2), \text{ or } H_4P_2O_6;$$

and we must suppose that the peroxide compound corresponding to phosphoric acid, if it be discovered, will have the following structure :—

$$(H_2PO_4)_2 \text{ or } H_4P_2O_8 = 2H_2O + 2PO_3.*$$

As far as is known at present, the highest form of peroxides is met with in the peroxide of uranium, UO_4, prepared by Fairley ;† while OsO_4 is the highest oxide giving salts. The line of argument which is inspired by the periodic law, so far from being weakened by the discovery of peroxides, is thus actually strengthened, and we must hope that a further exploration of the region under consideration will confirm the applicability to chemistry generally of the principles deduced from the periodic law.

Permit me now to conclude my rapid sketch of the oxygen compounds by the observation that the periodic law is especially brought into evidence in the case of the oxides which constitute the immense majority of bodies at our disposal on the surface of the earth.

The oxides are evidently subject to the law, both as regards their chemical and their physical properties, especially if we take into account the cases of polymerism which are so obvious when comparing CO_2 with Si_nO_{2n}. In order to prove this I give the densities s and the specific volumes v of the higher oxides of two short periods. To render comparison easier, the oxides are all represented as of the form R_2O_n. In the column headed Δ the differences are given between the volume of the oxygen compound and that of the parent element, divided by n, that is, by the number of atoms of oxygen in the compound :—‡

* In this sense, oxalic acid, $(COOH)_2$, also corresponds to carbonic acid, $OH(COOH)$, in the same way that dithionic acid corresponds to sulphuric acid, and subphosphoric acid to phosphoric ; therefore, if a peroxide, corresponding to carbonic acid, be obtained, it will have the structure of $(HCO_3)_2$, or $H_2C_2O_6 = H_2O + C_2O_5$. So also lead must have a real peroxide, Pb_2O_5.

† The compounds of uranium prepared by Fairley seem to me especially instructive in understanding the peroxides. By the action of hydrogen peroxide on uranium oxide, UO_3, a peroxide of uranium, $UO_4 4H_2O$, is obtained ($U = 240$) if the solution be acid; but if hydrogen peroxide act on uranium oxide in the presence of caustic soda, a crystalline deposit is obtained, which has the composition $Na_4UO_8 4H_2O$, and evidently is a combination of sodium peroxide, Na_2O_2, with uranium peroxide, UO_4. It is possible that the former peroxide, $UO_4 4H_2O$, contains the elements of hydrogen peroxide and uranium peroxide, U_2O_7, or even $U(OH)_6H_2O_2$, like the peroxide of tin recently discovered by Spring, which has the constitution $Sn_2O_5H_2O_2$.

‡ Δ thus represents the average increase of volume for each atom of oxygen con-

	s.	v.	Δ.		s.	v.	Δ.
Na_2O	2·6	24	−22	K_2O	2·7	35	−55
Mg_2O_2	3·6	22	−3	Ca_2O	3·15	36	−7
Al_2O_3	4·0	26	+1·3	Sc_2O_3	3·86	35	0
Si_2O_4	2·65	45	5·2	Li_2O_4	4·2	38	+5
P_2O_5	2·39	59	6·2	V_2O_5	3·49	52	6·7
S_2O_6	1·96	82	8·7	Cr_2O_6	2·74	73	9·5

I have nothing to add to these figures, except that like relations appear in other periods as well. The above relations were precisely those which made it possible for me to be certain that the relative density of eka-silicon oxide would be about 4·7; germanium oxide, actually obtained by Winckler, proved, in fact, to have the relative density 4·703.

The foregoing account is far from being an exhaustive one of all that has already been discovered by means of the periodic law telescope in the boundless realms of chemical evolution. Still less is it an exhaustive account of all that may yet be seen, but I trust that the little which I have said will account for the philosophical interest attached in chemistry to this law. Although but a recent scientific generalisation, it has already stood the test of laboratory verification and appears as an instrument of thought which has not yet been compelled to undergo modification; but it needs not only new applications, but also improvements, further development, and plenty of fresh energy. All this will surely come, seeing that such an assembly of men of science as the Chemical Society of Great Britain has expressed the desire to have the history of the periodic law described in a lecture dedicated to the glorious name of Faraday.

SECTION B.

CHEMISTRY.

OPENING ADDRESS BY PROF. WILLIAM RAMSAY, PH.D.,
LL.D., SC.D., F.R.S., PRESIDENT OF THE SECTION.

An Undiscovered Gas.

A SECTIONAL address to members of the British Association falls under one of three heads. It may be historical, or actual, or prophetic ; it may refer to the past, the present, or the future. In many cases, indeed in all, this classification overlaps. Your former Presidents have given sometimes a historical introduction, followed by an account of the actual state of some branch of our science, and, though rarely, concluding with prophetic remarks. To those who have an affection for the past, the historical side appeals forcibly ; to the practical man, and to the investigator engaged in research, the actual, perhaps, presents more charm ; while to the general public, to whom novelty is often more of an attraction than truth, the prophetic aspect excites most interest. In this address I must endeavour to tickle all palates ; and perhaps I may be excused if I take this opportunity of indulging in the dangerous luxury of prophecy, a luxury which the managers of scientific journals do not often permit their readers to taste.

The subject of my remarks to-day is a new gas. I shall describe to you later its curious properties; but it would be unfair not to put you at once in possession of the knowledge of its most remarkable property—it has not yet been discovered. As it is still unborn, it has not yet been named. The naming of a new element is no easy matter. For there are only twenty-six letters in our alphabet, and there are already over seventy elements. To select a name expressible by a symbol which has not already been claimed for one of the known elements is difficult, and the difficulty is enhanced when it is at the same time required to select a name which shall be descriptive of the properties (or want of properties) of the element.

It is now my task to bring before you the evidence for the existence of this undiscovered element.

It was noticed by Döbereiner, as long ago as 1817, that certain elements could be arranged in groups of three. The choice of the elements selected to form these triads was made on account of their analogous properties, and on the sequence of their atomic weights, which had at that time only recently been discovered. Thus calcium, strontium, and barium formed such a group; their oxides, lime, strontia, and baryta are all easily slaked, combining with water to form soluble lime-water, strontia-water, and baryta-water. Their sulphates are all sparingly soluble, and resemblance had been noticed between their respective chlorides and between their nitrates. Regularity was also displayed by their atomic weights. The numbers then accepted were 20, 42·5, and 65; and the atomic weight of strontium, 42·5, is the arithmetical mean of those of the other two elements, for $(65+20)/2 = 42·5$. The existence of other similar groups of three was pointed out by Döbereiner, and such groups became known as "Döbereiner's triads."

Another method of classifying the elements, also depending on their atomic weights, was suggested by Pettenkofer, and afterwards elaborated by Kremers, Gladstone, and Cooke. It consisted in seeking for some expression which would represent the differences between the atomic weights of certain allied elements. Thus, the difference between the atomic weight of lithium, 7, and sodium, 23, is 16; and between that of sodium and of potassium, 39, is also 16. The regularity is not always so conspicuous; Dumas, in 1857, contrived a somewhat complicated expression which, to some extent, exhibited regularity in the atomic weights of fluorine, chlorine, bromine, and iodine; and also of nitrogen, phosphorus, arsenic, antimony and bismuth.

The upshot of these efforts to discover regularity was that, in 1864, Mr. John Newlands, having arranged the elements in eight groups, found that when placed in the order of their atomic weights, "the eighth element. starting from a given one, is a kind of repetition of the first, like the eighth note of an octave in music." To this regularity he gave the name "The Law of Octaves."

The development of this idea, as all chemists know, was due to the late Prof. Lothar Meyer, of Tübingen, and to Prof. Mendeléeff, of St. Petersburg. It is generally known as the "Periodic Law." One of the simplest methods of showing this arrangement is by means of a cylinder divided into eight segments by lines drawn parallel to its axis; a spiral line is then traced round the cylinder, which will, of course, be cut by these lines eight times at each revolution. Holding the cylinder vertically, the name and atomic weight of an element is written at each intersection of the spiral with a vertical line, following the numerical order of the atomic weights. It will be found, according to Lothar Meyer and Mendeléeff, that the elements grouped down each of the vertical lines form a natural class; they possess similar properties, form similar compounds, and exhibit a graded relationship between their densities, melting-points, and many of their other properties. One of these vertical columns, however, differs from the others, inasmuch as on it there are three groups, each consisting of three elements with approximately equal atomic weights. The elements in question are iron, cobalt, and nickel; palladium, rhodium, and ruthenium; and platinum, iridium, and osmium. There is apparently room for a fourth group of three elements in this column, and it may be a fifth. And the discovery of such a group is not unlikely, for when this table was first drawn up Prof. Mendeléeff drew attention to certain gaps, which have since been filled up by the discovery of gallium, germanium, and others.

The discovery of argon at once raised the curiosity of Lord Rayleigh and myself as to its position in this table. With a density of nearly 20, if a diatomic gas, like oxygen and nitrogen, it would follow fluorine in the periodic table; and our first idea was that argon was probably a mixture of three gases, all of which possessed nearly the same atomic weights, like iron, cobalt, and nickel. Indeed, their names were suggested, on this supposition, with patriotic bias, as Anglium, Scotium, and Hibernium! But when the ratio of its specific heats had, at least in our opinion, unmistakably shown that it was molecularly monatomic, and not diatomic, as at first conjectured, it was necessary to believe that its atomic weight was 40, and not 20, and that it followed chlorine in the atomic table, and not fluorine. But here arises a difficulty. The atomic weight of chlorine is 35·5, and that of potassium, the next element in order in the table, is 39·1; and that of argon, 40, follows, and does not precede, that of potassium, as it might be expected to do. It still remains possible that argon, instead of consisting wholly of monatomic molecules, may contain a small percentage of diatomic molecules; but the evidence in favour of this supposition is, in my opinion, far from strong. Another possibility is that argon, as at first conjectured, may consist of a mixture of more than one element; but, unless the atomic weight of one of the elements in the supposed mixture is very high, say 82, the case is not bettered, for one of the elements in the supposed trio would still have a higher atomic weight than potassium. And very careful experiments, carried out by Dr. Norman Collie and myself, on the fractional diffusion of argon, have disproved the existence of any such element with high atomic weight in argon, and, indeed, have practically demonstrated that argon is a simple substance, and not a mixture.

The discovery of helium has thrown a new light on this subject. Helium, it will be remembered, is evolved on heating certain minerals, notably those containing uranium; although it appears to be contained in others in which uranium is not present, except in traces. Among these minerals are clèveite, monazite, fergusonite, and a host of similar complex mixtures, all containing rare elements, such as niobium, tantalum, yttrium, cerium, &c. The spectrum of helium is characterised by a remarkably brilliant yellow line, which had been observed as long ago as 1868 by Profs. Frankland and Lockyer in the spectrum of the sun's chromosphere, and named "helium" at that early date.

The density of helium proved to be very close to 2·0, and, like argon, the ratio of its specific heat showed that it, too, was a monatomic gas. Its atomic weight therefore is identical with its molecular weight, viz. 4·0, and its place in the periodic table is between hydrogen and lithium, the atomic weight of which is 7·0.

The difference between the atomic weights of helium and argon is thus 36, or 40 − 4. Now there are several cases of such a difference. For instance, in the group the first member of which is fluorine we have—

Fluorine...	...	...	...	19	16·5
Chlorine..	...	...	...	35·5	19·5
Manganese	...	...	...	55	

In the oxygen group—

Oxygen ...	...	...	...	16	16
Sulphur ...	...	...	...	32	20·3
Chromium	...	...	...	52·3	

In the nitrogen group—

Nitrogen	...	...	...	14	17
Phosphorus	...	...	...	31	20·4
Vanadium	..	...	...	51·4	

And in the carbon group—

Carbon ...	...	...	...	12	16·3
Silicon ...	...	...	...	28·3	19·8
Titanium	...	...	...	48·1	

These instances suffice to show that approximately the differences are 16 and 20 between consecutive members of the corresponding groups of elements. The total differences between the extreme members of the short series mentioned are—

Manganese − Fluorine	...	...	36
Chromium − Oxygen...	...	...	36·3
Vanadium − Nitrogen	...	...	37·4
Titanium − Carbon ...	...	...	36·1

This is approximately the difference between the atomic weights of helium and argon, 36.

There should, therefore, be an undiscovered element between helium and argon, with an atomic weight 16 units higher than than of helium, and 20 units lower than that of argon, namely 20. And if this unknown element, like helium and argon, should prove to consist of monatomic molecules, then its density should be half its atomic weight, 10. And pushing the analogy still further, it is to be expected that this element should be as indifferent to union with other elements as the two allied elements.

My assistant, Mr. Morris Travers, has indefatigably aided me in a search for this unknown gas. There is a proverb about looking for a needle in a haystack ; modern science, with the aid of suitable magnetic appliances, would, if the reward were sufficient, make short work of that proverbial needle. But here is a supposed unknown gas, endowed no doubt with negative properties, and the whole world to find it in. Still, the attempt had to be made.

We first directed our attention to the sources of helium— minerals. Almost every mineral which we could obtain was heated in a vacuum, and the gas which was evolved examined. The results are interesting. Most minerals give off gas when heated, and the gas contains, as a rule, a considerable amount of hydrogen, mixed with carbonic acid, questionable traces of nitrogen, and carbonic oxide. Many of the minerals, in addition, gave helium, which proved to be widely distributed, though only in minute proportion. One mineral—malacone— gave appreciable quantities of argon ; and it is noteworthy that argon was not found except in it (and, curiously, in much larger amount than helium), and in a specimen of meteoric iron. Other specimens of meteoric iron were examined, but were found to contain mainly hydrogen, with no trace of either argon or helium. It is probable that the sources of meteorites might be traced in this manner, and that each could be relegated to its particular swarm.

Among the minerals examined was one to which our attention had been directed by Prof. Lockyer, named eliasite, from which he said that he had extracted a gas in which he had observed spectrum lines foreign to helium. He was kind enough to furnish us with a specimen of this mineral, which is exceedingly rare, but the sample which we tested contained nothing but undoubted helium.

During a trip to Iceland in 1895, I collected some gas from the boiling springs there ; it consisted, for the most part, of air, but contained somewhat more argon than is usually dissolved when air is shaken with water. In the spring of 1896 Mr. Travers and I made a trip to the Pyrenees to collect gas from the mineral springs of Cauterets, to which our attention had been directed by Dr. Bouchard, who pointed out that these gases are rich in helium. We examined a number of samples from the various springs, and confirmed Dr. Bouchard's results, but there was no sign of any unknown lines in the spectrum of these gases. Our quest was in vain.

We must now turn to another aspect of the subject. Shortly after the discovery of helium, its spectrum was very carefully examined by Profs. Runge and Paschen, the renowned spectroscopists. The spectrum was photographed, special attention being paid to the invisible portions, termed the " ultra-violet" and "infra-red." The lines thus registered were found to have a harmonic relation to each other. They admitted of division into two sets, each complete in itself. Now, a similar process had been applied to the spectrum of lithium and to that of sodium, and the spectra of these elements gave only one series each. Hence, Profs. Runge and Paschen concluded that the gas, to which the provisional name of helium had been given, was, in reality, a mixture of two gases, closely resembling each other in properties. As we know no other elements with atomic weights between those of hydrogen and lithium, there is no chemical evidence either for or against this supposition. Prof. Runge supposed that he had obtained evidence of the separation of these imagined elements from each other by means of diffusion ; but Mr. Travers and I pointed out that the same alteration of spectrum, which was apparently produced by diffusion, could also be caused by altering the pressure of the gas in the vacuum tube ; and shortly after Prof. Runge acknowledged his mistake.

These considerations, however, made it desirable to subject helium to systematic diffusion, in the same way as argon had been tried. The experiments were carried out in the summer of 1896 by Dr. Collie and myself. The result was encouraging. It was found possible to separate helium into two portions of different rates of diffusion, and consequently of different density by this means. The limits of separation, however, were not very great. On the one hand, we obtained gas of a density close on 2·0 ; and on the other, a sample of density 2·4 or thereabouts. The difficulty was increased by the curious behaviour, which we have often had occasion to confirm, that helium possesses a rate of diffusion too rapid for its density. Thus, the density of the lightest portion of the diffused gas, calculated from its rate of diffusion, was 1·874 ; but this corresponds to a real density of about 2·0. After our paper, giving an account of these experiments, had been published, a German investigator, Herr A. Hagenbach, repeated our work and confirmed our results.

The two samples of gas of different density differ also in other properties. Different transparent substances differ in the rate at which they allow light to pass through them. Thus, light travels through water at a much slower rate than through air, and at a slower rate through air than through hydrogen. Now Lord Rayleigh found that helium offers less opposition to the passage of light than any other substance does, and the heavier of the two portions into which helium had been split offered more opposition than the lighter portion. And the retardation of the light, unlike what has usually been observed, was nearly proportional to the densities of the samples. The spectrum of these two samples did not differ in the minutest particular ; therefore it did not appear quite out of the question to hazard the speculation that the process of diffusion was instrumental, not necessarily in separating two kinds of gas from each other, but actually in removing light molecules of the same kind from heavy molecules. This idea is not new. It had been advanced by Prof. Schützenberger (whose recent death all chemists have to deplore), and later, by Mr. Crookes, that what we term the atomic weight of an element is a mean ; that when we say the atomic weight of oxygen is 16, we merely state that the average atomic weight is 16 ; and it is not inconceivable that a certain number of molecules have a weight somewhat higher than 32, while a certain number have a lower weight.

We therefore thought it necessary to test this question by direct experiment with some known gas ; and we chose nitrogen, as a good material with which to test the point. A much larger and more convenient apparatus for diffusing gases was built by Mr. Travers and myself, and a set of systematic diffusions of nitrogen was carried out. After thirty rounds, corresponding to 180 diffusions, the density of the nitrogen was unaltered, and that of the portion which should have diffused most slowly, had there been any difference in rate, was identical with that of the most quickly diffusing portion—*i.e.* with that of the portion which passed first through the porous plug. This attempt, therefore, was unsuccessful ; but it was worth carrying out, for it is now certain that it is not possible to separate a gas of undoubted chemical unity into portions of different density by diffusion. And these experiments rendered it exceedingly improbable that the difference in density of the two fractions of helium was due to separation of light molecules of helium from heavy molecules.

The apparatus used for diffusion had a capacity of about two litres. It was filled with helium, and the operation of diffusion was carried through thirty times. There were six reservoirs, each full of gas, and each was separated into two by diffusion. To the heavier portion of one lot, the lighter portion of the next was added, and in this manner all six reservoirs were successfully passed through the diffusion apparatus. This process was carried out thirty times, each of the six reservoirs having had its gas diffused each time, thus involving 180 diffusions. After this process, the density of the more quickly diffusing gas was reduced to 2·02, while that of the less quickly diffusing had increased to 2·27. The light portion on re-diffusion hardly altered in density, while the heavier portion, when divided into three portions by diffusion, showed a considerable difference in density between the first third and the last third. A similar set of operations was carried out with a fresh quantity of helium, in order to accumulate enough gas to obtain a sufficient quantity for a second series of diffusions. The more quickly diffusing portions of both gases were mixed and rediffused. The density of the lightest portion of these gases was 1·98 ; and after other 15 diffusions, the density of the lightest portion had not decreased. The end had been reached ; it was not possible to obtain a lighter portion by diffusion. The density of the main body of this gas is therefore 1·98 ; and its refractivity, air being taken as unity, is 0·1245. The spectrum of this portion does not differ in any respect from the usual spectrum of helium.

As re-diffusion does not alter the density or the refractivity of this gas, it is right to suppose that either one definite element has now been isolated; or that if there are more elements than one present, they possess the same, or very nearly the same, density and refractivity. There may be a group of elements, say three, like iron, cobalt, and nickel; but there is no proof that this idea is correct, and the simplicity of the spectrum would be an argument against such a supposition. This substance, forming by far the larger part of the whole amount of the gas, must, in the present state of our knowledge, be regarded as pure helium.

On the other hand, the heavier residue is easily altered in density by re-diffusion, and this would imply that it consists of a small quantity of a heavy gas mixed with a large quantity of the light gas. Repeated re-diffusion convinced us that there was only a very small amount of the heavy gas present in the mixture. The portion which contained the largest amount of heavy gas was found to have the density 2·275, and its refractive index was found to be 0·1333. On re-diffusing this portion of gas until only a trace sufficient to fill a Plücker's tube was left, and then examining the spectrum, no unknown lines could be detected, but, on interposing a jar and spark gap, the well-known blue lines of argon became visible; and even without the jar the red lines of argon, and the two green groups were distinctly visible. The amount of argon present, calculated from the density, was 1·64 per cent., and from the refractivity 1·14 per cent. The conclusion had therefore to be drawn that the heavy constituent of helium, as it comes off the minerals containing it, is nothing new, but, so far as can be made out, merely a small amount of argon.

If, then, there is a new gas in what is generally termed helium, it is mixed with argon, and it must be present in extremely minute traces. As neither helium nor argon has been induced to form compounds, there does not appear to be any method, other than diffusion, for isolating such a gas, if it exists, and that method has failed in our hands to give any evidence of the existence of such a gas. It by no means follows that the gas does not exist; the only conclusion to be drawn is that we have not yet stumbled on the material which contains it. In fact, the haystack is too large and the needle too inconspicuous. Reference to the periodic table will show that between the elements aluminium and indium there occurs gallium, a substance occurring only in the minutest amount on the earth's surface; and following silicon, and preceding tin, appears the element germanium, a body which has as yet been recognised only in one of the rarest of minerals, argyrodite. Now, the amount of helium in fergusonite, one of the minerals which yields it in reasonable quantity, is only 33 parts by weight in 100,000 of the mineral; and it is not improbable that some other mineral may contain the new gas in even more minute proportion. If, however, it is accompanied in its still undiscovered source by argon and helium, it will be a work of extreme difficulty to effect a separation from these gases.

In these remarks it has been assumed that the new gas will resemble argon and helium in being indifferent to the action of reagents, and in not forming compounds. This supposition is worth examining. In considering it, the analogy with other elements is all that we have to guide us.

We have already paid some attention to several triads of elements. We have seen that the differences in atomic weights between the elements fluorine and manganese, oxygen and chromium, nitrogen and vanadium, carbon and titanium, is in each case approximately the same as that between helium and argon, viz. 36. If elements further back in the periodic table be examined, it is to be noticed that the differences grow less, the smaller the atomic weights. Thus, between boron and scandium, the difference is 33; between beryllium (glucinum) and calcium, 31; and between lithium and potassium, 32. At the same time, we may remark that the elements grow liker each other, the lower the atomic weights. Now, helium and argon are very like each other in physical properties. It may be fairly concluded, I think, that in so far they justify their position. Moreover, the pair of elements which show the smallest difference between their atomic weights is beryllium and calcium; there is a somewhat greater difference between lithium and potassium. And it is in accordance with this fragment of regularity that helium and argon show a greater difference. Then again, sodium, the middle element of the lithium triad, is very similar in properties both to lithium and potassium; and we might, therefore, expect that the unknown element of the helium series should closely resemble both helium and argon.

Leaving now the consideration of the new element, let us turn our attention to the more general question of the atomic weight of argon, and its anomalous position in the periodic scheme of the elements. The apparent difficulty is this: The atomic weight of argon is 40; it has no power to form compounds, and thus possesses no valency; it must follow chlorine in the periodic table, and precede potassium; but its atomic weight is greater than that of potassium, whereas it is generally contended that the elements should follow each other in the order of their atomic weights. If this contention is correct, argon should have an atomic weight smaller than 40.

Let us examine this contention. Taking the first row of elements, we have:

Li = 7, Be = 9·8, B = 11, C = 12, N = 14, O = 16, F = 19, ? = 20.

The differences are:

$$2\cdot8,\ 1\cdot2,\ 1\cdot0,\ 2\cdot0,\ 2\cdot0,\ 3\cdot0,\ 1\cdot0.$$

It is obvious that they are irregular. The next row shows similar irregularities. Thus:

(? = 20), Na = 23, Mg = 24·3, Al = 27, Si = 28, P = 31, S = 32, Cl = 35·5, A = 40.

And the differences:

$$3\cdot0,\ 1\cdot3,\ 2\cdot7,\ 1\cdot0,\ 3\cdot0,\ 1\cdot0,\ 3\cdot5,\ 4\cdot5.$$

The same irregularity might be illustrated by a consideration of each succeeding row. Between argon and the next in order, potassium, there is a difference of −0·9; that is to say, argon has a higher atomic weight than potassium by 0·9 unit; whereas it might be expected to have a lower one, seeing that potassium follows argon in the table. Further on in the table there is a similar discrepancy. The row is as follows:

Ag = 108, Cd = 112, In = 114, Sn = 119, Sb = 120·5, Te = 127·7, I = 127.

The differences are:

$$4\cdot0,\ 2\cdot0,\ 5\cdot0,\ 1\cdot5,\ 7\cdot2,\ -\ 0\cdot7.$$

Here, again, there is a negative difference between tellurium and iodine. And this apparent discrepancy has led to many and careful redeterminations of the atomic weight of tellurium. Prof. Brauner, indeed, has submitted tellurium to methodical fractionation, with no positive results. All the recent determinations of its atomic weight give practically the same number, 127·7.

Again, there have been almost innumerable attempts to reduce the differences between the atomic weights to regularity, by contriving some formula which will express the numbers which represent the atomic weights, with all their irregularities. Needless to say, such attempts have in no case been successful. Apparent success is always attained at the expense of accuracy, and the numbers reproduced are not those accepted as the true atomic weights. Such attempts, in my opinion, are futile. Still, the human mind does not rest contented in merely chronicling such an irregularity; it strives to understand why such an irregularity should exist. And, in connection with this, there are two matters which call for our consideration. These are: Does some circumstance modify these "combining proportions" which we term "atomic weights"? And is there any reason to suppose that we can modify them at our will? Are they true "constants of nature," unchangeable, and once for all determined? Or are they constant merely so long as other circumstances, a change in which would modify them, remain unchanged?

In order to understand the real scope of such questions, it is necessary to consider the relation of the "atomic weights" to other magnitudes, and especially to the important quantity termed "energy."

It is known that energy manifests itself under different forms, and that one form of energy is quantitatively convertible into another form, without loss. It is also known that each form of energy is expressible as the product of two factors, one of which has been termed the "intensity factor," and the other the "capacity factor." Prof. Ostwald, in the last edition of his "Allgemeine Chemie," classifies some of these forms of energy as follows:

Kinetic energy is the product of Mass into the square of velocity.

Linear	,,	,,	Length into force.
Surface	,,	,,	Surface into surface tension.
Volume	,,	,,	Volume into pressure.
Heat	,,	,,	Heat capacity (entropy) into temperature.
Electrical	,,	,,	Electrical capacity into potential.
Chemical	,,	,,	"Atomic weight" into affinity.

In each statement of factors, the "capacity factor" is placed first, and the "intensity factor" second.

In considering the "capacity factors," it is noticeable that they may be divided into two classes. The two first kinds of energy, kinetic and linear, are *independent of the nature of the material* which is subject to the energy. A mass of lead offers as much resistance to a given force, or, in other words, possesses as great inertia as an equal mass of hydrogen. A mass of iridium, the densest solid, counterbalances an equal mass of lithium, the lightest known solid. On the other hand, surface energy deals with molecules, and not with masses. So does volume energy. The volume energy of two grammes of hydrogen, contained in a vessel of one litre capacity, is equal to that of thirty-two grammes of oxygen at the same temperature, and contained in a vessel of equal size. Equal masses of tin and lead have not equal capacity for heat ; but 119 grammes of tin has the same capacity as 207 grammes of lead ; that is, equal atomic masses have the same heat capacity. The quantity of electricity conveyed through an electrolyte under equal difference of potential is proportional, not to the mass of the dissolved body, but to its equivalent ; that is, to some simple fraction of its atomic weight. And the capacity factor of chemical energy is the atomic weight of the substance subjected to the energy. We see, therefore, that while mass or inertia are important adjuncts of kinetic and linear energies, all other kinds of energy are connected with atomic weights, either directly or indirectly.

Such considerations draw attention to the fact that quantity of matter (assuming that there exists such a carrier of properties as we term "matter") need not necessarily be measured by its inertia, or by gravitational attraction. In fact the word "mass" has two totally distinct significations. Because we adopt the convention to measure quantity of matter by its mass, the word "mass" has come to denote "quantity of matter." But it is open to any one to measure a quantity of matter by any other of its energy factors. I may, if I choose, state that those quantities of matter which possess equal capacities for heat are equal ; or that "equal numbers of atoms" represent equal quantities of matter. Indeed, we regard the value of material as due rather to what it can do, than to its mass ; and we buy food, in the main, on an atomic, or perhaps, a molecular basis, according to its content of albumen. And most articles depend for their value on the amount of food required by the producer or the manufacturer.

The various forms of energy may therefore be classified as those which can be referred to an "atomic" factor, and those which possess a "mass" factor. The former are in the majority. And the periodic law is the bridge between them ; and yet, an imperfect connection. For the atomic factors, arranged in the order of their masses, display only a partial regularity. It is undoubtedly one of the main problems of physics and chemistry to solve this mystery. What the solution will be is beyond my power of prophecy ; whether it is to be found in the influence of some circumstance on the atomic weights, hitherto regarded as among the most certain "constants of nature" ; or whether it will turn out that mass and gravitational attraction are influenced by temperature, or by electrical charge, I cannot tell. But that some means will ultimately be found of reconciling these apparent discrepancies, I firmly believe. Such a reconciliation is necessary, whatever view be taken of the nature of the universe and of its mode of action ; whatever units we may choose to regard as fundamental among those which lie at our disposal.

In this address I have endeavoured to fulfil my promise to combine a little history, a little actuality, and a little prophecy. The history belongs to the Old World ; I have endeavoured to share passing events with the New ; and I will ask you to join with me in the hope that much of the prophecy may meet with its fulfilment on this side of the ocean.

NOTES.

WE are glad to learn that Lord Armstrong, who has for the past few days been suffering from a slight sunstroke, is now much better. Dr. Gibb, of Newcastle, who was hastily summoned to Bamburgh Castle on Sunday, anticipates that if the progress is maintained his lordship will be quite well again by the end of the week.

LIEUT. DE GERLACHE'S expedition to the Antarctic regions left Antwerp on Monday on board the steamer *Belgica*.

PROF. CORFIELD has been elected an honorary member of the Royal Society of Public Health of Belgium, of which he has been a corresponding member for some years.

A SPECIAL number of the *Rendiconti della R. Accademia dei Lincei* announces the award of the following prizes, besides others for essays of a literary character :—The Royal Prize for physics to Prof. Adolfo Bartoli, of Padua, for his two monographs on the specific heat of water between the temperatures of $0°$ and $35°$, and on the heat of the sun, and for other investigations. For the Ministerial Prize for physical and chemical science eight competitors entered, and the judges have awarded a prize of 1000 lire to Prof. Carlo Bonacini, of Modena, for his essays on orthochromatic and colour photography, and on the reflection and other properties of Röntgen rays ; also awards of 250 lire each to Prof. Carlo Cattaneo, of Turin, for his notes on the conductivity of electrolytes and on the velocity of ions, and to Prof. Pietro Bartolotti for chemical investigations relating to the compound Rottlerine and other derivatives.

Science of August 6 prints a long article, by Mr. Cyrus Adler, on the movement towards an international catalogue of scientific works, and reprints the official reports of the proceedings of the conference held at the Royal Society a year ago. The report which the American delegates, Prof. Simon Newcomb and Dr. John S. Billings, presented to the Secretary of State, was, in accordance with their suggestion, referred to the Secretary of the Smithsonian Institution for his views as to the propriety and feasibility of the work proposed being undertaken by that Institution, and as to the probable cost. After considering the matter, Mr. S. P. Langley replied that if the work should be assigned to the Institution, a grant of not less than ten thousand dollars per annum would be required to carry it out. This reply and the documents to which it refers were transmitted to the U.S. Senate and House of Representatives towards the close of last year ; and though no result has yet been reached, it is hoped that Congress will give support to the proposal, so that when the time comes the funds needed for cataloguing the scientific publications of the United States will be granted.

THE meeting of the French Association for the Advancement of Science, which opened at St. Étienne on August 5, was concluded on Saturday last. At the opening of the meeting, the President, M. Marey, gave an address on "La méthode graphique et les sciences experimentales." The address is printed in the *Revue Scientifique* for August 7. Next year's meeting will be at Nantes, while in 1899 the congress will be held at Boulogne, in order that visits may be exchanged with members of the British Association at Dover.

Section V

THE REVIVAL OF PROUT'S HYPOTHESIS

Dumas was the doyen of French chemists, responsible for bringing some order into the organic realm. In 1851, at the meeting of the British Association, he spoke in favour of Prout's hypothesis; he later elaborated the comparisons between families of elements and the homologous series of organic radicals. His Faraday Lecture of 1869 on this topic was reprinted in the first series of *Classical Scientific Papers, Chemistry*. Before *Chemical News* and *Nature* began, reports of the more informal discussions at the British Association and the full texts of some Presidential Addresses were carried in the magazine *The Athenaeum*.

J. H. Gladstone was a chemist famous in his day; with the logician Augustus de Morgan he demonstrated how unlikely it was that by chance the atomic weights of so many elements should be nearly whole numbers. In an address to the British Association in 1883, he discussed Prout's hypothesis at length. Lyon Playfair had in 1845 become Professor of Chemistry at the Royal School of Mines in London; in 1858 he had gone to Edinburgh as Professor of Chemistry; in 1865 he became a Member of Parliament and embarked upon a successful public career. He had played an important part in organising the Great Exhibition of 1851. His paper continues the Proutian trend.

Thomas Sterry Hunt was a pupil of J. B. Stallo, the American philosopher of science who began as a devotee of *Naturphilosophie* but later came to hold views close to those of Ernst Mach. Some of Hunt's papers show the dynamical chemical speculations of disciples of Schelling and Hegel; in the papers reprinted here it is the evolutionary turn which he gave which is interesting. His speculations about the ether resemble somewhat the latest views of Mendeléeff; a quantitative attempt to synthesise such ideas and the hypothesis that elements were like organic radicals was made by Thomas Carnelley, Professor of Chemistry at St. Andrew's. His ether with negative weight is a strange item to find in the chemistry of the late nineteenth century; and though he proposed it to the British Association it does not appear to have found any converts.

Lord Rayleigh was the successor of Clerk Maxwell, and the predecessor of J. J. Thomson at the Cavendish Laboratory at Cambridge. He was famous for the precision of his measurements; and to test Prout's Hypothesis was therefore a reasonable objective for him. Out of this research came the discovery of argon; and Rayleigh and Ramsay received Nobel Prizes for Physics and Chemistry in 1904. In 1887 Roscoe, in another useful survey of the history and present state of chemistry, referred to the mystery to be solved in connexion with Prout's hypothesis; and compared the Periodic Table with the genealogical tables of Francis Galton, the pioneer in the study of heredity.

A new turn was given by Lockyer, the editor of *Nature* and the founder of the Science Museum, South Kensington, and one of the most daring and controversial physicists of the time. To explain anomalies in spectra from the Sun, in which only some of the spectral lines of certain elements could be discerned, he propounded the hypothesis that under extreme conditions elements could become dissociated. And to this he added the working hypothesis that the elements were all, in fact, compound bodies, stable under ordinary conditions. He was enthusiastically supported by Crookes, whose own researches on the rare earth metals convinced him that they were not really distinct elements at all, but arbitrarily named points on a continuum of atoms of varying weights. His suggestion of 'meta-elements' is not unlike the later theory of isotopes; but it was invoked to explain different phenomena, and had less experimental basis.

Finally we come into the story of radioactivity, with the first transmutation observed by Ramsay and Soddy, and Ramsay's prediction for the use of nuclear energy. The atomic models of J. J. Thomson, Rutherford and Bohr had the simplicity for which the Proutians had sought; and the concept of atomic number and the theory of isotopes helped to explain how it is that many of the atomic weights are close to whole numbers, and why iodine and tellurium have to go into the Table the wrong way round if they are to be in the right families. These researches also settled the number of rare earth elements that there were; Crookes' difficulties had arisen from the extreme difficulty of separating any one completely from the others. And, at last, the study of radioactive processes began to cast light on the origin of the Sun's energy, and on how the heavier elements really have in a sense evolved from hydrogen.

FURTHER READING

W. H. Brock, 'Lockyer and the Chemists: the First Dissociation Hypothesis', *Ambix*, XVI (1969), 81–99.
D. M. Knight, 'Steps towards a Dynamical Chemistry', *Ambix*, XIV (1967), 179–97.
A. Romer (ed.), *The Discovery of Radioactivity and Transmutation*, New York, 1964.
J. B. Stallo, *The Concepts and Theories of Modern Physics*, ed. P. W. Bridgman, Cambridge, Mass., 1960.
M. W. Travers, *A Life of Sir William Ramsay*, London, 1956.
S. Wright (ed.), *Classical Scientific Papers, Physics*, London, 1964.

FRIDAY.

SECTION B.—CHEMISTRY—INCLUDING ITS APPLICATION TO AGRICULTURE AND THE ARTS.

DR. FARADAY read extracts from a letter by Prof. Bergeman, of Bonn, and exhibited a specimen of Orangite, the mineral affording the new metal Donarium, which was stated as easily to be reduced by potassium and sodium, in preference to reduction by the gas hydrogen; and as fear had been entertained for the perfect preservation of the properties of the new metal if exposed to the chances of a sea voyage, as it readily tarnishes and oxidizes, the discoverer had thought it more expedient to send a specimen of the hydrated oxide of Donarium.—In allusion to the statement that Prof. Rose had repeated and confirmed these results, Dr. FARADAY expressed an opinion that chemists had of late years viewed with regret the increase in the number of metals, and hoped that the day was not far distant when some of the metals would afford honour to chemists by new modes of investigation leading to their decomposition.

'Observations on Atomic Volumes and Atomic Weights, with considerations on the probability that certain bodies now considered as elementary may be decomposed,' by Prof. DUMAS.—Prof. Dumas alluded to the solubility of some substances and the insolubility of others, giving many instances of the difference of this quality in regard to solution in water, sulphuric and strong acids, and referred to Berthollet's views and experiments on this subject. The measure or volume of bodies he thought might be represented with as much facility as the weight: thus, for example, magnesia and sulphuric acid may have their volumes numerically expressed before and after combination, and also graphically by lines. Magnesia with sulphuric acid showed a certain degree of condensation, lime a greater condensation, and barytes the greatest condensation; and these he could represent and reason on as well by lines of different lengths as by figures or by words. The degree of condensation (however expressed) had also relation to the quality or degree of solubility. Thus, sulphate of magnesia was very soluble, sulphate of lime but little soluble, and the greatly condensed sulphate of baryta was insoluble. He then pursued the analogy with the chlorides, comparing the chloride of sodium with the extreme case of the chloride of silver. After graphically expressing the solubility of bases with sulphuric acid by lines, he proceeded to show that the relative volumes of the elements chlorine, bromine, and iodine could be perfectly represented by lines equal in length. Prof. Dumas said that when a number of metals are represented by lines, at first they seem in confusion, and it would appear like an impossibility to arrange them in a system of lines to permit their relations to appear; but when considered in relation to the substitution of one property for another, or of the substitution of one substance for another in groups, then their arrangement became easy. And here we may remark, that Prof. Dumas had not previously prepared diagrams or tables, but covered a large black board with lines, figures, and formulæ, to follow his train of reasoning,—and symbols, volumes, and names were rapidly produced and as rapidly effaced to illustrate the Professor's views of the laws of the substitution of one body for another in a compound. Prof. Dumas gave many examples of groups of bodies, such as the alkalies, earths, &c., arranged in the order of their affinities. He called attention in the Triad groups, to the intermediate body having most of its qualities intermediate with the properties of the extremes, and also that the atomic or combining number was also of the middle term, exactly half of the extremes added together; thus, sulphur 16, selenium 40, and tellurium 64. Half of the extremes give 40, the number for the middle term. Chlorine 35, bromine 80, and iodine 125. Or the alkalies, lithia, soda, and potassa, or earths, lime, strontia, and baryta, afford, with many others, examples of this coincidence; hence the suggestion, that in a series of bodies, if the extremes were known by some law, intermediate bodies might be discovered; and in the spirit of these remarks, if bodies are to be transformed or decomposed into others the suggestion of suspicion is thrown upon the possibility of the intermediate body being composed of the extremes of the series, and transmutable changes thus hoped for. Prof. Dumas then showed that in the metals similar properties are found to those of non-metallic bodies; alluding to the possibility that metals that were similar in their relations, and which may be substituted one for the other in certain compounds, might also be found *transmutable* the one into the other. He then took up the inorganic bodies where substitutions took place which he stated much resembled the metals. After discussing groups in triads, Prof. Dumas alluded to the ideas of the ancients of the transmutation of metals and their desire to change lead into silver and mercury into gold; but these metals do not appear to have the requisite similar relations to render these changes possible. He then passed to the changes of other bodies,—such as the transmutation of diamonds into black lead under the voltaic arc. After elaborate reasoning and offering many analogies from the stores of chemical analysis, Prof. Dumas expressed the idea that the law of the substitution of one body for another in groups of compounds might lead to the transformation of one group into another at will; and we should endeavour to devise means to divide the molecules of one body of one of these groups into two parts, and also of a third body, and then unite them, and probably the intermediate body might be the result. In this way, if bodies of similar properties and often associated together were transmutable one into the other, then by changes portions of one might often, if not always, be associated with the other. Thus, in nature when chlorine occurred, iodine and bromine might also be found, and always would be if they were transmutable the one into the other. Cobalt is thus mysteriously associated with nickel, iron with manganese, sulphur with selenium, &c. In the arts during operations when certain radicles were produced, analogous ones were found constantly to be associated. In the distillation of brandy, oil of wine is always an associated result.

DR. FARADAY expressed his hope that Prof. Dumas was setting chemists in the right path; and although conversationally acquainted with the subject, yet he had been by no means prepared for the multitude of analogies pointed out.—Mr. GROVE spoke of the importance of the view; as, by knowing the extreme compounds, it might serve as a guide in experiments and as a check to the results. He adverted to the allotropic condition of substances when their principal characters were changed but their chemical qualities were unaltered; thus, carbon in the state of diamond had a change of property so complete that it had one of the properties of metals given or transferred to it by its conducting power for electricity under these conditions, and its other forms were states resistant to electric passage. He thought this fact of certain bodies having two sets of physical properties with greatly differing character might, with this law of the substitution of one set of chemical qualities for another in a compound group, give the hope of the great realization of some of the ideas embodied in the views of the possible transformation of one body at will so as to possess the properties of others.—Prof. WILLIAMSON, Dr. ANDERSON, and Dr. GLADSTONE remarked on these analogies,—and referred to the groups of bodies of similar characters, but whose history was difficult or inexplicable. Thus, the metals of the platina group of bodies, the red states of phosphorus and of sulphur, the carrying of certain of these properties into the sulphurets of phosphorus, and the unsatisfactory history of bodies like the phosphates, might be rendered clear in future researches by the ideas resulting from numerous examples of the triad groups alluded to by Prof. Dumas.

THE
LONDON, EDINBURGH AND DUBLIN
PHILOSOPHICAL MAGAZINE
AND
JOURNAL OF SCIENCE.

[FOURTH SERIES.]

MAY 1853.

L. *On the Relations between the Atomic Weights of analogous Elements.* By J. H. GLADSTONE, *Ph.D.**

CHEMISTS who have turned their attention to the series of numbers representing the atomic weights of the elementary bodies, have frequently remarked curious relations between them. It is between similar elements that these numerical relations occur; and to such an extent is this the case, that Berzelius, after mentioning numerous instances, says, " We see that bodies which present the same properties up to a certain point have certain relations between their atomic weights†."

To illustrate this statement, to show the extent of its truth, and to draw certain analogical inferences tending to the proper understanding of such a fact, are the objects of this communication.

The following is a list of the atomic weights arranged from the lowest to the highest, and thus without any reference to chemical relationship. The numbers adopted are those given in the last volume of Liebig's *Jahresbericht*.

1	Hydrogen.	14	Nitrogen.
4·7	Glucinum.	16	Sulphur.
6	Carbon.	19	Fluorine.
6·5	Lithium.	20	Calcium.
8	Oxygen.	21·3	Silicon.
10·9	Boron.	22·4	Zirconium.
12	Magnesium.	23	Sodium.
13·7	Aluminium.	25	Titanium.

* Communicated by the Author.
† *Traité de Chimie*, vol. iv.

26·7	Chromium.		58	Tin.
27·6	Manganese.		59·6	Thorinum.
28	Iron.		60	Uranium.
29·5	Cobalt.		64·2	Tellurium.
29·6	Nickel.		68·5	Barium.
31	Phosphorus.		68·6	Vanadium.
31·7	Copper.		75	Arsenic.
32·6	Zinc.		80	Bromine.
35·5	Chlorine.		92	Tungsten.
39·2	Potassium.		98·7	Platinum.
39·5	Selenium.		99	Iridium.
43·8	Strontium.		99·6	Osmium.
46	Molybdenum.		100	Mercury.
47	Cerium.		103·7	Lead.
47	Lanthanium.		108·1	Silver.
50	Didymium.		127·1	Iodine.
52·2	Rhodium.		129	Antimony.
52·2	Ruthenium.		184	Tantalum.
53·3	Palladium.		197	Gold.
56	Cadmium.		208	Bismuth.

If we glance at this list we notice some peculiarities, but no very striking ones. We might ask, for instance, Why should there be so many elements congregated about No. 28; and, again, about 52? Why should there be only one atomic weight between 80 and 99, and then a group of four?

The following letter, kindly sent me by Professor De Morgan, will give the data for calculating the probabilities of this. I introduce it on account of its applicability, not only to this particular case, but to others which will occur in these observations.

" Univ. Coll., Lond., Dec. 18, 1852.

" DEAR SIR,—The following, though but an imperfect view of the whole question, will be enough, I think, for your purpose. I send formula and all, that who likes may verify it.

" If there be n numbers, each of which may be drawn at any trial, and all equally likely, and if the following denominations be used,

$$P = \left(1 - \frac{1}{n}\right)^m$$

$$Q = m \cdot \left(1 - \frac{1}{n}\right)^{m-1} \cdot \frac{1}{n}$$

$$R = m \frac{m-1}{2} \left(1 - \frac{1}{n}\right)^{m-2} \frac{1}{n^2}$$

$$S = m \frac{m-1}{2} \frac{m-2}{3} \left(1 - \frac{1}{n}\right)^{m-3} \frac{1}{n^3}, \text{ &c.}$$

Then, speaking of one assigned number, the chance that that number shall not appear in m trials, is P; that it shall appear once and once only, is Q; twice and twice only, is R; and so on. Further, the chance that it shall appear once or more is $1-P$. That it shall appear twice or more, the chance is $1-(P+Q)$. Three times or more, $1-(P+Q+R)$; and so on.

"For calculation,

$$Q = \frac{mP}{n-1}$$

$$R = \frac{(m-1)Q}{2(n-1)}$$

$$S = \frac{(m-2)R}{3(n-1)};$$

and so on.

"Let there be 100 numbers, and 60 trials to be made. I find

$$P = \cdot54716$$
$$Q = \cdot33161$$
$$R = \cdot09881$$
$$S = \cdot01929$$
$$T = \cdot00278$$
$$U = \cdot00031$$
$$P+Q+R+S+T+U = \cdot99996$$
$$1\cdot00000$$

$\cdot00004$. Chance of six or more of a given number.

"It is then 99996 to 4, or 24999 to 1, against the appearance of a predicted number six or more times.

"Now suppose the question to be what is the chance that *some one* number, not named, shall occur six or more times; that is, either the one named in the last case, or some other? This is a much more complicated question, but it is certain that 100 times the chance in the last is too great. Now $\cdot00004 \times 100 = \cdot004$, which is too much decidedly. Consequently 996 to 4, or 249 to 1, are too small odds to lay against the appearance of some one number six or more trials.

"That is, you may lay more than 250 to 1 that of all the numbers, *no one* will occur six or more times in 60 trials.

"I am, dear Sir,
"Yours faithfully,
"*Dr. Gladstone.*" "A. De Morgan."

Reverting to the list of elements given above, we certainly

Y 2

find no recurrence of certain numbers, or other peculiarity, sufficiently striking to warrant us in drawing any inference; but let us arrange the elements according to their chemical relations, and the case will be entirely altered. Any arrangement of the elements is attended with difficulty: I shall not attempt to form one of my own, as it would be open to the objection that my mind had been biassed by dwelling upon the numerical relations; but I shall adopt that given in Gmelin's Handbook of Chemistry at the commencement of vol. ii.

O	N	H
F Cl Br I		L Na K
S Se Te		Mg Ca Sr Ba
P As Sb	G Er Y Tr Ce Di La	
C B Si	Zr Th Al	
Ti Ta Nb Pe W	Sn Cd Zn	
Mo V Cr	U Mn Co Ni Fe	
Bi Pb Ag Hg Cu		
Os Ru Ir R Pt Pd Au		

Even here many elements are grouped together which have but a faint chemical resemblance.   Thus tin has little in common with cadmium and zinc, or mercury with copper; fluorine is very different from the other halogens; magnesium can scarcely rank with the metals of the alkaline earths; whilst late researches have shown the strict isomorphism of chromium with manganese or iron.

If we substitute the equivalent numbers in this arrangement, the slightest glance will make us acquainted with many remarkable resemblances.   Decimals are omitted for the sake of brevity.

8	14	1
19, 35, 80, 127		6, 23, 39
16, 39, 64		12, 20, 44, 68
31, 75, 129	5, —, —, —, 47, 50, 47	
6, 11, 21	22, 60, 14	
25, 184, —, —, 92	58, 56, 33	
46, 69, 27	60,	28, 29, 30, 28
208, 104, 108,	100, 32	
100, 52, 99,	52,	99, 53, 197.

Looking more closely into this arrangement of numbers, we shall find the observation of Berzelius borne out in every instance, but one, of a well-defined chemical group.

These numerical relations are of three kinds. The atomic weights of analogous elements may be the same; or may be in multiple proportion; or may differ by certain increments.

Of the first class we remark the strictly analogous metals,—chromium 26·7, manganese 27·6, iron 28, cobalt 29·5, and nickel 29·6. Then a double group of the platinum ore metals :—palladium 53·3, rhodium, 52·2, and ruthenium 52·2; and also platinum 98·7, iridium 99, and osmium 99·6. We are tempted to add to this—mercury 100. Again, in the mineral cerite we find together—cerium 47, lanthanium 47, and didymium 50. It has been remarked, not only that the metals of each of these groups have similar properties and weights, but that they are found associated together in nature. The question has often been put,—Would more accurate determinations show these atomic weights to be not nearly but exactly the same? It may be doubted. Yet it ought to be remembered that these numbers are the actual results of experiment, and are not controled by any theory, as is always the case with organic compounds.

As to the second class of numerical relations among atomic weights, namely, multiple proportions, Who has failed to remark that the platinum group has double the atomic weight of the palladium group, and that gold 197 is again the double of platinum? These two pairs have frequently been noticed—boron 10·9, and silicon 21·3; oxygen 8, and sulphur 16. We now come to a large group, those metals whose oxides principally affect an acid character, being also insoluble in water. The highest of these in weight is tantalum 184; half 184 is 92—the equivalent of tungsten; half 92 is 46—the equivalent of molybdenum; and half 46 is 23—just below the recognized equivalent of titanium 25*. Three times 23 is 69: now 68·6 is the equivalent of vanadium, being intermediate between tungsten and molybdenum. Tin has certain claims to be grouped along with the same elements; its equivalent is 58: now 57·5 would be two and a half times 23, and intermediate between molybdenum and vanadium. Taking 11·5 as the basis number of this series, we have—

		Received equiv.
Titanium . .	$2 \times 11\cdot5 = 23$	25
Molybdenum .	$4 \times 11\cdot5 = 46$	46
Tin	$5 \times 11\cdot5 = 57\cdot5$	58
Vanadium . .	$6 \times 11\cdot5 = 69$	68·6
Tungsten . .	$8 \times 11\cdot5 = 92$	92
Tantalum . .	$16 \times 11\cdot5 = 184$	184

* 23·6 according to Mosander.

Silicon is certainly very similar to the first of this group, titanium, but its atomic number 14·2 (reckoning silica to be SiO^2) is unconformable.

These metals, whose numbers have a multiple relationship, are not remarkable for being found together in nature.

The third kind of relationship is where an element, having properties intermediate between those of two other elements, has the intermediate atomic weight. Four instances of this have been noticed*. We observe all four groups in the arrangement of Gmelin given above.   They are—

The metals of the alkalies :—
> Lithium, 6·5.   Sodium, 23.   Potassium, 39·2.

The metals of the alkaline earths :—
> Calcium, 20.   Strontium, 43·8.   Barium, 68·5.

The halogens :—
> Chlorine, 35·5.   Bromine, 80.   Iodine, 127·1.

Sulphur and its congeners :—
> Sulphur, 16.   Selenium, 39·5.   Tellurium, 64·2.

The members of the last two groups generally occur together in nature.

There are certain analogies which may perhaps lead us to some understanding of these facts.

First, in the case where there is the same atomic weight. If the allotropism of an element were carried through all its compounds, we should have what occurs in the iron and similar series.   The only partial instances of this which I remember are the sulphides of phosphorus, as remarked by Berzelius, and silicic acid, the two conditions of which bear a striking analogy to the two allotropic forms of the elementary silicon itself.

Secondly, in the case of multiple atomic weights, is there not something analogous to the polymerizing of which we have many examples in organic chemistry ? or to the modifications of a metal, such as mercury, where we have 100 or 200 parts combining with 1 equivalent of chlorine to form a salt ? the difference in this latter case being, that when we regenerate the metal, from whatever source, it is always the same mercury.

Thirdly, in the case where an element of intermediate properties has an intermediate atomic weight. I regard this as strictly analogous to the series of homologous bodies so common in organic chemistry.   My meaning may be best explained perhaps by a reference to the quasi-metals, or compound hydrogens. We have hydrogen 1, methyle 15, æthyle 29, amyle 71 : the com-

---

* See Gmelin's Handbook of Chemistry, part 1. Dumas also brought forward some speculations on these groups at the Ipswich Meeting of the British Association.

pounds of these bodies differ progressively in properties—the boiling-point for instance—and they occur together in the processes of their preparation. Methyle is intermediate in chemical characters between hydrogen and æthyle, and has the intermediate atomic weight $\frac{1+29}{2}=15$. If we did not know in what respect the one quasi-metal differed from another, we should have a series of bodies precisely analogous to the metals of the alkalies: but we do know it; we know that methyle is hydrogen plus a certain increment $C^2 H^2$; æthyle is hydrogen plus twice $C^2 H^2$, &c. The general expression for any such homologous series is, taking x as the increment,—

$$a; \quad a+x; \quad a+2x; \quad a+3x, \&c.$$

This will equally apply to any other cases of the addition of increments—the conjugate organic acids, such as formo-benzoic acid, or the series phosphoric acid PO^5, 71, azophosphoric acid $P^2 NO^5$, 116, and deutazophosphoric acid $P^3 N^2 O^5$, 161. Now it is precisely in like manner that I regard these series of elements: I view sodium as lithium plus a certain increment; in fact, $Na = L + x$, and $K = L + 2x$. Lithium is here the starting-point, the "hydrogen" of the series; and so in like manner are calcium, chlorine, and sulphur. We do not know what the increments are, but we know their atomic weights.

In the lithium series it is 16·3
...   calcium   ...   24·2
...   chlorine   ...   45·8
...   sulphur   ...   24·1

It is remarkable that the increments of the calcium and sulphur series are the same in weight—24, and that the increment of the lithium series should be almost exactly two-thirds of that number.

Why should this numerical relation always give us triads? As yet we have no instance of a fourth member of one of these series, unless indeed we view the titanium series in this light; but the advance of science may furnish us with such instances if my theory of increments be correct. Whether any element not contained in these groups have the atomic weight of some other plus a certain increment, cannot be known until we have a third member of the series to prove the fact. Thus we can only speculate upon the curious circumstance, that zinc and cadmium—two similar metals occurring together in nature—have atomic weights differing almost by the remarkable number 24. Zn 32·6 + 23·4 = 56 at. wt. of Cd.

I believe that, with one exception, every well-defined group of elements has been considered under these three classes. That exception is in the case of arsenic and antimony, which are unquestionably analogous, phosphorus being also closely allied to them: the three are placed together in Gmelin's arrangement, but I see no ratio between their numbers 31, 75, 129. Schrœtter's reduction of the equivalent of phosphorus has prevented our considering antimony, 129, as four times that element. It is, however, double tellurium, 64·2, which is in some respects an analogous body.

There are several elements, such as bismuth, which have no very evident analogues; and of others which are similar we cannot speak, because we are ignorant of their atomic weights. These are pelopium and niobium, and yttrium erbium and terbium.

Alumina is usually classed with the earths; but its compounds are strictly isomorphous with those of the sesquioxides of the iron group. The equivalent of aluminium, 13·7, happens to be half of theirs. Again, glucina is certainly an analogous earth; viewing it also as a sesquioxide, the equivalent of glucinum is 7, which is half that of aluminium.

Some of the properties of lead would ally it to the family of the alkaline earths, but there is no apparent numerical relation: in other points it resembles silver, with which it is so generally found; their respective equivalents are 103·7 and 108·1, but neither the resemblance of chemical properties nor of number is very close. I would rather not consider it an instance of the general law.

Whatever may be thought of some of the speculations towards the close of this paper, these numerical relations are indisputable facts. That we should frequently find relations between 56 numbers drawn at hazard from a range little exceeding 100 might be predicted from the laws of probability, but that this should be to a considerable extent coincident with chemical relationship is not probable. Still more is it against all probability that, by mere chance, whenever, with one exception, close analogy of properties exists, there exists also numerical relationship; and although we cannot now see the precise reason of this, we can scarcely imagine that the intimate constitution of these related elementary bodies will long remain an unfruitful field of investigation.

THE CHEMICAL NEWS.

Vol. II. No. 27. — June 9, 1860.

TO OUR READERS.

THE commencement of our second volume affords the opportunity for saying a few words on the present position and future prospects of the CHEMICAL NEWS. Originally started to supply a want which had long been felt and expressed in scientific circles here in London, this journal has attained a circulation on the continent of Europe and in America, as well as in Great Britain, which we believe was never reached by any other periodical devoted to the same department of knowledge. This flattering result of our labours we accept as a proof of success, and also as a stimulus to future exertions.

In no science is the necessity for a journal, published at short intervals, so clearly seen, and the advantages of one so distinctly felt, as in Chemistry. So extensive is the science, and so numerous are the experimenters engaged in its pursuit, that it may be said no day passes without the discovery of some new compound, or the registration of some new fact. To record these, and thus rapidly diffuse a knowledge of scientific progress, not only throughout Great Britain, but over the rest of the world, has now been recognised by scientific men as the especial province of the CHEMICAL NEWS.

But it is not only science in the abstract which demands the attention of the journalist. The applications of Chemistry to the arts and manufactures are so numerous as almost alone to require a special journal for their development and discussion. In consequence of the extended reports we have hitherto given of the meetings of the learned societies the room for technical articles has necessarily been at times limited. The demands on our space having lately become so numerous, we intend in future to reserve verbatim reports of lectures only for the most important, believing that our readers would in most instances be as well satisfied with a condensed abstract. This will leave us more room for articles of a practical character on various subjects of commercial and technical interest.

The permanent value of a work which contains so great a number of facts distributed over so large a number of pages, necessarily depends on the perfection of the Index, and consequently the greatest pains have been taken to render that of the volume just completed as copious and perfect as possible. It has occupied the unremitting attention of the Editor for many weeks past, and it is confidently hoped that it will in great measure tend to raise the CHEMICAL NEWS from the ephemeral position of a mere periodical into that of a standard work of reference.

SCIENTIFIC AND ANALYTICAL CHEMISTRY.

Note on the Numerical Relations between the Specific Gravities of the Diamond, Graphite, and Charcoal Forms of Carbon and its Atomic Weight, by Dr. LYON PLAYFAIR, C.B. F.R.S.

[*Read before the Royal Society of Edinburgh.*]

RECENT researches have shown that there is an intimate relation between the specific gravities and atomic weights or equivalents of solid or liquid bodies. This relation is not so simple as that which prevails in regard to the volumes and combining numbers of gaseous bodies, and yet it is sufficiently marked to indicate many important chemical analogies. The formula for eliciting these relations is—

$$\frac{E}{d} = V,$$

in which E is the equivalent, d the specific gravities, and V the atomic volume.

It is to be borne in mind, that the unities or starting-points for specific gravities and for atomic weights are essentially distinct. In the first case, the weights of the bodies are compared with the weight of an equal bulk of water; in the second instance, the combining numbers refer to a unit weight of hydrogen. Nevertheless, the relations observed between the specific gravities and the atomic weights are well marked in bodies of a like character.

It has always been considered interesting to examine these relations in regard to carbon, which has three well characterised allotropic forms. The atomic volumes obtained by the above formula show no satisfactory relations between the numbers obtained for each of the states in which the element presents itself.

Before we examine them in another way, it is desirable to obtain a mean specific gravity for the diamond, graphite, and charcoal, as the recorded results of experiment show a considerable variation.

1. **Diamond.**—The specific gravity of this gem is generally stated in elementary works to range from 3·5 to 3·55; but these numbers do not represent the mean of recorded experiments, as will be seen by the following table:—

Diamond in Hunterian Museum, Glasgow	3·53	Thomson.[1]
Specific gravity, as stated by Mohs	3·52	Mohs.[2]
Brazilian diamond	3·44	Brisson.[3]
Another variety of the same	3·52	
Mean specific gravity of a "beautiful collection of diamonds"	3·48	Lowry.[4]
"Star of the south"	3·53	Dufrenoy & Halphin.[5]

1 *Thomson's Mineralogy*, vol. i. p. 46.
2 *Mohs' Mineralogy*, vol. ii. p. 306.
3 *Brisson*, as quoted by *Böttger, Specifisches Gewicht*, p. 32.
4 *Lowry*, as quoted by *Thomson's Mineralogy*, vol. i. p. 46.
5 *Dufrenoy, Comptes-Rendus*, vol. xl. p. 3.

Borneo diamond . . .	3·49	Grailich.[6]
Do. do. compact . . .	3·41	} Rivot.[7]
Do. do. do. .	3·25	
Diamond used in Jacquelain's experiments .	3·33	Jacquelain.[8]
Specific gravity, as given by Henry .	3·55	Henry.[9]
Well crystallised Brazilian diamond, weighing 0·5761 gramme in the Edinburgh Museum . .	3·48	Playfair.[10]
Mean sp. gr.	3·461	

If we reject the second Borneo diamond of Rivot, which has too low a specific gravity, we have a mean sp. gr. of 3·48, which is the same number as that found by Wilson Lowry for the mean specific gravity of "his beautiful collection of crystallised diamonds." (*Thomson's Mineralogy*, vol. i. p. 46.)

It is to be expected that the experimental determination of the specific gravity of diamonds should be rather above than below the truth; for we are aware that they all leave a minute quantity of ash on burning, and that this ash, according to Petzhold, contains silica and iron.

2. **Graphite.**—This variety of carbon is often impure, being not unfrequently contaminated with upwards of five per cent. of earthy impurities. Recorded specific gravities upon such impure specimens are of no value for the mean result as regards pure graphite. The following determinations are all those which I can find upon specimens which have been chemically examined to establish their purity :—

Natural graphite . . .	2·27	Regnault.[11]
Do. . . .	2·25	} Schrader.[12]
Do. . . .	2·32	
Graphite of iron furnaces . .	2·33	Karsten.[13]
Natural graphite, in fine crystalline plates .	2·14	} Breithaupt.[13]
Do. do. another variety	2·22	
Do. do. do. .	2·23	Kengott.[14]
Natural graphite . .	2·50	{ Pelouze and Fremy.[15]
Gas carbon graphite . . .	2·35	Graham.
Mean sp. gr. .	2·29	

It would have been interesting to have added to this list a determination of the specific gravity of Brodie's purified Ceylon graphite; but its minute division causes the air to adhere to it so tenaciously that I have failed in getting any correct determinations of its density.

3. **Charcoal.**—There are comparatively few determinations of the specific gravity of pure charcoal. It is in fact not so easy to obtain this substance. A specimen of charcoal from pure sugar, repeatedly calcined, and treated with chlorine to remove the last traces of hydrogen, and again calcined, gave me the sp. gr. 1·80; but bubbles of air still adhered to it, although it was kept for several hours under a good air-pump. The following determinations are those recorded :—

Pure lamp-black . .	1·78	Baudrimont.[16]
Fibrous gas coke . .	1·76	Colquhoun.[17]
Compact gas carbon .	2·08	Baudrimont.[18]
Powdered coke (mean) .	1·80	Regnault.[19]
Charcoal from alcohol .	2·10	Scholtz.[20]
Charcoal from sugar .	1·80	Playfair.[21]
Pure charcoal, without pores	1·84	Griffith.[22]
Mean sp. gr. .	1·88	

4. From the preceding data we take the mean specific gravity of the three varieties of carbon to be as follows :—

	Mean Sp. Gr.
Diamond . . .	3·48 or 3·461
Graphite . . .	2·29
Charcoal . . .	1·88

5. We have now to consider whether these numbers stand in any simple relation to their atomic weight. The formula

$$\frac{E}{d} = V$$

gives the following atomic volumes, taking $C = 12$.

	Atomic Volumes.
Diamond . . .	3·44
Graphite . . .	5·24
Charcoal . . .	6·38

These numbers do not bear to each other any simple relation.

6. If we now take the atomic weight of carbon ($C = 12$), and then extract from it its square, cube, and fourth roots, numbers are obtained which bear a striking approximation to the mean specific gravity of the three forms of carbon :—

	Roots.		Sp. Gr.
1 -	$\sqrt[2]{12} = 3·464$	-	Diamond, 3·48 or 3·46
2 -	$\sqrt[3]{12} = 2·289$	-	Graphite, 2·29
3 -	$\sqrt[4]{12} = 1·865$	-	Charcoal, 1·88

In other words, if we raise the specific gravity of diamond to its second power, that of graphite to its third power, and that of charcoal to its fourth power, we obtain numbers closely approaching in each case to 12, the atomic weight of carbon.

Diamond . . .	$3·48^2 = 12·11$
Graphite . . .	$2·29^3 = 12·00$
Charcoal . . .	$1·88^4 = 12·49$

These approximations are remarkable, and the relations of the numbers are natural and simple. The differences between the mean experimental numbers and the corresponding roots of the atomic weight of carbon are not so great as the differences observed in the specific gravities of the same form of carbon.

7. It may be useful to condense into the form of a table the previous observations :—

Forms of Carbon.	Experiment.		Calculations.	
	Sp. Gr.	Power	C	Roots.
Diamond .	3·46 or 3·48	$3·48^2 = 12·11$		$\sqrt[2]{12} = 3·464$
Graphite .		$2·29^3 = 12·00$		$\sqrt[3]{12} = 2·289$
Charcoal .		$1·88^4 = 12·49$		$\sqrt[4]{12} = 1·865$

These relations appear to be so simple that it is scarcely possible to conceive that they may not have been described before; but I have been unable to find such descriptions. The nearest approach to it which I know is the fact that Mr. Hawksley, the engineer, stated to me, many years since, that he had brought under the attention of the late Mr. Cooper the relation which seemed to subsist between the specific gravities of silver and gold and their atomic weights, this being approx-

6 *Grailich Bull. Geol.* [2] vol. xiii. p. 542.
7 *Rivot, Ann. des Mines*, vol. xiv. p. 423.
8 *Jacquelain, Ann. de Ch. et Phys.* [2] vol. xx. p. 459.
9 *Henry's Mineralogy*, vol. iv. p. 19.
10 Experiment made for this paper.
11 *Regnault, Ann. de Ch. et. Phys.* vol. lvi. p. 37.
12 *Schrader, Annals of Philosophy*, vol. i. p. 299.
13 As quoted in *Bottger's Specifisches Gewicht*.

14 *Kengott, Wien Akad.* vol. xiii. p. 469.
15 *Traité de Chimie*, vol. v. p. 518.
16 *Baudrimont, Traité de Chimie*, vol. i. p. 511.
17 *Colquhoun's Annals of Philosophy* [2] vol. xii. p. 1.
18 *Baudrimont, Traité de Chimie*, vol. i. p. 514.
19 *Regnault, Traité de Chimie*, vol. i. p. 369.
20 *Böttger Specifisches Gewicht.*
21 Experiment recorded above.
22 *Böttger Specifisches Gewicht.*

matively the square root of their atomic weights, or of multiples of these numbers. But I cannot find any record either of Mr. Hawksley's or Mr. Cooper's views on the subject.

PHARMACY, TOXICOLOGY, &c.

On the Chemical Reactions of Strychnia, by H. LETHEBY, *M.B. &c. Professor of Chemistry and Toxicology in the College of the London Hospital.*

THE medico-legal chemistry of strychnia has been very fully investigated during the last few years, and Dr. Wormley's paper on the chemical reactions of the alkaloid is a valuable *resumé* of the subject.[1] It will be noticed, however, that his results are always obtained by adding the tests to a known quantity of the *pure* alkaloid, a condition which is not at once secured in toxicological research. Experience, therefore, has shown that some of the tests must be applied with certain precautions, or the results will be very unsatisfactory. Those who are conversant with the practice of the matter will not agree with Dr. Wormley that the colour test, for example, is to be employed in the way described by him. "We have succeeded best," he says, "by placing the strychnia, or a drop of the solution evaporated to dryness, in a watch glass, and by its side a drop of sulphuric acid into which a fragment of bichromate of potash was introduced, and stirred until it imparted a yellow colour, then by inclining the watch glass the coloured sulphuric acid was allowed to flow over the strychnia.[2]" This mode of experimenting is dangerous in every way, for, in the first place, if the bichromate is in excess, and the quantity of strychnia small, the colour is so evanescent as to be uncertain, and, secondly, if organic matter be present, the colour will be masked; and, thirdly, if a nitrate or chloride is with the strychnia, the colour is not produced at all; and, fourthly, if strychnia be entirely absent, but sugar and bile, or piperine, or any of those substances be present which give a purple or red reaction with sulphuric acid, a fallacious result will be obtained; and lastly, if the result fails, the whole of the strychnia is lost, and the inquiry brought to an unsatisfactory conclusion.

So strongly have I felt these difficulties, that in my paper on the *Medico-legal Chemistry of Strychnia*, published in the *Lancet* of June and July 1856,[3] I have endeavoured to guard the operator against them by describing a process which generally ensures success. It is as follows:—"First place the strychnia, or the suspected matter, on a clean white plate; then touch it with a small drop of concentrated sulphuric acid (the acid should be free from nitric acid); stir it about with a glass rod, so as to mix the strychnia very perfectly with the acid; allow it to remain in this state for a few minutes, and if the strychnia be pure there will be no discoloration.[4] Then cautiously add the reagent, namely, peroxide of lead, bichromate of potash, or peroxide of manganese, taking care not to add too much of it; in fact, it is best done by dropping the powder into the oil of vitriol and strychnia from the point of a penknife. Lastly, either incline the plate so that the acid may gently flow over the powder, or else with great caution stir the powder about with the point of a glass rod. In this way the colour is always sure to be brought out, and, as far as I

know, it is not to be confounded with the reaction of any other substance. Indeed, the only thing which approaches it in appearance is the dirty violet colour which is occasioned by morphia and its salts when they are treated in the same way. As to the so-called fallacies to the test, namely, salicine, bile, sugar, pyroxanthine, piperine, resinous matters, and many other things, it must be manifest that they are not fallacies when the test is properly performed, for all these compounds acquire their colour directly the sulphuric acid is added to them, and *before* the other reagent is applied.

"Of all the substances which have been proposed for thus developing the tints with strychnia, bichromate of potash is assuredly the worst, for—

"1st. It is itself coloured by the acid, and may thus complicate the result.

"2nd. It will not act when organic matter is present, as for example, the vegetable acids,—citric and tartaric, cream of tartar, tartar-emetic, potassio-tartrate of soda, the residue of an effervescing draught, sugar, gum, and even a little morphia.

"3rd. It will not act when nitre, nitric acid, or common salt are present with the strychnia.

"4th. It is of all the tests the least delicate: for while the peroxide of manganese, or the peroxide of lead, will discover the presence of the $\frac{1}{20000}$th of a grain of strychnia, the bichromate will not act well with less than the $\frac{1}{2000}$th of a grain.

"It is true that by means of the process which I shall hereafter detail for extracting the alkaloid, none of those impurities will be present; yet, in making a comparison of the respective values of the several tests, it is right to know that the bichromate of potash reaction is the least satisfactory.[5]"

The true cause of its want of delicacy is the rapidity with which it oxydises the alkaloid; and therefore the peroxides of manganese and lead, because of their evolving oxygen slowly, are more delicate and suitable to the purpose, for their reactions take time to develop and the colour is far more enduring. Finding that the *modus operandi* of the test was the action of nascent oxygen upon the strychnia, it occurred to me that the galvanic current might be used instead of the oxygen compound.

"The mode of applying the galvanic test is as follows:—Place a drop of a solution of strychnia (say of one part of the alkaloid in 10,000, or even 20,000, of water) into a cup-shaped depression made in a piece of platinum foil. Allow the fluid to evaporate, and, when dry, moisten the spot with a drop of concentrated sulphuric acid. Connect the foil with the positive pole of a single cell of Grove's or Smee's battery, and then touch the acid with a platinum terminal of the negative pole. In an instant the violet colour will flash out, and, on removing the pole from the acid, the tint will remain."[6]

By this mode of proceeding the colour is perfectly under control, and the test is free from every known source of fallacy. I have lately applied it in the recognition of strychnia in the urine of a girl poisoned at Peterborough, and the results were very satisfactory. They were confirmed by the physiological action of the poison, obtained from the urine, on frogs.

Lastly, I would remark that the carbazotic acid, and iodine tests, which Dr. Wormley says he has not seen described, are fully discussed in my paper, and drawings of the crystalline precipitates are given.[7]

---

[1] CHEMICAL NEWS, pp. 218 and 242.   [2] *Ibid.* p. 243.

[3] See *Lancet*, June 28, 1856, and July 12, 1856.

[4] The process to be followed for rendering the strychnia pure is detailed at page 37 of the *Lancet* for July 12, 1856; it consists in treating the impure strychnia with concentrated sulphuric acid until all the impurities are destroyed and then extracting with chloroform or ether.

[5] *Lancet*, June 28, 1856, p. 708.   [6] *Ibid.*

[7] *Ibid.* p. 707, and July 12, p. 37.

PROCEEDINGS OF SOCIETIES.

ROYAL INSTITUTION.
Friday, May 31, 1867.
On the Chemistry of the Primeval Earth*, by T. STERRY
HUNT, M.A., F.R.S.

THE subject of my lecture this evening, as has been
announced, is the Chemistry of the Primeval Earth. The
natural history of our planet, to which we give the name
of geology, is, necessarily, a very complex science, including, as it does, the concrete sciences of mineralogy, of
botany, and zoology, and the abstract sciences of chemistry
and physics. These latter sustain a necessary, and very
important relation, to the whole process of development of
our earth from its earliest ages, and we find that the same
chemical laws which have presided over its changes, apply
also to those of extra-terrestrial matter. Recent investigations show the presence in the sun, and even in the fixed
stars—suns of other systems—the same chemical elements
as in our own planet. The spectroscope, that marvellous
instrument, has, in the hands of modern investigators,
thrown new light upon the composition of the farthest
bodies of the universe, and has made clear many points
which the telescope was impotent to resolve. The results
of extra-terrestrial spectroscopic research have lately been
set forth in an admirable manner by one of its most successful students, Mr. Huggins. We see by its aid matter
in all its stages, and trace the process of condensation and
the formation of worlds. It is long since Herschel, the
first of his illustrious name, conceived the nebulæ, which
his telescope could not resolve, to be the uncondensed
matter from which worlds are made. Subsequent astro-
nomers, with more powerful glasses, have been able to
show that many of these nebulæ are really groups of stars,
and thus a doubt was thrown over the existence in space
of nebulous luminous matter: but the spectroscope has
now placed the matter beyond doubt. We thus find in the
heavens planets, bodies like our earth, shining only by
reflected light; suns, self-luminous, radiating light from
solid matter; and, moreover, true nebulæ, or masses of
luminous gaseous matter. These three forms represent
three distinct phases in the condensation of the primeval
matter, from which our own and other planetary systems
have been formed.

This nebulous matter is conceived to be so intensely
heated as to be in the state of true gas or vapour, and, for
this reason, feebly luminous when compared with the sun.
It would be out of place, on the present occasion, to discuss
the detailed results of spectroscopic investigation, or the
beautiful and ingenious methods by which modern science
has shown the existence in the sun, and in many other
luminous bodies in space, of the same chemical elements
that are met with in our earth, and even in our own bodies;
realising, in a most literal manner, the genial intuition of
the poet, who

> "Sees alike in stars and flowers a part
> Of the self-same universal being
> That is throbbing in his mind and heart."

Calculations based on the amount of light and heat
radiated from the sun show that the temperature which
reigns at its surface is so great that we can hardly form an
adequate idea of it. Of the chemical relations of such
intensely heated matter modern chemistry has made known
to us some curious facts, which help to throw light on the
constitution and luminosity of the sun. Heat, under
ordinary conditions, is favourable to chemical combination, but a higher temperature reverses all affinities. Thus,
the so-called noble metals, gold, silver, mercury, &c., unite
with oxygen and other elements; but these compounds are
decomposed by heat, and the pure metals are regenerated.
A similar reaction was many years since shown by Mr.

* Reported specially for this paper, and revised by the author.

Grove with regard to water, whose elements—oxygen and
hydrogen—when mingled and kindled by flame, or by the
electric spark, unite to form water, which, however, at a
much higher temperature, is again resolved into its component gases. Hence, if we had these two gases existing
in admixture at a very high temperature, cold would
actually effect their combination precisely as heat would
do if the mixed gases were at the ordinary temperature,
and literally it would be found that "frost performs the
effect of fire." The recent researches of Henry Ste.-Claire
Deville and others go far to show that this breaking up
of compounds, or dissociation of elements by intense heat,
is a principle of universal application; so that we may
suppose that all the elements, which make up the sun or
our planet, would, when so intensely heated as to be in that
gaseous condition which all matter is capable of assuming, be uncombined—that is to say, would exist together
in the condition of what we call chemical elements, whose
further dissociation in stellar or nebulous masses may even
give us evidence of matter still more elemental than that
revealed by the experiments of the laboratory, where we
can only conjecture the compound nature of many of the
so-called elementary substances.

The sun, then, is to be conceived as an immense mass
of intensely heated gaseous and dissociated matter, so
condensed, however, that, notwithstanding its excessive
temperature, it has a specific gravity not much below
that of water; probably offering a condition analogous
to that which Cagniard de la Tour observed for volatile bodies when submitted to great pressure at temperatures much above their boiling point. The radiation
of heat, going on from the surface of such an intensely
heated mass of uncombined gases, will produce a superficial cooling, which will permit the combination of certain
elements and the production of solid or liquid particles,
which, suspended in the still dissociated vapours, become
intensely luminous and form the solar photosphere. The
condensed particles, carried down into the intensely heated
mass, again meet with a heat of dissociation, so that the
process of combination at the surface is incessantly renewed, while the heat of the sun may be supposed to be
maintained by the slow condensation of its mass; a diminution by $\frac{1}{1000}$th of its present diameter being sufficient,
according to Helmholtz, to maintain the present supply of
heat for 21,000 years.

This hypothesis of the nature of the sun and of the
luminous process going on at its surface is the one
lately put forward by Faye, and, although it has met
with opposition, appears to be the one which accords
best with our present knowledge of the chemical and
physical conditions of matter, such as we must suppose it to exist in the condensing gaseous mass, which
according to the nebular hypothesis, should form the
centre of our solar system. Taking this, as we have
already done, for granted, it matters little whether we
imagine the different planets to have been successively
detached as rings during the rotation of the primal mass,
as is generally conceived, or whether we admit with
Chacornac a process of aggregation, or concretion, operating
within the primal nebular mass, resulting in the production of sun and planets. In either case we come to the
conclusion that our earth must at one time have been in
an intensely heated gaseous condition, such as the sun
now presents, self-luminous, and with a process of condensation going on at first at the surface only, until by
cooling it must have reached the point where the gaseous
centre was exchanged for one of combined and liquefied
matter.

Here commences the chemistry of the earth, to the
discussion of which the foregoing considerations have
been only preliminary. So long as the gaseous condition
of the earth lasted, we may suppose the whole mass to
have been homogeneous; but when the temperature became
so reduced that the existence of chemical compounds at

the centre became possible, those which were most stable at the elevated temperature then prevailing, would be first formed. Thus, for example, while compounds of oxygen with mercury or even with hydrogen could not exist, oxides of silicon, aluminium, calcium, magnesium, and iron might be formed and condense in a liquid form at the centre of the globe. By progressive cooling, still other elements would be removed from the gaseous mass, which would now become the atmosphere of the non-gaseous nucleus. We may suppose an arrangement of the condensed matters at the centre according to their respective specific gravities, and thus the fact that the density of the earth as a whole is about twice the mean density of the matters which form its solid surface. Metallic or metalloidal compounds of elements grouped differently from any compounds known to us, and far more dense, may exist in the centre of the earth.

The process of combination and cooling having gone on, until those elements which are not volatile in the heat of our ordinary furnaces, were condensed into a liquid form, we may here inquire what would be the result, upon the mass, of a further reduction of temperature. It is generally assumed that in the cooling of a liquid globe of mineral matter, congelation would commence at the surface, as in the case of water; but water offers an exception to most other liquids, inasmuch as it is denser in the liquid than in the solid form. Hence ice floats on water, and freezing water becomes covered with a layer of ice, which protects the liquid below. With most other matters, however, and notably with the various mineral and earthy compounds analogous to those which may be supposed to have formed the fiery-fluid earth, numerous and careful experiments show that the products of solidification are much denser than the liquid mass; so that solidification would have commenced at the centre, whose temperature would thus be the congealing point of these liquid compounds. The important researches of Hopkins and Fairbairn on the influence of pressure in augmenting the melting point of such compounds as contract in solidifying, are to be considered in this connexion.

It is with the superficial portions of the fused mineral mass of the globe that we have now to do, since there is no good reason for supposing that the deeply seated portions have intervened in any direct manner in the production of the rocks which form the superficial crust. This, at the time of its first solidification, presented probably an irregular, diversified surface, from the result of contraction of the congealing mass, which at last formed a liquid bath of no great depth, surrounding the solid nucleus. It is to the composition of this crust that we must now direct our attention, since therein would be found all the elements (with the exception of such as were still in the gaseous form) now met with in the known rocks of the earth. This crust is now everywhere buried beneath its own ruins, and we can only, from chemical considerations, attempt to reconstruct it. If we consider the conditions through which it has passed, and the chemical affinities which must have come into play, we shall see that they are just what would now result if the solid land, sea, and air were made to react upon each other under the influence of intense heat. To the chemist it is at once evident that from this would result the conversion of all carbonates, chlorides, and sulphates into silicates, and the separation of the carbon, chlorine, and sulphur in the form of acid gases, which, with nitrogen, watery vapour, and a probable excess of oxygen, would form the dense primeval atmosphere. The resulting fused mass would contain all the bases as silicates, and must have much resembled in composition certain furnace slags or volcanic glasses. The atmosphere, charged with acid gases which surrounded this primitive rock, must have been of immense density. Under the pressure of such a high barometric column, condensation would take place at a temperature much above the present boiling point of water, and the depressed

portions of the half-cooled crust would be flooded with a highly heated solution of hydrochloric acid, whose action in decomposing the silicates is easily intelligible to the chemist. The formation of chlorides of the various bases, and the separation of silica, would go on until the affinities of the acid were satisfied, and there would be a separation of silica taking the form of quartz, and the production of a sea-water holding in solution, besides the chlorides of sodium, calcium, and magnesium, salts of aluminium and other metallic bases. The atmosphere, being thus deprived of its volatile chlorine and sulphur compounds, would approximate to that of our own time, but differ in its greater amount of carbonic acid.

We next enter into the second phase in the action of the atmosphere upon the earth's crust. This, unlike the first, which was subaqueous, or operative only on the portion covered with the precipitated water, is sub-aerial, and consists in the decomposition of the exposed parts of the primitive crust under the influence of the carbonic acid and moisture of the air, which would convert the complex silicate of the crust into a silicate of alumina, or clay, while the separated lime, magnesia, and alkalies, being converted into carbonates, would be carried down into the sea in a state of solution. The first effect of these dissolved carbonates would be to precipitate the dissolved alumina and the heavy metals, after which would result a decomposition of the chloride of calcium of the sea-water, resulting in the production of carbonate of lime or limestone, and chloride of sodium or common salt. This process is one still going on at the earth's surface, slowly breaking down and destroying the hardest rocks, and, aided by mechanical processes, transforming them into clays; although the action, from the comparative rarity of carbonic acid in the atmosphere, is less energetic than in earlier times, when the abundance of this gas and a higher temperature, favoured the chemical decomposition of the rocks. But now, as then, every clod of clay formed from the decay of a crystalline rock corresponded to an equivalent of carbonic acid abstracted from the atmosphere, and equivalents of carbonate of lime and common salt formed from the chloride of calcium of the sea-water.

It is very instructive, in this connexion, to compare the composition of the waters of the modern ocean with that of the sea in ancient times, whose composition we learn from the fossil sea-waters which are still to be found in certain regions, imprisoned in the pores of the older stratified rocks. These are vastly richer in salts of lime and magnesia than those of the present sea, from which have been separated, by chemical processes, all the carbonate of lime of our limestones, with the exception of that derived from the sub-aerial decay of calcareous silicates belonging to the primitive crust.

The gradual removal, in the form of carbonate of lime, of the carbonic acid from the primeval atmosphere, has been connected with great changes in the organic life of the globe. The air was doubtless at first unfit for the respiration of warm-blooded animals, and we find the higher forms of life coming gradually into existence as we approach the present period of a purer air. Calculations lead us to conclude that the amount of carbon thus removed in the form of carbonic acid has been so enormous, that we must suppose the earlier forms of air-breathing animals to have been peculiarly adapted to live in an atmosphere which would probably be too impure to support modern reptilian life. The agency of plants in purifying the primitive atmosphere was long since pointed out by Brongniart, and our great stores of fossil fuel have been derived from the decomposition, by the ancient vegetation, of the excess of carbonic acid of the early atmosphere, which through this agency was exchanged for oxygen gas. In this connexion the vegetation of former periods presents the curious phenomenon of plants, allied to those now growing beneath the tropics, formerly flourishing within the polar circles. Many ingenious hypotheses have been proposed to account for the warmer climate of earlier times, but are at best unsatisfactory, and it appears to me that the true solution of the problem may be found in the constitution of the early atmosphere, when considered in the light of Dr. Tyndall's beautiful researches on radiant heat. He has found that the presence of a few hundredths of carbonic acid gas in the atmosphere, while offering almost no obstacle to the passage of the solar rays, would suffice to prevent almost entirely the loss by radiation of obscure heat, so that the surface of the land beneath such an atmosphere would become like a vast orchard-house, in which the conditions of climate, necessary to a luxuriant vegetation, would be extended even to the polar regions. This peculiar condition of the early atmosphere cannot fail to have influenced in many other ways the processes going on at the earth's surface. To take a single example : one of the processes by which gypsum may be produced at the earth's surface involves the simultaneous production of carbonate of magnesia. This, being more soluble than the gypsum, is not always now found associated with it, but we have indirect evidence that it was formed, and subsequently carried away, in the case of many gypsum deposits whose thickness indicates a long continuance of the process, under conditions much more perfect and complete than we can attain under our present atmosphere. While studying this reaction I was led to inquire whether the carbonic acid of the earlier periods might not have favoured the formation of gypsum, and I found, by repeating the experiments in an artificial atmosphere impregnated with carbonic acid, that such was really the case. We may thence conclude that the peculiar composition of the primeval atmosphere, was the essential condition under which the great deposits of gypsum, generally associated with magnesian limestones, were formed.

The reactions of the atmosphere which we have considered, would have the effect of breaking down and disintegrating the surface of the primeval globe, covering it everywhere with beds of stratified rock of mechanical or of chemical origin. These would now so deeply cover the partially cooled surface that the amount of heat escaping from below is inconsiderable, although in earlier times it was very much greater, and the increase of temperature met with in descending into the earth must have been many times more rapid than now. The effect of this heat upon the buried sediments would be to soften them, producing new chemical reactions between their elements, and converting them into what are known as crystalline or metamorphic rocks, such as gneiss, greenstone, granite, &c. We are often told that granite is the primitive rock or substratum of the earth, but this is not only unproved, but extremely improbable. As I endeavoured to show in the early part of this lecture, the composition of this primitive rock, now everywhere hidden, must have been very much like that of a slag or lava, and there are excellent chemical reasons for maintaining that granite is in every case a rock of sedimentary origin—that is to say, it is made up of materials which were deposited from water like beds of modern sand and gravel, and includes in its composition quartz, which, so far as we know, can only be generated by aqueous agencies, and at comparatively low temperatures.

The action of heat upon many buried sedimentary rocks, however, not only softens or melts them, but gives rise to a great disengagement of gases, such as carbonic and hydrochloric acids, and sulphur compounds, all results of the reaction of the elements of sedimentary rocks, heated in presence of the water which everywhere filled their pores. In the products thus generated we have a rational explanation of the chemical phenomena of volcanos, which are vents through which these fused rocks and confined gases find their way to the surface of the earth. In some cases, as where there is no disengagement of gases, the fused or half-fused rocks solidify *in situ*, or in rents or fissures in the overlying strata, and constitute eruptive or plutonic rocks like granite and basalt.

This theory of volcanic phenomena was put forward in germ by Sir John F. W. Herschel thirty years since, and, as I have during the past few years endeavoured to show, it is the one most in accordance with what we know both of the chemistry and the physics of the earth. That all volcanic and plutonic phenomena have their seat in the deeply buried and softened zone of sedimentary deposits of the earth, and not in its primitive nucleus, accords with the conclusions already arrived at relative to the solidity of that nucleus ; and also with the remarkable mathematical and astronomical deductions of the late Mr. Hopkins, of Cambridge, based upon the phenomena of precession and nutation ; those of Archdeacon Pratt ; and those of Professor Thompson on the theory of the tides ; all of which lead to the same conclusion—namely, that the earth, if not solid to the centre, must have a crust several hundred miles in thickness, which would practically exclude it from any participation in the plutonic phenomena of the earth's surface, except such as would result from its high temperature communicated by conduction to the sedimentary strata reposing upon it.

The old question between the plutonists and the neptunists, which divided the scientific world in the last generation, was, in brief, this—whether fire or water had been the great agent in giving origin and form to the rocks of the earth's crust. While some maintained the direct igneous origin of such rocks as gneiss, mica-schist, and serpentine, and ascribed to fire the filling of metallic veins, others—the neptunian school—were disposed to shut their eyes to the evidences of igneous action on the earth, and even sought to derive all rocks from a primal aqueous magma. In the light of the exposition which I have laid before you this evening, we can, I think, render justice to both of these opposing schools. We have seen how actions dependent on water and acid solutions have operated on the primitive plutonic mass, and how the resulting aqueous sediments, when deeply buried, come again within the domain of fire, to be transformed into crystalline and so-called plutonic or volcanic rocks.

The scheme which I have endeavoured to put before you in the short time allotted, is, as I have endeavoured to show, in strict conformity with known chemical laws and the facts of physical and geological science. Did time permit, I would gladly have attempted to demonstrate at greater length its adaptation to the explanation of the origin of the various classes of rocks, of metallic veins and deposits, of mineral springs, and of gaseous exhalation. I shall not, however, have failed in my object, if, in the hour which we have spent together, I shall have succeeded in showing that chemistry is able to throw a great light upon the history of the formation of our globe, and to explain in a satisfactory manner some of the most difficult problems of geology ; and I feel that there is a peculiar fitness in bringing such an exposition before the members of this Royal Institution, which has been for so many years devoted to the study of pure science, and whose glory it is, through the illustrious men who have filled, and those who now fill, its professorial chairs, to have contributed more than any other school in the world to the progress of modern chemistry and physics.

CELESTIAL CHEMISTRY FROM THE TIME OF NEWTON.*

By T. STERRY HUNT, LL.D., F.R.S.

THE late W. Vernon Harcourt, in 1845,† called attention to the remarkable perception of great chemical truths which is apparent in the Queries appended to the third book of Newton's " Optics," as well as in his Hypothesis touching Light and Colour. With regard to the latter, Harcourt then remarked. " It has, I think, scarcely been quoted, except by Dr. Young, and its existence is but little known, even among the best-informed scientific men." The essay in question was read before the Royal Society, December 9th and 16th, 1675, but remained unpublished till 1757, when Birch, at that time Secretary to the Society, printed it, not without verbal inaccuracies, in the third volume of his " History of the Royal Society "—a work intended to serve as Supplement to the *Philosophical Transactions* up to that date. In 1846, at the suggestion of Harcourt, the Hypothesis of Newton was again printed in the *L. E. and D. Philosophical Magazine* (vol. xxix.), and it subsequently appeared in the Appendix to the first volume of Brewster's " Memoirs of Sir Isaac Newton," in 1855.

The time has come for further inquiries into the science of Newton, and I shall endeavour to show that a careful examination of the writings of our great natural philosopher, in the light of the scientific progress of the last generation, renders still more evident the wonderful prevision of him who already, two centuries since, had anticipated most of the recent speculations and conclusions regarding cosmic chemistry.

As an introduction to the inquiries before us, and in order to show the real significance of the speculations of Newton, it will be necessary to review, somewhat at length, the history of certain views enunciated almost simultaneously by the late Sir Benjamin Brodie, of Oxford, and the present writer, and subsequently developed and extended by the latter. In Part I. of his " Calculus of Chemical Operations," read before the Royal Society, May 3, 1866, and published in the *Philosophical Transactions* for that year, Brodie was led to assume the existence of certain ideal elements. These, he said, " though now revealed to us through the numerical properties of chemical equations only as *implicit and dependent existences*, we cannot but surmise may sometimes become, or may in the past have been, *isolated and independent existences*." Shortly after this publication, in the spring of 1867, I spent several days in Paris with the late Henri Sainte-Claire Deville, repeating with him some of his remarkable experiments in chemical dissociation, the theory of which we then discussed in its relations to Faye's solar hypothesis. From Paris, in the month of May, I went, as the guest of Brodie, for a few days to Oxford, where I read for the first time and discussed with him his essay on the " Calculus of Chemical Operations," in which connection occurred the very natural suggestion that his ideal elements might perhaps be liberated in solar fires, and thus be made evident to the spectroscope. I was then about to give, by invitation, a lecture before the Royal Institution on " The Chemistry

of the Primeval Earth, which was delivered May 31, 1867. A stenographic report of the lecture, revised by the author, was published in the CHEMICAL NEWS of June 21, 1867, and in the *Proceedings of the Royal Inetitution.* Therein I considered the chemistry of nebulæ, sun, and stars in the combined light of spectroscopic analysis and Deville's researches on dissociation, and concluded with the generalisation that the " breaking-up " of compounds, or dissociation of elements, by intense heat, is a principle of universal application, so that we may suppose that all the elements which make up the sun, or our planet, would, when so intensely heated as to be in the gaseous condition which all matter is capable of assuming, remain uncombined,—that is to say, would exist together in the state of chemical elements,—whose further dissociation in stellar or nebulous masses may even give us evidence of matter still more elemental than that revealed in the experiments of the laboratory, where we can only conjecture the compound nature of many of the so-called elementary substances."

The importance of this conception, in view of subsequent discoveries in spectroscopy and in stellar chemistry, has been well set forth by Lockyer in his late lectures on Solar Physics,* where, however, the generalisation is described as having been first made by Brodie in 1867. A similar but later enunciation of the same idea by Clerk-Maxwell is also cited by Lockyer. Brodie, in fact, on the 6th of June, one week after my own lecture, gave a lecture on Ideal Chemistry before the Chemical Society of London, published in the CHEMICAL NEWS of June 14th, in which, with regard to his ideal elements, in further extension of the suggestion already put forth by him in the extract above given from his paper of May 6, 1866, he says " We may conceive that in remote ages the temperature of matter was much higher than it is now, and that these other things [the ideal elements] existed in the state of perfect gases—separate existences—uncombined." He further suggested, from spectroscopic evidence, that it is probable that " we may one day, from this source, have revealed to us independent evidence of the existence of these ideal elements in the sun and stars."

During the months of June and July, 1867, I was absent on the Continent, and this lecture of Brodie's remained wholly unknown to me until its republication in 1880, in a separate form, by its author,† with a preface, in which he pointed out that he had therein suggested the probable liberation of his ideal elements in the sun, referring at the same time to his paper of 1866, from which we have already quoted the only expression bearing on the possible independence of these ideal elements somewhere in time or in space.

The above statements are necessary in order to explain why it is that I have made no reference to Sir Benjamin Brodie on the several occasions on which, in the interval between 1867 and the present time, I have reiterated and enforced my views on the great significance of the hypothesis of celestial dissociation as giving rise to forms of matter more elemental than any known to us in terrestrial chemistry. The conception, as at first enunciated in somewhat different forms alike by Brodie and myself, was one to which we were both naturally, one might say inevitably, led by different paths from our respective fields of speculation, and which each might accept as in the highest degree probable, and make, as it were, his own. I write, therefore, in no spirit of invidious rivalry with my honoured and lamented friend, but simply to clear myself from the charge, which might otherwise be brought against me, of having, on various occasions within the past fourteen years, put forth and enlarged upon this conception without mentioning Sir Benjamin Brodie, whose only publication on the subject, so far as I am aware, was his lecture of 1867, unknown to me until its reprint in 1880.

It was at the grave of Priestley, in 1874, that I for the second time considered the doctrine of celestial dissocia-

---

* Read before the Cambridge Philosophical Society, November 28, 1881, and reprinted from its *Proceedings.*
† *L. E. and D. Phil. Mag.*, III., xxviii., 106 and 478; also xxix.. 185.

* *Nature*, August 25, 1881, vol. xxiv., p. 396.
† " Ideal Chemistry, a Lecture. Macmillan, 1880

tion, commencing with an account of the hypothesis put forward by F. W. Clarke, of Cincinnati, in January, 1873,* to explain the growing complexity which is observed when we compare the spectra of the white, yellow, and red stars ; in which he saw evidence of a progressive evolution of chemical species, by a stoichiogenic process, from more elemental forms of matter. I then referred to the further development of this view by Lockyer, in his communication to the French Academy of Sciences, in November of the same year, wherein he connected the successive appearance in celestial bodies of chemical species of higher and higher vapour-densities with the speculations of Dumas and Pettenkofer as to the composite nature of the chemical elements.† I then quoted from my lecture of 1867 the language already cited, to the effect that dissociation by intense heat in stellar worlds might give us more elemental forms of matter than any known on earth, and further suggested that the green line in the spectrum of the solar corona, which had been supposed to indicate a hitherto unknown substance, may be due to a " more elemental form of matter, which, though not seen in the nebulæ, is liberated by the intense heat of the solar sphere, and may possibly correspond to the primary matter conjectured by Dumas, having an equivalent weight one-fourth that of hydrogen." The suggestion of Lavoisier, that " hydrogen, nitrogen, and oxygen, with heat and light, might be regarded as simpler forms of matter from which all others are derived," was also noticed in connection with the fact that the nebulæ, which we conceive to be condensing into suns and planets, have hitherto shown evidences only of the presence of the first two of these elements, which, as is well known, make up a large part of the gaseous envelope of our planet, in the forms of air and aqueous vapour. With this I connected the hypothesis that our atmosphere and ocean are but portions of the universal medium which, in an attenuated form, fills the interstellar spaces ; and further suggested, as " a legitimate and plausible speculation," that " these same nebulæ and their resulting worlds may be evolved by a process of chemical condensation from this universal atmosphere, to which they would sustain a relation somewhat analogous to that of clouds and rain to the aqueous vapour around us.‡

These views were reiterated in the preface to a second edition of my " Chemical and Geological Essays," in 1878 ; and again before the British Association for the Advancement of Science, at Dublin,‖ and before the French Academy of Sciences in the same year.§ They were still further developed in an essay on the " Chemical and Geological Relations of the Atmosphere " (published in the *American Journal of Science* for May, 1880), in which attention was called to the important contribution to the subject by Mr. Lockyer in his ingenious and beautiful spectroscopic studies, the results of which are embodied in his " Discussion of the Working Hypothesis that the so-called Elements are Compound Bodies," communicated to the Royal Society, December 12, 1878. It was then remarked that the already noticed " speculation of Lavoisier is really an anticipation of that view to which spectroscopic study has led the chemists of to-day ;" while it was said that the hypothesis put forth by the writer in 1874, " which seeks for a source of the nebulous matter itself, is perhaps a legitimate extension of the nebular hypothesis."

To show the connection of the above views with the philosophy of Newton, it now becomes necessary to give some account of the conception of the universal distribution of matter throughout space, both as regards its dynamical relations and its chemical composition. Passing

over the speculations of the Greek physiologists, we com to the controversies on this subject in the seventeenth century, and find, in apparent opposition to the plenum maintained by Descartes and his followers, the teaching of Newton that " the heavens are void of all sensible matter." This statement is, however, qualified elsewhere by his assertion, that " to make way for the regular and lasting movements of the planets and comets, it is necessary to empty the heavens of all matter, except perhaps some very thin vapours, steams, and effluvia arising from the atmospheres of the earth, planets, and comets, and from such an exceedingly rare ethereal medium as we have elsewhere described," &c. (" Optics," Book III., Query 28).

In order to understand fully the views of Newton on this subject, it is necessary to compare carefully his various utterances, including the Hypothesis, in 1675, the first edition of the " Principia," in 1687, the second edition, in 1713, and the various editions of the " Optics." This work appeared in 1704, the third book, with its appended queries, having, according to its author's preface, been " put together out of scattered papers " subsequent to the publication of the first edition of the " Principia." The Latin translation of the " Optics," by Dr. Clarke, which was published in 1706, and the second English edition, in 1718, contain successive additions to these queries, which are indicated in the notes to Horsley's edition of the works of Newton, and are important in this connection. From a collation of all these we learn how the conceptions of the Hypothesis took shape, were reinforced, and in great part incorporated in the " Principia."

In the Hypothesis he imagines " an ethereal medium much of the same constitution with air, but far rarer, subtler, and more elastic." " But it is not to be supposed that this medium is one uniform matter, but composed partly of the main phlegmatic body of ether, partly of other various ethereal spirits, much after the manner that air is compounded of the phlegmatic body of air intermixed with various vapours and exhalations." Newton further suggests in his Hypothesis that this complex spirit or ether, which, by its elasticity, is extended throughout all space, is in continual movement and interchange. " For Nature is a perpetual circulatory worker, generating fluids out of solids, and solids out of fluids, fixed things out of volatile, and volatile out of fixed, subtile out of gross, and gross out of subtile ; some things to ascend and make the upper terrestrial juices, rivers, and the atmosphere, and by consequence others, to descend for a requital to the former. And as the earth, so perhaps may the sun imbibe this spirit copiously, to conserve his shining, and keep the planets from receding farther from him ; and they that will may also suppose that this spirit affords or carries with it thither the solary fuel and material principle of life, and that the vast ethereal spaces between us and the stars are for a sufficient repository for this food of the sun and planets."

The language of this last sentence, in which his late biographer, Sir David Brewster, regards Newton as " amusing himself with the extravagance of his speculations," at which " we may be allowed to smile,"* was not apparently regarded as unreasonable by its author when, more than ten years later, he quoted it in the postscript of his letter to Halley, dated Cambridge, June 20, 1686. The views therein contained, with the single exception of the suggestion regarding gravitation, have not wanted advocates in our own time, and many of them were embodied in the " Principia," which Newton was then engaged in writing.

But this was not all : Newton saw in the cosmic circulation, and the mutual convertibility of rare and dense forms of matter, a universal law, and rising to a still bolder conception, which completes his Hypothesis of the Universe, adds : " Perhaps the whole frame of Nature may be nothing but various contextures of some certain ethereal spirits or vapours, condensed, as it were, by precipitation,

* Clarke, "Evolution and the Spectroscope," *Popular Science Monthly*, New York, vol. ii., p. 32.
† Lockyer, *Comptes Rendus*, November 3, 1873.
‡ "A Century's Progress in Theoretical Chemistry ; being an Address at Northumberland, Penn., July 31, 1874"; *Amer. Chemist*, vol. v., pp. 46—61, and *Popular Science Monthly*, vi., p. 420.
‖ *Nature*, August 29, 1878, vol. xviii., p. 475.
§ *Comptes Rendus*, Sept. 23, 1878, vol. xxxviii., p. 452.

* Brewster's "Memoirs of Newton," vol. i. pp. 121 and 404.

much after the same manner that vapours are condensed into water, or exhalations into grosser substances, though not so easily condensible; and after condensation wrought into various forms, at first by the immediate hand of the Creator, and ever since by the power of Nature, which, by virtue of the command ' Increase and multiply,' became a complete imitator of the copy set her by the great Protoplast. Thus, perhaps, may all things be originated from ether."

(To be continued.)

PROCEEDINGS OF SOCIETIES.

PHYSICAL SOCIETY.
Annual General Meeting, Saturday, February 11th, 1882.

Prof. W. GRYLLS ADAMS in the Chair.

THE PRESIDENT read the Report of the Council for the past year, from which it appeared that in this, the tenth year of the Society, it was in a highly satisfactory condition and numbered 331 members. Sir Charles Wheatstone's papers had been published, Dr. Joules's were soon to be so, and delegates from the Society had taken part in the Electrical Congress at Paris, the Lightning Rod Committee, &c.

The TREASURER (Dr. Atkinson) read the Audited Report of the financial state of the Society: and the following officers were, after a ballot, declared elected for the ensuing year:—

President—Prof. R. B. Clifton, F.R.S.
Vice-Presidents—Sir W. Thomson (past President), Prof. G. C. Foster, Prof. F. Fuller, Dr. J. Hopkinson, Lord Rayleigh.
Secretaries—Prof. A. W. Reinold, Prof. W. Chandler Roberts.
Treasurer—Dr. E. Atkinson.
Demonstrator—Prof. F. Guthrie.
Other Members of Council—Prof. W. G. Adams, Prof. W. E. Ayrton, Mr. Shellford Bidwell, Mr. Walter Baily, Prof. J. A. Fleming, Mr. R. J. Lecky, Dr. Hugo Müller, Prof. Osborne Reynolds, Prof. A. W. Rücker.
Honorary Member—Prof. G. Quincke.

Votes of thanks were then passed to the Lords Commissioners of the Committee of Council on Education for the use of the Meeting Hall, to the past President (Sir W. Thomson), to the Secretaries, the Treasurer, and Demonstrator, as well as to the Auditors, Mr. Shellford Bidwell and Mr. E. Rigg.

Prof. ADAMS then resolved the meeting into an ordinary one, and called Prof. Clifton to the Chair.

Dr. C. R. ALDER WRIGHT, F.R.S., then read a paper "*On the Relation between the Electromotive Force of a Daniel Element and the Chemical Affinity involved in its Action.*" The author has investigated the causes which lead to a fall of E.M.F. in a Daniell cell when in action. He found the amount of fall for increasing current densities and plotted it in a curve. The fall was slight when pure commercial or amalgamated zinc, or zinc coated with a film of copper, was employed. Amalgamated copper plates gave more rapid rates of fall than electro-coated ones. Dilute sulphuric acid round the zinc gave a less rapid fall than sulphate of zinc solution round it. In all cases no appreciable fall was noticed when the current did not exceed 8 micro-Ampères per square centimetre of plate surface. With four to six times this density a decrease of E.M.F. from 0·5 to 1 per cent resulted; and with currents exceeding 3000 micro-Ampères in density per square centimetre of surface the fall exceeded 10 per cent. A series of experiments were made to determine the fall due to change in the density of the solution by migration of the ions, causing a stronger zinc and a weaker copper solution. This showed that with nearly saturated zinc sulphate solution (sp. gr. 1·4) and very dilute copper sulphate solution, the maximum fall in E.M.F. is developed, and is less than 0·04 volt. Hence the total fall in E.M.F. due to migration of the ions when moderately strong currents pass, is only a fraction of the total fall. It follows that the energy due to the actions taking place in the cell, although wholly manifested as electric action expressible in Volt-Coulombs when the current is very small, is not wholly so manifested when the current is stronger. The author expresses this idea by calling the energy manifested in electric action *adjuvant* and the remainder as *non-adjuvant*. He finds that the major part of the latter energy is absorbed in actions having their seat at the surface of the copper plate, and the rest in actions at the surface of the zinc plate. It is transformed into heat according to Joule's law. As a subsidiary result it appears that the E.M.F. of a Daniell cell with zinc and copper sulphate solutions of equal specific gravity, or pure amalgamated zinc plate, and either a freshly deposited copper or an amalgamated copper plate, is a standard subject to less departure from the E.M.F. of other Daniell cells than the Clark's standard elements, which appear to vary one from another. On the other hand, a Clark cell keeps sensibly constant to its original value (if properly set up) during a period of months or years at a constant temperature, whereas a Daniell standard falls from its original value after a few hours, or days at most.

NOTICES OF BOOKS.

The Practice of Commercial Organic Analysis. By ALFRED H. ALLEN, F.I.C., F.C.S., Lecturer on Chemistry at the Sheffield School of Medicine, &c. Volume II., Hydrocarbons; Fixed Oils and Fats; Sugars; Starch and its Isomers; Alkaloids and Organic Bases, &c. London: J. and A. Churchill.

MANY of our readers will be glad to hear of the appearance of this long-promised volume. The first part of Mr. Allen's work so clearly supplied a want, that the necessity for a book treating of previously-omitted portions of the subject has been the more felt. The volume just published is half as large again as Volume I., and is based on a similar plan. In his preface the author expresses regret that the growth of the subject-matter has compelled him to omit several important sections, but a feeling that it was better to ignore a subject entirely than to treat it inadequately had caused him to postpone their consideration till a demand should arise for an additional volume or a second edition of those already published. On the subjects treated in the volume under review Mr. Allen writes fully and in many cases exhaustively, though it is only to be expected that in a book dealing with such a variety of material those sections should be most elaborate which treat of bodies in the examination of which the author has had special experience.

In the chapter on "Hydrocarbons" the author gives a useful table showing the various organic products respectively obtained by the distillation of coal, bituminous shale, peat, wood, and petroleum. Full details are given of the methods of assaying petroleum and shale products, including lubricating oils, vaselene, and paraffin wax. Turpentine, essential oils, and rosin oil are described in the section on "Terpenes," and then follow valuable and suggestive sections on "Benzene and its Homologues," "Naphthalene," and "Anthracene." The descriptions of the methods of assaying commercial benzols and coaltar naphthas, and of examining crude anthracene deserve special mention, as they are by far the most complete account yet published, and ought to render the book indispensable to tar distillers and managers of gas works.

CELESTIAL CHEMISTRY FROM THE TIME OF NEWTON.*

By T. STERRY HUNT, LL.D., F.R.S.

(Concluded from p. 76.)

IF now we look to the third book of the " Principia, we shall find in proposition 41 the remarkable chemical argument by which Newton was led to regard the interstellar ether as affording " the material principle of life " and " the food of planets." Considering the exhalations from the tails of comets, he supposes that the vapours thus derived, being rarefied, dilated, and spread through the whole heavens, are by gravity brought within the atmospheres of the planets, where they serve for the support of vegetable life. Inasmuch, moreover, as all vegetation is supported by fluids, and subsequently by decay is, in part, changed into solids, by which the mass of the earth is augmented, he concludes that if these essential matters were not supplied from some external source they must continually decrease, and at last fail. This vital and subtle part of our atmosphere, so important, though small in amount, he then supposed might come from the tails of comets.†

This appeared in the first edition of the " Principia," in 1687. It was not until later that the conception of exhalations from other celestial bodies took shape in the mind of Newton, as we may learn from the " Optics." Thus, in the first edition of this work, in Query 11, the sun and fixed stars are spoken of as great earths, intensely heated, and surrounded with dense atmospheres which, by their weight, condense the exhalations arising from these hot bodies. To this Query is added, in 1706, the suggestion that the weight of such an atmosphere " may hinder the globe of the sun from being diminished except by the emission of light ;" while in the second English edition, in 1718, we find a further addition, in the words " and a very small quantity of vapours and exhalations." A similar change of view appears in the query now numbered 28, wherein we read of " places [almost] destitute of matter," and also that " the sun and planets gravitate towards each other without [dense] matter between." In these quotations the two words in brackets are wanting in the edition of 1706, and first appear in that of 1718; while the language which we have in a previous page quoted from this same Query is found in the edition of 1706.

The Queries now numbered 17-24 appeared for the first time in the edition of 1718, and herein we find, in 18, the ethereal medium spoken of as being " by its elastic force expanded through all the heavens." Of this medium, " which fills all space adequately," he asks, " may not its resistance be so small as to be inconsiderable," and scarcely to make any sensible alteration in the movements of the planets ?* This complex ether of the interstellar space was thus, in the opinion of Newton, made up in part of matter common to the planetary and stellar atmospheres, the origin and importance of which is concisely stated in the paragraph which appears for the first time in 1713, in the second edition of the " Principia," in the third book, at the end of proposition 42, here much augmented. In this statement, which serves to supplement and complete that already made in 1687, in proposition 41, we read that the vapours which arise alike from the sun, the fixed stars, and the tails of comets, may by gravity fall into the atmospheres of the planets, and there be condensed, and pass into the form of salts, sulphurs (i. e., combustible matters), tinctures, clay, sand, coral and other terrestrial substances.†

The conception of Newton, who—while rejecting alike the plenum of the Cartesians, with its vortices and an absolute vacuum—imagined space to be filled with an exceedingly attenuated matter, through which a free circulation of gaseous substances might take place between distant worlds, has found favour among modern thinkers, who seem to have been ignorant of his views. Sir William Grove, in 1842, suggested that the medium of light and heat may be " a universally diffused matter ; and subsequently, in 1843, in the chapter on Light in his " Essay on the Correlation of Physical Forces," concluded with regard to the atmospheres of the sun and the planets, that there is no reason " why these atmospheres should not be, with reference to each other, in a state of equilibrium. Ether, which term we may apply to the highly attenuated matter existing in the interplanetary spaces, being an expansion of some or all of these atmospheres, or of the more volatile portions of them, would thus furnish matter for the transmission of the modes of motion which we call light, heat, &c. ; and possibly minute portions of the atmospheres may, by gradual accretions and subtractions, pass from planet to planet, forming a link of material communication between the distant monads of the universe. Subsequently, in his address as President of the British Association for the Advancement of Science, in 1866, Grove further suggested that this diffused matter may become a source of solar heat," inasmuch as the sun may condense gaseous matter as it travels in space, and so heat may be produced."

Humboldt, also, in his " Cosmos," considers the existence of a resisting medium in space, and says " of this impeding ethereal and cosmical matter," it may be supposed

---

* Read before the Cambridge Philosophical Society, November 28, 1881, and reprinted from its *Proceedings*.

† " Vapor enim in spatiis illis liberrimis perpetuo rarescit, ac dilatatur. Quâ ratione fit ut cauda omnis ad extremitatem superiorem latior sit quâm juxta capita cometae. Eâ autem rarefactione vaporem perpetuo dilatatum diffundi tandem et spargi par coelos universos, deinde paulatim in planetas per gravitatem suam attrahi et cum eorum atmosphaeris misceri, rationi consentaneum videtur. Nam quemadmodum maria ad constitutionem Terrae hujus omnino requiruntur, idque ut ex iis per calorem Solis vapores copiosè satis excitentur, qui vel in nubes coacti decidant in pluviis, et Terram omnem ad procreationem vegetabilium irrigent et nutriant ; vel in frigidis montium verticibus condensati (ut aliqui cum ratione philosophantur) decurrant in fontes et flumina : sic ad conservationem marium et humorum in planetis requiri videntur cometae, ex quorem exhalationibus et vaporibus condensatis, quicquid liquoris per vegetationem et putrefactionem consumitur et in Terram aridam convertitur, continuo suppleri et refici possit. Nam vegetabilia omnia ex liquoribus omnino crescunt, dein magnâ ex parte in Terram aridam per putrefactionem abeunt, et limus ex liquoribus putrefactis perpetuo decidit. Hinc moles Terrae aridae indies augetur, et liquores, nisi aliunde augmentum sumerent, perpetuo decresere deberent, ac tandem deficere. Porro suspicor spiritum illum, qui aëris nostri pars minima est, sed subtillissima et optima, et ad rerum omnium vitam requiritur, ex cometis praecipue venire."—*Newton*, " Principia," lib. III., prop. xli.

---

* Compare this with Prop. x., Book III. of the " Principia."

† " Vapores autem, qui ex Sole et stellis fixis et caudis cometarum oriuntur, incidere possunt per gravitatem suam in atmosphaeras planetarum, e, ibi condensari et converti in aquam et spiritos humidos et subinde per lentem calorem in sales, et sulphura, et tincturas, et limum, et lutem, et argillam, et arenam, et lapides, et corolla, et substantias a ias terrestres paulatim migrare."—*Newton*, " Principia," lib. III., prop.

that it is in motion, that it gravitates, notwithstanding its great tenuity, that it is condensed in the vicinity of the great mass of the sun, and that it may include exhalations from comets ; in which connection he quotes from the 42nd proposition of the third book of the " Principia." He further speaks comprehensively of the " vaporous matter of the incommensurable regions of space, whether, scattered without definite limits, it exists as a cosmical ether, or is condensed in nebulous masses and becomes comprised among the agglomerated bodies of the universe."* Humboldt also cites in this connection a suggestion made by Arago in the " Annuaire du Bureau des Longitudes " for 1842, as to the possibility of determining, by a comparison of its refractive power with that of terrestrial gases, the density of " the extremely rare matter occupying the regions of space."†

In 1854 Sir William Thomson published his note " On the Possible Density of the Luminiferous Ether,"‡ wherein he remarks " that there must be a medium of material communication throughout space to the remotest visible body is a fundamental conception of the undulatory theory of light. Whether or no this medium is (as appears to me most probable) a continuation of our own atmosphere, its existence cannot be questioned." He then attempts to fix an inferior limit to the density of the luminiferous medium in interplanetary space by considering the mechanical value of sunlight, as deduced from the value of solar radiation and the mechanical equivalent of the thermal unit. He concludes " that the luminiferous medium is enormously denser than the continuation of the terrestrial atmosphere would be in interplanetary space if rarefied according to Boyle's law always, and if the earth were at rest in a state of constant temperature, with an atmosphere of the actual density at its surface." The earth itself in moving through space " cannot displace less than 250 pounds of matter."

In 1870 W. Mattieu Williams published his very ingenious work entitled " The Fuel of the Sun," in which, apparently without any knowledge of what had been written before with regard to an interstellar medium, he attempts to find therein the source of solar heat—the " solary fuel " of Newton. To quote his own language— " The gaseous ocean in which we are immersed is but a portion of the infinite atmosphere that fills the whole solidity of space, that links together all the elements of the universe, and diffuses among them light and heat, and all the other physical and vital forces which heat and light are capable of generating " (*loc. cit.*, p. 5).

Since the days of Newton, however, no one had hitherto considered the interstellar matter from a chemical point of view. In 1874, as already shown, the writer had, in extension of the conception of Humboldt that its condensation gives rise to nebulæ, ventured the suggestion that from an ethereal medium having the same composition as our own atmosphere, the chemical elements of the sun and the planets have been evolved, in accordance with the views of Brodie, Clarke, and Lockyer, by a stoichiogenic process ; so that in the language of Newton's Hypothesis, " all things may be originated from ether."

It was not, however, until 1878 that, from a consideration of the chemical processes which have gone on at the earth's surface within recorded geological time, I was led to another step in this enquiry. That all the de-oxidised carbon found in the earth's crust in the forms of coal and graphite, as well as that existing in a diffused state, as bituminous or carbonaceous matter, has come, through vegetation, from atmospheric carbonic acid, appears certain. To the same source we must ascribe the carbonic acid of all the limestones which, since the dawn of life on our earth, have been deposited from its waters. It is through the sub-aërial decay of crystalline silicated rocks, and the direct formation of carbonate of lime, or of carbonate of magnesia and alkalies which have reacted on the calcium salts of the primæval ocean, that all limestones and dolomites have been generated. These, apart from the coaly matter, hold, locked up and withdrawn from the aërial circulation, an amount of carbonic acid which may be probably estimated at not less than 200 atmospheres equal in weight to our own. That this amount, or even a thousandth part of it, could have existed at any one time in our terrestrial atmosphere since the beginning of life on our planet is inconceivable, and that it could be supplied from the earth's interior is an hypothesis equally untenable.

I was therefore led to admit for it an extra-terrestrial source, and to maintain that the carbonic acid has thence gradually come into our atmosphere to supply the deficiences created by chemical processes at the earth's surface. Since similar processes are even now removing from our atmosphere this indispensable element, and fixing it in solid forms, it follows that except volcanic agency, which can only restore a portion of what was primarily derived from the atmosphere, there are on earth, besides organic decay, only the artificial processes of human industry which can furnish carbonic acid ; so that but for a supply of this gas from the interstellar spaces now, as in the past, vegetation, and consequently animal life itself, would fail and perish from the earth for want of this " food of planets."

Such were the conclusions, based on an induction from the facts of modern chemistry and geology, which I enunciated in my papers in 1878 and 1880, already quoted in the first part of this essay. I was at that time unacquainted with the Hypothesis of Newton, and with his remarkable reasoning contained in the 41st proposition of the third book of the " Principia," in which he, so far as was possible with the chemical knowledge of his time, anticipated my own argument, and showed how and in what manner the interstellar ether may really afford the " food of planets," and, in a sense, " the material principle of life."

I have thus endeavoured to bring before the Philosophical Society of Cambridge a brief history of the development of this conception of an interstellar medium, and to show that the thought of two centuries has done little more than confirm the almost forgotten views of Newton. It is with feelings of peculiar gratification that I have been able to indite these pages within the very walls of the college in which our great philosopher lived and laboured, and where, combining all the science of his time with a foresight which seems well-nigh divine, he was enabled, in the words of the poet, " to think again the great thought of the creation."

Alkali, &c., Works Regulation Act, 1881.—The following circular has been sent by the Secretary of the Local Government Board to the owner, lessee, occupier, or other person carrying on any work to which the Alkali, &c., Works Regulation Act, 1881, applies:—Sir,—I am directed by the Local Government Board to refer to their Circular of the 31st of December last, in which they drew your attention to the requirements of the Alkali, &c., Works Regulation Act, 1881, with respect to the registration of the several kinds of Works to which the Act applies. I am directed to remind you that, under Subsection (3) of Section 11 of the Act, application for Certificates of Registration is required to be made in the month of January or February in every year ; and that, under Subsection (6), the owner of any work which is carried on after the 1st day of April, 1882, without being registered, exposes himself to a penalty not exceeding Five Pounds for every day during which it is so carried on. If, therefore, your Works come within the Act, it is necessary that an application for a Certificate of Registration in respect of them should be made to the Board in the prescribed form, of which a copy is enclosed, *without any further delay.*—I am, Sir, your obedient servant, JOHN LAMBERT, Secretary.

---

* " Cosmos," Otté's translation, Harper's ed., vol, i., pp. 82, 86.
† *Ibid.*, vol. iii., p. 40.
‡ *Trans. Roy. Soc. Edinburgh*, vol. xxi., part I. ; and *Ph Mag.*, 1855, vol. ix., p. 36.

THE CHEMICAL NEWS.

VOL. XLVI. No. 1188.

ADDRESS TO THE MATHEMATICAL AND PHYSICAL SCIENCE SECTION OF THE BRITISH ASSOCIATION.

By Right Hon. LORD RAYLEIGH, M.A., F.R.S , F.R.A.S.

IN common with some of my predecessors in this chair, I recognise that probably the most useful form which a presidential address could take, would be a summary of the progress of physics, or of some important branch of physics, during recent years. But the difficulties of such a task are considerable, and I do not feel myself equal to grappling with them. The few remarks which I have to offer are of a general, I fear it may be thought of a commonplace, character. All I can hope is that they may have the effect of leading us into a frame of mind suitable for the work that lies before us.

The diversity of the subjects which come under our notice in this section, as well as of the methods by which alone they can be adequately dealt with, although a sign of the importance of our work, is a source of considerable difficulty in the conduct of it. From the almost inevitable specialisation of modern science, it has come about that much that is familiar to one member of our section is unintelligible to another, and that details whose importance is obvious to the one fail altogether to rouse any interest in the mind of the other. I must appeal to the authors of papers to bear this difficulty in mind, and to confine within moderate limits their discussion of points of less general interest.

Even within the limits of those departments whose foundation is evidently experimental, there is room, and indeed necessity, for great variety of treatment. One class of investigators relies mainly upon reiterated appeals to experiment to resolve the questions which appear still to be open, while another prefers, with Thomas Young, to base its decisions as far as possible upon deductions from experiments already made by others. It is scarcely necessary to say that in the present state of science both methods are indispensable. Even where we may fairly suppose that the fundamental principles are well established, careful and often troublesome work is necessary to determine with accuracy the constants which enter into the expression of natural laws. In many cases the accuracy desirable, even from a practical point of view, is hard to attain. In many others, where the interest is mainly theoretical, we cannot afford to neglect the confirmations which our views may derive from the comparison of measurements made in different fields and in face of different experimental difficulties. Examples of the interdependence of measurements apparently distinct will occur to every physicist. I may mention the absolute determinations of electrical resistance, and of the amounts of heat developed from electrical and mechanical work, any two of which involve also the third, and the relation of the velocity of sound to the mechanical and thermal properties of air.

Where a measurement is isolated, and not likely to lead to the solution of any open question, it is doubtless possible to spend upon it time and attention that might with advantage be otherwise bestowed. In such a case we may properly be satisfied for a time with work of a less severe and accurate character, knowing that with the progress of knowledge the way is sure to be smoothed both by a better appreciation of the difficulties involved, and by the invention of improved experimental appliances. I hope I shall not be misunderstood as underrating the importance of great accuracy in its proper place if I express the opinion that the desire for it has sometimes had a prejudicial effect. In cases where a rough result would have sufficed for all immediate purposes, no measurement at all has been attempted, because the circumstances rendered it unlikely that a high standard of precision could be attained. Whether our aim be more or less ambitious, it is important to recognise the limitations to which our methods are necessarily subject, and as far as possible to estimate the extent to which our results are uncertain. The comparison of estimates of uncertainty made before and after the execution of a set of measurements may sometimes be humiliating, but it is always instructive.

Even when our results show no greater discrepancies than we were originally prepared for, it is well to err on the side of modesty in estimating their trustworthiness. The history of science teaches only too plainly the lesson that no single method is absolutely to be relied upon, that sources of error lurk where they are least expected, and that they may escape the notice of the most experienced and conscientious worker. It is only by the concurrence of evidence of various kinds and from various sources that practical certainty may at last be attained, and complete confidence justified. Perhaps I may be allowed to illustrate my meaning by reference to a subject which has engaged a good deal of my attention for the last two years—the absolute measurement of electrical resistance. The unit commonly employed in this country is founded upon experiments made about twenty years ago by a distinguished committee of this Association, and was intended to represent an absolute resistance of 10^9 C.G.S., *i.e.*, one ohm. The method emplyed by the committee at the recommendation of Sir W. Thomson (it had been originally proposed by Weber) consisted in observing the deflection from the magnetic meridian of a needle suspended at the centre of a coil of insulated wire, which formed a closed circuit, and was made to revolve with uniform and known speed about a vertical axis. From the speed and deflection in combination with the mean radius of the coil and the number of its turns, the absolute resistance of the coil, and thence of any other standard, can be determined.

About ten years later Kohlrausch attacked the problem by another method, which it would take too long to explain, and arrived at the result that the B.A. unit was equal to 1·02 ohms—about two per cent too large. Rowland, in America, by a comparison between the steady battery current flowing in a primary coil with the transient current developed in a secondary coil when the primary current is reversed, found that the B.A. unit was 0·991 ohm. Lorentz, using a different method again, found 0·980, while H. Weber, from distinct experiments, arrived at the conclusion that the B.A. unit was correct. It will be seen that the results obtained by these highly competent observers range over about four per cent. Two new determinations have lately been made in the Cavendish laboratory at Cambridge, one by myself with the method of the revolving coil, and another by Mr. Glazebrook, who used a modification of the method followed by Rowland, with the result that the B.A. unit is 0·986 ohm. I am now engaged upon a third determination, using a method which is a modification of that of Lorentz.

In another important part of the field of experimental science, where the subject-matter is ill understood, and the work is qualitative rather than quantitative, success depends more directly upon sagacity and genius. It must be admitted that much labour spent in this kind of work is ill-directed. Bulky records of crude and uninterpreted observations are not science, nor even in many cases the raw material out of which science will be constructed. The door of experiment stands always open ; and when the question is ripe, and the man is found, he will nine times out of ten find it necessary to go through the work again. Observations made by the way, and under favourable conditions, may often give rise to valuable suggestions, but these must be tested by experiment, in which the conditions are simplified to the utmost, before they can lay claim to acceptance.

When an unexpected effect is observed, the question will arise whether or not an explanation can be found upon admitted principles. Sometimes the answer can be quickly given; but more often it will happen that an assertion of what *ought* to have been expected can only be made as the result of an elaborate discussion of the circumstances of the case, and this discussion must generally be mathematical in its spirit, if not in its form. In repeating, at the beginning of the century, the well-known experiment of the inaudibility of a bell rung *in vacuo,* Leslie made the interesting observation that the presence of hydrogen was inimical to the production of sound, so that not merely was the sound less in hydrogen than in air of equal pressure, but that the actual addition of hydrogen to rarefied air caused a diminution in the intensity of sound. How is this remarkable fact to be explained? Does it prove that, as Herschel was inclined to think, a mixture of gases of widely different densities differs in its acoustical properties from a single gas? These questions could scarcely be answered satisfactorily but by a mathematical investigation of the process by which vibrations are communicated from a vibrating solid body to the surrounding gas. Such an investigation, founded exclusively upon principles well established before the date of Leslie's observation, was undertaken years afterwards by Stokes, who proved that what Leslie observed was exactly what ought to have been expected. The addition of hydrogen to attenuated air increases the wave-length of vibrations of given pitch, and consequently the facility with which the gas can pass round the edge of the bell from the advancing to the retreating face, and thus escape those rarefactions and condensations which are essential to the formation of a complete sound wave. There remains no reason for supposing that the phenomenon depends upon any other elements than the density and pressure of the gaseous atmosphere, and a direct trial, *e.g.,* a comparison between air and a mixture of carbonic anhydride and hydrogen of like density, is almost superfluous.

Examples such as this, which might be multiplied *ad libitum,* show how difficult it often is for an experimenter rightly to interpret his results without the aid of mathematics. It is eminently desirable that the experimenter himself should be in a position to make the calculations, to which his work gives occasion, and from which in return he would often receive valuable hints for further experiment. I should like to see a course of mathematical instruction arranged with especial reference to physics, within which those whose bent was plainly towards experiment might, more or less completely, confine themselves. Probably a year spent judiciously on such a course would do more to qualify the student for actual work than two or three years of the usual mathematical curriculum. On the other side, it must be remembered that the human mind is limited, and that few can carry the weight of a complete mathematical armament without some repression of their energies in other directions. With many of us difficulty of remembering, if not want of time for acquiring, would impose an early limit. Here, as elsewhere, the natural advantages of a division of labour will assert themselves. Innate dexterity and facility in contrivance, backed by unflinching perseverance, may often conduct to successful discovery or invention a man who has little taste for speculation; and on the other hand the mathematician, endowed with genius and insight, may find a sufficient field for his energies in interpreting and systematising the work of others.

The different habits of mind of the two schools of physicists sometimes lead them to the adoption of antagonistic views on doubtful and difficult questions. The tendency of the purely experimental school is to rely almost exclusively upon direct evidence, even when it is obviously imperfect, and to disregard arguments which they stigmatise as theoretical. The tendency of the mathematician is to overrate the solidity of his theoretical structures, and to forget the narrowness of the experimental foundation upon which many of them rest.

By direct observation, one of the most experienced and successful experimenters of the last generation convinced himself that light of definite refrangibility was capable of further analysis by absorption. It has happened to myself, in the course of measurements of the absorbing power of various media for the different rays of the spectrum, to come across appearances at first sight strongly confirmatory of Brewster's views, and I can therefore understand the persistency with which he retained his opinion. But the possibility of further analysis of light of definite refrangibility (except by polarisation) is almost irreconcilable with the wave theory, which on the strongest grounds had been already accepted by most of Brewster's contemporaries; and in consequence his results, though urgently pressed, failed to convince the scientific world. Further experiment has fully justified this scepticism, and in the hands of Airy, Helmholtz, and others, has shown that the phenomena by which Brewster was misled can be explained by the unrecognised intrusion of diffused light. The anomalies disappear when sufficient precaution is taken that the refrangibility of the light observed shall really be definite.

On similar grounds undulationists early arrived at the conviction that physically light and invisible radiant heat are both vibrations of the same kind, differing merely in wave-length; but this view appears to have been accepted slowly, and almost reluctantly, by the experimental school.

When the facts which appear to conflict with theory are well defined and lend themselves easily to experiment and repetition, there ought to be no great delay at arriving at a judgment. Either the theory is upset, or the observations, if not altogether faulty, are found susceptible of another interpretation. The difficulty is greatest when the necessary conditions are uncertain, and their fulfilment rare and uncontrollable. In many such cases an attitude of reserve, in expectation of further evidence, is the only wise one. Premature judgments err perhaps as much on one side as on the other. Certainly in the past many extraordinary observations have met with an excessive incredulity. I may instance the fire-balls which sometimes occur during violent thunderstorms. When the telephone was first invented, the early reports of its performances were discredited by many on quite insufficient grounds.

It would be interesting, but too difficult and delicate a task, to enumerate and examine the various important questions which remain still undecided from the opposition of direct and indirect evidence. Merely as illustrations I will mention one or two in which I happen to have been interested. It has been sought to remedy the inconvenience caused by excessive reverberation of sound in cathedrals and other large unfurnished buildings by stretching wires overhead from one wall to another. In some cases no difference has been perceived, but in others it is thought that advantage has been gained. From a theoretical point of view it is difficult to believe that the wires could be of service. It is known that the vibrations of a wire do not communicate themselves in any appreciable degree directly to the air, but require the intervention of a sounding-board, from which we may infer that vibrations in the air would not readily communicate themselves to stretched wires. It seems more likely that the advantage supposed to have been gained in a few cases is imaginary than that the wires should really have played the part attributed to them.

The other subject on which, though with diffidence, I should like to make a remark or two, is that of Prout's law, according to which the atomic weights of the elements, or at any rate of many of them, stand in simple relation to that of hydrogen. Some chemists have reprobated strongly the importation of *à priori* views into the consideration of the question, and maintain that the only numbers worthy of recognition are the immediate results of experiment. Others, more impressed by the argument that the close approximations to simple numbers cannot be merely fortuitous, and more alive to the inevitable imperfections of our measurements, consider that the experimental evidence against the simple numbers is of a very slender character, balanced, if not outweighed, by the *à priori* argument in favour of simplicity. The subject is eminently one for further experiment; and as it is now engaging the attention of chemists, we may look forward to the settlement of the question by the present generation. The time has perhaps come when a redetermination of the densities of the principal gases may be desirable—an undertaking for which I have made some preparations.

If there is any truth in the views that I have been endeavouring to impress, our meetings in this section are amply justified. If the progress of science demands the comparison of evidence drawn from different sources, and fully appreciated only by minds of different order, what may we not gain from the opportunities here given for public discussion, and, perhaps more valuable still, private interchange of opinion? Let us endeavour, one and all, to turn them to the best account.

375

SECTION B

CHEMICAL SCIENCE

OPENING ADDRESS BY J. H. GLADSTONE, PH.D., F.R.S.,
V.P.C.S., PRESIDENT OF THE SECTION.

A SECTIONAL address usually consists either of a review of the work done in the particular science during the past year, or of an exposition of some branch of that science to which the speaker has given more especial attention. I propose to follow the latter of these practices, and shall ask the indulgence of my brother chemists while I endeavour to place before them some thoughts on the subject of Elements.

Though theoretical and practical chemistry are now intertwined, with manifest advantage to each, they appear to have been far apart in their origin. Practical chemistry arose from the arts of life, the knowledge empirically and laboriously acquired by the miner and metallurgist, the potter and the glassworker, the cook and the perfumer. Theoretical chemistry derived its origin from cosmogony. In the childhood of the human race the question was eagerly put, "By what process were all things made?" and some of the answers given started the doctrine of elements. The earliest documentary evidence of the idea is probably contained in the Shoo King, the most esteemed of the Chinese classics for its antiquity. It is an historical work, and comprises a document of still more venerable age, called "The Great Plan, with its Nine Divisions,"

which purports to have been given by Heaven to the Great Yu, to teach him his royal duty and "the proper virtues of the various relations." Of course there are wide differences of opinion as to its date, but we can scarcely be wrong in considering it as older than Solomon's writings. The First Division of the Great Plan relates to the Five Elements. "The first is named Water; the second, Fire; the third, Wood; the fourth, Metal; the fifth, Earth. The nature of water is to soak and descend; of fire, to blaze and ascend; of wood, to be crooked and to be straight; of metal, to obey and to change; while the virtue of the earth is seen in seed-sowing and ingathering. That which soaks and descends becomes salt; that which blazes and ascends becomes bitter; that which is crooked and straight becomes sour; that which obeys and changes becomes acrid; and from seed-sowing and ingathering comes sweetness."[1]

A similar idea of five elements was also common among the Indian races, and is stated by Professor Rodwell to have been in existence before the fifteenth century B.C., but, though the number is the same, the elements themselves are not identical with those of the ancient Chinese classic; thus, in the Institutes of Menu, the "subtle ether" is spoken of as being the first created, from which, by transmutation, springs air, whence, by the operation of a change, rises light or fire; from this comes water, and from water is deposited earth. These five are curiously correlated with the five senses, and it is very evident that they are not looked upon as five independent material existences, but as derived from one another. This philosophy was accepted alike by Hindoos and Buddhists. It was largely extended over Asia, and found its way into Europe. It is best known to us in the writings of the Greeks. Among these people, however, the elements were reduced to four—fire, air, earth, and water—though Aristotle endeavoured to restore the "blue ether" to its position as the most subtle and divine of them all. It is true that the fifth element, or "quinta essentia," was frequently spoken of by the early chemists, though the idea attaching to it was somewhat changed, and the four elements continued to retain their place in popular apprehension, and still retain it even among many of the scholars who take degrees at our universities. The claim of wood to be considered an element seems never to have been recognised in the West, unless, indeed, we are to seek this origin for the choice of the word ὕλη to signify that original chaotic material out of which, according to Plato and his school, all things were created.[2] The idea also of a primal element, from which the others, and everything else, were originated, was common in Greece, the difficulty being to decide which of the four had the greatest claim to this honour. Thales, as is well known, in the sixth century B.C. affirmed that water was the first principle of things; but Anaxamenes afterwards looked upon air, Heraclytus upon fire, and Theracleides on earth, as the primal element. This notion of elements, however, was essentially distinct from our own. It was always associated with the idea of the genesis of matter rather than with its ultimate analysis, and the idea of *simple* as contrasted with *compound* bodies probably never entered into the thoughts of the contending philosophers.

The modern idea appears to have had a totally different origin, and we must again travel back to China. There, also in the sixth century B.C., the great philosopher Lao-tse was meditating on the mysteries of the world and the soul, and his disciples founded the religion of Taou. They were materialists; nevertheless they believed in a "finer essence," or spirit, that rises from matter, and may become a star; thus they held that the souls of the five elements, water, metal, fire, wood, and earth, arose and became the five planets. These speculations naturally led to a search after the sublimated essences of things, and the means by which this immortality might be secured. It seems that at the time of Tsin-she-hwang, the builder of the Great Wall, about two centuries before Christ, many romantic stories were current of immortal men inhabiting islands in the Pacific Ocean. It was supposed that in these magical islands was found the "herb of immortality" growing, and that it gave them

[1] Quoted from the translation by the Rev. Dr. Legge. In that most obscure classic. the "Yi-King," fire and water. wind and thunder, the ocean and the mountains, appear to be recognised as the elements.
[2] Students of the Apocrypha will remember the expression in the Book of Wisdom, xi. 17, ʽἡ παντοδύναμός σου χείρ καὶ κτίσασα τὸν κόσμον ἐξ ἀμόρφου ὕλης' ('Thy Almighty hand, that made the world of matter without form'). The same book contains two allusions to the ordinary elements, vii. 17, and xix. 18 to 20. The word στοιχεῖον is used in the New Testament only in a general sense (2 Pet. iii. 10), or in its more popular meaning of the first steps in knowledge.

exemption from the lot of common mortals. The emperor determined to go in search of these islands, but some untoward event always prevented him.[1]

Some two or three centuries after this a Taouist, named Weipahyang, wrote a remarkable book called "The Uniting Bond." It contains a great deal about the changes of the heavenly bodies, and the mutual relation of heaven and men; and then the author proceeds to explain some transformations of silver and water. About elixir he tells us, "What is white when first obtained becomes red after manipulation on being formed into the elixir" ("tan," meaning red or elixir). "That substance, an inch in diameter, consists of the black and the white, that is, water and metal combined. It is older than heaven and earth. It is most honourable and excellent. Around it, like a wall, are the sides of the cauldron. It is closed up and sealed on every side, and carefully watched. The thoughts must be undisturbed, and the temper calm, and the hour of its perfection anxiously waited for. The false chemist passes through various operations in vain. He who is enlightened expels his evil passions, is delighted morning and night, forgets fame and wealth, comprehends the true objects of life, and gains supernatural powers. He cannot then be scorched by fire, nor drowned in water, &c., &c. . . . The cauldron is round like the full moon, and the stove beneath is shaped like the half-moon. The lead ore is symbolised by the White Tiger; and it, like metal amongst the elements, belongs to the West. Mercury resembles the sun, and forms itself into sparkling globes; it is symbolised by the Blue Dragon belonging to the East, and it is assigned to the element wood. Gold is imperishable. Fire does not injure its lustre. Like the sun and moon, it is unaffected by time. Therefore the elixir is called 'the Golden Elixir.' Life can be lengthened by eating the herb called Hu ma; how much more by taking the elixir, which is the essence of gold, the most imperishable of all things! The influence of the elixir, when partaken of, will extend to the four limbs; the countenance will become joyful; white hair will be turned black; new teeth will grow in the place of old ones, and age at once become youth. . . . Lead ore and mercury are the bases of the process by which the elixir is prepared; they are the hinge upon which the principles of light and darkness revolve."

This description suggests the idea that the elixir of the Taouists was the red sulphide of mercury—vermilion—for the preparation of which the Chinese are still famous. That Weipahyang believed in his own philosophy is testified by a writer named Ko-hung, who, about a century afterwards, wrote the lives of celebrated Taouists. He tells how the philosopher, after preparing the elixir, took it, with his disciples, into a wood, and gave it first to his dog, then took it himself, and was followed by one of his pupils. They all three died, but, it appears, rose to life again, and to immortality. This brilliant example did not remain without imitators; indeed, two emperors of the Tang family are said to have died from partaking of the elixir. This circumstance diminished its popularity, and alchemy ceased to be practised in the Celestial Empire.

At the beginning of the seventh century the doctrine of Lao-tse was in great favour at the Chinese Court; learning was encouraged, and there was much enterprise. At the same time the disciples of Mohammed carried their arms and his doctrines over a large portion of Asia, and even to the Flowery Land. Throughout the eighth century there were frequent embassies between eastern and western Asia, wars with the Caliphs, and even a matrimonial alliance. We need not wonder, therefore, that the teachings of the Taouist alchemists penetrated westward to the Arabian philosophers. It was at this period that Yeber-Abou-Moussah-Djaferal-Sofé, commonly called Geber, a Sabæan of great knowledge, started what to the West was a new philosophy about the transmutation of metals, the Philosopher's Stone, and the Elixir of Life; and this teaching was couched in highly poetic language, mixed with astrology and accompanied by religious directions and rites. He held that all metals were composed of mercury, sulphur, and arsenic, in various proportions, and that the noblest metal could be procured only by a very lengthy purification. It was in the salts of gold and silver that he looked for the Universal Medicine. Geber himself was an experimental philosopher, and the belief in transmutation led to the acquirement of a considerable amount of chemical knowledge amongst the alchemists of Arabia and Europe. This

gradually brought about a conviction that the three reputed elementary bodies, mercury, sulphur, and salt or acid, were not really the originators of all things. There was a transition period, during which the notion was itself suffering a transmutation. The idea became gradually clearer that all material bodies were made up of certain constituents, which could not be decomposed any further, and which, therefore, should be considered as elementary. The introduction of quantitative methods compelled the overthrow of mediæval chemistry, and led to the placing of the conception of simple and compound bodies upon the foundation of scientific fact. Lavoisier, perhaps, deserves the greatest credit in this matter, while the labours of the other great chemists of the eighteenth and the beginning of the nineteenth centuries were in a great measure directed to the analysis of every conceivable material, whether solid, liquid, or gaseous. These have resulted in the table of so-called elements, now nearly seventy in number, to which fresh additions are constantly being made.

Of this ever-growing list of elements not one has been resolved into simpler bodies for three-quarters of a century; and we, who are removed by two or three generations from the great builders of our science, are tempted to look upon these bodies as though they were really simple forms of matter, not only unresolved, but unresolvable. The notation we employ favours this view and stamps it upon our minds.

Is it, however, a fact that these reputed elements are really simple bodies? or, indeed, are they widely different in the nature of their constitution from those bodies which we know to be chemical compounds? Thus, to take a particular instance, are fluorine, chlorine, bromine, and iodine essentially distinct in their nature from the compound halogens, cyanogen, sulphocyanogen, ferricyanogen, &c.? Are the metals lithium, sodium, and potassium essentially distinct from such alkaline bases as ammonium, ethylamine, di-ethylamine, &c.? No philosophical chemist would probably venture to answer this question categorically with either "yes" or "no." Let us endeavour to approach it from three different points of attack—(1) the evidence of the spectroscope, (2) certain peculiarities of the atomic weights, and (3) specific refraction.

1. *The Spectroscope.*—It was at first hoped that the spectroscope might throw much light upon the nature of elements, and might reveal a common constituent in two or more of them; thus, for instance, it was conceivable that the spectrum line of bromine or iodine vapour might consist of the rays given by chlorine *plus* some others. All expectations of this have hitherto been disappointed; yet, of the other hand, it must not be supposed that such a result disproves the compound nature of elements, for as investigation proceeds it becomes more and more clear that the spectrum of a compound is not made up of the spectra of its component parts.

Again, the multiplicity of rays given out by some elements, when heated, in a gaseous condition, such as iron, has been supposed to indicate a more complex constitution than in the case of those metals, such as magnesium, which give a more simple spectrum. Yet it is perfectly conceivable that this may be due to a complexity of arrangement of atoms all of the same kind.

Again, we have changes of a spectrum at different temperatures; new rays appear, others disappear; or even there occurs the very remarkable change from a fluted spectrum to one of sharp lines at irregular intervals, or to certain recurring groups of lines. This, in all probability, does arise from some redistribution, but it may be a redistribution in a molecular grouping of atoms of the same kind, and not a dissociation or rearrangement of dissimilar atoms.

A stronger argument has been derived from the revelations of the spectroscope in regard to the luminous atmospheres of the sun. There we can watch the effect of heat enormously transcending that of our hottest furnaces, and of movements compared with which our hurricanes and whirlwinds are the gentlest of zephyrs. Mr Lockyer, in studying the prismatic spectra of the luminous prominences or spots of the sun, has frequently observed that on certain days certain lines, say of the iron spectrum are non-existent and on other days certain other lines disappear, and that in almost endless variety; and he has also remarked that occasionally certain lines of the iron spectrum will be crooked or displaced, thus showing the vapour to be in very rapid motion, while others are straight, and therefore comparatively at rest. Now, as a gas cannot be both at rest and in motion at the same time and the same place, it seems very clear that the two sets of lines must originate in two distinct layers of atmosphere, one above the other, and Mr. Lockyer's conclusion is

---

[1] Nearly all the statements relating to this Taouist alchemy are derived from the writings of the Rev. Joseph Edkins, of Pekin, and the matter is treated in greater detail in an article on the "Birth of Alchemy," in the "Argonaut," vol iii P I

that the iron molecule was dissociated by heat, and that it[s] different constituents, on account of their different volatility, or some other cause, had floated away from one another. This seems to me the easiest explanation of the phenomenon; and, as dissociation by heat is a very common occurrence, there is no *a priori* improbability about it. But we are not shut up to it, for the different layers of atmosphere are certainly at different temperatures, and most probably of different composition. If they are of different temperatures, the variations of the spectrum may only be an extreme case of what must be acknowledged by every one more or less—that bodies emit, or cease to emit, different rays as their temperature increases, and notably when they pass from the liquid to the gaseous condition. And again, if the composition of the two layers of atmosphere be different, we have lately learnt how profoundly the admixture of a foreign substance will sometimes modify a luminous spectrum.

2. *Peculiarities of Atomic Weights.*—At the meeting of this Association at Ipswich, in 1851, M. Dumas showed that in several cases analogous elements form groups of three, the middle one of which has an atomic weight intermediate between those of the first and third, and that many of its physical and chemical properties are intermediate also. During the discussion upon his paper, and subsequently,[1] attention was drawn to the fact that this is not confined to groups of three, but that there exist many series of analogous elements having atomic weights which differ by certain increments, and that these increments are in most cases multiples of 8. Thus we have lithium, 7; sodium, 23, *i.e.* 7 + 16; potassium, 39, *i.e.* 7 + (16 × 2); and the more recently discovered rubidium, 85, *i.e.* 7 + (16 × 5) nearly; and cæsium, 133, *i.e.* 7 + (16 × 8) nearly. This is closely analogous to what we find in organic chemistry, where there are series of analogous bodies playing the part of metals, such as hydrogen, methyl, ethyl, &c., differing by an increment which has the atomic weight 14, and which we know to be CH_2. Again, there are elements with atomic weights nearly the same or nearly multiples of one another, instances of which are to be found in the great platinum group and the great cerium group.[2] This suggests the analogy of isomeric and polymeric bodies. There is also this remarkable circumstance: the various members of such a group as either of those just mentioned are found together at certain spots on the surface of the globe, and scarcely anywhere else. The chemist may be reminded of how in the dry distillation of some organic body he has obtained a mixture of polymerised hydrocarbons, and may perhaps be excused if he speculates whether in the process of formation of the platinum or the cerium group, however and whenever it took place, the different elements had been made from one another and imperfectly polymerised.

But this is not the largest generalisation in regard to the peculiarities of these atomic weights. Newlands showed that, by arranging the numbers in their order, the octaves presented remarkable similarities, and, on the same principle, Mendeléeff constructed his well-known table. I may remind you that in this table the atomic weights are arranged in horizontal and vertical series, those in the vertical series differing from one another, as a rule, by the before-mentioned multiples of 8—namely 16, 16, 24, 24, 24, 24, 32, 32—the elements being generally analogous in their atomicity and in other chemical characters. Attached to the elements are figures, representing various physical properties, and these in the horizontal series appear as periodic functions of the atomic weights. The table is incomplete, especially in its lower portions, but, with all its imperfections and irregularities, there can be no doubt that it expresses a great truth of nature. Now, if we were to interpolate the compound bodies which act like elements—methyl, 15; ammonium, 18; cyanogen, 26—into Mendeléeff's table, they would be utterly out of place, and would upset the order both of chemical analogy and of the periodicity of the physical properties.

3. *Specific Refraction.*—The specific refraction has been determined for a large majority of the elements, and is a very fundamental property, which belongs to them apparently in all their combinations, so long at least as the atomicity[3] is unchanged. If the figures representing this property be inserted into Mendeléeff's table, we find that in the vertical columns the

figures almost invariably decrease as the atomic weights increase. If, however, we look along the horizontal columns, or better still if we plot the figures in the table by which Lothair Meyer has shown graphically that the molecular volume is a periodic function of the atomic weights, we shall see that they arrange themselves in a series of curves similar to but not at all coincident with his. The observations are not so complete or accurate as those of the molecular volumes, but they seem sufficient to establish the fact, while the points of the curves would appear to be, not the alkaline metals, as in Meyer's diagram, but hydrogen, phosphorus and sulphur, titanium and vanadium, selenium, antimony. Now, if we were to insert the specific refractions of cyanogen, ammonium, and methyl into this table, we should again show that it was an intrusion of strangers not in harmony with the family of elements.

But there is another argument to be derived from the action of light. The refraction equivalent of a compound body is the sum of the refraction equivalents of its compounds; and, if there is anything known for certain in the whole subject, it is that the refraction equivalent of an organic compound advances by the same quantity (7·6) for every increment of CH_2. If, therefore, the increment between the different members of a group of analogous elements, such as the alkaline metals, be of the same character, we may expect to find that there is a regular increase of the refraction equivalent for each addition of 16. But this is utterly at variance with fact: thus, in the instance above quoted, the refraction equivalent of lithium being 3·8, that of sodium is 4·8, of potassium 8·1, of rubidium 14·0, and of cæsium about 13·7. Neither does the law obtain in those series in which the increment is not a multiple of 8, as in the case of the halogens, where the increment of atomic weight is 45, and the refraction equivalents are chlorine 9·9, bromine 15·3, and iodine 24·5.

The refraction equivalents of isomeric bodies are generally identical, and the refraction equivalents of polymeric bodies are in proportion to their atomic weights. Among the groups of analogous elements of the same, or nearly the same, atomic weight we do find certain analogies: thus cobalt and nickel are respectively 10·8 and 10·4, while iron and manganese are respectively 12·0 and 12·2. But, as far as observation has gone at present, we have reason to conclude that, if metals stand to one another in the ratio of 2 : 1 in atomic weight, their refraction equivalents are much nearer together than that; while, on the other hand, the equivalent of sulphur, instead of being the double of that of oxygen, is at least five times as great.

The general tendency of these arguments is evidently to show that the elementary radicals are essentially different from the compound radicals, though their chemical functions are similar.

There remains still the hypothesis that there is a "primordial element," from which the others are derived by transmutation. With the sages of Asia it was the "blue ether," with Thales water, with Dr. Prout hydrogen. The earlier views have passed away, and the claims of hydrogen are being fought out on the battle-field of atomic weights and their rigorous determination.

There does not appear to be any argument which is fatal to the idea that two or more of our supposed elements may differ from one another rather in form than in substance, or even that the whole seventy are only modifications of a prime element; but chemical analogies seem wanting. The closest analogy would be if we could prepare two allotropic conditions of some body, such as phosphorus or cyanogen, which should carry their allotropism into all their respective compounds, no compound of the one form being capable of change into a compound of the other. Our present knowledge of allotropism, and of variations in atomicity, affords little, if any, promise of this.

The remarkable relations between the atomic weights of the elements, and many peculiarities of their grouping, force upon us the conviction that they are not separate bodies created without reference to one another, but that they have been fashioned or built up from one another, according to some general plan. This plan we may hope gradually to understand better, but if we are ever to transform one of these supposed elements into another, or to split up one of them into two or three dissimilar forms of matter, it will probably be by the application of some method of analysis hitherto unknown.

Nothing can be of greater promise than the discovery of new methods of research; hence I need make no apology to others who have lately done excellent work in chemistry if I single out the Bakerian Lecture of this year, by Mr. Crookes, on "Radiant Matter Spectroscopy." It relates to the prismatic analysis, not of the light transmitted or absorbed in the ordinary way by a solid or liquid, nor of that given out by incandescent gas, but the

---

[1] "Phil. Mag.," May, 1853.
[2] Another curious instance is the occurrence of nickel and cobalt in all meteoric irons, with occasionally chromium or manganese, the atomic weights and other properties of which are very similar.
[3] This exception includes not merely such changes as that from a ferrous to a ferric salt, but the different ways in which the carbon is combined in such bodies as ethene, benzene, and pyrene.

analysis of the fluorescence that manifests itself in certain bodies when they are exposed to an electric discharge in a highly exhausted vacuum. He describes, in an interesting and even amusing manner, his three years' quest after the origin of a certain citron band, which he observed in the spectrum of the fluorescence of many substances, till he was led into that wonderful labyrinth of uncertain elements which are found together in samarskite, and eventually he proved the appearance to be due to yttrium. As the test is an extremely delicate one, he has obtained evidence of the very general dissemination of that element, in very minute quantities—and not always very minute —for the polypes that built up a certain pink coral were evidently able to separate the earth from the sea water, as their calcareous secretion contained about ½ per cent. of yttrium. We have reason to hope that this is only the first instalment of discoveries to be made by this new method of research.

I cannot conclude without a reference to the brightening prospects of technical chemistry in this country. I do not allude to the progress of any particular industry, but to the increased facilities for the education of those engaged in the chemical manufactures. First as to the workpeople. Hitherto the young artisan has had little opportunity of learning at school what would be of the greatest service to him in his after career. The traditions of the Middle Ages were all in favour of literary culture for the upper classes, and the education suited for these has been retained in our schools for the sons of the people. It is true that some knowledge of common things has been given in the best schools, and the Education Department has lately encouraged the teaching of certain sciences in the upper standards. In the Mundella Code, however, which came into operation last year, "elementary science" may receive a grant in all the classes of a boys' or girls' school, and in the suggested scheme there is mentioned simple lessons on "the chemical and physical principles involved in one of the chief industries of England, among which Agriculture may be reckoned," while "Chemistry" is inserted among "the specific subjects of instruction" that may be given to the older children. It is impossible, as yet, to form an estimate of the extent to which managers and teachers have availed themselves of this permission, for the examinations of Her Majesty's inspectors under the new code have only just commenced ; but one of the best of the Board Schools in London has just passed satisfactorily in chemistry, both with boys and girls. I trust that in those parts of the country where chemical industries prevail, chemistry may be largely taken up in our elementary schools.

The great deficiency in our present educational arrangements is the want of the means of teaching a lad who has just left the common school the principles of that industry by which he is to earn his livelihood. The more purely scientific chemistry, however, may be learnt by him now in those evening classes which may be formed under the Education Department, as well as in those that have long been established under the Science and Art Department. The large amount of attention that is now being given to the subject of technical education is creating in our manufacturing centres many technical classes and colleges for students of older growth.

As to inventors and the owners of our chemical factories, in addition to the Chemical Society and the Chemical Institute, there has recently been founded the Society of Chemical Industry. It came into existence with much promise of success ; at the close of its second year it numbered 1400 members ; it has now powerful sections in London, Manchester, Liverpool, Newcastle, and Birmingham ; and it diffuses information on technical subjects in a well-conducted monthly journal.

May the abstract science and its useful applications ever prove helpful to one another, and become more and more one chemistry for the benefit of mankind.

———

Time and Longitude

I HAVE been much amused at the questions on the above (NATURE, vol. xviii. p. 40), by Mr. Latimer Clark, and the answer (p. 66) by my old friend Capt. J. P. Maclear; the numbers of NATURE for May having only just reached my "out-of-the-world" residence. I suspect Mr. L. C. has had in his mind what I have often had, and with which I have frequently puzzled some "unco guid" Sabbatarians! If it is such a deadly sin to work on Sunday, one or the other of A and B coming, one from the east, the other from the west, of 180° meridian, must, if he continues his daily avocations, be in a bad way! Some of our people in Fiji are in this unenviable position, as the line of 180° passes through Loma-Loma!

I went from Fiji to Tonga in H.M.S. *Nymph*, and arrived at our destination on *Sunday*, according to our reckoning from Fiji, but *Monday*, according to the proper computation west from Greenwich. We, however, found the natives all keeping Sunday. On my asking the missionaries about it they told me that the missions to that group and the "navigators," having all come from the *eastward*, had determined to observe their *seventh* day, as usual, so as not to subject the natives to any future puzzle, and agreed to put the dividing line further off, between them and Hawaii, somewhere in the broad ocean, where there were no metaphysical natives or "intelligent Zulus" to cross-question them!   E. L. LAYARD
British Consulate, Noumea, New Caledonia

Hereditary Transmission

I HAVE perused with interest Mr. Edmund Watt's account of the six-fingered family in Dominica, as it recalls to my memory a family showing precisely the same peculiarities in Ceylon, at Point Pedro, the most northerly point of the island, where, twenty-six years ago, I was magistrate.

A family quarrel came before me, and I found, to my great astonishment, that plaintiff and defendant, and all the witnesses, had six fingers on each hand and six toes on each foot! The additional finger or toe was, in each instance, a "little finger" (or toe) inserted in the side of the hand or foot, quite loosely, adhering to the skin, and not part of the skeleton. It might easily have been excised with a pair of ordinary scissors. The parties were all closely related—brothers and sisters, uncles and aunts, nephews, nieces, and cousins—they must have had a common progenitor. It would be easy, and most interesting, to ascertain if any of the family now exist, and, if so, if the

supplementary finger has been transmitted to the present generation. A note to the "Resident Magistrate," Point Pedro, would, I hope, produce a reply. If any of the family of my old clerk, Mr. Dehoedt, survive, they would recollect the fact. I think the party came from Panditerripu.   E. L. LAYARD
British Consulate, Noumea

"Survival of the Fittest"

IN NATURE, vol. xix. p. 155, Mr. S. F. Clarke's observations on the cannibal habits so rapidly developed by the larvæ of the New England salamanders are cited in illustration of the survival of the fittest. The fact that similar tendencies are invariably betrayed very early in life by theyoung of the common Mexican Axolotl (*Siredon mexicanum*), numbers of which are annually hatched out in the Brighton Aquarium, may perhaps be of interest. Many of the smaller and weaker ones are bodily devoured by their stronger brethren of the same brood, an inclination which is so marked that systematic over-feeding is resorted to in order to arrest the diminution in the number of specimens.
Brighton, December 27, 1878   A. CRANE

Shakespeare's Colour-Names

IN the very interesting articles and correspondence which you have published on the subject of colour-blindness, it is rather surprising that no one has referred to a passage which, if taken alone, would appear to show that Shakespeare did not know the difference between green and blue. In "Romeo and Juliet" (Act iii., Scene 5), the Nurse says to Juliet, speaking of Paris :—

> "Oh, he's a lovely gentleman;
> Romeo's a dish-clout to him; an eagle, madam,
> Hath not so green, so quick, so fair an eye
> As Paris hath."

What is here called a green eye is evidently what we call a blue one. But Iago ("Othello," Act iii., Scene 3) calls jealousy a "green-eyed monster," using the expression "green-eyed" as a modern might use it, and meaning something very unlike "blue-eyed." These instances appear only to show that in the language of Shakespeare's time the names of colours were used somewhat vaguely.   JOSEPH JOHN MURPHY
Old Forge, Dunmurry, co. Antrim, December 23

DISCUSSION OF THE WORKING HYPOTHESIS THAT THE SO-CALLED ELEMENTS ARE COMPOUND BODIES[1]

I.

IT is known to many Fellows of the Society that I have for the last four years been engaged upon the preparation of a map of the solar spectrum on a large scale, the work including a comparison of the Fraunhofer lines with those visible in the spectrum of the vapour of each of the metallic elements in the electric arc.

To give an idea of the thoroughness of the work, at all events in intention, I may state that the complete spectrum of the sun, on the scale of the working map, will be half a furlong long; that to map the metallic lines and purify the spectra in the manner which has already been described to the Society, more than 100,000 observations have been made and about two thousand photographs taken.

In some of these photographs we have vapours compared with the sun; in others vapours compared with each other; and others again have been taken to show which lines are long and which are short in the spectra.

I may state in way of reminder that the process of purification consisted in this : When, for instance, an impurity of manganese was searched for in iron, if the longest line of Mn was absent, the short lines must also be absent on the hypothesis that the elements are elementary; if the longest line were present, then the impurity was traced down to the shortest line present.

The Hypothesis that the Elements are Simple Bodies does not include all the Phenomena

The final reduction of the photographs of all the metallic elements in the region 39-40—a reduction I

[1] Paper read at the Royal Society, December 12, by J. Norman Lockyer, F.R.S.

began in the early part of the present year, and which has taken six months, summarised all the observations of metallic spectra compared with the Fraunhofer lines accumulated during the whole period of observation. Now this reduction has shown me that the hypothesis that identical lines in different spectra are due to im-

TABLE I.—FINAL REDUCTION—IRON.

Intensity in Sun.	Wave-length and length of line.	Coincidences with Short Lines.
1	39 / 0600 / 2	U/3, Zr/5, Yt/4
3	0622 / 4	Va/4
2	0920 / 3	Va/2, Ba/3
3	1010 / 4	Va/4, Pt/3
2	1648 / 2	Co/3
2	1755 / 3	Mn/3, Ce/4
2	1835 / 4	Os/2
1	2700 / 1	Va/2
1	2950 / 1	Mo/3
3	3023 / 4	Ce/4
5	3435 / 4	U/2
3	3475 / 2	Ba/2, Rh/2
3	3628 / 3	Ta/3
2	3975 / 2	Co/3
3	4026 / 3	Va/5
3	4422 / 4	Mo/3
3	4720 / 2	Yt/5, Th/3
2	5012 / 2	Di/2
2	5160 / 2	Ce/3, Ru/3
2	5210 / 3	W/3
2	5423 / 4	U/3, Mo/3, W/4
3	6215 / 3	Yt/5, Ce/3, Di/2
2	6571 / 2	Zr/2
3	6662 / 2	Th/1
1	7555 / 3	Os/2, Ta/4, Cr/2
3	7578 / 4	Di/2
2	7685 / 2	Va/4
2	8083 / 2	Ti/1
1	8320 / 1	Cr/3
3	9520 / 3	Ru/3
2	9750 / 2	Mo/3

TABLE II.—FINAL REDUCTION—TITANIUM.

Intensity in Sun.	Wave-length and length of line.	Coincidences with Short Lines.
1	39 / 0000	Zr/4
4	3 / 0048	Th/4
5	1040 / 5	Mn/4 Ce/5 Di/3
2	1360 / 3	Va/4
5	1915 / 8	Ce/4
4	2050 / 3	U/3 ...... La/3
3	2368 / 2	Va/3
3	3718 / 5	Th/4 .... Ce/4
2	4775 / 1	Fe/2
2	5722 / 1	Zr/1 .... Rh/3
4	6175 / 2	U/3
3	6335 / 2	Di/3 .... Ta:/5
2	8083 / 1	Fe/2
3	8152 / 2	Mo/3
1	8922 / 1	Mn/4 .... Cr/4
2	9798 / longest	Va / longest

purities is not sufficient. I shall show in detail in a subsequent paper the hopeless confusion in which I have been landed. I limit myself on the present occasion to giving tables showing how the hypothesis deals with the spectra of iron and titanium.

We find short line coincidences between many metals the impurities of which have been eliminated or in which the freedom from mutual impurity has been demonstrated by the absence of the longest lines.

Evidences of Celestial Dissociation

It is five years since I first pointed out that there are many facts and many trains of thought suggested by solar and stellar physics which point to another hypothesis—namely, *that the elements themselves, or at all events some of them, are compound bodies.*

In a letter written to M. Dumas, December 3, 1873, and printed in the *Comptes Rendus*, I thus summarised a memoir which has since appeared in the *Philosophical Transactions.*

"Il semble que plus une étoile est chaude plus son spectre est simple, et que les éléments métalliques se font voir dans l'ordre de leurs poids atomiques.[1]

"Ainsi nous avons :

"1. Des étoiles très-brillantes où nous ne voyons que l'hydrogène, *en quantité énorme*, et le magnésium ;

"2. Des étoiles plus froides, comme notre Soleil, où nous trouvons :

$$H + Mg + Na$$
$$H + Mg + Na + Ca, Fe, . . . ;$$

dans ces étoiles, pas de métalloïdes ;

"3. Des étoiles plus froides encore, dans lesquelles

[1] This referred to the old numbers in which Mg = 12, Na = 23.

tous les éléments métalliques sont ASSOCIÉS, où leurs lignes ne sont plus visibles, et où nous n'avons que les spectres des métalloïdes et des composés.

"4. Plus une étoile est âgée, *plus l'hydrogène libre disparaît ;* sur la terre, nous ne trouvons plus d'hydrogène en liberté.

"Il me semble que ces faits sont les preuves de plusieurs idées émises par vous. J'ai pensé que nous pouvions imaginer une '*dissociation céleste*,' qui continue le travail de nos fourneaux, et que les métalloïdes sont des composés qui sont dissociés par la température solaire, pendant que les éléments métalliques monatomiques, dont les poids atomiques sont les moindres, son précisément ceux qui résistent, même à la température des étoiles les plus chaudes."

Before I proceed further, I should state that while observations of the sun have since shown that calcium should be introduced between hydrogen and magnesium for that luminary, Dr. Huggins' photographs have demonstrated the same fact for the stars, so that in the present state of our knowledge, independent of all hypotheses, the facts may be represented as follows, the symbol indicating the spectrum in which the lines are visible.

	Lines of	Metalloids
Hottest Stars	H + Ca + Mg	
Sun ...	H + Ca + Mg + Na + Fe	
Cooler Stars	— — Mg + Na + Fe + Bi + Hg	
Coolest ...	— — — — — —	Fluted bands of

Following out these views, I some time since communicated a paper to the Society on the spectrum of calcium, to which I shall refer more expressly in the sequel.

Differentiation of the Phenomena to be observed on the Two Hypotheses

When the reductions of the observations made on metallic spectra, on the hypothesis that the elements were really elementary, had landed me in the state of utter confusion to which I have already referred, I at once made up my mind to try the other hypothesis, and therefore at once sought for a critical differentiation of the phenomena on the two hypotheses.

Obviously the first thing to be done was to inquire whether one hypothesis would explain these short line coincidences which remained after the reduction of all the observations on the other. Calling for simplicity' sake the short lines common to many spectra *basic lines*, the new hypothesis, to be of any value, should present us with a state of things in which basic molecules representing bases of the so-called elements should give us their lines, varying in intensity from one condition to another, the *conditions* representing various compoundings.

Suppose A to contain B as an impurity and as an element, what will be the difference in the spectroscopic result?

A in both cases will have a spectrum of its own;

B as an impurity will add its lines according to the amount of impurity, as I have shown in previous papers.

B as an element will add its lines according to the amount of dissociation, as I have also shown.

The difference in the phenomena, therefore, will be that, with gradually increasing temperature, the spectrum of A *will fade*, if it be a compound body, as it will be increasingly dissociated, and it *will not* fade if it be a simple one.

Again, on the hypothesis that A is a compound body, that is, one compounded of at least two similar or dissimilar molecular groupings, then the longest lines at one temperature will not be the longest at another, the whole fabric of "impurity elimination," based upon the assumed single molecular grouping, falls to pieces, and the origin of the basic lines is at once evident.

This may be rendered clearer by some general considerations of another order.

General Considerations

Let us assume a series of furnaces A . . . D, of which A is the hottest.

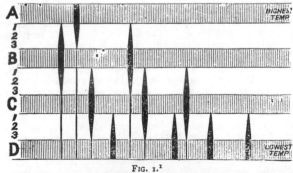

FIG. 1.[1]

Let us further assume that in A there exists a substance *a* by itself competent to form a compound body *β* by union with itself or with something else when the temperature is lowered.

Then we may imagine a furnace B in which this compound body exists alone. The spectrum of the compound *β* would be the only one visible in B, as the spectrum of the assumed elementary body *a* would be the only one visible in A.

A lower temperature furnace C will provide us with

[1] The figures between the hypothetical spectra point to the gradual change as the spectrum is observed near the temperature of each of the furnaces.

a more compound substance *γ*, and the same considerations will hold good.

Now if into the furnace A we throw some of this doubly compounded body *γ* we shall get at first an integration of the three spectra to which I have drawn attention; the lines of *γ* will first be thickest, then those of *β*, and finally *a* would exist alone, and the spectrum would be reduced to one of the utmost simplicity.

This is not the only conclusion to be drawn from these considerations. Although we have by hypothesis *β*, *γ*, and *δ* all higher, that is, more compound forms of *a*, and although the strong lines in the diagram may represent the true spectra of these substances in the furnaces B, C, and D, respectively, yet, in consequence of incomplete dissociation, the strong lines of *β* will be seen in furnace C, and the strong lines of *γ* will be seen in furnace D, *all as thin lines*. Thus, although in C we have no line which is not represented in D, the intensities of the lines in C and D are entirely changed.

In short, the line of *a* strong in A is *basic* in B, C, and D, the lines of *β* strong in B are *basic* in C and D, and so on.

I have prepared another diagram which represents the facts on the supposition that the furnace A, instead of having a temperature sufficient to dissociate *β*, *γ*, and *δ* into *a* is far below that stage, although higher than B.

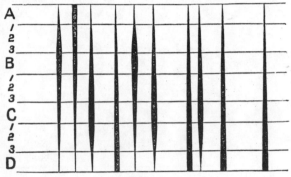

FIG. 2.

It will be seen from this diagram that then the only difference in the spectra of the bodies existing in the four furnaces would consist merely in the relative thicknesses of the lines. The spectrum of the substances as they exist in A would contain as many lines as would the spectrum of the substances as they exist in D; each line would in turn be *basic in the whole series of furnaces* instead of in one or two only.

Application of these General Considerations to Impurity Elimination

Now let us suppose that in the last diagram (Fig. 2) the four furnaces represent the spectra of say, iron, broken up into different finenesses by successive stages of heat. It is first of all abundantly clear that the relative thicknesses of the iron lines observed will vary according as the temperature resembles that of A, B, C, or D. The positions in the spectra will be the same, but the intensities will vary; this is the point. *The longest lines, represented in the diagram by the thickest ones, will vary as we pass from one temperature to another.* It is on this ground that I have before stated that the whole fabric of impurity elimination must fall to pieces on such an hypothesis. Let us suppose, for instance, that manganese is a compound of the form of iron represented in furnace B, with something else; and suppose again that the photograph of iron which I compare with manganese represents the spectrum of the vapour at the temperature of the furnace D. To eliminate the impurity of iron in manganese, as I have eliminated it, we begin the search by looking for the longest and strongest lines shown in the photograph of iron, in the photograph of manganese taken under the same conditions. I do not find these lines. I say, therefore, that there is no impurity of iron in manganese, but although the longest iron lines are not there, some of the fainter basic ones are. This I hold to be the explanation of the apparent confusion in which we are landed on the supposition that the elements are elementary.

Application of these Considerations to Known Compounds

Now to apply this reasoning to the dissociation of a known compound body into its elements—

A compound body, such as a salt of calcium, has as definite a spectrum as a simple one; but while the spectrum of the metal itself consists of lines, the number and thickness of some of which increase with increased quantity, the spectrum of the compound consists in the main of channelled spaces and bands, which increase in like manner.

In short, the molecules of a simple body and a compound one are affected in the same manner by quantity in so far as their spectra are concerned; *in other words, both spectra have their long and short lines*, the lines in the spectrum of the element being represented by bands or fluted lines in the spectrum of the compound; and in each case the greatest simplicity of the spectrum depends upon the smallest quantity, and the greatest complexity (a continuous spectrum) upon the greatest.

The heat required to act upon such a compound as a salt of calcium so as to render its spectrum visible, dissociates the compound according to its volatility; the number of true metallic lines which thus appear is a measure of the quantity of the metal resulting from the dissociation, and as the metal lines increase in number, the compound bands thin out.

I have shown in previous papers how we have been led to the conclusion that binary compounds have spectra of their own, and how this idea has been established by considerations having for a basis the observations of the long and short lines.

It is absolutely similar observations and similar reasoning which I have to bring forward in discussing the compound nature of the chemical elements themselves.

In a paper communicated to the Royal Society in 1874, referring, among other matters, to the reversal of some lines in the solar spectrum, I remarked [1]:—

"It is obvious that greater attention will have to be given to the precise *character* as well as to the position of each of the Fraunhofer lines, in the thickness of which I have already observed several anomalies. I may refer more particularly at present to the two H lines 3933 and 3968 belonging to calcium, which are much thicker in all photographs of the solar spectrum [I might have added that they were by far the thickest lines in the solar spectrum] than the largest calcium line of this region (4226·3), this latter being invariably thicker than the H lines in all photographs of the calcium spectrum, and remaining, moreover, visible in the spectrum of substances containing calcium in such small quantities as not to show any traces of the H lines.

"How far this and similar variations between photographic records and the solar spectrum are due to causes incident to the photographic record itself, or to variations in the intensities of the various molecular vibrations under solar and terrestrial conditions, are questions which up to the present time I have been unable to discuss."

An Objection Discussed

I was careful at the very commencement of this paper to point out that the conclusions I have advanced are based upon the analogies furnished by those bodies which, by common consent and beyond cavil and discussion are compound bodies. Indeed, had I not been careful to urge this point the remark might have been made that the various changes in the spectra to which I shall draw

attention are not the results of successive dissociations, but are effects due to putting the same mass into different kinds of vibration or of producing the vibration in different ways. Thus the many high notes, both true and false, which can be produced out of a bell with or without its fundamental one, might have been put forward as analogous with those spectral lines which are produced at different degrees of temperature with or without the line, due to each substance when vibrating visibly with the lowest temperature. To this argument, however, if it were brought forward, the reply would be that it proves too much. If it demonstrates that the *h* hydrogen line in the sun is produced by the same molecular grouping of hydrogen as that which gives us two green lines only when the weakest possible spark is taken in hydrogen inclosed in a large glass globe, it also proves that calcium is identical with its salts. For we can get the spectrum of any of the salts alone without its common base, calcium, as we can get the green lines of hydrogen without the red one.

I submit, therefore, that the argument founded on the overnotes of a sounding body, such as a bell, cannot be urged by any one who believes in the existence of any compound bodies at all, because there is no spectroscopic break between acknowledged compounds and the supposed elementary bodies. The spectroscopic differences between calcium itself at different temperatures is, as I shall show, as great as when we pass from known compounds of calcium to calcium itself. There is a perfect continuity of phenomena from one end of the scale of temperature to the other.

Inquiry into the Probable Arrangement of the Basic Molecules

As the results obtained from the above considerations seemed to be so far satisfactory, inasmuch as they at once furnished an explanation of the *basic lines* actually observed, the inquiry seemed worthy of being carried to a further stage.

The next point I considered was to obtain a clear mental view of the manner in which, on the principle of evolution, various bases might now be formed, and then become basic themselves.

It did not seem unnatural that the bases should increase their complexity by a process of continual multiplication, the factor being 1, 2, or even 3, if conditions were available under which the temperature of their environment should decrease, as we imagined it to do from the furnace A down to furnace D. This would bring about a condition of molecular complexity in which the proportion of the molecular weight of a substance so produced in a combination with another substance would go on continually increasing.

Another method of increasing molecular complexity would be represented by the addition of molecules of different origins. Representing the first method by $A + A$, we could represent the second by $A + B$. A variation of the last process would consist in a still further complexity being brought about by the addition of another molecule of B, so that instead of $(A + B)_2$ merely, we should have $A + B_2$.

Of these three processes the first one seemed that which it was possible to attack under the best conditions, because the consideration of impurities was eliminated; the prior work has left no doubt upon the mind about such and such lines being due to calcium, others to iron, and so forth. That is to say, they are visible in the spectra of these substances as a rule. The inquiry took this form: Granting that these lines are special to such and such a substance, does each become basic in turn as the temperature is changed?

I therefore began the search by reviewing the evidence concerning calcium and seeing if hydrogen, iron, and lithium behaved in the same way.

[1] *Phil. Trans.*, vol. clxiv. part 2, p. 807.

(To be continued.)

DISCUSSION OF THE WORKING HYPOTHESIS THAT THE SO-CALLED ELEMENTS ARE COMPOUND BODIES[1]

II.

Application of the above Views to Calcium, Iron, Lithium, and Hydrogen

Calcium

IT was in a communication to the Royal Society made now some time ago (*Proc.*, vol. xxii. p. 380, 1874), that I first referred to the possibility that the well-known line-spectra of the elementary bodies might not result from the vibration of similar molecules. I was led to make the remark in consequence of the differences to which I have already drawn attention in the spectra of certain elements as observed in the spectrum of the sun and in those obtained with the ordinary instrumental appliances.

Later (*Proc.* Roy. Soc., No. 168, 1876) I produced evidence that the molecular grouping of calcium which, with a small induction-coil and small jar, gives a spectrum with its chief line in the blue, is nearly broken up in the sun, and quite broken up in the discharge from a large coil and jar, into another or others with lines in the violet.

I said "another," or "others," because I was not then able to determine whether the last-named lines proceeded from the same or different molecules; and I added that it was possible we might have to wait for photographs of the spectra of the brighter stars before this point could be determined. I also remarked that this result enabled us to fix with very considerable accuracy the electric dissociating conditions which are equivalent to that degree of dissociation at present at work in the sun.

In Fig. 3 I have collected several spectra copied from photographs in order that the line of argument may be grasped.

First we see what happens to the non-dissociated and the dissociated chloride. Next we have the lines with a weak voltaic arc, the single line to the right (W L

FIG. 3.—The blue end of the spectrum or calcium under different conditions. 1. Calcium is combined with chlorine (CaCl₂). When the temperature is low, the compound molecule vibrates as a whole, the spectrum is at the red end, and no lines of calcium are seen. 2. The line of the metal seen when the compound molecule is dissociated to a slight extent with an induced current. 3. The spectrum of metallic calcium in the electric arc with a small number of cells. 4. The same when the number of cells is increased. 5. The spectrum when a coil and small jar are employed. 6. The spectrum when a large coil and large jar are used. 7. The absorption of the calcium vapour in the Sun.

4226·3) is much thicker than the two lines (W L 3933 and 3968) to the left, and reverses itself.

We have next calcium exposed to a current of higher tension. It will be seen that here the three lines are almost equally thick, and all reverse themselves.

Now it will be recollected, that in the case of known compounds the band structure of the true compounds is reduced as dissociation works its way, and the spectrum of each constituent element makes its appearance. If in 3 we take the wide line as representing the banded spectrum of the compound, and the thinner ones as representing the longest elemental lines making their appearance as the result of partial dissociation, we have, by hypothesis, an element behaving like a compound.

If the hypothesis be true, we ought to be able not only to obtain, with lower temperatures, a still greater preponderance of the single line, *as we do;* but with higher temperatures, a still greater preponderance of the double ones, *as we do.*

I tested this in the following manner :—Employing

photography, because the visibility of the more refrangible lines is small, and because a permanent record of an experiment, free as it must be from all bias, is a very precious thing.

Induced currents of electricity were employed in order that all the photographic results might be comparable.

To represent the lowest temperature I used a small induction coil and a Leyden jar only just large enough to secure the requisite amount of photographic effect. To represent the highest, I used the largest coil and jar at my disposal. The spark was then taken between two aluminium electrodes, the lower one cup-shaped, and charged with a salt of calcium.

In the figure I give exact copies of the results obtained. It will be seen that with the lowest temperature only the single line (2) and with the highest temperature only the two more refrangible lines (6) are recorded on the plate.

This proved that the intensity of the vibrations was quite changed in the two experiments.

Perhaps it may not be superfluous here to state the reasons which induced me to search for further evidence in the stars.

[1] Paper read at the Royal Society, December 12, 1878, by J. Norman Lockyer, F.R.S. Continued from p. 201.

It is abundantly clear that if the so-called elements, or more properly speaking their finest atoms—those that give us line spectra—are really compounds, the compounds must have been formed at a very high temperature. It is easy to imagine that there may be no superior limit to temperature, and therefore no superior limit beyond which such combinations are possible, because the atoms which have the power of combining together at these transcendental stages of heat do not exist as such, or rather they exist combined with other atoms, like or unlike, at all lower temperatures. Hence association will be a combination of more complex molecules as temperature is reduced, and of dissociation, therefore, with increased temperature there may be no end.

That is the first point.

The second is this :—

We are justified in supposing that our "calcium," once formed, is a distinct entity, whether it be an element or not, and therefore, by working at it alone, we should never know whether the temperature produces a single simpler form or more atomic condition of the same thing, or whether we actually break it up into X + Y, because neither X nor Y will ever vary.

But if calcium be a product of a condition of relatively lower temperature, then in the stars, hot enough to enable its constituents to exist uncompounded, we may expect these constituents to vary in quantity ; there may be more of X in one star and more of Y in another ; and if this be so, then the H and K lines will vary in thickness, and the extremest limit of variation will be that we shall only have H representing, say X in one star, and only have K representing, say Y in another. Intermediately between these extreme conditions we may have cases in which, though both H and K are visible, H is thicker in some and K is thicker in others.

Prof. Stokes was good enough to add largely to the value of my paper as it appeared in the *Proceedings* by appending a note pointing out that "When a solid body such as a platinum wire, traversed by a voltaic current, is heated to incandescence, we know that as the temperature increases not only does the radiation of each particular refrangibility absolutely increase, but the proportion of the radiations of the different refrangibilities is changed, the proportion of the higher to the lower increasing with the temperature. It would be in accordance with analogy to suppose that as a rule the same would take place in an incandescent surface, though in this case the spectrum would be discontinuous instead of continuous. Thus if A, B, C, D, E denote conspicuous bright lines of increasing refrangibility in the spectrum of the vapour, it might very well be that at a comparatively low temperature A should be the brightest and the most persistent ; at a higher temperature, while all were brighter than before, the relative brightness might be changed, and C might be the brightest and the most persistent, and at a still higher temperature E."

On these grounds Prof. Stokes, while he regarded the facts I mentioned as evidence of the high temperature of the sun, did not look upon them as *conclusive* evidence of the dissociation of the molecule of calcium.

Since that paper was sent in, however, the appeal to the stars to which I referred in it has been made, and made with the most admirable results, by Dr. Huggins.

The result of that appeal is that the line which, according to Prof. Stokes' view, should have prevailed over all others, as Sirius is acknowledged to be a hotter star than our sun, is that, if it exists at all in the spectrum, it is so faint that it was not recognised by Dr. Huggins in the first instance.

In Sirius, indeed, the H line due to one molecular grouping of calcium is as thick as are the hydrogen lines as mapped by Secchi, while the K line, due to another molecular grouping, which is equally thick in the spectrum of the sun, has not yet made its appearance.

In the sun, where it is as thick as H, the hydrogen lines have vastly thinned.

While this paper has been in preparation, Dr. Huggins has been good enough to communicate to me the results of his most important observations, and I have also had an opportunity of inspecting several of the photographs which he has recently taken. The result of the recent work has been to show that H and *h* are of about the same breadth in Sirius. In *a* Aquilæ while the relation of H to *h* is not greatly changed, a distinct approach to the solar condition is observed, K being now unmistakably present, although its breadth is small as compared with that of H. I must express my obligations to Dr. Huggins for granting me permission to enrich my paper by reference to these unpublished observations. His letter, which I have permission to quote, is as follows :—

" It may be gratifying to you to learn that in a photograph I have recently taken of the spectrum of *a* Aquilæ there is a line corresponding to the more refrangible of the solar H lines [that is K], but about half the breadth of the line corresponding to the first H lines.

" In the spectra of *a* Lyræ and Sirius the second line is absent."

Prof. Young's observations of the chromospheric lines, to which I shall afterwards refer, give important evidence regarding the presence of calcium in the chromosphere of the sun. He finds that the H and K lines of calcium are strongly reversed in every important spot, and that in solar storms H has been observed injected into the chromosphere seventy-five times, and K fifty times, while the blue line at W. L. 4226·3, the all-important line at the arc-temperature, was only injected thrice.

Further, in the eclipse observed in Siam in 1875, the H and K lines left the strongest record in the spectrum of the chromosphere, while the line near G in a photographic region of much greater intensity was not recorded at all. In the American eclipse of the present year the H and K lines of calcium were distinctly visible at the base of the corona, in which for the first time the observers could scarcely trace the existence of any hydrogen.

To sum up, then, the facts regarding calcium, we have first of all the H-line differentiated from the others

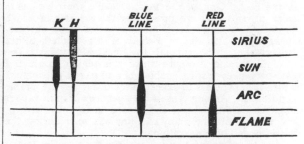

Fig. 4.—The Molecular Groupings of Calcium.

by its almost solitary existence in Sirius. We have the K-line differentiated from the rest by its birth, so to speak, in *a* Aquilæ, and the thickness of its line in the sun, as compared to that in the arc. We have the blue line differentiated from H and K by its thinness in the solar spectrum while they are thick, and by its thickness in the arc while they are thin. We have it again differentiated from them by its absence in solar storms in which they are almost universally seen, and finally, by its absence during eclipses, while the H and K lines have been the brightest seen or photographed. Last stage of all, we have calcium, distinguished from its salts by the fact that the blue line is only visible when a high temperature is employed, each salt having a definite spectrum of its own, in which none of the lines to which I have drawn attention appear, so long as the temperature is kept below a certain point.

Iron

With regard to the iron spectrum I shall limit my remarks to that portion of it visible on my photographic plates between H and G. It may be described as a very complicated spectrum so far as the number of lines is concerned in comparison with such bodies as sodium and potassium, lead, thallium, and the like, but unlike them again it contains no one line which is clearly and unmistakably reversed on all occasions. Compared, however, with the spectrum of such bodies as cerium and uranium the spectrum is simplicity itself.

Now among these lines are two triplets, two sets of three lines each, giving us beautiful examples of those repetitions of structure in the spectrum which we meet with in the spectra of almost all bodies, some of which have already been pointed out by Mascart, Cornu, and myself. Now the facts indicate that these two triplets are not due to the vibration of the same molecular grouping which gives rise to most of the other lines. They are as follows. In many photographs in which iron has been compared with other bodies, and in others again in which iron has been photographed as existing in different degrees of impurity in other bodies, these triplets have been seen almost alone, and the relative intensity of them, as compared with the few remaining lines, is greatly changed. In this these photographs resemble one I took three years ago, in which a large coil and jar were employed instead of the arc, which necessitated an exposure of an hour instead of two minutes. In this the triplet near G is very marked, the two adjacent lines more refrangible near it, which are seen nearly as strong as the triplet itself in some of the arc photographs I possess, are only very faintly visible, while dimmer still are seen the lines of the triplet between H and *h*.

There is another series of facts in another line of work. In solar storms, as is well known, the iron lines sometimes make their appearance in the chromosphere. Now, if we were dealing here with one molecular grouping, we should expect the lines to make their appearance in the order of their lengths, and we should expect the shortest lines to occur less frequently than the longest ones. Now, precisely the opposite is the fact. One of the most valuable contributions to solar physics that we possess is the memoir in which Prof. C. A. Young records his observation of the chromospheric lines, made on behalf of the United States Government, at Sherman, in the Rocky Mountains. The glorious climate and pure air of this region, to which I can personally testify, enabled him to record phenomena which it is hopeless to expect to see under less favourable conditions. Among these were injections of iron vapour into the chromosphere, the record taking the form of the number of times any one line was seen during the whole period of observation.

Now two very faint and short lines close to the triplet near G were observed to be injected thirty times, while one of the lines of the triplet was only injected twice.

The question next arises, Are the triplets produced by one molecular grouping or by two? This question I also think the facts help us to answer. I will first state by way of reminder that in the spark photograph the more refrangible triplet is barely visible, while the one near G is very strong. Now if one molecular grouping alone were in question this relative intensity would always be preserved however much the absolute intensity of the compound system might vary, but if it is a question of two molecules we might expect that in some of the regions open to our observation we should get evidence of cases in which the relative intensity is reversed or the two intensities are assimilated. What might happen does happen; the relative intensity of the two triplets in the spark photograph is grandly reversed in the spectrum of the sun. The lines barely visible in the spark photograph are among the most prominent in the solar spectrum, while the triplet which is strong in that photograph is represented by Fraunhofer lines not half so thick. Indeed, while the hypothesis that the iron lines in the region I have indicated are produced by the vibration of one molecule does not include all the facts, the hypothesis that the vibrations are produced by at least three distinct molecules includes all the phenomena in a most satisfactory manner.

Lithium

Before the maps of the long and short lines of some of the chemical elements compared with the solar spectra, which were published in the *Phil. Trans.* for 1873, "Plate IX.," were communicated to the Society, I very carefully tested the work of prior observers on the non-coincidence of the red and orange lines of that metal with the Fraunhofer lines, and found that neither of them were strongly if at all represented in the sun, and this remark also applies to a line in the blue at wavelength 4,603.

The photographic lithium line, however, in the violet, has a strong representative among the Fraunhofer lines.

Applying, therefore, the previous method of stating the facts, the presence of this line in the sun differentiates it from all the others. For the differentiation of the red and yellow lines I need only refer to Bunsen's spectral analytical researches, which were translated in the *Phil. Mag.*, December, 1875.

In Plate IV. two spectra of the chloride of lithium are given, one of them showing the red line strong and the yellow one feeble, the other showing merely a trace of the red line, while the intensity of the yellow one is much increased, and a line in the blue is indicated. Another notice of the blue line of lithium occurs in a discourse by Prof. Tyndall, reprinted in the *Chemical News*, and a letter of Dr. Frankland's to Prof. Tyndall, dated November 7, 1861. This letter is so important for my argument, that I reprint it entire from the *Philosophical Magazine*, vol. xxii. p. 472 :—

" On throwing the spectrum of lithium on the screen yesterday, I was surprised to see a magnificent blue band. At first I thought the lithic chloride must be adulterated with strontium, but on testing it with Steinheil's apparatus it yielded normal results without any trace of a blue band. I am just now reading the report of your discourse in the *Chemical News*, and I find that you have noticed the same thing. Whence does this blue line arise? Does it really belong to the lithium, or are the carbon points or ignited air guilty of its production? I find there blue bands with common salt, but they have neither the definiteness nor the brilliancy of the lithium band. When lithium wire burns in air it emits a somewhat crimson light; plunge it into oxygen, and the light changes to bluish white. This seems to indicate that a high temperature is necessary to bring out the blue ray."

" POSTSCRIPT, Nov. 22, 1861.—I have just made some further experiments on the lithium spectrum, and they conclusively prove that the appearance of the blue line depends entirely on the temperature. The spectrum of lithic chloride, ignited in a Bunsen's burner flame, does not disclose the faintest trace of the blue line; replace the Bunsen's burner by a jet of hydrogen (the temperature of which is higher than that of the Bunsen's burner) and the blue line appears, faint, it is true, but sharp and quite unmistakable. If oxygen now be slowly turned into the jet, the brilliancy of the blue line increases until the temperature of the flame rises high enough to fuse the platinum, and thus put an end to the experiment."

These observations of Profs. Tyndall and Frankland differentiate this blue line from those which are observed at low temperatures. The line in the violet to which I have already referred, is again differentiated from all the rest by the fact that it is the only line in the spectrum of the sun which is strongly reversed, so far as our present knowledge extends. The various forms of lithium, therefore, may be shown in the following manner.

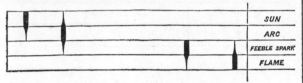

		SUN
		ARC
		FEEBLE SPARK
		FLAME

FIG. 5.—The Molecular Groupings of Lithium.

It is remarkable that in the case of this body which at relatively low temperature goes through its changes, its compounds are broken up at the temperature of the Bunsen burner. The spectrum, *e.g.* of the chloride, so far as I know, has never been seen.

Hydrogen

All the phenomena of variability and inversion in the order of intensity presented to us in the case of calcium can be paralleled by reference to the knowledge already acquired regarding the spectrum of hydrogen. Dr. Frankland and myself were working together on the subject in 1869. In that year (*Proc.*, No. 112) we pointed out that the behaviour of the *h* line was *hors ligne*, and that the whole spectrum could be reduced to one line, F.

"1. The Fraunhofer line on the solar spectrum, named *h* by Ångström, which is due to the absorption of hydrogen, is not visible in the tubes we employ with low battery and Leyden-jar power ; it may be looked upon, therefore, as an indication of relatively high temperature. As the line in question has been reversed by one of us in the spectrum of the chromosphere, it follows that the chromosphere, when cool enough to absorb, is still of a relatively high temperature.

"2. Under certain conditions of temperature and pressure, the very complicated spectrum of hydrogen is reduced in our instrument *to one line in the green* corresponding to F in the solar spectrum."

As in the case of calcium also, solar observation affords us most precious knowledge. The *h* line was missing from the protuberances in 1875, as will be shown from the accompanying extract from the Report of the Eclipse Expedition of that year :—

"During the first part of the eclipse two strong protuberances close together are noticed ; on the limb towards the end these are partially covered, while a series of protuberances came out at the other edge. The strongest of these protuberances are repeated three times, an effect of course of the prism, and we shall have to decide if possible the wave-lengths corresponding to the images. We expect *à priori* to find the hydrogen lines represented. We know three photographic hydrogen lines : F, a line near G, and *h*. F is just at the limit of the photographic part of the spectrum, and we find indeed images of protuberances towards the less refrangible part at the limit of photographic effect. For, as we shall show, a continuous spectrum in the lower parts of the corona has been recorded, and the extent of this continuous spectrum gives us an idea of the part of the spectrum in which each protuberance line is placed. We are justified in assuming, therefore, as a preliminary hypothesis, that the least refrangible line in the protuberance shown on the photograph is due to F, and we shall find support of this view in the other lines. In order to determine the position of the next line the dispersive power of the prism was investigated. The prism was placed on a goniometer table in minimum deviation for F, and the angular distance between F and the hydrogen line near G, *i.e.*, Hγ, was found, as a mean of several measurements, to be 3'. The goniometer was graduated to 15", and owing to the small dispersive power, and therefore

relatively great breadth of the slit, the measurement can only be regarded as a first approximation. Turning now again to our photographs, and calculating the angular distance between the first and second ring of protuberances, we find that distance to be 3' 15". We conclude, therefore, that this second ring is due to hydrogen. We, therefore, naturally looked for the third photographic hydrogen line, which is generally called *h*, but we found no protuberance on our photographs corresponding to that wave-length. Although this line is always weaker than Hγ, its absence on the photograph is rather surprising, if it be not due to the fact that the line is one which only comes out at a high temperature. This is rendered likely by the researches of Frankland and Lockyer (*Proc. Roy. Soc.*, vol. xvii. p. 453).

"We now turn to the last and strongest series of protuberances shown on our photographs. The distance between this series and the one we have found reason for identifying with Hγ is very little greater than that between Hβ and Hγ. Assuming the distances equal, we conclude that the squares of the inverse wave-lengths of the three series are in arithmetical progression. This is true as a first approximation. We then calculated the wave-length of this unknown line, and found it to be approximately somewhat smaller than 3,957 tenth-metres. No great reliance can be placed, of course, on the number, but it appears that the line must be close to the end of the visible spectrum.

"In order to decide if possible what this line is due to, we endeavoured to find out both by photography and fluorescence whether hydrogen possesses a line in that part of the spectrum. We have not at present come to any definite conclusion. In vacuum tubes prepared by Geissler containing hydrogen, a strong line more refrangible than H is seen, but these same tubes show between Hγ and Hδ, other lines known not to belong to hydrogen, and the origin of the ultra-violet line is therefore difficult to make out. We have taken the spark in hydrogen at atmospheric pressures, as impurities are easier to eliminate, but a continuous spectrum extends over the violet and part of the ultra-violet, and prevents any observation as to lines. We are going on with experiments to settle this point.

"Should it turn out that the line is not due to hydrogen, the question will arise what substance it is due to. It is a remarkable fact that the calculated wave-length comes very close to H. Young has found that these calcium lines are always reversed in the penumbra and immediate neighbourhood of every important sun-spot, and calcium must therefore go up high into the chromosphere. We draw attention to this coincidence, but our photographs do not allow us to draw any certain conclusions.

"At any rate, it seems made out by our photographs that the photographic light of the protuberances is in great part due to an ultra-violet line which does not certainly belong to hydrogen. The protuberances as photographed by this ultra-violet ray seem to go up higher than the hydrogen protuberances, but this may be due to the relative greater length of the line."

In my remarks upon calcium I have already referred to the fact that the line which our observation led us to believe was due to calcium in 1875, was traced to that element in this year's eclipse. The observations also show the curious connection that, at the time when the hydrogen lines were most brilliant in the corona, the calcium lines were not detected ; next, when the hydrogen lines, being still brilliant, the *h* line was not present (a condition of things which, in all probability, indicated a reduction of temperature), calcium began to make itself unmistakably visible ; and finally, when the hydrogen lines are absent, H and K become striking objects in the spectrum of the corona.

To come back to *h*, then, I have shown that Dr.

Frankland and myself, in 1869, found that it only made its appearance when a high tension was employed. We have seen that it was absent from among the hydrogen lines during the eclipse of 1875.

I have now to strengthen this evidence by the remark that it is always the shortest line of hydrogen in the chromosphere.

I now pass to another line of evidence.

I submit to the Society a photograph of the spectrum of indium, in which, as already recorded by Thalèn, the strongest line is one of the lines of hydrogen (h), the other line of hydrogen (near G) being absent. I have observed the C line in the spark produced by the passage of an induced current between indium poles in dry air.

As I am aware how almost impossible it is to render air perfectly dry, I made the following differential experiment. A glass tube with two platinum poles about half an inch apart was employed. Through this tube a slow current of air was driven after passing through a U-tube one foot high, containing calcic chloride, and then through sulphuric acid in a Wolff's bottle. The spectrum of the spark passing between the platinum electrodes was then observed, a coil with five Grove cells and a medium-sized jar being employed. Careful notes were made of the brilliancy and thickness of the hydrogen lines as compared with those of air. This done, a piece of metallic indium, which was placed loose in the tube, was shaken so that one part of it rested against the base of one of the poles, and one of its ends at a distance of a little less than half an inch from the base of the other pole. The spark then passed between the indium and the platinum. The red and blue lines of hydrogen were then observed both by my friend Mr. G. W. Hemming, Q.C., and myself. Their brilliancy was most markedly increased. This unmistakable indication of the presence of hydrogen, or rather of that form of hydrogen which gives us the h line alone *associated into* that form which gives us the blue and red lines, showed us that in the photograph we were not dealing with a physical coincidence, but that in the arc this special form of hydrogen had really been present ; that it had come from the indium, and that it had registered itself on the photographic plate, although ordinary hydrogen persistently refuses to do so. Although I was satisfied from former experiments that occluded hydrogen behaves in this respect like ordinary hydrogen, I begged my friend Mr. W. C. Roberts, F.R.S., chemist to the Mint, to charge a piece of palladium with hydrogen for me. This he at once did, and I take this present opportunity to express my obligation to him. I exhibit to the Society a photograph of this palladium and of indium side by side. It will be seen that one form of hydrogen in indium has distinctly recorded itself on the plate, while that in palladium has not left a trace. I should add that the palladium was kept in a sealed tube till the moment of making the experiment, and that special precautions were taken to prevent the two pieces between which the arc was taken becoming unduly heated.

To sum up, then, the facts with regard to hydrogen ; we have h differentiated from the other lines by its appearance alone in indium ; by its absence during the eclipse of 1875, when the other lines were photographed ; by its existence as a short line only in the chromosphere of the sun, and by the fact that in the experiments of 1869 a very high temperature was needed to cause it to make its appearance.

With regard to the isolation of the F line I have already referred to other experiments in 1869, in which Dr. Frankland and myself got it alone.[1] I exhibit to the Society a globe containing hydrogen which gives us the F line without either the red or the blue one.

[1] See also Plücker, *Phil. Trans.*, 1865, part 1, p. 21.

The accompanying drawing shows how these lines are integrated in the spectrum of the sun.

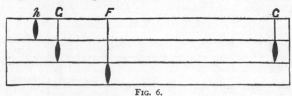

FIG. 6.

I have other evidence which, if confirmed, leads to the conclusion that the substance which gives us the non-reversed line in the chromosphere and the line at 1474 of Kirchhoff's scale, termed the coronal line, are really other forms of hydrogen. One of these is possibly more simple than that which gives us h alone, the other more complex than that which gives us F alone. The evidence on this point is of such extreme importance to solar physics, and throws so much light on star structure generally, that I am now engaged in discussing it and shall therefore reserve it for a special communication.

In the meantime I content myself by giving a diagram in which I have arranged the various groupings of hydrogen as they appear to exist, from the regions of highest to those of lowest temperature in our central luminary.

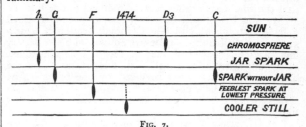

FIG. 7.

Summation of the above Series of Facts

I submit that the facts above recorded are easily grouped together, and a perfect continuity of phenomena established on the hypothesis of successive dissociations analogous to those observed in the cases of undoubted compounds.

The other Branches of the Inquiry

When we pass to the other possible evolutionary processes to which I have before referred, and which I hope to discuss on a future occasion, the inquiry becomes much more complicated by the extreme difficulty of obtaining pure specimens to work with, although I should remark that in the working hypothesis now under discussion the cause of the constant occurrence of the same substance as an impurity in the same connection is not far to seek. I take this opportunity of expressing my obligations to many friends who have put themselves to great trouble in obtaining specimens of pure chemicals for me during the whole continuance of my researches. Among these I must mention Dr. Russell, who has given me many specimens prepared by the lamented Matthiessen, as well as some of cobalt and nickel prepared by himself ; Prof. Roscoe, who has supplied me with vanadium and cæsium alum ; Mr. Crookes, who has always responded to my call for thallium ; Mr. Roberts, chemist to the Mint, who has supplied me with portions of the gold and silver trial plates and some pieces of palladium ; Dr. Hugo Müller, who has furnished me a large supply of electrolitically-deposited copper ; Mr. Holtzman, who has provided me with cerium, lanthanum, and didymium prepared by himself ; Mr. George Matthey, of the well-known metallurgical firm of Johnson and Matthey, who has provided me with magnesium and aluminium of marvellous purity ; while to

Mr. Valentin, Mr. Mellor, of Salford, and other friends, my thanks are due for other substances.

I have already pointed out that a large portion of the work done in the last four years has consisted in the elimination of the effects of impurities. I am therefore aware of the great necessity for caution in the spectroscopic examination of various substances. There is, however, a number of bodies which permit of the inquiry into their simple or complex nature being made in such a manner that the presence of impurities will be to a certain extent negligable. I have brought this subject before the Royal Society at its present stage, in the hope that possibly others may be induced to aid inquiry in a region in which the work of one individual is as a drop in the ocean. If there is anything in what I have said, the spectra of all the elementary substances will require to be re-mapped, and re-mapped from a new standpoint; further, the arc must replace the spark, and photography must replace the eye. A glance at the red end of the spectrum of almost any substance incandescent in the voltaic arc in a spectroscope of large dispersion, and a glance at the maps prepared by such eminent observers as Huggins and Thalén, who have used the coil, will give an idea of the mass of facts which have yet to be recorded and reduced before much further progress can be made.

In conclusion I would state that only a small part of the work to which I have drawn attention is my own. In some cases I have merely, as it were, codified the work done by other observers in other countries. With reference to that done in my own laboratory, I may here repeat what I have said before on other occasions, that it is largely due to the skill, patience, and untiring zeal of those who have assisted me. The burthen of the final reduction, to which I have before referred, has fallen to Mr. Miller, my present assistant; while the mapping of the positions and intensities of the lines was done by Messrs. Friswell, Meldola, Ord and Starling, who have successively filled that post.

I have to thank Corporal Ewings, R.E., for preparing the various diagrams which I have submitted to the notice of this Society.

THE CHEMICAL NEWS.

VOL. LIII. No. 1375.

SUGGESTIONS AS TO THE CAUSE OF THE PERIODIC LAW,
AND THE
NATURE OF THE CHEMICAL ELEMENTS.*

By THOS. CARNELLEY, D.Sc.,

Professor of Chemistry in University College, Dundee.

THE truth of the Periodic Law, as enunciated by Newlands, Mendeljeff, and Lothar Meyer, is now generally allowed by most chemists. Nevertheless but little has been done towards attaining a reasonable explanation of the Law.

The object of the present paper, therefore, is to offer a few suggestions on this subject.

The Periodic Law not only throws considerable light on many relationships and points of analogy which had previously been observed, but has also led to the discovery of numerous others of both theoretical and practical importance, and has opened up views of the elements and their compounds which have been of great service in attaining a correct knowledge of their mutual relationships.

Granting the truth of the Periodic Law, we cannot help theorising as to its cause, and thence, by a natural step, as to the nature of the elements themselves. Even long before the discovery of the law many chemists—more especially Gladstone, Cooke, Pettenkofer, Odling, Kremers, and Dumas—had pointed out certain numerical relationships existing between the atomic weights of bodies belonging to a given group, and had thence been induced to seek for explanations of such relationships. These chemists hence supposed that the elements belonging to the several natural groups were not primary, but were made up of two or more simpler elements. These considerations, as to the compound nature of the elements, were very ably discussed, more especially by Dumas and Gladstone (*Phil. Mag.* (4), v., 313; *Comptes Rendus*, xlv., 709, 731; xlvi., 951; xlvii., 1026). A perusal of their papers on this subject is extremely interesting in view of the more recent discoveries of Newlands, Meyer, and Mendeljeff, and show how nearly the law was detected prior to its first enunciation by Newlands in his "Law of Octaves."

The views of Dumas and Gladstone may be expressed somewhat as follows :—The atomic weights of any group of closely allied elements, when written down in numerical order, form an arithmetical progression represented by $a + x d$, in which a is the first term and d the common difference. Thus in the case of the magnesium group we have—

	Be	Mg	Ca	Sr	Ba
$a = 8$ ..	8	24	40	88	136
$d = 16$ ..	a	$a+d$	$a+2d$	$a+5d$	$a+8d$

Now this relationship, as pointed out by both Gladstone and Dumas, is exactly similar to that which holds in the case of the homologous series of organic chemistry. For example :—

Methyl	$CH_3 = 15 = a$
Ethyl	$C_2H_5 = 29 = a+d$
Propyl	$C_3H_7 = 43 = a+2d$
	&c.

In other groups of elements, however, the relations are not so simple, and it is necessary to assume such a complex expression as—

$$a + x d + x d' + x d'' + \&c. \quad \text{Thus :—}$$

---

*A Paper read before the British Association, Aberdeen Meeting.

		F	Cl	Br	I
$a = 19$ ..		F	Cl	Br	I
$d = 16\cdot5$ ..		19	35·5	80	127
$d' = 28$ ..		a	$a+d$	$a+2d+d'$	$2a+2d+2d'$

The conclusions of these chemists, however, were more or less fragmentary, and referred only to particular and isolated groups of elements. At that time the Periodic Law had not been propounded, and more general conclusions were therefore impossible, for the *general* application of the above facts would have necessarily implied a discovery of the Law prior to that made by Newlands.

In the light of the Periodic Law I wish to make a general extension of the fragmentary conclusions of Dumas. But I propose to arrive at this extension from a totally different standpoint and in a totally different way. It is an extension to which I have been led quite independently of the previous suggestions of Dumas, Gladstone, and others. In brief, what I wish to do is to bring into juxtaposition the Periodic Law, and an extended generalisation of the analogy of the elements to the alcohol radicles. At the same time I should wish my conclusions to be considered rather as *endeavours* after truth than the truth itself.

These conclusions are based on the relationships which I have observed to obtain between certain physical properties and the atomic weights of the elements and those of their compounds.

The relationships to which I refer have in part been published in various periodicals, but I wish more especially to direct attention to three of these, viz. :—

1. In this paper (*Phil. Mag.* (5), xviii., 1) a careful comparison was made of the melting- and boiling-points, and heats of formation, of the normal *halogen* compounds of the *elements*, and it was shown that certain well-defined relationships existed between these properties,—relationships which were dependent in a marked manner on the atomic weight of both the positive element and the halogen.

2. This paper (*Phil. Mag.* (5), xx., 259) dealt in a similar manner with some of the physical properties of the normal *alkyl* compounds (methides, ethides, &c.) of the elements, and it was shown that the *same relationships* obtained as in the case of the corresponding halogen compounds.

3. In this paper (*Ibid.* (5), xx., 497) it was shown that both the normal halogen and normal alkyl compounds of the hydrocarbon radicles* exhibit (with one exception, see Table I.) relationships similar to those of the corresponding compounds of the elements.

The relationships which were found to exist among all the above four classes of compounds were very numerous, but there were ten which were especially of importance ; and for the purpose of giving some idea of the weight-of the evidence which these relationships furnish, it will be well to state, in the following table, the number of cases in which they have been applied in each of the above classes of compounds, and the number of exceptions :—

	No. of cases in which the rules have been applied.	Number of exceptions.
I. Halogen compounds of the elements	3248	180 = 5·5%
II. Alkyl ditto..	942	54 = 5·7 ,,
III. Halogen compounds of the hydrocarbon radicles	877	11 = 1·2 ,,
IV. Alkyl ditto..	1110	60 = 5·4 ,,
Total	6177	305 = 5·0 ,,

For the reasons pointed out in my first paper (*q. v.*) 5 per cent of exceptions is very small, and is due chiefly

---

* For the definition of these radicles see Table IX.

to errors in the experimental melting- and boiling-points; thus no less than forty-five of the above exceptions are due to the boiling-point of propane being taken too high.

Examples of the kind of relationships referred to are given in the following Tables (I. to VII.). The arrows show the direction in which the melting- or boiling-point alters with the atomic weight. In the case of the halogen compounds of the elements, both calculated (*Phil. Mag.* (5), xviii., 1) and experimental data are given, in order that the Tables may be more complete.*

* So far as the determinations of the molecular magnetic rotations of the halogen compounds of the hydrocarbon radicles, made by Dr Perkin (*Chem. Journ.*, 1884; *Trans.*, 577), will allow a conclusion to be drawn, these physical constants also exhibit relationships similar to those given below in the case of the melting- and boiling-points. If this be true it is very probable that the molecular magnetic rotations of the halogen compounds of the elements would behave in a similar manner.

TABLE I.—Relation 1 (*Periodicity*).

X =	Cl (M.-pt.)	Br (B.-pt.)	C_2H_5 (B.-pt.)	Cl (B.-pt.)	Br (B.-pt.)	CH_3 (B.-pt.)	= X
↑ LiX_1	870	820	—	—	—	—	—
BeX_2	874	874	459	—	—	—	—
BX_3	210	271	369	—	—	—	—
CX_4	231	364 (?)	—	350	462	282	CX_4 ↑
NX_3	200	244	362	334	424	256	$(CH)X_3$
OX_2	199	237	308	315	370	245	$(CH_2)X_2$ ↓
FX_1	189	218	283	249	278	Gas	$(CH_3)X_1$
↑ NaX_1	1045	981	—	—	—	—	—
MgX_2	981	971	—	460 (?)	—	—	$(C_2)X_2$
AlX_3	340	366	467	360	435	309	$(C_2H)X_3$ ↓
SiX_4	205	260	425	411	488	325	$(C_2H_2)X_4$
PX_3	200	252	401	367	462	293	$(C_2H_3)X_3$ ↑
SX_2	199	240	364	345	393	265	$(C_2H_4)X_2$
ClX_1	198	229	285	285	312	245	$(C_2H_5)X_1$
↑ CuX_1	707	777	—	—	—	—	$(C_3H)X_1$
ZnX_2	535	667	391	351	—	314	$(C_3H_2)X_2$
GaX_3	346	378	—	402	457	343	$(C_3H_3)X_3$ ↓
$EkaSiX_4$	—	—	—	447	513	360	$(C_3H_4)X_4$
AsX_3	244	295	432	415	476	330	$(C_3H_5)X_3$ ↑
SeX_2	237	276	381	366	411	299	$(C_3H_6)X_2$
BrX_1	229	251	312	315	339	265	$(C_3H_7)X_1$
↑ AgX_1	724	700	—	—	—	323	$(C_4H_3)X_1$
CdX_2	814	844	—	—	—	355	$(C_4H_4)X_2$
InX_3	377	408	—	—	—	358	$(C_4H_5)X_3$ ↓
SnX_4	254	303	454	—	—	380	$(C_4H_6)X_4$
SbX_3	345	363	432	—	487	364	$(C_4H_7)X_3$ ↑
TeX_2	337 (?)	354 (?)	373	386	431	333	$(C_4H_8)X_2$
IX_1	298	309	345	339	361	299	$(C_4H_9)X_1$
↑ AuX_1	—	—	—	—	—	323	$(C_5H_5)X_1$
HgX_2	561	517	432	419	—	360	$(C_5H_6)X_2$
TlX_3	477	508	—	473	—	387	$(C_5H_7)X_3$ ↓
PbX_4	338	372	473	508	548	404	$(C_5H_8)X_4$
BiX_3	503	480	—	454	—	381	$(C_5H_9)X_3$
—	—	—	—	417	453	365	$(C_5H_{10})X_2$
—	—	—	—	371	391	333	$(C_5H_{11})X_1$

TABLE II.   *Relation 2 (Boiling-points).*

X =	Cl	Br	I	CH_3	C_2H_5	a-C_3H_7
SnX_4	389	474	568	351	454	498
SbX_3	496	549	693	354	432	—
AlX_3	453	533	623	403	467	543
HgX_2	576	592	622	367	432	464
IX_1	373	390	473	317	345	375
$(C_3H_4)X_4$	447	513	—	360	—	—
$(CH)X_3$	334	424	—	256	370	—
$(C_3H_3)X_3$	402	457	—	343	—	—
$(C_2H_4)X_2$	345	393	a.453	274	344	398
$C_5H_{11}X_1$	371	391	416 →	333	365	398

TABLE III.   *Relation 3 (Boiling-points).*

X =	Cl	Br	I	CH_3	C_2H_5	Cl	Br	CH_3	C_2H_5	= X
PX_3	351	444	575	314	401	334	424	256	370	CHX_3
AsX_3	405	493	677	—	432	367	462	293	—	$(C_2H_3)X_3$
SbX_3	496	549	693	354	432	415	476	330	—	$(C_3H_5)X_3$
↓ BiX_3	703	749	—	—	—	—	487	364	—	$(C_4H_7)X_3$ ↓
—	—	—	—	—	—	454	—	381	—	$(C_5H_9)X_3$

TABLE IV.   *Relation 4.*

	Br—Cl M.-pt.	I—Br M.-pt.	Et—Me B.-pt.	Pra—Et B.-pt.	Br—Cl B.-pt.	I—Br B.-pt.	Et—Me B.-pt.	Pra—Et B.-pt.	
Na	−64	−80	—	—	—	—	—	—	(C_3H)i
Mg	−10	−75	—	—	—	—	—	—	(C_3H_2)ii
Al	26	32	64	56	55	—	—	—	(C_3H_3)iii
Si	55	133	122	61	66	—	—	—	(C_3H_4)iv
P	52	76	87	—	61	—	—	—	(C_3H_5)iii
S	41	75	54	41	45	49	60	51	(C_3H_6)ii
Cl	31	69	36	34	24	29	37	33	(C_3H_7)i

TABLE V.—*Relations 5 and 6 (Boiling-points).*

	Br—Cl	I—Br	Et—Me	Pra—Me	Br—Cl	I—Br	Et—Me	Pra—Me	Normal primary.
Diff. of at. wts.	44·5	47	14	28	44·5	47	14	28	
Cl	46	87	36	70	29	37	—	—	CH_3
Br	45	59	35	67	27	33	—	—	C_2H_5
I	17	83 (?)	28	58	25	31	37	70	C_3H_7
Si	95	137	122	183	22	30	33	60	C_4H_9
Sn	86	94	103	147	22	27	27	54	C_5H_{11}
Pb	—	—	40	—	22	26	27	51	C_6H_{13}
Zn	19	—	72	113	20	22	24	48	C_7H_{15}
Hg	16	30	65	97	20	—	24	46	C_8H_{17}

TABLE VI.—*Relations 7a and 7b (Boiling-points).*

(I.)	Diff. of at. wts. of elements in (I.).	Cl	Br	I	CH_3	C_2H_5	$a.C_3H_7$	Diff. of at. wts. of elements in (I.a).	(I.a.)
As—P	44	54	49	102*(?)	82 (?)	19	—	89	Pb—Sn
Sb—As	47	91	56	16	48	29	12	90	Sn—Si
Sb—P	89	145	105	118*	130	48	—	179	Pb—Si
Bi—Sb	90	207	200	171	28	27	25	44·5	Br—Cl
Bi—As	135	298	256	187	40	33	31	47·0	I—Br
Bi—P	179	352	305	289	68	60	56	91·5	I—Cl
C_4H_9—C_3H_7	14	32	29	28	37	33	27	14	C_4H_9—C_3H_7
C_6H_{13}—C_4H_9	28	55	55	51	60	54	51	28	C_6H_{13}—C_4H_9
C_6H_{13}—C_3H_7	42	87	84	79	97	87	78	42	C_6H_{13}—C_3H_7
C_8H_{17}—C_4H_9	56	105	103	91	111	102	97	56	C_8H_{17}—C_4H_9
C_8H_{17}—C_3H_7	70	137	132	119	148	135	124	70	C_8H_{17}—C_3H_7
C_8H_{17}—C_2H_5	84	171	164	149	177	172	156	84	C_8H_{17}—C_2H_5
C_8H_{17}—CH_3	98	207	198	179	—	201	193	98	C_8H_{17}—CH_3

* Boiling-point of PI_3 too low and boiling-point of AsI_3 too high.

TABLE VII.—*Relations 8 and 9 (Boiling-points).*

(I.)	Diff. of at. wts. of elements in (I.).	Cl	Br	I	CH_3	C_2H_5	$a.C_3H_7$
P—I	−96	−22	+54	+102	−3	+56	—
As—I	−52	+32	103	204	—	87	—
P—Br	−49	65	113	185	+37	89	—
Sb—I	−7	123	159	220	37	87	—
P—Cl	−4·5	111	158	202	65	116	—
As—Br	−5·0	119	162	287	—	120	—
As—Cl	+39·5	165	207	304	—	147	—
Sb—Br	40·0	210	218	303	77	120	—
Sb—Cl	84·5	256	263	320	105	147	—
Bi—I	83·0	330	359	—	—	—	—
Bi—Br	130·0	417	418	474	—	—	—
Bi—Cl	174·5	463	463	491	—	—	—
CH_2—C_8H_{17}	−99	−141	−106	−41	−177	−135	−97
C_2H_4—C_8H_{17}	−85	−99	−71	—	−148	−102	−70
CH_2—C_6H_{13}	−71	−91	−58	−1	−126	−87	−51
C_4H_8—C_8H_{17}	−57	—	—	—	−78	−48	−22
CH_2—C_4H_9	−43	−36	−3	+50	—	−33	0
C_2H_4—C_4H_9	−29	+6	+32	—	−37	0	+27
C_2H_4—C_3H_7	−15	38	61	—	0	+33	54
C_3H_6—C_3H_7	−1	74	94	125	+37	60	78
C_3H_6—C_2H_5	+13	108	126	155	—	97	111
C_5H_{10}—C_3H_7	27	—	—	—	94	111	123
C_5H_{10}—C_2H_5	41	—	—	—	126	148	156
C_5H_{10}—CH_3	55	—	—	—	—	177	193

(To be continued).

THE CHEMICAL NEWS.

VOL. LIII. No. 1376.

SUGGESTIONS AS TO THE CAUSE OF THE PERIODIC LAW,
AND THE
NATURE OF THE CHEMICAL ELEMENTS.*

By THOS. CARNELLEY, D.Sc.,
Professor of Chemistry in University College, Dundee.

(Continued from p. 159.)

Now what conclusions are to be drawn from such facts as those represented in the above tables? In addition to the general conclusion *that the melting- and boiling-points of both organic and inorganic compounds obey the same rules*, we may also infer—

That the elements as a whole are analogous to the hydrocarbon radicles (see Tables IX. and XI.), *having a similar function in their several compounds, and most probably a somewhat similar chemical constitution, or, shortly, they are analogous in both form and function.*

This analogy will be best exhibited in Tables VIII. and IX.

Taking first the points of *resemblance* :—

1. The properties of the elements and their compounds and those of the hydrocarbon radicles and their compounds are strictly periodic, a period in each case consisting (if we neglect group VIII.) of seven members.

2. As shown above the melting and boiling-points of both the halogen and alkyl compounds of either the elements of Table VIII. or of the hydrocarbon radicles in Table IX. follow the same rules; that is to say, if we make any systematic arrangement of the compounds of the Elements in Table VIII., and a corresponding systematic arrangement of the compounds of the radicles in Table IX., then the numerical order of the physical properties (as represented by the melting and boiling-points) will be the same in the two cases.

3. The atomicity in each series, in the case of both elements and hydrocarbon radicles, increases up to the middle group and then diminishes or increases up to the 7th or last group (*cf.* differences *infra*).

4. The compounds of the elements are strictly analogous in form to those of the hydrocarbon radicles, as witness the following examples :—(See next page.)

5. Mendeljeff in his original paper on the Periodic Law (*Annalen*, Suppl., viii., 133) draws attention to the fact that the elements of the first or Li-series, and also part of the Na-series, are to some extent different from those of the other series, and the elements comprising them are hence termed by him " typical elements." The exceptional character of these series he compares with the usual and well-known exceptional character of the

TABLE VIII.—*Natural Classification of the Elements.* (After Mendeljeff).

Groups	I.	II.	III.	IV.	V.	VI.	VII.	
Series	Monads.	Dyads.	Triads.	Tetrads.	Triads or Pentads.	Dyads or Hexads.	Monads or Heptads.	Group VIII.
0.	$H=1$							
1.	$Li=7$	$Be=9$	$B=11$	$C=12$	$N=14$	$O=16$	$F=19$	
2.	$Na=23$	$Mg=24$	$Al=27$	$Si=28$	$P=31$	$S=32$	$Cl=35\cdot5$	
3.	$K=39$	$Ca=40$	$Sc=44$	$Ti=48$	$V=51$	$Cr=52$	$Mn=54$	{ $Fe=56, Co=59, Ni=59,$ ($Cu=63$).
4.	$Cu=63$	$Zn=65$	$Ga=69$	$EkaSi=72$ (?)	$As=75$	$Se=79$	$Br=80$	
5.	$Rb=85$	$Sr=87$	$Yt=89$	$Zr=90$	$Nb=94$	$Mo=96$	—	{ $Ru=104, Rh=104, Pd=106$ ($Ag=108$).
6.	$Ag=108$	$Cd=112$	$In=113$	$Sn=118$	$Sb=120$	$Te=125$	$I=127$	
7.	$Cs=133$	$Ba=137$	$La=138$	$Ce=140$	$Di=142$	$Tb=149$ (?)	—	—, —, —, (——).
8.	—	—	—	—	$Er=166$ (?)	—	—	
9.	—	—	$Yb=173$	—	$Ta=182$	$W=184$	—	{ $Os=193, Ir=193, Pt=195,$ ($Au=197$).
10.	$Au=197$	$Hg=200$	$Tl=204$	$Pb=207$	$Bi=208$	—	—	
11.	—	—	—	$Th=234$	—	$U=240$	—	—, —, —, (——).

TABLE IX.—*Classification of the Hydrocarbon Radicles.*

Groups	I.	II.	III.	IV.	V.	VI.	VII.	
Series	Monads or Heptads.	Dyads or Hexads.	Triads or Pentads.	Tetrads.	Triads.	Dyads.	Monads.	
0.							$H=1$	
1.	—	—	—	CX_4	$(CH)X_3$	$(CH_2)X_2$	$(CH_3)X$	
2.	—	$(C_2)X_2$	$(C_2H)X_3$	$(C_2H_2)X_4$	$(C_2H_3)X_3$	$(C_2H_4)X_2$	$(C_2H_5)X$	$X=$any monad element
3.	$(C_3H)X$	$(C_3H_2)X_2$	$(C_3H_3)X_3$	$(C_3H_4)X_4$	$(C_3H_5)X_3$	$(C_3H_6)X_2$	$(C_3H_7)X$	or radical, *e.g.*, Cl,
4.	$(C_4H_3)X$	$(C_4H_4)X_2$	$(C_4H_5)X_3$	$(C_4H_6)X_4$	$(C_4H_7)X_3$	$(C_4H_8)X_2$	$(C_4H_9)X$	Br, I, CH_3, C_2H_5, &c.
5.	$(C_5H_5)X$	$(C_5H_6)X_2$	$(C_5H_7)X_3$	$(C_5H_8)X_4$	$(C_5H_9)X_3$	$(C_5H_{10})X_2$	$(C_5H_{11})X$	
6.	$(C_6H_7)X$	$(C_6H_8)X_2$	$(C_6H_9)X_3$	$(C_6H_{10})X_4$	$(C_6H_{11})X_3$	$(C_6H_{12})X_2$	$(C_6H_{13})X$	
&c.	&c.	&c.	&c.	&c.	&c.	&c.	&c.	

The hydrocarbon radicles in the table may be represented by the general formula $C_n H_{2n+(2-x)}$, in which $n =$ the series, and $x=(8-a)$, a being the group to which the element belongs.

It will be seen at a glance that there is a great resemblance between the two tables both in structure and arrangement; but there are nevertheless certain points of difference which it will be well to mention in detail along with the points of resemblance.

first and second members of a homologous series of organic chemistry.*

* Compare Perkin's paper on " Magnetic Rotation " (*Jour. Chem. Soc. Trans.*, 1884, p. 552; CHEM. NEWS., xlix., 285), in which he says "that any body to be the first member of a homologous series must contain the CH_2-group, because it is by the introduction of additional members of this group that we form the series. Thus, propionic acid should be the first member of the homologous series of fatty acids. Formic and acetic acids do not contain the CH_2-group, and they are well known to possess properties which are not altogether comparable with those of the other fatty acids."

* A Paper read before the British Association, Aberdeen Meeting.

I.	II.	III.	IV.	V.	VI.	VII.
Na(OH). Sodium hydrate.	Mg(OH)$_2$. Magnesium hydrate.	Al(OH)$_3$. Aluminium hydrate.	Si(OH)$_4$. Silicic acid.	P(OH)$_3$. Phosphorous acid.	S(OH)$_2$. Hyposulphurous acid.	Cl(OH). Hypochlorous acid.
C$_3$H(OH). ?	C$_3$H$_2$(OH)$_2$. ?	C$_3$H$_3$(OH)$_3$. Trioxymethylene.	C$_5$H$_4$(OH)$_4$. Propylphycite.	C$_3$H$_5$(OH)$_3$. Glycerol.	C$_3$H$_6$(OH)$_2$. Propylene glycol.	C$_3$H$_7$(OH).* Propyl alcohol.
NaCl. Na chloride.	MgCl$_2$. Mg chloride.	AlCl$_3$. Al chloride.	SiCl$_4$. Si tetrachloride	PCl$_3$. P trichloride.	SCl$_2$. Sulphur dichloride.	ClCl. Chlorine chloride.
(C$_3$H)Cl. ?	(C$_3$H$_2$)Cl$_2$. Dichlorallylene.	(C$_3$H$_3$)Cl$_3$. Trichlorpropylene.	(C$_3$H$_4$)Cl$_4$. Tetrachlorpropane.	(C$_3$H$_5$)Cl$_3$. Trichlorpropane.	(C$_3$H$_6$)Cl$_2$. Dichlorpropane.	(C$_3$H$_7$)Cl.* Propyl chloride.
NaCH$_3$. ?	Mg(CH$_3$)$_2$. ?	Al(CH$_3$)$_3$. Al methide.	Si(CH$_3$)$_4$. Si methide.	P(CH$_3$)$_3$. Trimethylphosphine.	S(CH$_3$)$_2$. Methyl sulphide.	Cl(CH$_3$). Methyl chloride.
C$_3$H(CH$_3$). ?	C$_3$H$_2$(CH$_3$)$_2$. Valerylene.	C$_3$H$_3$(CH$_3$)$_3$. Hexylene.	C$_3$H$_4$(CH$_3$)$_4$. Heptane.	C$_3$H$_5$(CH$_3$)$_3$. Hexane.	C$_3$H$_6$(CH$_3$)$_2$. Pentane.	C$_3$H$_7$(CH$_3$).* Butane.
Na$_2$O. Sodium oxide.	MgO. Mg oxide.	Al$_2$O$_3$. Al oxide.	SiO$_2$. Si dioxide.	P$_2$O$_3$. P trioxide.	[SO]. ?	Cl$_2$O. Cl monoxide.
(C$_3$H)$_2$O. ?	(C$_3$H$_2$)O. ?	(C$_3$H$_3$)$_2$O$_3$. ?	(C$_3$H$_4$)O$_2$. ?	(C$_3$H$_5$)$_2$O$_3$. Glyceric ether.	(C$_3$H$_6$)O. Propylene oxide.	(C$_3$H$_7$)$_2$O.* Propyl oxide.

* Most probably the series of hydrocarbon radicles, as represented here and in Table IX., should be inverted so that Groups VII., VI., and V. would become I., II., and III., respectively. With regard to this point, *vide infra.* If such an inverted arrangement be adopted, then glycerol, propylene glycol, and propyl alcohol would be represented as analogous to Al, Mg, and Na hydrates respectively, an analogy which would be truer than the one shown here.

This therefore furnishes another point of analogy between the elements of Table VIII. and the radicles of Table IX.

6. The numerous cases in which, in many reactions, the hydrocarbon radicles act in a manner strictly analogous to that of the elements are too well known to need further particularisation in detail.

Let us now consider the points in which the elements of Table VIII. differ from the radicles of Table IX.

1. Though it is true that both the elements and the hydrocarbon radicles are strictly periodic in their properties and in those of their compounds, and that the periods are of the same length, consisting of seven members in each case, yet the nature of the periodicity (at least in so far as the melting- and boiling-points are concerned) is not exactly the same in both. In the case of the halides of the elements the melting- and boiling-points *diminish as we pass from the first to the seventh group*, whereas in the case of both the halides and alkides of the hydrocarbon radicles these properties *increase up to the fourth group, and then diminish to the seventh*; whilst the alkides of the elements partake of the features of each, since they increase up to the third and then diminish to the seventh group. The alteration in the order of size of these physical properties is therefore the same in all cases from the fourth to the seventh group inclusive, but different from the first to the third group. This difference and resemblance is shown in the following Table, in which the numbers referring to the halides have been obtained by adding the boiling-points (or melting-points) of the chlorides of the corresponding members of each series together, the same operation being repeated with the bromides and iodides. The sum of the boiling-points (or melting-points) of the chlorides is then added to that of the bromides and iodides, and the mean taken. A similar operation gives the average mean boiling-points for the alkides (methides, ethides, propides). As there are five series under consideration, each number is the mean of $5 \times 3 = 15$ separate data: whilst in the case of the halides of the elements each number is the mean of 18 data, since in this case we are dealing with six series.

The above facts are rendered still more evident by Diagrams I., II., and III., of which I. shows the *average mean* curves of (1) the melting-points of the halides (Cl, Br, and I) of the elements; (2) the boiling-points of the alkides (Me, Et, Pr) of the elements; (3) the boiling-points of the halides (Cl, Br, I) of the hydrocarbon radicles; (4) the boiling-points of the alkides (methides only) of the same. The curve is constructed by taking the average mean temperatures of melting or boiling (see Table X.) as ordinates, and the atomicity of the elements or hydrocarbon radicles as abscissæ.

Diagrams II. and III. represent the same facts, but in greater detail. Diagram II. gives the *mean* curves of— (1) the alkides (Me, Et, Pr) of the elements*; (2) the halides (Cl, Br, I) of the hydrocarbon radicles; (3) the

* That is, of elements of Series 1, 2. 4, 6, and 10, these being the only elements which combine with alcohol radicles.

TABLE X.

	ELEMENTS.		HYDROCARBON RADICLES.	
	Halides. (Cl, Br, I). Melting-points.*	*Alkides.* (Me, Et, Pr^a). Boiling-points.	*Halides.* (Cl, Br, I). Boiling-points.	*Alkides.* (Methides only) † Boiling-points.
Monads	870	—		289
Dyads	847	409	425	311
Triads	341	461	478	333
Tetrads	326	411	530	350
Triads	318	404	482	325
Dyads	303	366	415	301
Monads	277	312	338	264

* Data not sufficiently complete for boiling-points.   † Data not sufficiently complete for ethides and propides.

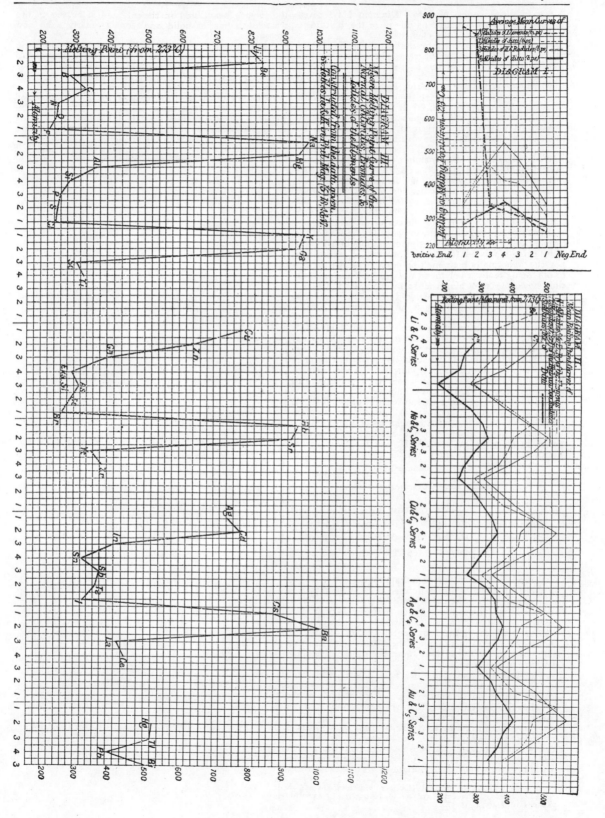

alkides (methides only) of ditto. Diagram III.* gives the mean curve of the melting-points of the chlorides, bromides, and iodides of the elements.

A very marked feature in the curves of the halides of the elements, as shown by Diagram III., and more especially by Diagram I., is the sudden and great rise in melting-point (the same also holds true for the boiling-point) which occurs at the positive end of each series as we pass from triad to dyad, and thence to monad, metals. What may be the cause of this it is very difficult to say. The only explanation which occurs to me is that it may possibly be due to polymerisation, in which case the molecules of the halides of the alkaline and alkaline-earth metals, also of Ag, Cu, Zn, Cd, &c., would not be really represented by their ordinary simple formulæ, but by some multiple thereof,—e.g., $(NaCl)_n$, $(KBr)_n$, &c. In this connection note the abnormally high boiling-point of C_2Cl_2 in the second series of the chlorides of the hydrocarbon radicles, and already referred to in my first paper, and shown in the diagram accompanying that paper. This compound stands on a part of the curve corresponding to that occupied by $MgCl_2$ in the curve of the halides of the elements, and its high boiling-point is undoubtedly due to its being a polymer $(C_2Cl_2)_n$. For a similar reason may not the abnormally high melting-point of $MgCl_2$ be due to its being really a polymer $(MgCl_2)_n$? This explanation, however, is not alogether satisfactory, for C_2Cl_2 is a non-saturated compound, and could hence easily undergo polymerisation, whilst $MgCl_2$ is, so far as known, a saturated compound.

2. A second point in which the elements of Table VIII. differ from the hydrocarbon radicles of Table IX. is that the groups in the former, though frequently, are not always numerically homologous, whereas in the latter case they are strictly so throughout. (If, however, the arrangement of the hydrocarbon radicles in Table XI. be adopted, this statement would require some modification.

3. In the case of the elements each group is divisible into two sub-groups, because, as pointed out by Mendeljeff, it is only the alternate members of a group which are strictly comparable with one another. The several series of elements in Table VIII. are therefore printed alternately in thin and thick type.

(To be continued).

NOTE ON THE
REACTION OF TIN WITH SULPHURIC AND NITRIĊ ACIDS.

By H. BASSETT.

THE formation of stannic sulphate by the action of sulphuric acid alone on tin requires, as is well known, a high temperature: the addition of a small quantity of nitric acid, which readily supplies the necessary oxygen, greatly facilitates the action, and was long ago applied in the manufacture of the ornamental articles of tin-plate known as "Moirée metallique," the mixed acids dissolving a film of tin from the surface and exposing the crystalline structure of the metal.

A mixture of 1 part sulphuric acid, 2 parts nitric acid, and 3 parts water, all by measure, has a remarkable solvent action on tin in the cold. A solid lump of the metal weighing 10 grms. dissolves completely in a few hours in 50 c.c. of this mixture, the containing vessel being immersed in cold water. Under these circumstances no red fumes are given off, but a steady evolution of nearly pure nitrous oxide goes on the whole time. The resulting solution of stannic sulphate is quite clear unless the temperature has been allowed to rise, in which case

it becomes beautifully opalescent, and if heated in the water-oven for an hour it becomes nearly solid and opaque.

On pouring the solution (still containing some nitric acid) into boiling water, the tin is completely precipitated in the form of gelatinous metastannic acid, which, after thorough washing with boiling water, dries up in the water-oven into semi-transparent lumps.

Attempts to utilise this reaction for the solution and analysis of tin alloys have not proved satisfactory: in presence of some metals a small quantity of tin appears to escape precipitation, while others show more or less tendency to go down with the precipitate.

---

* The separate curves for the chlorides, bromides, and iodides respectively of the elements have already been given in an earlier paper (Phil. Mag., v., 18, 1).

SUGGESTIONS AS TO THE CAUSE OF THE PERIODIC LAW,
AND THE
NATURE OF THE CHEMICAL ELEMENTS.*

By THOS. CARNELLEY, D.Sc.,
Professor of Chemistry in University College, Dundee.

(Continued from p. 172.)

As illustrations of this difference between odd and even series we have the following among others ; (a) It is only the elements printed in thick type (including also the typical elements beginning with Li) which combine with alcohol radicles, those printed in thin type being devoid of that property.† (b) Elements printed in thin type are paramagnetic, whilst those in thick type are diamagnetic.‡

(c) Elements in thick type are generally easily reducible and frequently occur in the free state in nature, or when combined are usually found as sulphides or double sulphides, and only in very few cases as oxides ; whereas elements in thin type are only reducible with great difficulty, and (except C, N, and O) *never* occur in the free state in nature, nor (except Mo, Mn, and O) as sulphides, but usually as oxides or double oxides, forming silicates, carbonates, sulphates, &c.§ (d) The colour relations,‖ also relations between the melting-points, boiling-points, heats of formation, &c.,¶ only hold good when alternate series are compared.

Now in the case of the hydrocarbon radicles of Table IX. no such difference in the proporties of alternate series can be detected, for the members of any one group are all mutually comparable, and are not divisible into two sub-groups as in the case of the elements in Table VIII. (cf. however, Baeyer's observation, *vide infra*). Here we have a marked difference, but may it not be due to the series of hydrocarbon radicles in Table IX. being either all odd or all even series, the alternate even or odd series not being represented in the latter table ? Such alternate series might be supposed to be generated as follows :—(See next column).

It will be observed that in the above table the system of odd and even series only becomes *complete* at series 10, though it commences *partially* at series 4. Now in the arrangement of the elements in Table VIII. the difference

Formic acid series.	Melting-point. Even.	Melting-point. Odd.	Oxalic acid series.	Melting-point. Even.	Melting-point. Odd.
CH_2O_2		+8·6			—
$C_2H_4O_2$	+17		$C_2H_2O_4$	200	
$C_3H_6O_2$		b. −21	$C_3H_4O_4$		132
$C_4H_8O_2$	+1		$C_4H_6O_4$	180	
$C_5H_{10}O_2$		b. −16	$C_5H_8O_4$		97
$C_6H_{12}O_2$	−2		$C_6H_{10}O_4$	148	
$C_7H_{14}O_2$		−10·5	$C_7H_{12}O_4$		103
$C_8H_{16}O_2$	+16		$C_8H_{14}O_4$	140	
$C_9H_{18}O_2$		+12	$C_9H_{16}O_4$		106
$C_{10}H_{20}O_2$	+30		$C_{10}H_{18}O_4$	127	
$C_{11}H_{22}O_2$		+28·5	$C_{11}H_{20}O_4$		108
$C_{12}H_{24}O_2$	+43·6		—		
$C_{13}H_{26}O_2$		+40·5	$C_{17}H_{32}O_4$		132
$C_{14}H_{28}O_2$	+53·8				
$C_{15}H_{30}O_2$		+51			
$C_{16}H_{32}O_2$	+62				
$C_{17}H_{34}O_2$		+60			
$C_{18}H_{36}O_2$	+69	—			
$C_{20}H_{40}O_2$	+75				
$C_{21}H_{42}O_2$		+72·5			
$C_{22}H_{44}O_2$	+78				

* A Paper read before the British Association, Aberdeen Meeting.
† Mendeljeff (*Annalen*, Suppl., viii., 133).
‡ Carnelley, *Ber.*, xii., 1958.
§ *Ibid.*, *Phil. Mag.*, (5), xviii., 197. *Ber.*, xvii., 2287.
‖ *Ibid.*, *Phil. Mag.*, (5), xviii., 132. *Ber.*, xvii., 2151.
¶ *Ibid.*, *Phil. Mag.*, (5), xviii., 6.

TABLE XI.

Groups Series.	I. Monads.	II. Dyads.	III. Triads.	IV. Tetrads.	V. Triads.	VI. Dyads.	VII. Monads.	VIII.
0.	—	—	—	—	—	—	$H=1$	
1.	—	—	—	$CX_4=12$	$(CH)X_3=13$	$(CH_2)X_2=14$	$(CH_3)X=15$	
2.	—	$C_2X_2=24$	$(C_2H)X_3=25$	$(C_2H_2)X_4=26$	$(C_2H_3)X_3=27$	$(C_2H_4)X_2=28$	$(C_2H_5)X=29$	
3.	$(C_3H)X=37$	$(C_3H_2)X_2=38$	$(C_3H_3)X_3=39$	$(C_3H_4)X_4=40$	$(C_3H_5)X_3=41$	$(C_3H_6)X_2=42$	$(C_3H_7)X=43$	
4.	$(C_4H_3)X=51$	$(C_4H_4)X_2=52$	$(C_4H_5)X_3=53$	$(C_4H_6)X_4=54$	$(C_4H_7)X_3=55$	$(C_4H_8)X_2=56$	$(C_4H_9)X=57$	
5.	$(C_5H_5)X=65$	$(C_5H_6)X_2=66$	$(C_5H_7)X_3=67$	$(C_5H_8)X_4=68$	$(C_5H_9)X_3=69$	$(C_5H_{10})X_2=70$	$(C_5H_{11})X=71$	
6.	$(C_6H_7)X=79$	$(C_6H_8)X_2=80$	$(C_6H_9)X_3=81$	$(C_6H_{10})X_4=82$	$(C_6H_{11})X_3=83$	$(C_6H_{12})X_2=84$	$(C_6H_{13})X=85$	
7.	$(C_7H_9)X=93$	$(C_7H_{10})X_2=94$	$(C_7H_{11})X_3=95$	$(C_7H_{12})X_4=96$	$(C_7H_{13})X_3=97$	$(C_7H_{14})X_2=98$	$(C_7H_{15})X=99$	
8.	$(C_8H_{11})X=107$	$(C_8H_{12})X_2=108$	$(C_8H_{13})X_3=109$	$(C_8H_{14})X_4=110$	$(C_8H_{15})X_3=111$	$(C_8H_{16})X_2=112$	$(C_8H_{17})X=113$	C_8H_1
9.	$(C_9H_{13})X=121$	$(C_9H_{14})X_2=122$	$(C_9H_{15})X_3=123$	$(C_9H_{16})X_4=124$	$(C_9H_{17})X_3=125$	$(C_9H_{18})X_2=126$	$(C_9H_{19})X=127$	C_9H_1; C_9H_2; C_9H_3
10.								$C_{10}H_1$; $C_{10}H_2$; $C_{10}H_3$; $C_{10}H_4$; $C_{10}H_5$
11.	&c.	&c.	&c.	&c.	&c.	&c.	&c.	
12.								
13.								
14.								
15.								
&c.								

DIAGRAM IV.

ELEMENTS.

(Left handed Spiral)

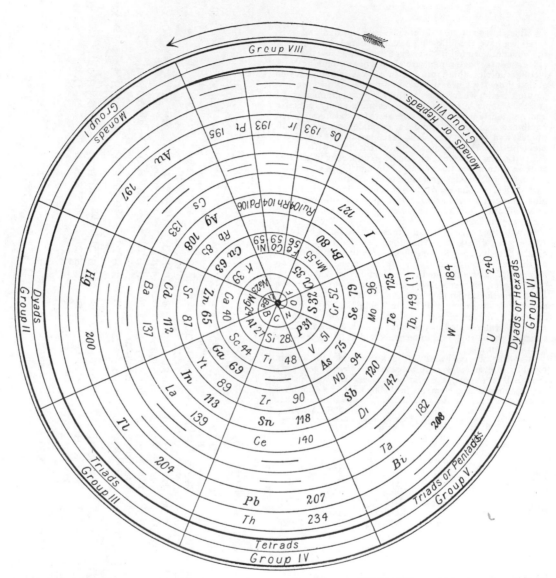

between odd and even series only becomes evident about the middle of series 3.*

It is also interesting to observe in this connection that Baeyer (*Ber.*, x., 1286) has shown that, if the melting-points of homologous series be compared, this property is materially influenced by *odd and even numbers* in such a way that those compounds containing an *even* number of carbon atoms melt higher than the neighbouring compounds containing an odd number (see Table on preceding page, col. 1).

4. The last point of difference in the complete analogy of the elements of Table VIII. to the hydrocarbon radicles of Tables IX. or XI. that it will be necessary to mention is what may be called " inversion of the series." What is meant by this will be better understood from an inspection of Tables VIII. and IX. than from a written explanation. As there arranged the atomic weights of both tables increase across the tables from Group I. to Group VII., indicating thereby that methyl, ethyl, propyl, &c., are analogous to the halogens. And further, Groups I. II. and III. of the elements are represented as usually exhibiting but one atomicity, whilst Groups V., VI., and VII. are represented as triads or pentads, dyads or hexads, monads or heptads, respectively. In the Table of the hydrocarbon radicles, on the other hand, the conditions,

* *Cf.* the examples which have been given above in illustration of the difference between odd and even series of elements.

DIAGRAM V.

HYDROCARBON RADICLES.

(Right handed Spiral)

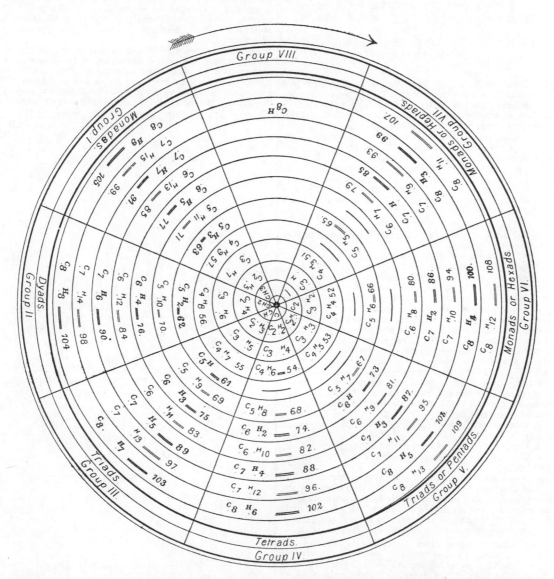

as regards to the atomicity of the several groups are reversed.

If, however, the hydrocarbon radicles in Table IX. be so arranged that the atomic weight increases across the table from Group VII. to Group I., instead of from Group I. to Group VII., then the condition of things as regards the atomicity of the several groups becomes the same in both tables. Under these circumstances, however, the alkyl radicles, methyl, ethyl, propyl, &c., are represented as analogous to the monad metals of Group I.

On the whole I am inclined to think that the analogy indicated by this latter arrangement is the truer one. It should be stated that whichever arrangement be adopted the relationships between the physical properties of the elements and their compounds and those of the hydrocarbon radicles referred to above in my previous papers, and the conclusions deduced therefrom, hold good in either case.

If the arrangement last proposed be adopted (*i.e.*, the one in which the atomic weights of the hydrocarbon radicles increase across the table from Group VII. to Group I.) then such an arrangement of the hydrocarbon radicles would bear the same relation to that of the elements in Table VIII. which a right-handed spiral bears to a left-handed spiral. Diagrams IV. and V. will make this clear. The spiral arrangement represented in IV. is in principle the same as that first adopted by Baumhauer (1870) and subsequently modified by Huth (1884). The latter determines the position of the elements in the spiral by measuring off distances on the radii of the spiral proportional to the respective atomic weights.

(To be continued).

REPORT OF COMMITTEE OF THE CHEMICAL SOCIETY OF WASHINGTON,
ON THE
"METHOD OF STATING RESULTS OF WATER ANALYSES."

THE Chemical Society of Washington, at the meeting of November 12th, 1885, appointed a Committee to consider the present state of water analysis, and to present a method of stating analyses adapted for general use, in order that those hereafter published may be readily compared with each other and with future work. This Committee reported February 11th, 1886, and was authorised to prepare an abstract for publication, in order to call the attention of chemists to the subject. The Society earnestly recommends the adoption of the scheme, which is herewith briefly presented. The full text of the Report will be published in the next *Bulletin* of the Society.

Abstract.

Water analyses are usually made to answer one of three questions, viz.,—

1st. Is the water useful medicinally?
2nd. Is it injurious to health? and
3rd. Is it suitable for manufacturing purposes?

Many books relating to water were published during the eighteenth century, but accurate chemical analysis was not attempted until about 1820. As the earlier analyses were isolated, rare, and made for special purposes, the form of the statement was of little importance if it was only intelligible. At the present time, however, water analyses are very numerous. An examination of about a thousand shows some forty-two methods of stating quantitative results, there being sometimes three different ratios in the report of one analysis. Such discrepancies render comparisons difficult and laborious. The various methods of statement may be classified under the following general forms :—

1. Grains per imperial gallon of 10 lbs., or 70,000 grains.

2. Grains per U.S., or wine gallon of 58,372 grains.

3. Decimally, as parts per 100, 1000, 100,000, or 1,000,000.

4. As so many grammes or milligrammes per litre.

The last two would be identical if all waters had the same density, but, as the densities of sea water, mineral waters, &c., are much above that of pure water, it is plain that the third and fourth modes are not comparable.

The Committee therefore unanimously recommends—

1st. That water analyses be uniformly reported according to the decimal system, in parts per million or milligrammes per kilogramme, with the temperature stated, and that Clark's scale of degree of hardness and all other systems be abandoned.

2nd. That all analyses be stated in terms of the radicals found.

3rd. That the constituent radicals be arranged in the order of the usual electro-chemical series, the positive radicals first.

4th. That the combination deemed most probable by the chemist should be stated in symbols as well as by name.

The abandonment of Clark's *scale* has been recommended by Wanklyn and Chapman, and the recommendation made by the Committee does not involve the disuse of his method, but merely the bringing of it into accord with the decimal system,—the changing from grains per gallon to milligrammes per kilogramme.

The last conclusion was deemed desirable from the frequent confusion in the statement of the iron salts and of the carbon oxides.

The Committee is unanimously of the opinion that analyses in the form recommended will prove quite as acceptable to boards of health, and to the public in general, for whom such analyses are often made, as if presented in the mixed and irregular forms commonly adopted.

The Committee also feels sure that people in general are better able to form a definite idea of the character of a water from a report stated in parts per 100, parts per 1,000,000, &c., than from one expressed as grains per gallon, the latter being a ratio wholly unfamiliar to any but those in the medical or pharmaceutical professions.

(Signed)   A. C. PEALE, M.D.
              WM. H. SEAMAN, M.D.
              CHAS. H. WHITE, M.D.

MANUFACTURE OF MINERAL COLOURS IN THE UNITED STATES.

THE following particulars, taken from the " Report of the United States' Geological Survey on the Mineral Resources of the States," have been forwarded to us by the author, Dr. Marcus Benjamin, F.C.S.

There are in the States 31 white-lead works, in all of which the so-called Dutch method is followed, the material used being pig-lead. The total produce during the year 1884 was about 65,000 tons.

A "sublimed lead" is made in Missouri by the direct oxidation of galena in a reverberatory furnace.

Zinc-white was manufactured in the same year to the extent of 12,000 to 15,000 tons. It is used not only as a colour but in the manufacture of india-rubber, in pottery, and in the paper trade.

Barium sulphate (heavy spar) was raised to between 25,000 and 30,000 tons. Barium compounds are used as paints under the names of blanc fixe, satin-white, &c., and in the form of peroxide for bleaching purposes. Barium sulphate, both the natural and the precipitated, is largely used as an adulterant.

Terra alba (ground gypsum) is imported from Nova Scotia, whilst a superior quality is brought from France. In addition to its legitimate use in making white pigments of a low grade it serves for adulterating a variety of commercial articles.

The quantity of red-lead produced in the United States could not be ascertained, but the imports at New York amounted to 198,588 lbs.

The American production of litharge is also an unknown quantity. The imports were only 54,183 lbs.

Concerning ochres it is said that with the possible exception of the deposits recently opened up near St. Louis, the American production is inferior to the imported qualities. "American ochres for the most part lack strength or tinting properties and require too much oil for grinding." The annual consumption in the United States is estimated at 10,000 tons, of which about 3000 tons are imported.

American umbers are inferior to those imported from Italy and Turkey. Siennas are found to a small extent in Virginia and Pennsylvania, but most of that used is imported from Italy.

There is no mention of lapis lazuli having been found in the United States, but there are two American manufactories of artificial ultramarine with a yearly output of 1400 tons.

Ground slate is used as a pigment to the extent of 2000 tons yearly, and occurs in four colours—green, red, slate, and drab.

SUGGESTIONS AS TO THE CAUSE OF THE PERIODIC LAW,
AND THE
NATURE OF THE CHEMICAL ELEMENTS.*

By THOS. CARNELLEY, D.Sc.,
Professor of Chemistry in University College, Dundee.

(Concluded from p. 186.)

General Conclusions.

A CAREFUL consideration of the points submitted above leads almost irresistibly to the conclusion *that the elements as a whole are analogous to the hydrocarbon radicles.* This is a conclusion which, if true, would further lead us to infer *that the elements are not elements in the strict sense of the term, but are in fact compound radicles, made up of* AT LEAST *two simple elements,* A *and* B.†

Generally speaking the elements, as I have endeavoured to show above, are analogous to hydrocarbon radicles, but differing from them in certain points, these differences depending most probably on the difference in the nature of one or more of the primary elements in the two cases. In making this analogy carbon and hydrogen are supposed to be primary elements.

Supposed Genesis of the Elements.

From the above point of view it should be possible to build up a series of compounds of two primary elements, A and B, corresponding to the—at present—so called elements. Such a system should fulfil the following conditions:—

(1.) The secondary elements generated from the two primary elements, A and B, must be capable of division into series and groups,—that is to say, they must exhibit the phenomenon of periodicity.

(2.) The several series must run in octaves.

(3.) The system must exhibit some feature corresponding to "odd and even series."

(4.) The atomic weights must increase across the system from the first to the seventh group,—*i. e.*, from the positive to the negative end of each series.

(5.) The atomicity (*i. e.*, valency) must increase from the first to the fourth or middle group, and then either increase or diminish to the seventh group.

6. It should exhibit some feature corresponding to the eighth group in Mendeljeff's classification of the elements.

(7.) The atomic weights in such a system should coincide with, or approximate to, the commonly received atomic weights of the elements.

Though up to the present I have not been able to draw out a system which will rigorously accord with all the above conditions, still the one given in the lower half of Diagram VI.‡ fulfils all except the last. This condition is indeed the most important and crucial, but nevertheless the system proposed gives atomic weights which, with but few exceptions, approximate closely to those usually taken as the true atomic weights.

---

* A Paper read before the British Association, Aberdeen Meeting.
† If composed of more than two simple elements, it is probable that such primary elements will be but few in number.
‡ The *upper* part of this diagram is, with one or two slight modifications, identical with that proposed by Bayley (*Phil. Mag.* (5), 13, 26

DIAGRAM SHOWING THE

		Diff.		Diff.	
K 39·0		46·3	Rb 85·3	47·5	Cs 132·8
Ca 40·0		47·4	Sr 87·4	49·4	Ba 136 8
Sc 44·0		45·8	Yt 89·8	48·2	La 138·0
Ti 48·0		42·0	Zr 90·0	50·0	Ce 140·0
V 51·3		42·5	Nb 93·8	48·2	Di 142
Cr 52·0		43·5	Mo 95·5		—
Mn 54·0		—	—		—
Fe 56·0		48·2	Ru 104·2		—
Co 58·9		45·3	Rh 104·2		—
Ni 58·9		47·1	Pd 106·0		—
Cu 63·2		44·8	Ag 108·0		—
Zn 64·9		47·1	Cd 112·0		—
Ga 68·9		44·7	In 113·6		—
—			Sn 117·8		—
As 74·9		45·3	Sb 120·2		—
Se 78·8		{46·2 / 49·2}	Te {125 / 128·0}		—
Br 79·8		47·2	I 127·0		—

Diff.

	Diff.	
Li 7·0	16	Na 23·0
Be 9·1	14·9	Mg 24·0
B 10·9	16·1	Al 27·0
C 12·0	16	Si 28·0
N 14·0	17	P 31·0
O 16·0	16	S 32·0
F 19·0	16·4	Cl 35·4

H

16 Difference

16, 17, 20, 20·3, 20, 19·6, 40·2, 40·9, 41·9, 43·9, 46·8, 44·4

GENERAL FORMULA = $A_nB_{2n+(2-x)}$.

n = Series. x = Group.

$A = 12$ and $B = -2$.

16 Difference

A_5B_{11} 38	48	$A_{11}B_{23}$ 86	48	$A_{17}B_{35}$ 134
A_5B_{10} 40	48	$A_{11}B_{22}$ 88	48	$A_{17}B_{34}$ 136
A_5B_9 42	48	$A_{11}B_{21}$ 90	48	$A_{17}B_{33}$ 138
A_5B_6 48	44	$A_{11}B_{20}$ 92	48	$A_{17}B_{32}$ 140
A_5B_5 50	44	$A_{11}B_{19}$ 94	48	$A_{17}B_{31}$ 142
A_5B_4 52	44	$A_{11}B_{18}$ 96	48	$A_{17}B_{30}$ 144
A_5B_3 54	44	$A_{11}B_{17}$ 98	48	$A_{17}B_{29}$ 146
$A_6^5B_8^2$ 56	48	$A_{12}^{11}B_{20}^{16}$ 104		—
$A_6^5B_7^1$ 58	48	$A_{12}^{11}B_{19}^{15}$ 106		—
A_6B_6 60	48	$A_{12}^{11}B_{18}^{14}$ 108		—
A_6B_5 62	48	$A_{12}B_{17}$ 110	48	$A_{18}B_{29}$ 158
A_6B_4 64	48	$A_{12}B_{16}$ 112	48	$A_{18}B_{28}$ 160
A_6B_3 66	48	$A_{12}B_{15}$ 114	48	$A_{18}B_{27}$ 162
A_6B_2 68[72]	48	$A_{12}B_{14}$ 116[120]	48	$A_{18}B_{26}$ 164
A_7B_3 78[74]	48	$A_{13}B_{15}$ 126[122]	48	$A_{19}B_{27}$ 174
A_7B_2 80[76]	48	$A_{13}B_{14}$ 128[124]	48	$A_{19}B_{26}$ 176
A_7B_1 82[78]	48	$A_{13}B_{13}$ 180[126]	48	$A_{19}B_{25}$ 178

	Diff.	
AB_3 6	16	A_3B_7 22
AB_2 8	16	A_3B_6 24
AB_1 10	16	A_3B_5 26
AB_0 12	16	A_3B_4 28
AB^*_{-1} 14	16	A_3B_3 30
AB^*_{-2} 16	16	A_3B_2 32
AB^*_{-3} 18	16	A_3B_1 34

H

16, 16, 20, 20, 20, 40, 40, 40(44), 48(44), 48(44), 48(44)

* $A = DB_4$ and $D = 2$.

SUPPOSED GENESIS OF THE ELEMENTS.

Diff.		Diff.				
	—		—	Monads.	Group I.	
	—		—	Dyads.	Group II.	
	—		—	Triads.	Group III.	
	—		Th 234	Tetrads.	Group IV.	a.
40·1	Ta 182·1		—	Triads or Pentads.	Group V.	
	W 184·0	56	U 240	Dyads or Hexads.	Group VI.	
	—		—	Monads or Heptads.	Group VII.	
	Os 198·5 (?)		—			
	Ir 192·7		—		Group VIII.	
	Pt 194·4		—			
	Au 196·2		—	Monads.	Group I. or IX.	
	Hg 200		—	Dyads.	Group II. or X.	
	Tl 203·7		—	Triads.	Group III. or XI.	
	Pb 206·5		—	Tetrads.	Group IV. or XII.	b.
	Bi 207·5		—	Triads or Pentads	Group V. or XIII.	
	—		—	Dyads or Hexads.	Group VI. or XIV.	
	—		—	Monads and Heptads.	Group VII. or XV.	

GENERAL FORMULÆ.

Diff.		Diff.			Possibly		
40	$A_{22}B_{45}$ 174	56	$A_{29}B_{59}$ 230	$(A_nB_{2n+1})^i$		Group I.	
40	$A_{22}B_{44}$ 176	56	$A_{29}B_{58}$ 232	$(A_nB_{2n})^{ii}$		Group II.	
40	$A_{22}B_{43}$ 178	56	$A_{29}B_{57}$ 234	$(A_nB_{2n-1})^{iii}$		Group III.	
40	$A_{22}B_{42}$ 180	56	$A_{29}B_{56}$ 236	$(A_nB_{2n-2})^{iv}$	$(A_nB_{2n-4})^{iv}$	Group IV.	a.
40	$A_{22}B_{41}$ 182	56	$A_{29}B_{55}$ 238	$(A_nB_{2n-3})^{iii, v}$	$(A_nB_{2n-5})^{v}$	Group V.	
40	$A_{22}B_{40}$ 184	56	$A_{29}B_{54}$ 240	$(A_nB_{2n-4})^{ii, vi}$	$(A_1B_{2n-6})^{vi}$	Group VI.	
40	$A_{22}B_{39}$ 186	56	$A_{29}B_{53}$ 242	$(A_nB_{2n-5})^{i, vii}$	$(A_nB_{2n-7})^{vii}$	Group VII.	
	$A_{23}^{22}B_{42}^{38}$ 192		—				
	$A_{23}^{22}B_{41}^{37}$ 194		—			Group VIII.	
	$A_{23}^{22}B_{40}^{36}$ 196		—				
40	$A_{23}B_{39}$ 198		—	$(A_nB_{2n-7})^i$		Group IX.	
40	$A_{23}B_{38}$ 200		—	$(A_nA_{2n-8})^{ii}$		Group X.	
40	$A_{23}B_{37}$ 202		—	$(A_nB_{2n-9})^{iii}$		Group XI.	
40	$A_{23}B_{36}$ 204[208]		—	$(A_nB_{2n-10})^{iv}$	$(A_nB_{2n-4})^{iv}$	Group XII.	b.
32	$A_{23}B_{35}$ 206[210]		—	$(A_nB_{2n-11})^{iii, v}$	$(A_nB_{2n-5})^{iii, v}$	Group XIII.	
32	$A_{23}B_{34}$ 208		—	$(A_nB_{2n-12})^{ii, vi}$	$(A_nB_{2n-6})^{ii, vi}$	Group XIV.	
32	$A_{23}B_{32}$ 210		—	$(A_nB_{2n-13})^{i, vii}$	$(A_nB_{2n-7})^{i, vii}$	Group XV.	

Of the atomic weights represented by this system—

    27 agree almost exactly with the actual numbers.
    19 are not more than 1 unit out.
    5   ,,   ,,   1·5 units out.
    6   ,,   ,,   2   ,,
    3   ,,   ,,   3   ,,
    1 (viz., Sb) is 5·8 units out.

    61 = Total.

Ag, Au, and Te are not included in the above, because the positions of Ag and Au are not known with certainty, for they may belong either to Group VIII., in which case the calculated numbers would be correct, or to Group I. (IX.), for which positions the calculated numbers are about 2 units too high. As regards Te, there are two numbers which claim to represent its atomic weight, viz., 125 (Brauner) and 128 (Wills, Berzelius, &c.), of which the former is most probably the true one.

The similarity between the upper and lower parts of the Diagram VI. is strikingly illustrated by the *differences* between the atomic weights of successive periods.

With regard to the *structure* of the system represented in the lower half of the diagram the following remarks are necessary:—

First. The binary elements of this system (except Ti, V, Cr, and Mn, the general formula for which would be $A_n B_{2n+(4-x)}$) may all be represented by the general formula $A_n B_{2n+(2-x)}$, in which $A = 12$ and $B = -2$, whilst $x = $ the group and n the homologous series to which any element belongs. A is supposed to be a tetrad element identical with carbon, and B a monad element.

Second. As represented in the diagram the atomic weights are all *even integers*. This is due to the value for B, which, for the sake of simplicity, has been taken as $= -2$; whereas, if the theory be correct, it should have a value somewhere between $-1·99$ and $-2·00$. Such a value would, on the whole, bring the calculated atomic weights more into accord with the actual numbers.

Third. The difference between what are commonly called odd and even series is indicated by the diagram to be due to the several series beginning with K, Rb, Cs, &c., consisting of *saturated* binary elements, whereas the series beginning with Cu, Ag, Au,* &c., consist of *unsaturated* binary elements. Thus the difference between K, on the one hand, and Cu on the other, is analogous to the difference between some such saturated formula as—

$$\begin{array}{l} CH_3 \\ | \\ CH_2 \\ | \\ CH_2 \\ | \\ CH_2 \\ | \\ CH_2- \\ \hline (C_5H_{11})- \end{array}$$

And some such unsaturated formula as—

Or

$$\begin{array}{l} CH \\ \| \\ CH \\ | \\ CH \\ \| \\ CH \\ | \\ CH \\ \| \\ C- \\ \hline \end{array} \qquad \begin{array}{l} CH_2 \\ \| \\ C \\ ||| \\ C \\ | \\ CH \\ \| \\ CH- \\ \hline \end{array}$$

$$(C_6H_5)-$$

Fourth. If $A = 12$, then   Nitrogen $= A B_{-1} = 14$
               Oxygen $= A B_{-2} = 16$
         and Fluorine $= A B_{-3} = 18$.

Now if A be really a primary element, and therefore indivisible, it would be impossible to take B from it, as indicated by the formula $A B_{-1}$ for N, and the two following formulæ for O and F. This difficulty would, however, be removed by substituting $D B_4$ for A throughout the whole scheme, D being a heptavalent element of atomic weight $= 20$. By this substitution neither the atomic weights nor the general structure of the diagram would be affected.

Fifth. If there be any truth in the above speculation as

---

* Assuming that these metals belong to Group I. (IX.).

---

to the genesis of the chemical elements, then the existence of elements of *identical atomic weights*, and isomeric with one another, would be possible. May not Ni and Co, Ru and Rh, Os and Ir, and some of the rare earth metals, be isomers in this sense? If this be so, it would remove one of the principal objections which have been raised against the truth of the Periodic Law.*

Sixth. In the scheme represented in Diagram VI. the chemical elements are all (except hydrogen) supposed to be composed of two simpler elements, viz., $A = 20$, and $B = -2$. Of these A is identical with the element carbon, whilst B is a substance of negative weight, possibly the ethereal fluid of space?

Supposing the theory advanced above to be true, then it is interesting to observe—

That whereas the hydrocarbons are compounds of carbon and hydrogen, the chemical elements would be compounds of carbon with ether (atomic weight $= -2$), the two sets of bodies being generated in an exactly analogous manner from their respective elements. There would hence be three primitive elements, viz., carbon, hydrogen, and ether.

Truly this is a bold, though most interesting, speculation,† and one which would need very ample confirmation —and no doubt some considerable modification—before it could be generally accepted by chemists. It is offered simply as a speculation, with the object of indicating new lines in which research is required, and as possibly throwing some light, however dim, on the cause of the Periodic Law and the nature of the chemical elements.

---

* In this connection compare the article on the Decomposition of Didymium, by Welsbach, in *Nature*, xxxii., 435; also CHEMICAL NEWS, 1885

† I have dealt with the question entirely from my own point of view, but there are many other circumstances which might be adduced in favour of the compound nature of the elements, including the results obtained by Abney (*Proc. Roy. Soc*, xxxi., 416; *Journ. Chem. Soc.*, xlii., 131) in regard to the infra-red absorption-spectra of certain carbon compounds, and the arguments advanced some time ago by Lockyer and others.

A joint discussion on gold and silver has been arranged between Sections C and F. As these discussions will be real, and as several eminent foreigners are expected to take part in them, the meeting on the whole promises to be lively.

The social distractions—*conversaziones*, receptions, dinners, and excursions—are perplexingly numerous. The hand-books for the excursions have been got up with much care and thoughtfulness. There is, indeed, a separate little hand-book for each excursion, the whole set being done up in a case. Another hand-book of about one hundred pages gives an epitome of the history, antiquities, meteorology, physiography, flora and fauna of Manchester and the district.

Thus, so far as the officials are concerned, everything has been done to make the Manchester meeting a success. At the present moment the weather is not quite what could be wished; it is raining hard, and the weather is oppressively sultry. We can only hope it will improve before active operations begin.

INAUGURAL ADDRESS BY SIR HENRY E. ROSCOE, M.P., D.C.L., LL.D., PH.D., F.R.S., V.P.C.S., PRESIDENT.

MANCHESTER, distinguished as the birthplace of two of the greatest discoveries of modern science, heartily welcomes to-day, for the third time, the members and friends of the British Association for the Advancement of Science.

On the occasion of our first meeting in this city in the year 1842, the President, Lord Francis Egerton, commenced his address with a touching allusion to the veteran of science, John Dalton, the great chemist, the discoverer of the laws of chemical combination, the framer of the atomic theory upon which the modern science of chemistry may truly be said to be based. Lord Francis Egerton said :—" Manchester is still the residence of one whose name is uttered with respect wherever science is cultivated, who is here to-night to enjoy the honours due to a long career of persevering devotion to knowledge, and to receive from myself, if he will condescend to do so, the expression of my own deep personal regret that increase of years, which to him up to this hour has been but increase of wisdom, should have rendered him, in respect of mere bodily strength, unable to fill on this occasion an office which in his case would have received more honour than it could confer. I do regret that any cause should have prevented the present meeting in his native town from being associated with the name "—and here I must ask you to allow me to exchange the name of Dalton in 1842 for that of Joule in 1887, and to add, again in the words of the President of the former year, that I would gladly have served as a doorkeeper in any house where Joule, the father of science in Manchester, was enjoying his just pre-eminence.

For it is indeed true that the mantle of John Dalton has fallen on the shoulders of one well worthy to wear it, one to whom science owes a debt of gratitude not less than that which it willingly pays to the memory of the originator of the atomic theory. James Prescott Joule it was who, in his determination of the mechanical equivalent of heat, about the very year of our first Manchester meeting, gave to the world of science the results of experiments which placed beyond reach of doubt or cavil the greatest and most far-reaching scientific principle of modern times ; namely, that of the conservation of energy. This, to use the words of Tyndall, is indeed a generalization of conspicuous grandeur fit to take rank with the principle of gravitation ; more momentous, if that be possible, combining as it does the energies of the material universe into an organic whole, and enabling the eye of science to follow the flying shuttles of the universal power as it weaves what the Erdgeist in " Faust " calls " the living garment of God."

It is well, therefore, for us to remember, in the midst of the turmoil of our active industrial and commercial life, that Manchester not only well represents the energy of England in these practical directions, but that it possesses even higher claims to our regard and respect as being the seat of discoveries of which the value not only to pure science is momentous, but which also lie at the foundation of all our material progress and all our industrial success. For without a knowledge of the laws of chemical combination all the marvellous results with which modern industrial chemistry has astonished the world could not have been achieved, whilst the knowledge of the quantitative relations existing between the several forms of energy, and the possibility of expressing their amount in terms of ordinary mechanics, are matters which now constitute the life-breath of every branch of applied science. For example, before Dalton's discovery every manufacturer of oil of vitriol—a substance now made each week in thousands of tons within a few miles of this spot—every manufacturer had his own notions of the quantity of sulphur which he ought to burn in order to make a certain weight of sulphuric acid, but he had no idea that only a given weight of sulphur can unite with a certain quantity of oxygen and of water to form the acid, and that an excess of any one of the component parts was not only useless but harmful. Thus, and in tens of thousands of other instances, Dalton replaced rule of thumb by scientific principle. In like manner the applications of Joule's determination of the mechanical equivalent of heat are even more general ; the increase and measurement of the efficiency of our steam-engines and the power of our dynamos are only two of the numerous examples which might be adduced of the practical value of Joule's work.

If the place calls up these thoughts, the time of our meeting also awakens memories of no less interest, in the recollection that we this year celebrate the Jubilee of Her Most Gracious Majesty's accession to the throne. It is right that the members of the British Association for the Advancement of Science should do so with heart and voice, for, although science requires and demands no royal patronage, we thereby express the feeling which must be uppermost in the hearts of all men of science, the feeling of thankfulness that we have lived in an age which has witnessed an advance in our knowledge of Nature, and a consequent improvement in the physical, and let us trust also in the moral and intellectual, well-being of the people hitherto unknown ; an age with which the name of Victoria will ever be associated.

To give even a sketch of this progress, to trace even in the merest outline the salient points of the general history of science during the fifty momentous years of Her Majesty's reign, is a task far beyond my limited powers. It must suffice for me to point out to you, to the best of my ability, some few of the steps of that progress as evidenced in the one branch of science with which I am most familiar, and with which I am more closely concerned, the science of chemistry.

In the year 1837 chemistry was a very different science from that existing at the present moment. Priestley, it is true, had discovered oxygen, Lavoisier had placed the phenomena of combustion on their true basis, Davy had decomposed the alkalies, Faraday had liquefied many of the gases, Dalton had enunciated the laws of chemical combination by weight, and Gay-Lussac had pointed out the fact that a simple volumetric relation governs the combination of the gases. But we then possessed no knowledge of chemical dynamics, we were then altogether unable to explain the meaning of the heat given off in the act of chemical combination. The atomic theory was indeed accepted, but we were as ignorant of the mode of action of the atoms and as incapable of explaining their mutual relationship as were the ancient Greek philosophers. Fifty years ago, too, the connexion existing between the laws of life, vegetable and animal, and the phenomena of inorganic chemistry, was ill understood. The idea that the functions of living beings are controlled by the same forces, chemical and physical, which regulate the changes occurring in the inanimate world, was then one held by only a very few of the foremost thinkers of the time. Vital force was a term in everyone's mouth, an expression useful, as Goethe says, to disguise our ignorance, for

" Wo die Begriffe fehlen,
Da tritt ein Wort zur rechten Zeit sich ein."

Indeed the pioneer of the chemistry of life, Liebig himself, cannot quite shake himself free from the bonds of orthodox opinion, and he who first placed the phenomena of life on a true basis cannot trust his chemical principles to conduct the affairs of the body, but makes an appeal to vital force to help him out of his difficulties ; as when in the body politic an unruly mob requires the presence and action of physical force to restrain it and to bring its members under the saving influence of law and order, so too, according to Liebig's views, in the body corporeal a continual conflict between the chemical forces and the vital power occurs throughout life, in which the latter, when it prevails, insures health and a continuance of life, but of which defeat insures disease or death. The picture presented to the student of to-day is a very different one. We now believe that no such conflict is possible, but that life is governed by chemical and physical forces, even though we cannot in every case explain its phenomena in terms of these forces ; that whether these tend to continue or to end existence depends upon their nature and amount, and that disease and death are as much a consequence of the operation of chemical and physical laws as are health and life.

Looking back again to our point of departure fifty years ago, let us for a moment glance at Dalton's labours, and compare his views and those of his contemporaries with the ideas which now prevail. In the first place it is well to remember that the keystone of his atomic theory lies not so much in the idea of the existence and the indivisible nature of the particles of matter—though this idea was so firmly implanted in his mind that, being questioned on one occasion on the subject, he said to his friend the late Mr. Ransome, "Thou knowst it must be so, for no man can split an atom "—as in the assumption that the weights of these particles are different. Thus whilst each of the ultimate particles of oxygen has the same weight as every other particle of oxygen, and each atom of hydrogen, for example, has the same weight as every other particle of hydrogen, the oxygen atom is sixteen times heavier than that of hydrogen, and so on for the atoms of every chemical element, each having its own special weight. It was this discovery of Dalton, together with the further one that the elements combine in the proportions indicated by the relative weights of their atoms or in multiples of these proportions, which at once changed chemistry from a qualitative to a quantitative science, making the old invocation prophetic, "Thou hast ordered all things in measure and number and weight."

The researches of chemists and physicists during the last fifty years have not only strengthened but broadened the foundations of the great Manchester philosopher's discoveries. It is true that his original numbers, obtained by crude and inaccurate methods, have been replaced by more exact figures, but his laws of combination and his atomic explanation of those laws stand as the great bulwarks of our science.

On the present occasion it is interesting to remember that within a stone's-throw of this place is the small room belonging to our Literary and Philosophical Society which served Dalton as his laboratory. Here, with the simplest of all possible apparatus—a few cups, penny ink-bottles, rough balances, and self-made thermometers and barometers—Dalton accomplished his great results. Here he patiently worked, marshalling facts to support his great theory, for as an explanation of his laborious experimental investigations the wise old man says : " Having been in my progress so often misled by taking for granted the results of others, I have determined to write as little as possible but what I can attest by my own experience." Nor ought we when here assembled to forget that the last three of Dalton's experimental essays—one of which, on a new method of measuring water of crystallization, contained more than the germ of a great discovery—were communicated to our Chemical Section in 1842, and that this was the last memorable act of his scientific life. In this last of his contributions to science, as in his first, his method of procedure was that which has been marked out as the most fruitful by almost all the great searchers after Nature's secrets ; namely, the assumption of a certain view as a working hypothesis, and the subsequent institution of experiment to bring this hypothesis to a test of reality upon which a legitimate theory is afterwards to be based. "Dalton," as Henry well says, " valued detailed facts mainly, if not solely, as the stepping-stones to comprehensive generalizations."

Next let us ask what light the research of the last fifty years has thrown on the subject of the Daltonian atoms : first, as regards their size ; secondly, in respect to their indivisibility and mutual relationships ; and thirdly, as regards their motions.

As regards the size and shape of the atoms, Dalton offered no opinion, for he had no experimental grounds on which to form it, believing that they were inconceivably small and altogether beyond the grasp of our senses aided by the most powerful appliances of art. He was in the habit of representing his atoms and their combinations diagrammatically as round disks or spheres made of wood, by means of which he was fond of illustrating his theory. But such mechanical illustrations are not without their danger, for I well remember the answer given by a pupil to a question on the atomic theory : "Atoms are round balls of wood invented by Dr. Dalton." So determinedly indeed did he adhere to his mechanical method of representing the chemical atoms and their combinations that he could not be prevailed upon to adopt the system of chemical formulæ introduced by Berzelius and now universally employed. In a letter addressed

to Graham in April 1837, he writes : " Berzelius's symbols are horrifying. A young student in chemistry might as soon learn Hebrew as make himself acquainted with them." And again : " They appear to me equally to perplex the adepts in science, to discourage the learner, as well as to cloud the beauty and simplicity of the atomic theory."

But modern research has accomplished, as regards the size of the atom, at any rate to a certain extent, what Dalton regarded as impossible. Thus in 1865, Loschmidt, of Vienna, by a train of reasoning which I cannot now stop to explain, came to the conclusion that the diameter of an atom of oxygen or nitrogen was $1/10,000,000$ of a centimetre. With the highest known magnifying power we can distinguish the $1/40,000$ part of a centimetre ; if now we imagine a cubic box each of whose sides has the above length, such a box when filled with air will contain from 60 to 100 millions of atoms of oxygen and nitrogen. A few years later William Thomson extended the methods of atomic measurement, and came to the conclusion that the distance between the centres of contiguous molecules is less than $1/5,000,000$ and greater than $1/1,000,000,000$ of a centimetre ; or, to put it in language more suited to the ordinary mind, Thomson asks us to imagine a drop of water magnified up to the size of the earth, and then tells us that the coarseness of the graining of such a mass would be something between a heap of small shot and a heap of cricket balls. Or again, to take Clifford's illustration, you know that our best microscopes magnify from 6000 to 8000 times ; a microscope which would magnify that result as much again would show the molecular structure of water. Or again, to put it in another form, if we suppose that the minutest organism we can now see were provided with equally powerful microscopes, these beings would be able to see the atoms.

Next, as to the indivisibility of the atom, involving also the question as to the relationships between the atomic weights and properties of the several elementary bodies.

Taking Dalton's aphorism, "Thou knowst no man can split an atom," as expressing the view of the enunciator of the atomic theory, let us see how far this idea is borne out by subsequent work. In the first place, Thomas Thomson, the first exponent of Dalton's generalization, was torn by conflicting beliefs until he found peace in the hypothesis of Prout, that the atomic weights of all the so-called elements are multiples of a common unit, which doctrine he sought to establish, as Thorpe remarks, by some of the very worst quantitative determinations to be found in chemical literature, though here I may add that they were not so incorrect as Dalton's original numbers.

Coming down to a somewhat later date, Graham, whose life was devoted to finding what the motion of an atom was, freed himself from the bondage of the Daltonian aphorism, and defined the atom not as a thing which cannot be divided, but as one which has not been divided. With him, as with Lucretius, as Angus Smith remarks, the original atom may be far down.

But speculative ideas respecting the constitution of matter have been the scientific relaxation of many minds from olden time to the present. In the mind of the early Greek the action of the atom as one substance taking various forms by unlimited combinations was sufficient to account for all the phenomena of the world. And Dalton himself, though upholding the indivisibility of his ultimate particles, says : " We do not know that any of the bodies denominated elementary are absolutely indecomposable." Again, Boyle, treating of the origin of form and quality, says : " There is one universal matter common to all bodies—an extended divisible and impenetrable substance." Then Graham in another place expresses a similar thought when he writes : " It is conceivable that the various kinds of matter now recognized as different elementary substances may possess one and the same ultimate or atomic molecules existing in different conditions of movement. The essential unity of matter is an hypothesis in harmony with the equal action of gravity upon all bodies."

What experimental evidence is now before us bearing upon these interesting speculations ? In the first place, then, the space of fifty years has completely changed the face of the inquiry. Not only has the number of distinct well-established elementary bodies increased from fifty-three in 1837 to seventy in 1887 (not including the *twenty* or more new elements recently said to have been discovered by Krüss and Nilson in certain rare Scandinavian minerals), but the properties of these elements have been studied, and are now known to us with a degree of precision then undreamt of. So that relationships existing between these bodies which fifty years ago were undiscernible are now clearly manifest, and it is to these relationships that I would for a moment ask your attention. I have already stated that Dalton measured the relative weights of the ultimate particles by assuming hydrogen as the unit, and that Prout believed that on this basis the atomic weights of all the other elements would be found to be multiples of the atomic weight of hydrogen, thus indicating that an intimate constitutional relation exists between hydrogen and all the other elements.

Since the days of Dalton and Prout the truth or otherwise of Prout's law has been keenly contested by the most eminent chemists of all countries. The inquiry is a purely experimental one, and only those who have a special knowledge of the difficulties which surround such inquiries can form an idea of the amount of labour and self-sacrifice borne by such men as Dumas, Stas, and Marignac in carrying out delicate researches on the atomic weights of the elements. What is, then, the result of these most laborious experiments? It is that, whilst the atomic weights of the elements are not exactly either multiples of the unit or of half the unit, many of the numbers expressing most accurately the weight of the atom approximate so closely to a multiple of that of hydrogen, that we are constrained to admit that these approximations cannot be a mere matter of chance, but that some reason must exist for them. What that reason is, and why a close approximation and yet something short of absolute identity exists, is as yet hidden behind the veil; but who is there that doubts that when this Association celebrates its centenary, this veil will have been lifted, and this occult but fundamental question of atomic philosophy shall have been brought into the clear light of day?

But these are by no means all the relationships which modern science has discovered with respect to the atoms of our chemical elements. So long ago as 1829 Döbereiner pointed out that certain groups of elements exist presenting in all their properties strongly marked family characteristics, and this was afterwards extended and insisted upon by Dumas. We find, for example, in the well-known group of chlorine, bromine, and iodine, these resemblances well developed, accompanied moreover by a proportional graduation in their chemical and physical properties. Thus, to take the most important of all their characters, the atomic weight of the middle term is the mean of the atomic weights of the two extremes. But these groups of triads appeared to be unconnected in any way with one another, nor did they seem to bear any relation to the far larger number of the elements not exhibiting these peculiarities.

Things remained in this condition until 1863, when Newlands threw fresh light upon the subject, showing a far-reaching series of relationships. For the first time we thus obtained a glance into the mode in which the elements are connected together, but, like so many new discoveries, this did not meet with the recognition which we now see it deserves. But whilst England thus had the honour of first opening up this new path, it is to Germany and to Russia that we must look for the consummation of the idea. Germany, in the person of Lothar Meyer, keeps, as it is wont to do, strictly within the limits of known facts. Russia, in the person of Mendelejeff, being of a somewhat more imaginative nature, not only seizes the facts which are proved, but ventures upon prophecy. These chemists, amongst whom Carnelley must be named, agree in placing all the elementary bodies in a certain regular sequence, thus bringing to light a periodic recurrence of analogous chemical and physical properties, on account of which the arrangement is termed the periodic system of the elements.

In order to endeavour to render this somewhat complicated matter clear to you, I may perhaps be allowed to employ a simile. Let us, if you please, imagine a series of human families: a French one, represented by Dumas; an English one, by name Newlands; a German one, the family of Lothar Meyer; and lastly, a Russian one, that of Mendelejeff. Let us next imagine the names of these chemists placed in a horizontal line in the order I have mentioned. Then let us write under each the name of his father, and again, in the next lower line, that of his grandfather, followed by that of his great-grandfather, and so on. Let us next write against each of these names the number of years which has elapsed since the birth of the individual. We shall then find that these numbers regularly increase by a definite amount, *i.e.* by the average age of a generation, which will be approximately the same in all the four families. Comparing the ages of the chemists themselves we shall observe certain differences, but these are small in comparison with the period which

has elapsed since the birth of any of their ancestors. Now each individual in this series of family trees represents a chemical element; and just as each family is distinguished by certain idiosyncrasies, so each group of the elementary bodies thus arranged shows distinct signs of consanguinity.

But more than this, it not unfrequently happens that the history and peculiarities of some member of a family may have been lost, even if the memory of a more remote and more famous ancestor may be preserved, although it is clear that such an individual must have had an existence. In such a case Francis Galton would not hesitate from the characteristics of the other members to reproduce the physical and even the mental peculiarities of the missing member; and should genealogical research bring to light the true personal appearance and mental qualities of the man, these would be found to coincide with Galton's estimate.

Such predictions and such verifications have been made in the case of no less than three of our chemical elements. Thus, Mendelejeff pointed out that if, in the future, certain lacunæ in his table were to be filled, they must be filled by elements possessing chemical and physical properties which he accurately specified. Since that time these gaps have actually been stopped by the discovery of gallium by Lecoq de Boisbaudron, of scandium by Nilson, and of germanium by Winkler, and their properties, both physical and chemical, as determined by their discoverers, agree absolutely with those predicted by the Russian chemist. Nay, more than this, we not unfrequently have had to deal with chemical foundlings, elements whose parentage is quite unknown to us. A careful examination of the personality of such waifs has enabled us to restore them to the family from which they have been separated by an unkind fate, and to give them that position in chemical society to which they are entitled.

These remarkable results, though they by no means furnish a proof of the supposition already referred to, viz. that the elements are derived from a common source, clearly point in this direction, and lend some degree of colour to the speculations of those whose scientific imagination, wearying of dry facts, revels in picturing to itself an elemental Bathybius, and in applying to the inanimate, laws of evolution similar to those which rule the animate world. Nor is there wanting other evidence regarding this inquiry, for here heat, the great analyzer, is brought into court. The main portion of the evidence consists in the fact that distinct chemical individuals capable of existence at low temperatures are incapable of existence at high ones, but split up into new materials possessing a less complicated structure than the original. And here it may be well to emphasize the distinction which the chemist draws between the atom and the molecule, the latter being a more or less complicated aggregation of atoms, and especially to point out the fundamental difference between the question of separating the atoms in the molecule and that of splitting up the atom itself. The decompositions above referred to are, in fact, not confined to compound bodies, for Victor Meyer has proved in the case of iodine that the molecule at high temperatures is broken to atoms, and J. J. Thomson has added to our knowledge by showing that this breaking up of the molecule may be effected not only by heat vibrations, but likewise by the electrical discharge at a comparatively low temperature.

How far, now, has this process of simplification been carried? Have the atoms of our present elements been made to yield? To this a negative answer must undoubtedly be given, for even the highest of terrestrial temperatures, that of the electric spark, has failed to shake any one of these atoms in two. That this is the case has been shown by the results with which spectrum analysis, that new and fascinating branch of science, has enriched our knowledge, for that spectrum analysis does give us most valuable aid in determining the varying molecular conditions of matter is admitted by all. Let us see how this bears on the question of the decomposition of the elements, and let us suppose for a moment that certain of our present elements, instead of being distinct substances, were made up of common ingredients, and that these compound elements, if I may be allowed to use so incongruous a term, are split up at the temperature of the electric spark into less complicated molecules. Then the spectroscopic examination of such a body must indicate the existence of these common ingredients by the appearance in the spark-spectra of these elements of identical bright lines. Coincidences of this kind have indeed been observed, but on careful examination these have been shown to be due either to the presence of some one of the other elements as an impurity, or to insufficient observational power. This absence of coincident lines admits, however, of two explanations—either that the elements are not decomposed at the temperature of the electric spark, or, what appears to me a much more improbable supposition, each one of the numbers of bright lines exhibited by every element indicates the existence of a separate constituent, no two of this enormous number being identical.

Terrestrial analysis having thus failed to furnish favourable evidence, we are compelled to see if any information is forthcoming from the chemistry of the sun and stars. And here I would remark that it is not my purpose now to dilate on the wonders which this branch of modern science has revealed. It is sufficient to remind you that chemists thus have the means placed at their disposal of ascertaining with certainty the presence of elements well known on this earth in fixed stars so far distant that we are now receiving the light which emanated from them perhaps even thousands of years ago.

Since Bunsen and Kirchhoff's original discovery in 1859, the labours of many men of science of all countries have largely increased our knowledge of the chemical constitution of the sun and stars, and to no one does science owe more in this direction than to Lockyer and Huggins in this country, and to Young in the New England beyond the seas. Lockyer has of late years devoted his attention chiefly to the varying nature of the bright lines seen under different conditions of time and place on the solar surface, and from these observations he has drawn the inference that the matching observed by Kirchhoff between, for instance, the iron lines as seen in our laboratories and those visible in the sun, has fallen to the ground. He further explains this want of uniformity by the fact that at the higher transcendental temperatures of the sun the substance which we know here as iron is resolved into separate components. Other experimentalists, however, while accepting Lockyer's facts as to the variations in the solar spectrum, do not admit his conclusions, and would rather explain the phenomena by the well-known differences which occur in the spectra of all the elements when their molecules are subject to change of temperature or change of position.

Further, arguments in favour of this idea of the evolution of the elements have been adduced from the phenomena presented by the spectra of the fixed stars It is well known that some of these shine with a white, others with a red, and others again with a blue light; and the spectroscope, especially under the hands of Huggins, has shown that the chemical constitution of these stars is different. The white stars, of which Sirius may be taken as a type, exhibit a much less complicated spectrum than the orange and the red stars; the spectra of the latter remind us more of those of the metalloids and of chemical compounds than of the metals. Hence it has been argued that in the white, presumably the hottest, stars a celestial d ssociation of our terrestrial elements may have taken place, whilst in the cooler stars, probably the red, combination even may occur. But even in the white stars we have no *direct* evidence that a decomposition of any terrestrial atom has taken place; indeed we learn that the hydrogen atom, as we know it here, can endure unscathed the inconceivably fierce temperature of stars presumably many times more fervent than our sun, as Sirius and Vega.

Taking all these matters into consideration, we need not be surprised if the earth-bound chemist should, in the absence of celestial evidence which is incontestable, continue, for the present at least, and until fresh evidence is forthcoming, to regard the elements as the unalterable foundation-stones upon which his science is based.

Pursuing another line of inquiry on this subject, Crookes has added a remarkable contribution to the question of the possibility of decomposing the elements. With his well-known experimental prowess, he has discovered a new and beautiful series of phenomena, and has shown that the phosphorescent lights emitted by certain chemical compounds, especially the rare earths, under an electric discharge in a high vacuum exhibit peculiar and characteristic lines. For the purpose of obtaining his material Crookes started from a substance believed by chemists to be homogeneous, such, for example, as the rare earth yttria, and succeeded by a long series of fractional precipitations in obtaining products which yield different phosphorescent spectra, although when tested by the ordinary methods of what we may term high temperature spectroscopy, they appear to be the one substance employed at the starting-point. The other touchstone by which the identity, or otherwise, of these various products

might be ascertained, viz. the determination of their atomic weights, has not, as yet, engaged Crookes's attention. In explanation of these singular phenomena, the discoverer suggests two possibilities. First, that the bodies yielding the different phosphorescent spectra are different elementary constituents of the substance which we call yttria Or, if this be objected to because they all yield the same spark-spectrum, he adopts the very reasonable view that the Daltonian atom is probably, as we have seen, a system of chemical complexity; and adds to this the idea that these complex atoms are not all of exactly the same constitution and weight, the differences, however, being so slight that their detection has hitherto eluded our most delicate tests, with the exception of this one of phosphorescence in a vacuum. To these two explanations, Marignac, in a discussion of Crookes's results, adds a third. It having been shown by Crookes himself that the presence of the minutest traces of foreign bodies produce remarkable alterations in the phosphorescent spectra, Marignac suggests that in the course of the thousands of separations which must be made before these differences become manifest, traces of foreign bodies may have been accidently introduced, or, being present in the original material, may have accumulated to a different extent in the various fractions, their presence being indicated by the only test by which they can now be detected. Which of these three explanations is the true one must be left to future experiment to decide.

We must now pass from the statics to the dynamics of chemistry; that is, from the consideration of the atoms at rest to that of the atoms in motion. Here, again, we are indebted to John Dalton for the first step in this direction, for he showed that the particles of a gas are constantly flying about in all directions; that is, that gases diffuse into one another, as an escape of coal gas from a burner, for example, soon makes itself perceptible throughout the room. Dalton, whose mind was constantly engaged in studying the molecular condition of gases. first showed that a light gas cannot rest upon a heavier gas as oil upon water, but that an interpenetration of each gas by the other takes place. It is, however, to Graham's experiments, made rather more than half a century ago, that we are indebted for the discovery of the law regulating these molecular motions of gases, proving that their relative rates of diffusion are inversely proportional to the square roots of their densities, so that oxygen being 16 times heavier than hydrogen, their relative rates of diffusion are 1 and 4.

But whilst Dalton and Graham indicated that the atoms are in a continual state of motion, it is to Joule that we owe the first accurate determination of the rate of that motion. At the Swansea meeting, in 1848, Joule read a paper before Section A on the "Mechanical Equivalent of Heat and on the Constitution of Elastic Fluids." In this paper Joule remarks that whether we conceive the particles to be revolving round one another according to the hypothesis of Davy, or flying about in every direction according to Herapath's view, the pressure of the gas will be in proportion to the *vis viva* of its particles. "Thus it may be shown that the particles of hydrogen at the barometrical pressure of 30 inches at a temperature of 60° must move with a velocity of 6225·54 feet per second in order to produce a pressure of 14·714 lbs. on the square inch," or, to put it in other words, a molecular cannonade or hailstorm of particles, at the above rate—a rate, we must remember, far exceeding that of a cannon ball—is maintained against the bounding surface.

We can, however, go a step further and calculate with Clerk Maxwell the number of times in which this hydrogen molecule, moving at the rate of 70 miles per minute, strikes against others of the vibrating swarm, and we learn that in one second of time it must knock against others no less than 18 thousand million times.

And here we may pause and dwell for a moment on the reflection that in Nature there is no such thing as great or small, and that the structure of the smallest particle, invisible even to our most searching vision, may be as complicated as that of any one of the heavenly bodies which circle round our sun.

But how does this wonderful atomic motion affect our chemistry? Can chemical science or chemical phenomena throw light upon this motion, or can this motion explain any of the known phenomena of our science? I have already said that Lavoisier left untouched the dynamics of combustion. He could not explain why a fixed and unalterable amount of heat is in most cases emitted, but in some cases absorbed, when chemical combination takes place. What Lavoisier left unexplained Joule has made clear. On August 25, 1843, Joule read a short communication, I am glad to remember, before the Chemical Section of our Association, meeting that year at Cork, containing an announcement of a discovery which was to revolutionize modern science. This consisted in the determination of the mechanical equivalent of heat, in proving by accurate experiment that by the expenditure of energy equal to that developed by the weight of 772 pounds falling through 1 foot at Manchester, the temperature of 1 pound of water can be raised 1° F. In other words, every change in the arrangement of the particles is accompanied by a definite evolution or an absorption of heat. In all such cases the molecular energy leaves the potential to assume the kinetic form, or *vice versâ*. Heat is evolved by the clashing of the atoms, and this amount is fixed and definite.

Thus it is to Joule we owe the foundation of chemical dynamics and the basis of thermal chemistry. As the conservation of mass or the principle of the indestructibility of matter forms the basis of chemical statics, so the principle of the conservation of energy [1] constitutes the foundation of chemical dynamics. Change in the form of matter and change in the form of energy are the universal accompaniments of every chemical operation. Here again it is to Joule we owe the proof of the truth of this principle in another direction, viz. that when electrical energy is developed by chemical change a corresponding quantity of chemical energy disappears. Energy, as defined by Maxwell, is the power of doing work, and work is the act of producing a configuration in a system in opposition to a force which resists that change. Chemical action produces such a change of configuration in the molecules. Hence, as Maxwell says, 'A complete knowledge of the mode in which the potential energy of a system varies with the configuration would enable us to predict every possible motion of the system under the action of given external forces, provided we were able to overcome the purely mathematical difficulties of the calculation." The object of thermal chemistry is to measure these changes of energy by thermal methods, and to connect these with chemical changes, to estimate the attractions of the atoms and molecules to which the name of chemical affinity has been applied, and thus to solve the most fundamental problem of chemical science. How far has modern research approached the solution of this most difficult problem? How far can we answer the question, What is the amount of the forces at work in these chemical changes? What laws govern these forces? Well, even in spite of the results with which recent researches, especially the remarkable ones of the Danish philosopher Thomsen have enriched us, we must acknowledge that we are yet scarcely in sight of Maxwell's position of successful prediction. Thermal chemistry, we must acknowledge, is even yet in its infancy; it is, however, an infant of sturdy growth, likely to do good work in the world, and to be a credit to him who is its acknowledged father, as well as to those who have so carefully tended it in its early years.

But recent investigation in another direction bids fair even to eclipse the results which have been obtained by the examination of thermal phenomena. And this lies in the direction of electrical chemistry. Faraday's work relating to conductivity of chemical substances has been already referred to, and this has been since substantiated and extended to pure substances by Kohlrausch. It has been shown, for example, that the resistance of absolutely pure water is almost an infinite quantity. But a small quantity of an acid, such as acetic or butyric acid, greatly increases the conductivity; but more than this, it is possible by determination of the conductivity of a mixture of water with these two acids to arrive at a conclusion as to the partition of the molecules of the water between the acids. Such a partition, however, implies a change of position, and therefore we are furnished with a means of recognizing the motion of the molecules in a liquid, and of determining its amount. Thus it has been found that the hindrance to molecular motion is more affected by the chemical character of the liquid than by physical characters such as viscosity. We have seen that chemical change is always accompanied by molecular motion, and further evidence of the truth of this is gained from the extraordinary chemical inactivity of pure unmixed substances. Thus pure anhydrous hydrochloric acid does not act upon lime, whereas the addition of even a trace of moisture sets up a most active chemical change, and hundreds of other examples of a similar kind might be stated. Bearing in mind that these pure anhy-

drous compounds do not conduct, we are led to the conclusion that an intimate relation exists between chemical activity and conductivity. And we need not stop here; for a method is indicated indeed by which it will be possible to arrive at a measure of chemical affinity from determination of conductivity. It has indeed been already shown that the rate of change in the saponification of acetic ether is directly proportional to the conductivity of the liquid employed.

Such wide-reaching inquiries into new and fertile fields, in which we seem to come into nearer touch with the molecular state of matter, and within a measurable distance of accurate mathematical expression, leads to confident hope that Lord Rayleigh's pregnant words at Montreal may ere long be realized : "It is from the further study of electrolysis that we may expect to gain improved views as to the nature of chemical reactions, and of the forces concerned in bringing them about ; and I cannot help thinking that the next great advance, of which we already have some foreshadowing, will come on this side."

There is, perhaps, no branch of our science in which the doctrine of the Daltonian atom plays a more conspicuous part than in organic chemistry or the chemistry of the carbon compounds, as there is certainly none in which such wonderful progress has been made during the last fifty years. One of the most striking and perplexing discoveries made rather more than half a century ago was that chemical compounds could exist which, whilst possessing an identical chemical composition, that is containing the same percentage quantity of their constituents, are essentially distinct chemical substances exhibiting different properties. Dalton was the first to point out the existence of such substances, and to suggest that the difference was to be ascribed to a different or to a multiple arrangement of the constituent atoms. Faraday soon afterwards proved that this supposition was correct, and the research of Liebig and Wöhler on the identity of composition of the salts of fulminic and cyanic acid gave further confirmation to the conclusion, leading Faraday to remark that "now we are taught to look for bodies composed of the same elements in the same proportion, but differing in their qualities, they may probably multiply upon us." How true this prophecy has become we may gather from the fact that we now know of thousands of cases of this kind, and that we are able not only to explain the reason of their difference by virtue of the varying position of the atoms within the molecule, but even to predict the number of distinct variations in which any given chemical compound can possibly exist. How large this number may become may be understood from the fact that, for example, one chemical compound, a hydrocarbon containing thirteen atoms of carbon combined with twenty-eight atoms of hydrogen, can be shown to be capable of existing in no less than 802 distinct forms.

Experiment in every case in which it has been applied has proved the truth of such a prediction, so that the chemist has no need to apply the cogent argument sometimes said to be used by experimentalists enamoured of pet theories, " When facts do not agree with theory, so much the worse for the facts"! This power of successful prediction constitutes a high-water mark in science, for it indicates that the theory upon which such a power is based is a true one.

But if the Daltonian atom forms the foundation of this theory, it is upon a knowledge of the mode of arrangement of these atoms and on a recognition of their distinctive properties that the superstructure of modern organic chemistry rests. Certainly it does appear almost to verge on the miraculous that chemists should now be able to ascertain with certainty the relative position of atoms in a molecule so minute that millions upon millions, like the angels in the schoolmen's discussion, can stand on a needle's point. And yet this process of orientation is one which is accomplished every day in our laboratories, and one which more than any other has led to results of a startling character. Still, this sword to open the oyster of science would have been wanting to us if we had not taken a step farther than Dalton did, in the recognition of the distinctive nature of the elemental atoms. We now assume on good grounds that the atom of each element possesses distinct capabilities of combination : some a single capability, others a double, others a triple, and others again a fourfold combining capacity. The germs of this theory of valency, one of the most fruitful of modern chemical ideas, were enunciated by Frankland in 1852, but the definite explanation of the linking of atoms, of the tetrad nature of the carbon atoms, their power of combination, and of the difference in structure between the fatty and aromatic series of compounds,

[1] " The total energy of any material system is a quantity which can neither be increased nor diminished by any action between the parts of the system, though it may be transformed into any of the forms of which energy is susceptible."—MAXWELL.

was first pointed out by Kekulé in 1857 ; though we must not forget that this great principle was foreshadowed so long ago as 1833 from a physical point of view by Faraday in his well-known laws of electrolysis, and that it is to Helmholtz, in his celebrated Faraday Lecture, that we owe the complete elucidation of the subject ; for, whilst Faraday has shown that the number of the atoms electrolytically deposited is in the inverse ratio of their valencies, Helmholtz has explained this by the fact that the quantity of electricity with which each atom is associated is directly proportional to its valency.

Amongst the tetrad class of elements, carbon, the distinctive element of organic compounds, finds its place ; and the remarkable fact that the number of carbon compounds far exceeds that of all the other elements put together receives its explanation. For these carbon atoms not only possess four means of grasping other atoms, but these four-handed carbon atoms have a strong partiality for each other's company, and readily attach themselves hand in hand to form open chains or closed rings, to which the atoms of other elements join to grasp the unoccupied carbon hand, and thus to yield a dancing company in which all hands are locked together. Such a group, each individual occupying a given position with reference to the others, constitutes the organic molecule. When, in such a company, the individual members change hands, a new combination is formed. And as in such an assembly the eye can follow the changing positions of the individual members, so the chemist can recognize in his molecule the position of the several atoms, and explain by this the fact that each arrangement constitutes a new chemical compound possessing different properties, and account in this way for the decompositions which each differently constituted molecule is found to undergo.

Chemists are, however, not content with representing the arrangement of the atoms in one plane, as on a sheet of paper, but attempt to express the position of the atoms in space. In this way it is possible to explain certain observed differences in isomeric bodies, which otherwise baffled our efforts. To Van t'Hoff, in the first instance, and more recently to Wislicenus, chemistry is indebted for work in this direction, which throws light on hitherto obscure phenomena, and points the way to still further and more important advances.

It is this knowledge of the mode in which the atoms in the molecule are arranged, this power of determining the nature of this arrangement, which has given to organic chemistry the impetus which has overcome so many experimental obstacles, and given rise to such unlooked-for results. Organic chemistry has now become synthetic. In 1837 we were able to build up but very few and very simple organic compounds from their elements ; indeed the views of chemists were much divided as to the possibility of such a thing. Both Gmelin and Berzelius argued that organic compounds, unlike inorganic bodies, cannot be built up from their elements. Organic compounds were generally believed to be special products of the so-called vital force, and it was only intuitive minds like those of Liebig and Wöhler who foresaw what was coming, and wrote in 1837 strongly against this view, asserting that the artificial production in our laboratories of all organic substances, so far as they do not constitute a living organism, is not only probable but certain. Indeed, they went a step farther, and predicted that sugar, morphia, salicine, will all thus be prepared ; a prophecy which, I need scarcely remind you, has been after fifty years fulfilled, for at the present time we can prepare an artificial sweetening principle, an artificial alkaloid, and salicine.

In spite of these predictions, and in spite of Wöhler's memorable discovery in 1828 of the artificial production of urea, which did in reality break down for ever the barrier of essential chemical difference between the products of the inanimate and the animate world, still, even up to a much later date, contrary opinions were held, and the synthesis of urea was looked upon as the exception which proves the rule. So it came to pass that for many years the artificial production of any of the more complicated organic substances was believed to be impossible. Now the belief in a special vital force has disappeared like the *ignis fatuus*, and no longer lures us in the wrong direction. We know now that the same laws regulate the formation of chemical compounds in both animate and inanimate nature, and the chemist only asks for a knowledge of the constitution of any definite chemical compound found in the organic world in order to be able to promise to prepare it artificially.

But the progress of synthetic organic chemistry, which has of late been so rapid, was made in the early days of the half-century

only by feeble steps and slow. Seventeen long years elapsed between Wöhler's discovery and the next real synthesis. This was accomplished by Kolbe, who in 1845 prepared acetic acid from its elements. But then a splendid harvest of results gathered in by chemists of all nations quickly followed, a harvest so rich and so varied that we are apt to be overpowered by its wealth, and amidst so much that is alluring and striking we may well find it difficult to choose the most appropriate examples for illustrating the power and the extent of modern chemical synthesis.

Next, as a contrast to our picture, let us for a moment glance back again to the state of things fifty years ago, and then notice the chief steps by which we have arrived at our present position. In 1837 organic chemistry possessed no scientific basis, and therefore no classification of a character worthy of the name. Writing to Berzelius in that year, Wöhler describes the condition of organic chemistry as one enough to drive a man mad. " It seems to me," says he, " like the tropical forest primæval, full of the strangest growths, an endless and pathless thicket in which a man may well dread to wander." Still clearances had already been made in this wilderness of facts. Berzelius in 1832 welcomed the results of Liebig and Wöhler's research on benzoic acid as the dawn of a new era ; and such it really was, inasmuch as it introduced a novel and fruitful idea—namely, the possibility of a group of atoms acting like an element by pointing out the existence of organic radicals. This theory was strengthened and confirmed by Bunsen's classical researches on the cacodyl compounds, in which he showed that a common group of elements which acts exactly as a metal can exist in the free state, and this was followed soon afterwards by isolation of the so-called alcohol radicals by Frankland and Kolbe. It is, however, to Schorlemmer that we owe our knowledge of the true constitution of these bodies, a matter which proved to be of vital importance for the further development of the science.

Turning our glance in another direction we find that Dumas in 1834 by this law of substitution threw light upon a whole series of singular and unexplained phenomena by showing that an exchange can take place between the constituent atoms in a molecule. Laurent indeed went farther, and assumed that a chlorine atom, for example, took up the position vacated by an atom of hydrogen and played the part of its displaced rival, so that the chemical and physical properties of the substitution-product were thought to remain substantially the same as those of the original body. A singular story is connected with this discovery. At a *soirée* in the Tuileries in the time of Charles X. the guests were almost suffocated by acrid vapours which were evidently emitted by the burning wax candles, and the great chemist Dumas was called in to examine into the cause of the annoyance. He found that the wax of which the candles were made had been bleached by chlorine, that a replacement of some of the hydrogen atoms of the wax by chlorine had occurred, and that the suffocating vapours consisted of hydrochloric acid given off during the combustion. The wax was as white and as odourless as before, and the fact of the substitution of chlorine for hydrogen could only be recognized when the candles were destroyed by burning. This incident induced Dumas to investigate more closely this class of phenomena, and the results of this investigation are embodied in his law of substitution. So far indeed did the interest of the French school of chemists lead them that some assumed that not only the hydrogen but also the carbon of organic bodies could be replaced by substitution. Against this idea Liebig protested, and in a satirical vein he informs the chemical public, writing from Paris under the *nom de plume* of S. Windler, that he has succeeded in substituting not only the hydrogen but the oxygen and carbon in cotton cloth by chlorine, and he adds that the London shops are now selling nightcaps and other articles of apparel made entirely of chlorine, goods which meet with much favour, especially for hospital use !

But the debt which chemistry, both inorganic and organic, thus owes to Dumas' law of substitution is serious enough, for it proved to be the germ of Williamson's classical researches on etherification, as well as of those of Wurtz and Hofmann on the compound ammonias, investigations which lie at the base of the structure of modern chemistry. Its influence has been, however, still more far-reaching, inasmuch as upon it depends in great measure the astounding progress made in the wide field of organic synthesis.

It may here be permitted to me to sketch in rough outline the principles upon which all organic syntheses have been effected. We have already seen that as soon as the chemical structure of a body has been ascertained its artificial preparation may be certainly anticipated, so that the first step to be taken is the study of the structure of the naturally occurring substance which it is desired to prepare artificially by resolving it into simpler constituents, the constitution of which is already known. In this way, for example, Hofmann discovered that the alkaloid coniine, the poisonous principle of hemlock, may be decomposed into a simpler substance well known to chemists under the name of pyridine. This fact having been established by Hofmann, and the grouping of the atoms approximately determined, it was then necessary to reverse the process, and, starting with pyridine, to build up a compound of the required constitution and properties, a result recently achieved by Ladenburg in a series of brilliant researches. The well-known synthesis of the colouring matter of madder by Graebe and Liebermann, preceded by the important researches of Schunck, and that of indigo by Baeyer, are other striking examples in which this method has been successfully followed.

Not only has this intimate acquaintance with the changes which occur within the molecules of organic compounds been utilized, as we have seen, in the synthesis of naturally occurring substances, but it has also led to the discovery of many new ones. Of these perhaps the most remarkable instance is the production of an artificial sweetening agent termed saccharin, 250 times sweeter than sugar, prepared by a complicated series of reactions from coal-tar. Nor must we imagine that these discoveries are of scientific interest only, for they have given rise to the industry of the coal-tar colours, the value of which is measured by millions sterling annually, an industry which Englishmen may be proud to remember was founded by our countryman Perkin.

Another interesting application of synthetic chemistry to the needs of every-day life is the discovery of a series of valuable febrifuges, amongst which I may mention antipyrin as the most useful. An important aspect in connexion with the study of these bodies is the physiological value which has been found to attach to the introduction of certain organic radicals, so that an indication is given of the possibility of preparing a compound which will possess certain desired physiological properties, or even to foretell the kind of action which such bodies may exert on the animal economy.

But it is not only the physiological properties of chemical compounds which stand in intimate relation with their constitution, for we find that this is the case with all their physical properties. It is true that at the beginning of our period any such relation was almost unsuspected, whilst at the present time the number of instances in which this connexion has been ascertained is almost infinite. Amongst these perhaps the most striking is the relationship which has been pointed out between the optical properties and chemical composition. This was in the first place recognized by Pasteur in his classical researches on racemic and tartaric acids in 1848 ; but the first to indicate a quantitative relationship and a connexion between chemical structure and optical properties was Gladstone in 1863. Great instrumental precision has been brought to bear on this question, and consequently most important practical applications have resulted. I need only refer to the well-known accurate methods now in every-day use for the determination of sugar by the polariscope, equally valuable to the physician and to the manufacturer.

But now the question may well be put, is any limit set to this synthetic power of the chemist? Although the danger of dogmatizing as to the progress of science has already been shown in too many instances, yet one cannot help feeling that the barrier which exists between the organized and unorganized worlds is one which the chemist at present sees no chance of breaking down.

It is true that there are those who profess to foresee that the day will arrive when the chemist, by a succession of constructive efforts, may pass beyond albumen, and gather the elements of lifeless matter into a living structure. Whatever may be said regarding this from other standpoints, the chemist can only say that at present no such problem lies within his province. Protoplasm, with which the simplest manifestations of life are associated, is not a compound, but a structure built up of compounds. The chemist may successfully synthetize any of its component molecules, but he has no more reason to look forward to the synthetic production of the structure than to imagine that

the synthesis of gallic acid leads to the artificial production of gall-nuts.

Although there is thus no prospect of our effecting a synthesis of organized material, yet the progress made in our knowledge of the chemistry of life during the last fifty years has been very great, and so much so indeed that the sciences of physiological and of pathological chemistry may be said to have entirely arisen within this period.

In the introductory portion of this address I have already referred to the relations supposed to exist fifty years ago between vital phenomena and those of the inorganic world. Let me now briefly trace a few of the more important steps which have marked the progress of this branch of science during this period. Certainly no portion of our science is of greater interest, nor, I may add, of greater complexity, than that which, bearing on the vital functions both of plants and of animals, endeavours to unravel the tangled skein of the chemistry of life, and to explain the principles according to which our bodies live, and move, and have their being. If, therefore, in the less complicated problems with which other portions of our science have to deal, we find ourselves, as we have seen, often far from possessing satisfactory solutions, we cannot be surprised to learn that with regard to the chemistry of the living body—whether vegetable or animal—in health or disease we are still farther from a complete knowledge of phenomena, even those of fundamental importance.

It is of interest here to recall the fact that nearly fifty years ago Liebig presented to the Chemical Section of this Association a communication in which, for the first time, an attempt was made to explain the phenomena of life on chemical and physical lines, for in this paper he admits the applicability of the great principle of the conservation of energy to the functions of animals, pointing out that the animal cannot generate more heat than is produced by the combustion of the carbon and hydrogen of his food.

"The source of animal heat," says Liebig, "has previously been ascribed to nervous action or to the contraction of the muscles, or even to the mechanical motions of the body, as if these motions could exist without an expenditure of force [equal to that] consumed in producing them." Again he compares the living body to a laboratory furnace in which a complicated series of changes occur in the fuel, but in which the end-products are carbonic acid and water, the amount of heat evolved being dependent, not upon the intermediate, but upon the final products. Liebig asked himself the question, Does every kind of food go to the production of heat ; or can we distinguish, on the one hand, between the kind of food which goes to create warmth, and, on the other, that by the oxidation of which the motions and mechanical energy of the body are kept up? He thought that he was able to do this, and he divided food into two categories. The starchy or carbohydrate food is that, said he, which by its combustion provides the warmth necessary for the existence and life of the body. The albuminous or nitrogenous constituents of our food, the flesh meat, the gluten, the casein out of which our muscles are built up, are not available for the purposes of creating warmth, but it is by the waste of those muscles that the mechanical energy, the activity, the motions of the animal are supplied. We see, said Liebig, that the Esquimaux feeds on fat and tallow, and this burning in his body keeps out the cold. The Gaucho, riding on the pampas, lives entirely on dried meat, and the rowing man and pugilist, trained on beefsteaks and porter, require little food to keep up the temperature of their bodies, but much to enable them to meet the demand for fresh muscular tissue, and for this purpose they need to live on a strongly nitrogenous diet.

Thus far Liebig. Now let us turn to the present state of our knowledge. The question of the source of muscular power is one of the greatest interest, for, as Frankland observes, it is the corner-stone of the physiological edifice and the key to the nutrition of animals.

Let us examine by the light of modern science the truth of Liebig's view—even now not uncommonly held—as to the functions of the two kinds of food, and as to the cause of muscular exercise being the oxidation of the muscular tissue. Soon after the promulgation of these views, J. R. Mayer, whose name as the first expositor of the idea of the conservation of energy is so well known, warmly attacked them, throwing out the hypothesis that all muscular action is due to the combustion of food, and not to the destruction of muscle, proving his case by showing that if the muscles of the heart be destroyed in doing mechanical work the heart would be burnt up in eight days! What does modern research say to this question? Can it be brought to the crucial test of experiment? It can; but how? Well, in the first place we can ascertain the work done by a man or any other animal; we can measure this work in terms of our mechanical standard, in kilogramme-metres or foot-pounds. We can next determine what is the destruction of nitrogenous tissue at rest and under exercise by the amount of nitrogenous material thrown off by the body. And here we must remember that these tissues are never completely burnt, so that free nitrogen is never eliminated. If now we know the heat-value of the burnt muscle, it is easy to convert this into its mechanical equivalent, and thus measure the energy generated. What is the result? Is the weight of muscle destroyed by ascending the Faulhorn or by working on the treadmill sufficient to produce on combustion heat enough when transformed into mechanical exercise to lift the body up to the summit of the Faulhorn or to do the work on the treadmill? Careful experiment has shown that this is so far from being the case that the actual energy developed is twice as great as that which could possibly be produced by the oxidation of the nitrogenous constituents eliminated from the body during twenty-four hours. That is to say, taking the amount of nitrogenous substance cast off from the body, not only whilst the work was being done but during twenty-four hours, the mechanical effect capable of being produced by the muscular tissue from which this cast-off material is derived would only raise the body half-way up the Faulhorn, or enable the prisoner to work half his time on the treadmill.

Hence it is clear that Liebig's proposition is not true. The nitrogenous constituents of the food do doubtless go to repair the waste of muscle, which, like every other portion of the body, needs renewal, whilst the function of the non-nitrogenous food is not only to supply the animal heat, but also to furnish, by its oxidation, the muscular energy of the body.

We thus come to the conclusion that it is the potential energy of the food which furnishes the actual energy of the body, expressed in terms either of heat or of mechanical work.

But there is one other factor which comes into play in this question of mechanical energy, and must be taken into account; and this factor we are as yet unable to estimate in our usual terms. It concerns the action of the mind upon the body, and, although incapable of exact expression, exerts none the less an important influence on the physics and chemistry of the body, so that a connexion undoubtedly exists between intellectual activity or mental work and bodily nutrition. In proof that there is a marked difference between voluntary and involuntary work, we need only compare the mechanical action of the heart, which never causes fatigue, with that of the voluntary muscles, which become fatigued by continued exertion. So, too, we know well that an amount of drill which is fatiguing to the recruit is not felt by the old soldier, who goes through the evolutions automatically. What is the expenditure of mechanical energy which accompanies mental effort, is a question which science is probably far removed from answering. But that the body experiences exhaustion as the result of mental activity is a well-recognized fact. Indeed, whilst the second law of thermodynamics teaches that in none of the mechanical contrivances for the conversion of heat into actual energy can such a conversion be complete, it is perhaps possible, as Helmholtz has suggested, that such a complete conversion may take place in the subtle mechanism of the animal organism.

The phenomena of vegetation, no less than those of the animal world, have, however, during the last fifty years been placed by the chemist on an entirely new basis. Although before the publication of Liebig's celebrated report on chemistry and its application to agriculture, presented to the British Association in 1840, much had been done, many fundamental facts had been established, still Liebig's report marks an era in the progress of this branch of our science. He not only gathered up in a masterly fashion the results of previous workers, but put forward his own original views with a boldness and frequently with a sagacity which gave a vast stimulus and interest to the questions at issue. As a proof of this I may remind you of the attack which he made on, and the complete victory which he gained over, the humus theory. Although Saussure and others had already done much to destroy the basis of this theory, yet the fact remained that vegetable physiologists up to 1840 continued to hold to the opinion that humus, or decayed vegetable matter, was the only source of the carbon of vegetation. Liebig, giving due consideration to the labours of Saussure, came to the con-

clusion that it was absolutely impossible that the carbon deposited as vegetable tissue over a given area, as for instance over an area of forest land, could be derived from humus, which is itself the result of the decay of vegetable matter. He asserted that the whole of the carbon of vegetation is obtained from the atmospheric carbonic acid, which, though only present in the small relative proportion of 4 parts in 10,000 of air, is contained in such absolutely large quantity that if all the vegetation on the earth's surface were burnt, the proportion of carbonic acid which would thus be thrown into the air would not be sufficient to double the present amount.

That this conclusion of Liebig's is correct needed experimental proof, but such proof could only be given by long-continued and laborious experiment, and this serves to show that chemical research is not now confined to laboratory experiments lasting perhaps a few minutes, but that it has invaded the domain of agriculture as well as of physiology, and reckons the periods of her observations in the field not by minutes, but by years. It is to our English agricultural chemists Lawes and Gilbert that we owe the complete experimental proof required. And it is true that this experiment was a long and tedious one, for it has taken forty-four years to give the definite reply. At Rothamsted a plot was set apart for the growth of wheat. For forty-four successive years that field has grown wheat without addition of any carbonized manure; so that the only possible source from which the plant could obtain the carbon for its growth is the atmospheric carbonic acid. Now, the quantity of carbon which on an average was removed in the form of wheat and straw from a plot manured only with mineral matter was 1000 pounds, whilst on another plot, for which a nitrogenous manure was employed, 1500 pounds more carbon was annually removed; or 2500 pounds of carbon are removed by this crop annually without the addition of any carbonaceous manure. So that Liebig's prevision has received a complete experimental verification.

May I without wearying you with experimental details refer for a moment to Liebig's views as to the assimilation of nitrogen by plants—a much more complicated and difficult question than the one we have just considered—and compare these with the most modern results of agricultural chemistry? We find that in this case his views have not been substantiated. He imagined that the whole of the nitrogen required by the plant was derived from atmospheric ammonia; whereas Lawes and Gilbert have shown by experiments of a similar nature to those just described, and extending over a nearly equal length of time, that this source is wholly insufficient to account for the nitrogen removed in the crop, and have come to the conclusion that the nitrogen must have been obtained either from a store of nitrogenous material in the soil or by absorption of free nitrogen from the air. These two apparently contradictory alternatives may perhaps be reconciled by the recent observations of Warington and of Berthelot, which have thrown light upon the changes which the so-called nitrogenous capital of the soil undergoes, as well as upon its chemical nature, for the latter has shown that under certain conditions the soil has the power of absorbing the nitrogen of the air, forming compounds which can subsequently be assimilated by the plant.

Touching us as human beings even still more closely than the foregoing, is the influence which chemistry has exerted on the science of pathology, and in no direction has greater progress been made than in the study of micro-organisms in relation to health and disease. In the complicated chemical changes to which we give the names of fermentation and putrefaction, the views of Liebig, according to which these phenomena are of a purely chemical character, have given way under the searching investigations of Pasteur, who established the fundamental principle that these processes are inseparably connected with the life of certain low forms of organisms. Thus was founded the science of bacteriology, which in Lister's hands has yielded such splendid results in the treatment of surgical cases; and in those of Klebs, Koch, William Roberts, and others, has been the means of detecting the cause of many diseases both in man and animals; the latest and not the least important of which is the remarkable series of successful researches by Pasteur into the nature and mode of cure of that most dreadful of maladies, hydrophobia. And here I may be allowed to refer with satisfaction to the results of the labours on this subject of a Committee the formation of which I had the honour of moving for in the House of Commons. These results confirm in every respect Pasteur's assertions, and prove beyond a doubt that the adoption of his method has prevented the occurrence of hydrophobia in a large proportion of persons bitten by rabid animals, who, if they had not been subjected to this treatment would have died of that disease. The value of his discovery is, however, greater than can be estimated by its present utility, for it shows that it may be possible to avert other diseases besides hydrophobia by the adoption of a somewhat similar method of investigation and of treatment. This, though the last, is certainly not the least of the debts which humanity owes to the great French experimentalist. Here it might seem as if we had outstepped the boundaries of chemistry, and have to do with phenomena purely vital. But recent research indicates that this is not the case, and points to the conclusion that the microscopist must again give way to the chemist, and that it is by chemical rather than by biological investigation that the causes of diseases will be discovered, and the power of removing them obtained. For we learn that the symptoms of infective diseases are no more due to the microbes which constitute the infection than alcoholic intoxication is produced by the yeast-cell, but that these symptoms are due to the presence of definite chemical compounds, the result of the life of these microscopic organisms. So it is to the action of these poisonous substances formed during the life of the organism, rather than to that of the organism itself, that the special characteristics of the disease are to be traced; for it has been shown that the disease can be communicated by such poisons in entire absence of living organisms.

If I have thus far dwelt on the progress made in certain branches of pure science it is not because I undervalue the other methods by which the advancement of science is accomplished, viz. that of the application and of the diffusion of a knowledge of Nature, but rather because the British Association has always held, and wisely held, that original investigation lies at the root of all application, so that to foster its growth and encourage its development has for more than fifty years been our chief aim and wish.

Had time permitted I should have wished to have illustrated this dependence of industrial success upon original investigation, and to have pointed out the prodigious strides which chemical industry in this country has made during the fifty years of Her Majesty's reign. As it is I must be content to remind you how much our modern life, both in its artistic and useful aspects, owes to chemistry, and, therefore, how essential a knowledge of the principles of the science is to all who have the industrial progress of the country at heart.

This leads me to refer to what has been accomplished in this country of ours towards the diffusion of scientific knowledge amongst the people during the Victorian era. It is true that the English people do not possess, as yet, that appreciation of the value of science so characteristic of some other nations. Up to very recent years our educational system, handed down to us from the Middle Ages, has systematically ignored science, and we are only just beginning, thanks in a great degree to the prevision of the late Prince Consort, to give it a place, and that but an unimportant one, in our primary and secondary schools or in our Universities. The country is, however, now awakening to the necessity of placing its house in order in this respect, and is beginning to see that if she is to maintain her commercial and industrial supremacy the education of her people from top to bottom must be carried out on new lines. The question as to how this can be most safely and surely accomplished is one of transcendent national importance, and the statesman who solves this educational problem will earn the gratitude of generations yet to come.

In conclusion, may I be allowed to welcome the unprecedentedly large number of foreign men of science who have on this occasion honoured the British Association by their presence, and to express the hope that this meeting may be the commencement of an international scientific organization, the only means nowadays existing, to use the words of one of the most distinguished of our guests, of establishing that fraternity among nations from which politics appear to remove us further and further by absorbing human powers and human work, and directing them to purposes of destruction. It would indeed be well if Great Britain, which has hitherto taken the lead in so many things that are great and good, should now direct her attention to the furthering of international organizations of a scientific nature. A more appropriate occasion than the present meeting could perhaps hardly be found for the inauguration of such a movement.

But whether this hope be realized or not, we all unite in that one great object, the search after truth for its own sake, and we all, therefore, may join in re-echoing the words of Lessing:—
" The worth of man lies not in the truth which he possesses, or believes that he possesses, but in the honest endeavour which he puts forth to secure that truth; for not by the possession of truth, but by the search after it are the faculties of man enlarged, and in this alone consists his ever-growing perfection. Possession fosters content, indolence, and pride. If God should hold in His right hand all truth, and in His left hand the ever-active desire to seek truth, though with the condition of perpetual error, I would humbly ask for the contents of the left hand, saying, ' Father, give me this; pure truth is only for Thee.' "

THE CHEMICAL NEWS.

VOL. LX. No. 1547.

ON

RECENT RESEARCHES
ON THE
RARE EARTHS
AS INTERPRETED BY THE
SPECTROSCOPE.*

By WILLIAM CROOKES, F.R.S., &c.

IF I name the Spectroscope as the most important scientific invention of the latter half of this century, I shall not fear to be accused of exaggeration. Photography has rendered vast services in recording astronomical and biological phenomena, and it even supplies us with indirect means of studying ray vibrations to which the human retina does not respond. The electro-acoustic devices of Edison and his co-workers permit almost magical communication between human beings. Ruhmkorff's coil and the Geissler tube have rendered notable service in physical investigation; and the electric lamp promises to aid in exploring the internal parts of living animals, as well as in studying the organic forms of the deep sea. But in the spectroscope we possess a power that enables us to peer into the very heart of Nature. In the extent of its grasp and the varied character of its applicability it surpasses the telescope and at least rivals the microscope. It enables the astronomer to defy immeasurable distance, and to study the physical condition and the chemical composition of the sun and the stars as if they were within touch, and even to ascertain the direction of their movements.

Without attempting to discuss the import of the results thus gained—which would lead us too far—I may point out that they overthrow a dogma concerning the classification of the sciences. It has been said that the simpler and more general sciences lend both doctrines and methods to the more complex and less general sciences, and that the latter give nothing in return. But we now see chemistry endowing astronomy with an original and fruitful method of research.

Turning to the very opposite extremity of the scientific hierarchy we find that to the biologist the spectroscope is of value in studying the relations of animal and vegetable fluids, and even of certain tissues. But this wonderful instrument is clearly destined to play its chief part in what is called terrestrial chemistry—the field where it has won the most signal triumphs.

It must be remarked, despite this vast range of applicability, a range sweeping through the whole universe and embracing all the four elements of antiquity; and despite the astonishing results already achieved and the prospect of greater revelations to come, that the spectroscope is still inadequately appreciated by professed men of science, and in consequence is, to a great extent, ignored by the "educated and intelligent public." In urging its more thorough recognition I do not advocate the formation of spectroscopic societies for the fragmentary study of everything that can be observed with a spectroscope. But I recommend researching chemists to appeal to this instrument wherever requisite and possible.

An elaborate spectroscopic study of the basic constituents of rare minerals from different localities would be of great value, and I would suggest that on all possible occasions meteorites should be submitted to careful spectroscopic analysis.

* Address delivered by the President of the Chemical Society at the Annual General Meeting March 21st, 1889.

I do not propose to discuss all the splendid achievements of the spectroscope in chemistry; nor its applications in ordinary analysis, qualitative and quantitative; nor the conduct of technical operations, such as the Bessemer process. I confine myself to the light thrown by the spectroscope upon the nature and the relations of our *elements*, real or supposed.

Though systematically employed by few experimentalists, the spectroscope has already led to the discovery of several hitherto unknown elements. In the early days of spectrum analysis attention was mainly concentrated on the flame spectra: that is, the bodies in question were vaporised and rendered luminous by the action of a flame, such as that of the Bunsen burner or of the oxyhydrogen jet. This procedure in the hands of Bunsen and Kirchhoff gave us cæsium and rubidium; afterwards, in my own hands, thallium; and in those of Reich and Richter, indium.

Then followed the production and examination of spark-spectra. The spark produced by means of the induction coil, especially when its energy is reinforced by the intercalation of a Leyden jar, volatilises and renders luminous minute portions of matter, solid, liquid, or gaseous, which may then be examined by the spectroscope. In this manner gallium was discovered in 1875 by Lecoq de Boisbaudran. In consequence of the sharpness and the well-marked character of these spark-spectra they are relied on by chemists as certain proof of the identity of any two elements which yield identical spectra.

Next was introduced the systematic study of the absorption-spectra seen when a beam of light is passed through certain transparent solids or through solutions of various substances. One of the earliest observers in this branch of spectroscopy was Dr. Gladstone, who, in 1858, read before this Society a paper on the absorption of light by various metallic salts, and gave the first description of the absorption-spectrum of didymium. This branch of spectroscopy has proved not less fruitful in the recognition of new metallic elements.

In the investigation of the rare earths my principal object has been to separate the true from the undemonstrated and spurious, verifying the true, rejecting the spurious, and reducing as far as possible the number of the doubtful. In the following table I have given a list of the so-called "rare-elements," with which for the last seven or eight years I have been specially occupied. Column 1 gives the names by which they are commonly known. Column 2 gives their atomic weights, &c. Column 3 shows in what manner they come under the domain of spectroscopy; and columns 4 and 5 notify the components or meta-elements into which some of these bodies have been decomposed in 1886 by myself, and in 1887 by Krüss and Nilson. In the first column I have exercised a judicial leniency in retaining candidates, for the sake possibly of old associations, when strict justice would have disestablished them. Thus, it may be doubted whether decipium, philippium, or gadolinium should have been retained. But since doubts have been cast on the integrity of nearly all the occupants of this column the line should not be drawn too strictly.

At first spectroscopic examination was applied directly to substances, natural or artificial, which had not undergone any special preparation. The idea next occurred of attempting to split up substances supposed to be simple into heterogeneous constituents before appealing to the spectroscope. The refined chemical processes used for this operation may be summarised under the name of fractionation, whether they be fractional precipitations, crystallisations, or decompositions. The essential principles of this processes were so fully discussed on the last occasion when I had the honour of addressing you that I need not further allude to them.

The Didymium Group.

A combination of such delicate and prolonged chemical

TABLE I.

	Atomic weight of Metal and Formula of Oxide.	Gives Spectrum by—	Component Meta-Elements according to—	
			Crookes (1886).	Nilson and Krüss (1887).
Didymium	Neodymium— 140·3, Nd_2O_3. / Praseodymium— 143·6, Pr_2O_3. / Unnamed.	Absorption.	$D\alpha$ $\lambda = 475$	$Di\alpha$ $Di\beta$ $Di\gamma$ $Di\delta$ $Di\epsilon$ $Di\eta$ $Di\theta$ $Di\iota$ $Di\chi$
Decipium.. ..				
Samarium ..	150·12, Sm_2O_3.	Absorption and phosphorescence.	$S\delta$ $G\epsilon$ $G\gamma$ $G\theta$	$Sm\alpha$ $Sm\beta$
Lanthanum ..	138. La_2O_3.	Phosphorescence.		
Erbium	166, Er_2O_3.	Absorption. and phosphorescence.	$\lambda\,550$ $\lambda\,493$	$Er\alpha$ $Er\beta$
Philippium ..	45—48, PpO.	Phosphorescence.		
Holmium.. ..		Absorption.		$X\alpha$ $X\beta$ $X\gamma$ $X\delta$
Thulium	170·7, Tm_2O_3.	Absorption.		$Tm\alpha$ $Tm\beta$
Dysprosium ..		Absorption. $\lambda\,457—448$ (1888)		$X\zeta$ $X\epsilon$ $X\eta$
Yttrium	88·9, Yt_2O_3.	Phosphorescence.	$G\alpha$ $G\beta$ $G\delta$ $G\zeta$ $G\eta$	$Z\alpha$ Tb Yt （Lecoq re Bois-baudran.）
Terbium	124·7, Tb_2O_3.			
Gadolinium (Yα)		Phosphorescence.	$G\beta$ $G\zeta$	
Ytterbium ..	173·01, Yb_2O_3.	Phosphorescence.		
Scandium ..	44·03, Sc_2O_3.			

processes with spectroscopic examination applied to bodies showing absorption spectra soon led to important discoveries. Fig. 1A* shows what I may call the normal didymium spectrum, as it was generally recognised down to the year 1878. Fig. 1 B shows the whole of the absorption-bands belonging to bodies subsequently separated from didymium by fractionation. When in 1878 the didymium from samarskite was examined by Delafontaine (*Compt. Rend.*, lxxxvii., 632; CHEM. NEWS, xxxviii., 223) he found it to differ somewhat from ordinary didymium as extracted from cerite and gadolinite, and by a series of chemical fractionations he succeeded in separating from it an earth which he called decipium, giving at least three absorption-bands, one having a wave-length of 416 ($1/\lambda^2$ 578) ; another narrower and stronger, at wave-length 478 ($1/\lambda^2$ 438), and a very faint "minimum of transmission" near the limit of the blue and green. Nine months later Lecoq de Boisbaudran (*Compt. Rend.*, lxxxix., 212 ; CHEM. NEWS, xl., 99) announced the discovery of samarium as a constituent of the didymium

from samarskite, giving a drawing of the decipium and samarium spectra to a common scale (Fig. 2), from which it is seen that samarium is characterised by the bands of Delafontaine's decipium together with two additional bands. Fig. 3 shows the samarium spectrum from the latest measurements.

Still didymium was not reduced to its ultimate simplicity. In 1885 Carl Auer (*Monatsh. Chem.*, vi., 477), by fractionally crystallising the mixed nitrates of ammonium, didymium, and lanthanum, showed it was thus possible to cleave didymium in a certain direction and separate it into two other bodies, one giving green salts and the other pink salts. Each of these has a characteristic absorption spectrum, the sum of the two sets of bands approximating to the old didymium spectrum. These bodies the discoverer has named respectively praseodymium and neodymium. The neodymium spectrum (Fig. 4), according to Auer, consists of the whole of the bands in the red, with part of the large one in the yellow ; it then misses all the green and blue, and takes in the second line in the violet. Fig. 5 is the spectrum of praseodymium, which I have also taken from Auer's description ;

* All the figures are drawn to the $1/\lambda^2$ scale.

Editorial Note. For technical reasons it has unfortunately not been possible to reproduce the plates relating to William Crookes' articles.

it takes the other part of the yellow band, and all the green and blue, except the second blue, which belongs to neodymium. Subtracting these two spectra from the old didymium spectrum (Fig. 1) you see there are still two bands left at λ 462 and 475 (1/λ² 465 and 443). Assuming that the argument from absorption-spectra is a legitimate one—and all recent research tends to show that if not quite trustworthy it is at all events a weighty one—the inference I draw from these results is that the old didymium still contains a third body distinct from neo- and praseodymium, to which one or both of these extra bands is due.

I must venture to lay especial emphasis on the words *in a certain direction.* Didymium in my own laboratory has undergone other cleavages, and I have not yet decided whether we shall have to recognise further decompositions of neodymium and praseodymium, or whether the original didymium is capable of being resolved differently according to the manner in which it is treated. Keeping the band in the orange always of the same strength, in many of the fractions of didymium from different sources the other bands of neo- and praseodymium are seen to vary from very strong almost to obliteration (CHEM. NEWS, liv., 27). In this way I have worked on the spectra of didymium from allanite, cerite, euxenite, fluocerite, gadolinite, hielmite, samarskite, yttrotitanite, &c., and the further I carry the examination the more the conclusion is forced upon me that didymium must not be regarded as compounded of two elements only, but rather as an aggregation of many closely allied bodies. Later researches of Krüss and Nilson have led them to the same conclusion.

When working in 1886 on the decomposition of the nitrate of didymium by heat, I found very decided indications of the possibility of depriving didymium of band after band until only the deep line in the blue λ 443 (1/λ² 509·6) is left (*Proc. Roy. Soc.*, xl., 503). I have provisionally named this single band element Dα. In this connection I would like to draw your attention to a few facts which have very recently come to light.

In some of my fractions of didymium the band λ 475 (1/λ² 443) intensifies in company with another band at λ 462 (1/λ² 465). Fig. 6 shows at A the group of blue bands as they are seen in a strong solution of didymium. It forms a well-marked set of four comparatively sharp lines, one at λ 462 (1/λ² 465) being the faintest. Under these lines I show (Fig. 6, B) the same group as seen in fractions of the same solution of didymium after disintegration has commenced. This fraction is very similar to Auer's praseodymium; and below this (Fig. 6, C) I again show the same group as seen in the didymium fractions most removed from the one last described. Here the two lines λ 475 (1/λ² 443) and λ 462 (1/λ² 465) have become very strong, while the other two have almost faded out. One of these, λ 475 (1/λ² 443), is included by Auer in the spectrum of his crude didymium, but he makes no further reference to it either in description or diagram. Of these lines, λ 462 (1/λ² 645), probably belongs to the samarium group, but the other, λ 475 (1/λ² 453), cannot belong to samarium, although it superposes on the most refrangible half of the broad and ill-defined samarium band.

In a paper read before the Royal Society dated June 9, 1886 (*Compt. Rend.*, cii., 1551), I gave an account of some observations I had made upon the line λ 475 (1/λ² 443), proving that it could be separated from the old didymium spectrum, and in conclusion said it "must be regarded as characteristic of a new body."

Subsequent to my paper Demarçay (*Ibid.*) drew attention to this line at λ 475 (1/λ² 443), and in 1887 (*Compt. Rend.*, civ., 580) he again returned to the subject, associating the two lines λ 475 (1/λ² 443) and λ 462 (1/λ² 465) as being due to the same element. With this opinion I cannot yet agree, for in many instances I have had fractions in which the relative intensities of the two are widely different. More recently Krüss and Nilson have ascribed this line to one of the constituents of

Soret's X or holmium, which gives a line falling nearly on the same place (CHEM. NEWS, lvi., 154, 173).

By examining the absorption-spectra of solutions of rare earths obtained from widely different sources, Krüss and Nilson (*Ber.*, xx., Part xii., 2134; and CHEM. NEWS, lvi., 74, 85, 135, 145, 154, 165, 172) came to the conclusion that the elements giving absorption-spectra, and known as didymium, samarium, holmium, thulium, erbium, and dysprosium, were not homogeneous, but that each one contained almost as many separate components as it produced bands of absorption.

They have discovered that in didymium obtained from some minerals one of the fainter lines of the normal didymium spectrum is strong, while others usually stronger are almost or quite absent; results to which I shall presently refer will show that this cannot be explained by dilution or concentration. In this way, by examining a great number of minerals, they found anomalies occurred in the case of almost each of the old didymium lines, and therefore decided, as above mentioned, that it is a compound body, capable of resolution into at least nine separate components.

Identical arguments are brought forward to prove that each of the other so-called elements, samarium, erbium, holmium, thulium, dysprosium, &c., are compounds of many closely allied bodies.

Krüss and Nilson, I believe, are pushing their investigations with the object of isolating the separate components of these different earths. They, however, question the possibility of resolving the erbia and didymia earths into their several ultimate constituents by a fractionated decomposition of the nitrates. In fact they assert that by means of the methods of separation at present known it would be almost impossible to completely isolate any single constituent of the mixed earths. They therefore propose, as I had previously done,[*] a method by which we may certainly arrive nearer to the mark and dispense with much tedious fractionation. If we examine the minerals which contain these rare earths we find they occur in very different states of mixture or combination. Sometimes many of the constituents which we wish to separate are conjointly present, and sometimes but few. The desired differentiation, in fact, has already been commenced by Nature. Krüss and Nilson, therefore, whichever ingredient they wish to separate, propose to operate on a mineral which contains that ingredient as far as possible in a state of isolation. In other words, they will take advantage of the work that Nature has already begun, and endeavour by refined chemical means to put the last finishing touches to her work. Thus they will be able to work with smaller quantities of primary material,—no small consideration in the case of some minerals,—and to obtain results in a shorter time. How widely the composition of one and the same mineral, as judged by our searching physical tests, may vary, will be seen from the following instances. Fergusonite from Arendal shows six of the bands of holmium, fergusonite from Ytterby four, and that from Hitterö only three. Moreover, the ingredient provisionally called Xα is to be found in the fergusonite from Ytterby, but not in that of Arendal and Hitterö.

The foundation for thus firmly declaring what I had previously ventured to infer, is the striking differences in the spectra given by several specimens of one earth, say didymium, when obtained from different sources.

We are anxiously waiting the results of this investigation, but although the paper quoted was published in July, 1887, no further communication has come from these illustrious workers.

Chemists recently have stated as proof of the existence of new elements the fact that certain bands of absorption, as seen in various fractions, "follow the same variations of intensity." Before deciding the question whether

---

[*] " Address to the Chemical Section of the British Association, Birmingham Meeting," CHEM. NEWS, liv., 123. " On the Fractionation of Yttria," CHEM. NEWS, liv., 157. *Proc. Roy. Soc.*, xl., 1886. 505

didymium is a homogeneous whole, or whether an argument in favour of its heterogeneity can be based on the fact that the absorption spectra of didymium from different minerals differ *inter se*, it was necessary to ascertain if the absorption-bands seen in its solutions, whatever the thickness of the layer, whether dilute or concentrated, followed the same variations, and also to ascertain the nature of these variations. To contribute to this inquiry I examined the absorption-spectrum of a solution of neutral didymium nitrate containing one part by weight of metal in 10 of water, as seen through a series of cells from 1 m.m. to 25 m.m. in thickness. For this work I used a new form of binocular spectroscope, fitted with a mechanical tracing arrangement, so that each spectrum can be automatically mapped on paper strips; from this set of tracings I have arranged the diagram (Fig. 7) now before you. It represents the bands of the normal didymium spectrum. The figures running up the side represent the thickness of the layer of solution observed, and they show at a glance the "life" of each of the bands. At the bottom, 25 m.m. thickness, all the known bands are visible, and they become fainter and die out in the particular order here given, some of them remaining visible almost to the end. For instance, almost as long as the deep line in the blue part of the spectrum at λ 443 ($1/\lambda^2$ 509), my Dα, can be distinguished, it is possible to see the group of three very narrow ones next to it. Two or three other less characteristic bands can be seen only when there is a very considerable depth of liquid; thus, the three lines in the red λ 636, 628, 622 ($1/\lambda^2$ 247, 253, 258), cannot be seen distinctly through less than 20 m.m. of this strength of solution.

Having ascertained in this series how the spectra varied in appearance with different thicknesses of the same solution (strength 1 of Di in 10 of water), I repeated the experiments, keeping the thickness of layer of solution constant, and diluting the standard solution of didymium so that the rays of light passed through the same quantity of metal as in the former series. The results in each case were practically identical; the differences being too slight to be detected in my apparatus. The spectrum exhibited, for instance, by 1 m.m. of the standard solution of didymium is found to be identical with the spectrum shown by the same solution diluted 20 times and viewed through a 20 m.m. cell.

It will be seen that in the case of 1 m.m., not only the line in the yellow at λ 582 ($1/\lambda^2$ 292)* is to be seen, but also two in the green at λ 525 ($1/\lambda^2$ 368) and λ 510 ($1/\lambda^2$ 382). Therefore, to get a more simple spectrum, I diluted the solution to 1 of didymium in 20 of liquid. I m.m. thickness of this shows little else than a broad faint trace of the line in the yellow, λ 582 ($1/\lambda^2$ 292).

(To be continued).

---

* The λ and $1/\lambda^2$ lengths given here are only approximate.

THE CHEMICAL NEWS.

VOL. LX. No. 1548.

ON

RECENT RESEARCHES

ON THE

RARE EARTHS

AS INTERPRETED BY THE

SPECTROSCOPE.*

By WILLIAM CROOKES, F.R.S., &c.

(Continued from p. 30).

In the year 1886 (*Proc. Roy. Soc.*, xl., 502, June 9, 1886; CHEM. NEWS, liv., 27, July 19, 1886) I demonstrated from experiments on the fractionation of didymium that this element was very probably a compound capable of being resolved into a number of constituents, each represented by a single band, like the cases of yttrium and samarium, which give bands by phosphorescence. In Krüss and Nilson's case, the batch of crude earth contained in any single mineral was examined as a whole without any attempt to separate the earths, and the composition of some of these minerals is extraordinarily complex, euxenite for example containing, after removal of the other metals, the rare bodies Ce, La, Di, Sm, Yt, Er, Tr, Ho, Tm, Th, De, Sc, Dy, Be, Nb. Ta.

In my own case the didymium earths, upon which I formed the "one band one element" theory (CHEM. NEWS, liv., 27) were in a much more simple state, as the whole of the yttrium group, including erbium, holmium, thulium, &c., and others, had been removed, and the earth under examination probably contained little besides didymium and lanthanum, with traces only of samarium, yttrium, and calcium. But not even here was the earth under examination in a state of even approximate purity.

There are at least two points in these researches that I must touch, since they illustrate the necessity of great caution in drawing conclusions from an examination of absorption-spectra. Paul Kiesewetter and Krüss (*Ber.*, xxi., 2310; CHEM. NEWS, lviii., 75, 91) have recently published a paper on this subject, although it goes no further, nor indeed so far as the previous communication of Krüss and Nilson. They have examined gadolinite, and find that some constituents of didymium and samarium are absent, notably those which produce the group of lines in the green to which I have already referred. In my own laboratory I have worked for the last two years almost exclusively upon the earths from gadolinite—of which I have obtained a large quantity from Fahlun—and there is not the shadow of a doubt that in my gadolinite earths the lines reported absent by Kiesewetter and Krüss are present in abundance.

Some hitherto unexplained condition doubtless rendered these lines invisible to Kiesewetter and Krüss; perhaps the presence of some other earths, or some condition of concentration or acidity. In the light of this knowledge I do not see how we can take the results of Krüss and Nilson or my own as final.

Concerning the influence of one body upon another little is yet known, but that little is of sufficient importance to make us very careful how we interpret absorption-spectra when uncorroborated by chemical results. Lecoq de Boisbaudran and Smith have pointed out some important modifications produced in absorption-spectra by an excess of acid (*Compt. Rend.*, lxxxviii., 1167), and later on Soret (*Compt. Rend.*, September 15, 1879) verified these observations. Brauner and others have recorded

experiments on mixing solutions of didymium and samarium; they find, in the case of a didymium solution showing the group of three bands, λ 476, 469, 428 ($1/\lambda^2$ 430·4, 441·3, 454·6), that by adding a dilute solution of samarium, all three of these bands vanish without any appearance of any of the samarium bands, until a certain proportion is reached, when the samarium bands gradually come in their place (Brauner, *Trans. Chem. Soc.*, 1883, xliii., 286). In my own experiments, I find that from a solution of erbia which originally shows no trace of the strong didymium band lying between λ 596 and 572 ($1/\lambda^2$ 281 and 305), appreciable quantities of didymium can certainly be squeezed out by fractionation.

Owing to its complicated nature, Kiesewetter and Krüss consider gadolinite an unfavourable source of didymium for these investigations, and recommend that a large quantity of earth from keilhauite should be systematically worked up, for the reason that keilhauite didymium is more simple in constitution.

The Erbium Group.

It is known that a certain oxide, ten years ago called erbia, and regarded as belonging to a simple elementary body, has been resolved by the investigations of Delafontaine, Marignac, Soret, Nilson, Clève, Brauner, and others into at least six distinct earths—three of them, scandia, ytterbia, and terbia, giving no absorption-spectra, whilst others, erbia (new), holmia, and thulia, give absorption-spectra.

In Fig. 8A I have represented the old erbium absorption-spectrum, as it was known down to the year 1878. In Fig. 8B, as in the case of didymium, I have given for diagrammatic purposes certain lines which, not seen in unfractionated erbia, belong to bodies capable of being separated from it by fractionation.

The first to announce that erbium was not a simple body was Delafontaine, who in 1878 (*Compt. Rend.*, lxxxvii., 559; CHEM. NEWS, xxxviii., 202) published an account of philippium, a yellow oxide characterised by a strong band in the violet, λ 400 to 405 ($1/\lambda^2$ 625, to 623), a broad, black absorption-band in the indigo-blue, λ about 450 ($1/\lambda^2$ 494), two rather fine bands in the green, and one in the red.

The history of philippium is curious, and I may perhaps be allowed to give it in some detail. A year after Delafontaine's discovery Soret (*Compt. Rend.*, lxxxix., 521; CHEM. NEWS, xl., 147) published a paper in which he declared that philippia was identical with his earth X. The next month, in a note on erbia, Clève (*Compt. Rend.*, lxxxix., 708; CHEM. NEWS, xl., 224) said he could not identify Soret's X with Delafontaine's philippia, as the latter was characterised by an absorption-band in the blue which occupied the same place as one of the erbia bands. In February, 1880 (*Compt. Rend.*, xc., 221; CHEM. NEWS, xli., 72), Delafontaine returned to the subject, enumerating 10 new earths in gadolinite and samarskite, viz., mosandra, philippia, decipia, scandia, holmia, thulia, samaria, and two others unnamed. He said that the properties of philippia were those of Soret's X and Clève's holmia, and proposed that the name "holmia," being a duplicate name for an already known earth, should be discarded in favour of philippia. In July, 1880 (*Compt. Rend.*, xci., 328; CHEM. NEWS, xlii., 185), Clève repeated his former statement that philippia was not the same body as Soret's X or holmia. Delafontaine next withdrew all he had said about the absorption-spectrum of philippium, and decided that it had no absorption-spectrum (*Archives de Genève*, [3], 999, 15). Finally, Roscoe (*Chem. Soc. Journ.*, xii., 277), in an elaborate chemical examination of the earth-metals in samarskite, proved that philippia was a mixture of yttria and terbia. From a prolonged chemical study of these earths I have since come to a similar conclusion; but a spectroscopic examination of the earth left on igniting some specially purified crystals of "philippium formate" tested in the radiant matter tube, has shown

* Address delivered by the President of the Chemical Society at the Annual General Meeting, March 21st, 1889.

me that in the separation of Delafontaine's philippium the yttria undergoes a partial fractionation, and three of its components or meta-elements, Gζ, Gδ, and Gβ, are present in great abundance, while others, Gα and Gη, are almost if not quite absent.

Shortly after the announcement of philippium, Soret (*Compt. Rend.*, lxxxvi.. 1062) described an earth which he provisionally called X. This was soon found to be identical with an earth subsequently discovered by Clève (*Compt. Rend.*, lxxxix., 479; CHEM. NEWS, xl., 125), and called by him holmia. Soret admitted the identity, and agreed to adopt Clève's name of holmia. Fig. 9 shows the absorption-spectrum of X or holmia. It consists of a very strong band in the extreme red, λ 804 ($1/\lambda^2$ 155), two characteristic bands in the orange and green, λ 640 and 536 ($1/\lambda^2$ 244 and 347), besides fainter lines in the more refrangible part of the spectrum, and a number of bands in the ultra-violet.

Simultaneously with the discovery of holmia, Clève announced the existence of a second earth from erbia, which he called thulia. Fig. 10 shows the absorption-spectrum of thulia. It consists of a very strong band in the red, λ 680 to 707 ($1/\lambda^2$ 216 to 200), and one in the blue, λ 464·5 ($1/\lambda^2$ 462). The residual erbia, after separation of these earths, gives a simpler absorption-spectrum, shown in Fig. 11.

In 1886 (*Compt. Rend.*, cii., 1003, 1005) Lecoq de Boisbaudran showed by fractional precipitation of Soret's X and by spectroscopic examination of the simple fractions, that this X, or holmium, consisted of at least two elements, one of which he named dysprosium, retaining the name of holmium for the residue left after deducting dysprosium. Fig. 11 shows the absorption-spectrum of dysprosium; it has four bands, λ 451·5, 475, 756·5, 427·5 ($1/\lambda^2$ 490·5, 443, 175, 547). The new holmium gives an absorption-spectrum shown in Fig. 13.

As regards dysprosium, I pointed out (*Proc. Soy. Soc.*, xl., 502), at the time it was first announced, that I had obtained a solution in which one of the bands claimed for dysprosium, the one at λ 451·5 ($1/\lambda^2$ 490·5), was very strong, while the others were absent. As M. de Boisbaudran associates the bands at λ 475 ($1/\lambda^2$ 443) and λ 451·5 ($1/\lambda^2$ 490·5) as both belonging to dysprosium, and as I have obtained an earth which gives λ 451·5 ($1/\lambda^2$ 490·5) strong with no trace of λ 475 ($1/\lambda^2$ 443), it is evident that the conclusion I arrived at in 1886 that dysprosium itself consisted of at least two simpler bodies—is correct.

The old spectrum of erbium (Fig. 8) shows two faint bands at λ 550 and λ 493 ($1/\lambda^2$ 331 and 409); the second being broader than the first. These bands do not occur in the spectra of holmium, thulium, dysprosium, or the new erbium. In a long-continued fractionation of the erbia group of earths, carried out with an ample supply of the old erbia, I find an earth giving these two bands concentrating at one end, and the bands become stronger. At the same time two other new bands are making their appearance. This, therefore, points to the existence of still another earth belonging to the erbium group.

Incandescence Spectra.

Another distinct method of spectrum analysis depends on the examination of the spectrum of the light emitted by a solid substance when raised to incandescence. Almost the only known example of this is obtained in the case of erbia (*ibid*, xl., 77, January 21, 1886). It is scarcely known that if erbia in the solid state is illuminated by the electric or other bright light, and examined in the spectroscope, it gives a spectrum of black lines and bands as distinct as the Frauenhofer lines. The spectrum of bright lines emitted when solid erbia is heated to incandescence in the blowpipe flame has been more often observed; in this case the lines come out luminous on a faintly continuous background, whilst the reflection spectrum just mentioned is composed of black lines sharply defined and much more luminous, upon a continuous spectrum. Holmia and thulia, components of

old erbia, possess a similar property (CHEM. NEWS, lvi., 165).

Phosphorescence Spectra.

I will now deal with phosphorescence spectra. Not a few chemists and physicists, conspicuous among whom is Ed. Becquerel, have carefully studied the phenomena of phosphorescence. Phosphorescence may be excited by elevation of temperature, by mechanical action, by electricity, and by exposure to the rays of the sun, and the light thus given off—for example, in the case of fluorspar, has been examined by means of the spectroscope. In my own spectroscopic research I have dealt with the phosphorescence occasioned by the impact of the molecules of radiant matter upon certain phosphorescent bodies, or what I have ventured to call molecular bombardment.

It is not necessary for me to describe the mode of procedure further than to say that the substance under examination is placed in a very high vacuum—a vacuum which varies in degree in the case of certain earths. Fig. 14 shows the form of radiant matter tube I prefer. In such a vacuum, when submitted to the action of the induction current, substances phosphoresce very differently from what they do when treated similarly at the ordinary pressure of the atmosphere. Under such circumstances the spectroscopic examination of matter affords what I have called the radiant matter test. The number of substances which are thus phosphorescent is very considerable. Glass of different kinds, according to its composition, phosphoresces with various colours. Phenakite (glucinium silicate) phosphoresces blue; spodumene (aluminium and lithium silicate) gives off a rich golden-yellow light, whilst the emerald phosphoresces crimson, and the diamond, being exceptionally sensitive and brilliant, throws off a bright, greenish white light.

The ruby, one of the minerals I examined earliest in this manner, glows with a rich, brilliant, red tone, quite independent, as regards its depth and intensity, of the colour of the stone as seen by daylight; the pale, almost colourless specimens, and the highly prized variety of the true "pigeon's blood," all phosphoresce with substantially the same colour.

There are several varieties of phosphorescence spectra, or rather, substances to be submitted to the radiant matter may be previously prepared in divers manner, so as to give modified results. Thus—

a. An earth alone—of course in the solid state—may be very strongly ignited, and, when cold, examined in the radiant matter tube. This method differs from the use of the spectrum of incandescent solids above noticed, solely in the final test of molecular bombardment. Of this variety of phosphorescence spectra, we may take, as examples, the phenomena yielded by alumina, yttria, didymia, and lanthana.

b. Another modification applicable in the treatment of the less pure earths takes as material the sulphates of the earths. The substance under examination is first treated with strong sulphuric acid; the excess of the acid is then removed by heat, and, lastly, the sulphate is heated to a point just short of driving off all the sulphuric acid. It is then placed in the radiant matter tube, and when the exhaustion has been carried as far as necessary, the induction current is passed through (*Phil. Trans.*, Part III., May, 1883). This method gives a spectrum of broad bands, easily recognised, but by no means easy to measure. The spectroscope used should have a low dispersive power, and should not have a very narrow slit. In the case of yttrium sulphate the bands are more analogous to the absorption-bands seen in solutions of didymium than to the lines given by spark spectra. If examined with a high magnifying power the outlines of the bands become less definite. The bands are seen much more sharply when the current first passes than after it has been passing for some time and the earth has become hot. On cooling, the bands again appear sharply defined.

This method of observing the constitution of the rare earths, duly aided by delicate and prolonged chemical processes, has permitted us to push our investigations further than had previously seemed practicable. It enables us to determine whether we have reached the end of our investigations—a consummation which had hitherto been vainly sought. It has enabled us to prove that yttrium, samarium, &c., are not simple, homogeneous bodies. But what of the constituents into which they have been thus resolved? Suppose we refine them down until each displays merely one spectral band—what then? Is each one of such bodies barely differentiated from its neighbours chemically or physically, entitled to rank as an element? If so, as I pointed out in the Address which I had the honour to deliver before you in March last, we shall have to deal with further perplexing questions, arising in part from the relation of such elements to the periodic system. In a discussion of the elements, not as yet published, Dr. Wundt maintains that their possible number cannot exceed 79. But I myself see no definite and sufficient reason for limitation to this number. If these bodies are not elementary, possessing as they do the properties commonly regarded as characteristic of an element, we must be prepared to show why not?

Whatever rank may ultimately be assigned to these substances, they must, for convenience sake, have names as soon as our knowledge of their properties is in a sufficiently advanced state to allow of their removal from the suspense account.

(To be continued).

THE CHEMICAL NEWS.

VOL. LX. No. 1549.

ON

RECENT RESEARCHES

ON THE

RARE EARTHS

AS INTERPRETED BY THE

SPECTROSCOPE.*

By WILLIAM CROOKES, F.R.S., &c.

Continued from p. 41).

The Yttrium Group.

YTTRIUM—the old yttrium—proves now to be not a simple element, but a highly complex substance. I have come to the conclusion that it may be split up certainly into five and probably into six constituents. Its phosphorescence spectrum is shown in Fig. 16, and the different bands are designated by certain letters. Fig. 17 shows the simple spectra of the separate components into which yttria can be separated by fractionation. If we take these constituents in the order of their approximate basicity — the chemical analogue of refrangibility — the lowest of these constituents gives a deep blue band, $G\alpha$; then follows a strong citron band, $G\delta$, which increases in sharpness until it may be called a line; then a red band, $G\zeta$; then a deeper red band, $G\eta$; and lastly a close pair of greenish blue lines, $G\beta$. Following these are frequently seen $G\varepsilon$, $G\gamma$, and $G\theta$, the yellow, green, and red components of samarium, Fig. 19.

As the result of many years' work and several thousand fractionations of old yttria, I exhibit the series of 19 phosphorescence spectra shown on Diagram 28. The centre spectrum, marked J, is approximately that given by the crude earth, although this differs slightly according to the mineral from which it is extracted. After a time, fractionation splits the earth J into two earths, I and K, giving slightly different spectra. Fractioning I gives H and J, whilst K on fractionation yields J and L. The state of separation to which the meta-elements of the samarskite earths may be brought after many years' work is seen by the series of 19 spectra selected to illustrate the progress of the work. It must not, however, be thought that there is so great a difference between any two adjacent spectra as is here shown. To make the diagram more accurately represent what actually occurs in the laboratory it would be necessary to place between each of these 19 spectra about 1000 intermediate spectra.

Beginning at the extreme red it will be seen that a strong band at λ 647 ($1/\lambda^2$ 239) is of maximum intensity from G to K, when it rapidly disappears and is not seen beyond C and N. The meta-element giving this band I have called $G\eta$. The next band in the red λ 639 ($1/\lambda^2$ 245), is at its maximum at A or even above, while it fades out between K and L. The next band at λ 619 ($1/\lambda^2$ 261) has its maximum between I and O, fading out rapidly below, but being more persistent above. The meta-element to which this band is due I have called $G\zeta$. I next come to an extremely sharp band at λ 609 ($1/\lambda^2$ 269), which appears to belong to an earth absent in gadolinite, and present in samarskite and a few other minerals. Its greatest brilliancy is between E and K, dying away rapidly on either side. This meta-element I have called $S\delta$. A double orange band follows, and there are indications that its two components are separable, although they must be very closely associated. The maximum

* Address delivered by the President of the Chemical Society at the Annual General Meeting March 21st, 1889.

brightness of the first component, λ 603 ($1/\lambda^2$ 275), extends from O to the top of the diagram. The second component, λ 597 ($1/\lambda^2$ 280), begins to lose brilliancy about G, and, like its companion, is at its greatest brightness at the highest spectrum shown on the diagram. This band has been almost isolated in a specimen of crude lanthana. I have called its meta-element $G\varepsilon$.

The next band in order is the citron or $G\delta$ band, λ 574 ($1/\lambda^2$ 303.5), the most prominent in the spectrum of old yttrium. This band extends with almost undiminished sharpness and brilliancy from G to S; above G it fades out rapidly, and from D upwards it is absent. Following the citron band is a double green band. This, like the orange band just described, is separable into two components. The first, at λ 568 ($1/\lambda^2$ 310), is almost absent in A, attains its maximum in about D, and fades out entirely at K. The second component of this green pair, λ 563 ($1/\lambda^2$ 315), is at its maximum at A and above, and extends only to H. The meta-element giving rise to this pair I have called $G\gamma$. Next in order comes a pair of bright green bands. The two components λ 550, 541 ($1/\lambda^2$ 331, 342) have not yet shown signs of dividing. The first appearance of this double band is at B, it attains its maximum at about E, and continues with almost undiminished brightness to Q. The body giving this double green band is the most persistent of all the meta-elements of these earths. I have provisionally called it $G\beta$.

After a dark interval a broad, hazy, double blue band is seen, having its centre at λ 482 ($1/\lambda^2$ 430.5). This band, first appearing at about F, increases in brightness down to the last fraction at S. Its meta-element is called $G\alpha$. Lastly, at λ 456 ($1/\lambda^2$ 481), is seen a deep violet band, coming in first at Q, and getting brighter as we get lower down the fractionations. This band is of extraordinary brilliancy in some samples of ytterbia, but as it is absent in a specimen of ytterbia from Nilson, said by him to be perfectly pure, it is probably due to some other new body which I have therefore named provisionally $S\gamma$.

Although, for economy of space, it is inconvenient to represent a longer series than is here shown, my fractionations have been pushed far beyond the limits shown on the diagram. Fractions above A and below S give ample evidence that the process of differentiation has not reached its utmost limit.

On the left of the diagram, Fig. 28, I have attached chemical symbols to some of the spectra. Thus the top spectrum, A, is the one shown by samarium. At D is the spectrum given by Marignac's $Y\alpha$ or gadolinium. At H is seen the spectrum of mosandrum, and at L is shown the spectrum given by what would ordinarily be called pure yttrium. A study of this diagram will, I think, convince any impartial observer that the lessons it conveys fully bear out my contention that samarium, gadolinium, mosandrum, and yttrium, are not actual chemical elements, but are compounded of certain simpler bodies which may conveniently be called "meta-elements."

A possible explanation of the existence and nature of the new bodies into which "old yttrium" has been split up, and of parallel cases which will doubtless be found on closer examination, is this. Our notions of a chemical element must be enlarged; hitherto the elemental molecule has been regarded as an aggregate of two or more atoms, and no account has been taken of the manner in which these atoms have been agglomerated. The structure of a chemical element is certainly more complicated than has hitherto been supposed. We may reasonably suspect that between the molecules we are accustomed to deal with in chemical reactions, and the component or ultimate atoms, there intervene sub-molecules, sub-aggregates of atoms, or meta-elements, differing from each other according to the position they occupy in the very complex structure known as "old yttrium."

The arguments in favour of the different theories are

as yet not unequally balanced. But the assumption of compound molecules will perhaps account for the facts, and thus legitimate itself as a good working hypothesis, whilst it does not seem so bold an alternative as the assumption of eight or nine new elements.

The history of the examination of the rare earths and of its results would lack completeness and intelligibility did I not refer to the views taken by my distinguished friend M. Lecoq de Boisbaudran.

In a communication to the Academy of Sciences this eminent chemist says (*Compt. Rend.*, c., 1437): "It is a singular fact that the position of the phosphorescence bands observed by Mr. Crookes with very pure compounds of yttrium are sufficiently near those which I, on my part, have obtained with the hydrochloric solutions of earths separated as far as possible from yttria, chemically as well as spectroscopically. My reversion spectrum cannot, I think, be attributed to yttrium, for on the one hand it is seen *brilliantly* with products which give no trace of yttrium rays with the direct spark, and on the other hand, I have found it impossible to obtain it sharply from certain earths extremely rich in yttria."

M. de Boisbaudran further states in a note :—"This spectrum (*i.e.*, the one which he had just described) is now recognised as being identical with that which is ascribed to pure yttria by Mr. Crookes, and which this *savant* obtained under experimental conditions very different from mine. Nevertheless, my latest observations, as well as earlier ones, lead to the conclusion that yttria is not the cause of the spectrum bands observed. In my fractionations the phosphorescence spectrum gradually gets weaker as I advance towards the yttria end. With almost pure yttria the phosphorescent bands show themselves faintly or not at all, whilst they are brilliant with the earths which do not give by the direct spark the rays of yttrium to any appreciable extent."

It will, I think, appear that the issue between myself and M. de Boisbaudran turns on the question : What is yttria? To what substance can this name be legitimately given? A short time ago the name conveyed to all chemists only one meaning, perfectly definite and undisputed. I have received specimens of yttria from M. de Marignac, considered by him to be purer than any which had ever been previously prepared ; from Professor Clève (called by him *purissimum*) ; and from M. de Boisbaudran, which he pronounces to be yttria "scarcely soiled by traces of other earths." Along with these I have specimens prepared by myself and purified up to the highest point known. A very short time ago every living chemist would have called these samples "yttria." Moreover, they all give my phosphorescence spectrum *in vaeuo* with such intensity that such phosphorescence cannot be rationally ascribed to slight traces of impurities.

This substance, like other chemists, I formerly called "yttria," but since its complex nature has been ascertained I speak of it as "*old* yttria."

M. de Boisbaudran, however, inadvertently uses the word in a manner not quite free from ambiguity. By "yttria," I repeat that I mean now, and have always meant, the yttria of Clève, of Marignac, and of all chemists up to the beginning of the year 1886, the yttria whose metallic base has the approximate atomic weight of 89. M. de Boisbaudran at one time writes as if our meanings were identical : "It is certain," he says, "that my earth, very rich in yttria, gives a beautiful spectrum in the vacuum tube." Again he informs me, "M. Becquerel has recently examined my earths A and B in ultra-violet light, and has obtained results analogous to yours ; that is to say, the earth rich in yttria has generally shone brighter than the other." The "earth A" here spoken of gives my phosphorescent spectrum with wonderful brilliancy, and is the one described above by M. de Boisbaudran as "yttria scarcely soiled by traces of other earths." This yttria, my distinguished friend goes on to say, "is the same as that of Clève and Marignac, varying

only in slight impurities, Marignac's being perhaps the purer." So far then we agree in the meaning to be attached to the word "yttria."

But at other times M. de Boisbaudran gives the name "yttria" to an earth of quite distinct properties, to an earth he has obtained in most minute quantities after months of reiterated fractionation, which gives *no* spectrum in the vacuum tube. Here, therefore, we have the name "yttria" applied to two distinct substances, the one giving a brilliant phosphorescence spectrum, and the other giving none at all, surely a misleading ambiguity. Moreover, I am constrained to question the propriety of my friend's use of the designation "yttria" in the latter sense. Quite recently (*Compt. Rend.*, cviii., 166, January 28, 1889) M. de Boisbaudran has defined his "yttria" as an earth, whether simple or complex, having a characteristic spark spectrum, but not giving a phosphorescence spectrum in the radiant matter tube or by his process of reversion, and having an atomic weight of nearly 89 for its metal.

Why should this non-phosphorescent earth deserve the name of elemental yttria? Why may we not just as well allot the name to some one or other of the phosphorescing earths which M. de Boisbaudran brands as impurities? It seems to me that to refine away the most characteristic attributes of a body, to call the *caput mortuum* by the original name, and pronounce everything else as "impurity," is a departure from the recognised principles of scientific reasoning and practice which it is a duty to protest against.

I have just mentioned that the earth heretofore called yttria and supposed to be simple has been split up into a number of simpler bodies. Now these constituents of the old yttria are not *impurities* in yttria any more than praseodymium and neodymium are impurities in didymium. They proceed from a real splitting up of the yttrium molecule into its components, and when this process is completed the old "yttria" has disappeared. If these newly discovered components on further examination should be found worthy to take the rank of elements, I think as first discoverer, I am entitled by the custom prevailing among men of science, to name them. For the present, and until their investigation is more advanced, I designate them by provisional symbols. One of the most distinct characteristics of "old yttria" is its very definite spark spectrum. To which of its components this spark spectrum belongs I am not yet able to say. It is possible that the particular component to which the spark spectrum is due yields no phosphorescent spectrum. It is also possible that the spark spectrum, like "old yttria," may prove to be compound, and then the well-known lines it contains will have to be shared between two or more of the newly-discovered bodies. I wish emphatically to re-state that at present no single component of old yttria can lawfully lay claim to what may be called the paternal name ; and it seems to me that in the present state of the question no one is entitled to call one of the new bodies "yttria," and to characterise the remainder as impurities.

I regret to add that a misunderstanding exists between M. de Boisbaudran and myself. In a memoir presented to the Academy of Sciences on January 28th of the present year my distinguished colleague shows he completely misapprehends my position, and he so far misinterprets me as to make me say that the gadolinia or Yα of M. de Marignac is a mixture of 61 parts of yttria and 39 parts of samaria !

But what is the fact? So far back as June 9th, 1886, I put it on record (*Proc. Roy. Soc.*, xl., 502) that "Yα is composed of the following band-forming bodies :— λ [541—549], [564], [597], [609], [619], together with a little samarium. Calling samarium an impurity, it is thus seen that gadolinium is composed of at least four simple bodies."

Again in a paper entitled "What is Yttria?" I wrote (CHEM. NEWS, liv., 39, July 23, 1886)—

" Gadolinium is composed of at least four simpler bodies, $G\beta$, $G\gamma$, $S\delta$, and $G\zeta$; the pair of green lines [λ 541 and λ 449, mean 545], being the strongest feature in its spectrum, may be characteristic of gadolinium."

" M. de Boisbaudran says :—' Mr. Crookes appears to attribute to gadolinium the double green band.' This is scarcely accurate; I did not ascribe the double green band to gadolinium, but finding by my test that the *so-called* gadolinium was a compound body, the earth $G\beta$ being its strongest component, I proposed to attach the name of gadolinium to $G\beta$ rather than give it a new name, and thus multiply names unnecessarily."

Thus in 1886 I corrected the error which M. de Boisbaudran now repeats.

Again, I said that the work of fractionating this mixture, "*for its completion*, would occupy a space of time in comparison with which the life of man is all too brief "—a statement transformed into the assertion that to separate yttria and samaria—a relatively easy task—would take more than a life-time.

M. de Boisbaudran's misconception is the more striking since in a foot-note he gives correctly my very words from my "Address to the Chemical Section of the British Association," in 1886, and from the "Genesis of the Elements." How he can have mistaken my meaning is a mystery, as he is an excellent English scholar. Certainly any reader of the *Comptes Rendus* who understands English will at once see that there is no foundation for M. de Boisbaudran's criticism.

(To be continued).

THE CHEMICAL NEWS.

VOL. LX. No. 1550.

ON

RECENT RESEARCHES

ON THE

RARE EARTHS

AS INTERPRETED BY THE

SPECTROSCOPE.*

By WILLIAM CROOKES, F.R.S., &c.

(Concluded from p. 53).

Action of Different Earths on Phosphorescence Spectra.

ANOTHER modification of phosphorescence spectroscopy is produced by the previous addition of other earths to the specially phosphorescent earths, and of the results of such addition some instances have already been mentioned. Lime exerts in this manner a remarkable action. By itself it phosphoresces with a continuous

The addition of lime also affords an argument for the compound nature of samaria, as it suppresses the sharp line $S\delta$, the most striking feature in the phosphorescence spectrum shown by pure samarium sulphate. On the other hand, an addition of old yttria deadens the other lines of samaria, but brings out the line $S\delta$ more strongly. Fig. 19 represents the spectra given by the meta-elements of samarium.

Lanthanum sulphate in the radiant matter tube phosphoresces with a reddish colour and has a broad hazy band in the orange, with a sharp line superposed. This line is identical with the line of $G\varepsilon$, one of the components of the phosphorescent spectrum of samarium. If lime is added to lanthanum, the phosphorescence changes its colour from red to yellow. Lime also brings out the lines of yttrium and samarium if they are present as impurities. When $G\delta$, $G\alpha$, and $G\beta$ are present in small proportions with lime, the lines $G\delta$ and $G\alpha$ become intensified, but a dark place appears in place of the green $G\beta$ band. Hence it appears that if only a small trace of $G\beta$ is found with lime and lanthanum, the green line is not only obliterated, but the quenching action suppresses that part of the continuous lime spectrum which has the same refrangibility as the $G\beta$ line, thus giving a black space in the spectrum.

A specimen of lanthana thrown out in a fractionation of

FIG. 22.

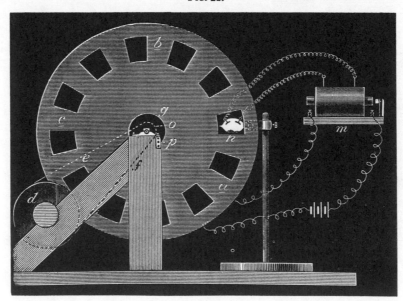

spectrum, after the fashion of yttria with a discontinuous spectrum. If, however, they are mixed, the phosphorescing energy of the lime does not extend over the whole spectrum, but concentrates itself on strengthening the yttria bands. These bands become broader, but also less definite, in proportion as the lime increases in quantity.

Lime also may be made to play a useful part in bringing out the phosphorescent bands of samaria. The bands are not so numerous as those of yttrium, but the contrasts are sharper. Examined with a somewhat broad slit, and disregarding the fainter bands, the samarium spectrum is seen to consist of three bright bands—red, orange, and green—nearly equidistant, the orange being brightest. With a narrower slit the orange and green bands are seen to be double. (Fig. 18).

* Address delivered by the President of the Chemical Society at the Annual General Meeting, March 21st, 1889.

didymia, examined in the radiant matter tube, phosphoresces of a yellow surface-colour and gives an extremely brilliant spectrum (Fig. 20). In the red there is a very fine and sharp line, somewhat like the alumina line, but nearer C of hydrogen. Then come a couple of misty red lines, the first apparently compound, then a hazy pair of green lines, and finally another wide apart pair of a bluish green; the first of these lines is intensely brilliant.

Interference of Phosphorescence Spectra.

I have already frequently noted the modification induced in the normal spectrum of one earth by the mixture of various quantities of others, when treated as anhydrous sulphates, and of these one of the most striking is seen in the case of a mixture of samaria with yttria, since the presence of even 40 per cent of yttria practically obliterates its spectrum. The most minute proportion of lime added to samaria causes the sharp line at $1/\lambda^2$ 269 in its spectrum to vanish, at the same time greatly intensifying the other bands. Strontia, baryta, glucina, thoria, magnesia, lanthana, alumina, and oxides of zinc, cadmium, lead, bismuth, and antimony, all give characteristic spectra with samaria, and have been fully described in my paper on that body (*Phil. Trans.*, Part II., 1885). The action of lime upon yttria is of great use in detecting very small quantities of this earth when in combination with other elements which otherwise would prevent its phosphorescence.

A recent discovery of some beautiful spectra given by the rare earths when their pure oxides are highly calcined, shows the remarkable changes produced in the spectra of these earths when two or more are observed in combination. It has likewise opened to me a wide field of investigation in the nature of the elements themselves. Alumina is especially active in inducing new spectra when mixed with the rare earths. I have given more than a twelvemonth to the exclusive study of alumina phosphorescence, and still the research is incomplete. But I have obtained some remarkable results. A moderate amount of fractionation has enabled me to penetrate below the surface of the red glow common to crude alumina, and to see traces of a most complicated sharp line spectrum. By pushing one particular process of fractionation to a considerable extent, I have obtained evidence of a body which is the cause of some of these lines. The spectrum described by me in 1887 (CHEM. NEWS, lvi., 62, 72) is one of great beauty, and a fair idea of it is given in Fig. 21. The new body is probably one of the rare elements or meta-elements closely connected with decipia, for I have reproduced the spectrum very fairly by adding decipia to alumina. Before arriving at definite conclusions much time must be devoted to the subject. Certain it is that this new earth is not yttria, erbia, samaria, didymia, lanthana, holmia, thulia, gadolinia, or ytterbia, the spectrum of each of these when mixed with alumina being very beautiful, but differing entirely from the decipia-alumina spectrum. Some of these new spectra are shown in Fig. 29.

The Phosphoroscope.

The phosphoroscope affords another method of verifying the simple or compound character of a substance. It is well known that the continuance of phosphorescence after the cessation of the exciting cause varies widely, from some hours, as in the case of the phosphorescent sulphides, to the fraction of a second in the case of uranium glass and quinine sulphate. On examining phosphorescent earths glowing in a vacuum tube under the action of the induction discharge, I found remarkable differences in the duration of this residual glow. Some of the earths, after the cessation of the current, remain luminous for an hour or more, whilst others cease to phosphoresce immediately on the stoppage of the current. Take the case of yttrium. As already stated, I succeeded in resolving this earth into several simpler bodies not equal in basicity. While seeking for further proof of the distinct character of these bodies I observed that the after-glow differed somewhat in colour from that which the earth exhibited whilst the current was still passing. Further, the spectrum of the after-glow seemed to show, so far as I could judge by the faint light, that some of the lines were missing. As this phenomenon indicated another difference among the components of yttrium, I examined them in an instrument similar to Becquerel's phosphoroscope, but acting electrically instead of by means of direct light.

Without describing this apparatus, shown in Fig. 22, I may mention some of the results obtained by its use. Under ordinary circumstances it is scarcely possible to perceive any phosphorescence in an earth until the vacuum is so high that the line spectrum of the residual gas begins to grow faint. Up to this point, the stronger light of the glowing gas overpowers the feeble glow of the phosphorescence. But in the phosphoroscope the light of the glowing gas lasts only for an inappreciable time, while that of the phosphorescent earth persists long enough to be distinctly observed. The different bands of the new constituents of yttria do not all appear at the same speed of rotation. At the lowest speed the double greenish blue band of Gβ is first seen, followed next by the dark blue band of Gα. As the velocity increases there follows the bright citron-yellow band of Gδ, and as the utmost speed approaches the red band of Gζ is seen, but with difficulty. If lanthanum sulphate along with a little lime is examined in the phosphoroscope the line of Gε is visible at the lowest speed; Gδ follows at an interval of 0·0035 second, and the Gα line immediately afterwards.

M. de Boisbaudran's Reversion Spectra.

Another modification of the phosphorescence process is afforded by the "reversion spectra" of M. Lecoq de Boisbaudran.

The following is the description of this process by M. Lecoq de Boisbaudran, read before the Academy of Sciences on June 8th, 1885 :—" When the electric spectrum of a solution *with a metallic base* is produced it is customary to make the outside platinum wire (whence the induction spark strikes) positive, the liquid consequently forming the negative pole. If the direction of the current is reversed the metallic rays (due to the free metal or to one of its compounds) are scarcely or not at all visible; at all events, so long as the exterior platinum wire now forming the negative pole is not coated with a deposit." Fig. 15 shows the arrangement. M. de Boisbaudran continues :—

" Having again taken up last year my researches on the rare earths belonging to the didymium and yttrium family, I had occasion to observe with many of my preparations the formation of spectrum bands, nebulous, but sometimes tolerably brilliant, having their origin in a thin layer of a beautiful green colour, which appeared at the surface of the liquid (a solution of a chloride) when it was rendered positive."

M. de Boisbaudran further adds :—" The production of my reversion spectrum appears to be analogous physically with the formation of the phosphorescence spectra obtained by Mr. Crookes at the negative pole in his high vacuum tubes containing certain compounds of yttria. The conditions of the two experiments are, however, practically speaking, very different."

By this method M. de Boisbaudran has discovered phosphorescent spectra, shown in Figs. 23 and 24, which he considers due to the presence of two earths, one of which, supposed to be new, he has provisionally named Zα, and another, also thought at first to be new, and therefore called Zβ, but since admitted by him to be terbia (*Compt. Rend.*, cviii., 167, Jan. 28, 1889). In the hands of so skilful an experimentalist as my accomplished friend, this method may give trustworthy indications, but the test is really beyond the range of practical analysis, owing to the difficulty of eliciting the phenomena. Unless the strength of the spark, the concentration and acidity of the solution, and the dispersive and magnifying power of the spectroscope bear a certain proportion to each other, the observer is likely to fail in seeing a spectrum even in solutions of earths which contain considerable quantities of Zα and terbia.

I have had not only the advantage of personal instruction in Paris from M. de Boisbaudran himself in the best methods of getting these reversion spectra, but I secured some of the identical earths which give these spectra most distinctly. Yet with all these advantages I have experimented for hours together without being able to see more than a feeble glimmer of the bands described by M. de Boisbaudran. Moreover, the bands of these "reversion" spectra, at their strongest, are but faint and hazy substitutes for the brilliant lines of yttria seen by the "bombardment" process, and they do not even in all cases agree with them in position. M. de Boisbaudran, speaking of the relative sensitiveness of our two processes, admits that the bombardment process *in vacuo* is incomparably more delicate than his reversion method. My own estimate of the relative sensitiveness of the two methods is in the proportion of at least 1 : 100. Though so accurate an observer, M. de Boisbaudran concluded apparently too hastily that two spectra are identical, when one of them has been measured only approximately, and contains bands in positions which are perfectly blank in the other.

The Phosphorescence of Alumina.

I now wish to draw attention to some recent researches on the phosphorescence spectrum given by alumina. So far back as 1859 Becquerel examined in his phosphoroscope pure alumina carefully prepared, and described it as glowing with a splendid red colour. He rendered his specimens phosphorescent by exposure to the sun, and made no use of the induction spark. As described by Becquerel (*Ann. Chem. Phys.*, lvii., 50, 1859), the spectrum of the red light emitted from alumina agrees with that of the ruby when submitted to the radiant matter test. It displays one intensely red line a little below the fixed line B in the spectrum, having a wave-length of about 689·5. There is a continuous spectrum beginning at about B and a few fainter lines beyond it, but in comparison with this red line the faint ones are so dim that they may be neglected. My latest observations in the vacuum tube prove this line to be double, the distance apart of the components being about half the distance separating the D lines (*Proc. Roy. Soc.*, xlii., 26, Dec. 30, 1886), their respective wave-lengths being 694·2 and 693·7 ($1/\lambda^2$ 207·5 and 207·8). This alumina spectrum is shown in Fig. 25.

The red phosphorescence of this alumina is exceedingly characteristic. M. de Boisbaudran (*Compt. Rend.*, ciii., 1107 ; civ., 330, 478, 554, 824) contends, however, that this red phosphorescence is due not to the alumina itself, but to an accompanying trace of chromium, $\frac{1}{1100}$th part of chromium being sufficient to give a splendid red phosphorescence, whilst even 1 part of chromic oxide in 10,000 will produce a very distinct rose colour. In testing this view I have purified alumina most carefully, so as to secure the absence of chromium, and on examining it in the radiant matter tube I have still obtained the characteristic phosphorescence and spectrum. I have then added to my purified alumina chromium in known varying proportions, but without finding any increase in the intensity of the phosphorescence. I fractionated my purified alumina by different methods and found that the substance which forms the crimson line becomes concentrated towards one end of the fractionations, whilst chromium concentrates at the other end. I have suggested four possible explanations of the phenomena—

1. The crimson line belongs to alumina, but it is liable to be masked or extinguished by some other earth, which accumulates towards one end of the fractionations.

2. The crimson line is not due to alumina, but to the presence of an accompanying earth, which accumulates towards the other end of the fractionations.

3. The crimson line belongs to alumina, but its development requires certain precautions to be taken in the duration and intensity of the ignition, and absolute freedom from alkaline and other bodies carried down by precipitated alumina, and difficult of removal by washing.

4. The earth alumina is a compound molecule, one only of its component sub-molecules giving the crimson line. If this hypothesis is correct, alumina must admit of being split up in a manner analogous to yttria.

Sharp Line Spectra with Phosphorescent Alumina.

About eighteen months ago M. de Boisbaudran published some results on "New Fluorescence with well-defined Spectral Rays" (*Compt. Rend.*, cv., 258). Having mixed alumina with 2 per cent of samaria and converted the mixture into sulphate, he heated it to a temperature

between the melting-point of copper and silver. This product, examined in the radiant matter tube, gives a faint fluorescence, the spectrum resembling the samarium-aluminium spectrum described by me in June, 1885 (*Phil. Trans.*, Part II., 1885, 712) (Fig. 26). That is, aluminium and samarium give a spectrum resembling the corresponding calcium-samarium one as to the red and the double orange, but having a very broad, rather faint, green band with a black division in the middle occupying the position of the bright green band of calcium-samarium.

On submitting this aluminium-samarium mixture to a very high temperature, M. de Boisbaudran finds that its spectrum alters greatly. In place of the three nebulous bands, there are now a number of sharp rays forming three groups corresponding respectively to each of the three diffused bands above described.

This spectrum seems to me closely to resemble the complicated system of sharp lines shown by alumina after moderate fractionation, to which I have already referred (Fig. 27, lower spectrum). Taking some of this alumina, I added to it one-fiftieth of its weight of samaria, and thereby obtained a spectrum similar to that described by M. de Boisbaudran (Fig. 27, upper spectrum). As I had suspected, the two spectra are almost identical ; the effect of the samaria is simply to intensify some of the lines and weaken others. But between this sharp line high temperature spectrum and the band spectrum (Fig. 26) given by the same mixture treated at a lower temperature I fail to see any such resemblance as could support the view that the groups correspond save that "those of the line spectra are less refrangible." The explanation of M. de Boisbaudran's result is simple. Both samarium sulphate and aluminium sulphate resist a red heat without their sulphuric acid being driven off. Aluminium sulphate does not phosphoresce, samarium sulphate does, therefore the mixture gives the samarium spectrum. But if the mixed sulphates are heated to the highest blow-pipe temperature both are decomposed, and there is left a mixture of samaria and alumina. Now, samaria by itself gives no phosphorescence spectrum, but alumina gives the new line spectrum I have described.

This method is applicable only when the earths are in a sufficiently high state of purity. The presence of exceedingly small traces of other matter may greatly modify the spectrum.

Conclusions.

During the course of the investigations—whose results are briefly summarised in the foregoing pages,—I have repeatedly had recourse to the balance, to ascertain how the atomic weights of the earths under treatment were varying. An atomic weight determination is valuable in telling when a stable molecular grouping is arrived at. During a fractionation, the atomic weight of the earth slowly rises or falls until it becomes stationary, after which no further fractionation of that lot *by the same process* makes it vary. Usually a result of this kind has been relied on as proof that the elementary stage has been reached. This constancy of atomic weight, however, only proves that the original body has been split up by the fractionating process into two molecular groupings capable of resisting further decomposition by that identical process ; but these groupings are not unlikely to break up when a different fractionating process is brought to bear on them, as I found in the separation of didymium and samarium when using dilute ammonia as the fractionating precipitant. In my paper on "Radiant Matter Spectroscopy" I said[*] :—"After a time a balance seems to be established between the affinities at work, when the earth would appear in the same proportion in the precipitate and the solution. At this stage they were thrown down by ammonia, and the precipitated earths set aside to be worked up by the fusion of their anhydrous nitrates so as to alter the ratio between them,

---

[*] Part II., Samarium," *Phil. Trans.*, Part II., 129, June 18, 1885.

when fractionation by ammonia could be again employed."

It is obvious that when the balance of affinities here spoken of was reached, the atomic weight of the mixture under treatment would have become constant, and no further fractionation would have caused the atomic weight to alter.

Atomic weight determinations are valuable in telling when the fractionating operation in use has effected all the separation it can: at this point it becomes constant. The true inference is, not that a new earth has been obtained, but simply that the fractionating operation requires changing for another, which will cleave the group of meta-elements in a different direction.

Meantime, I have kept strictly in view the question, What is an element, and how shall it be recognised when met ?

On this subject I beg to submit the following considerations, which, primarily referring to didymium, may at any moment apply to other cases :—

Neodymium and praseodymium are simply the products into which didymium is split up by one particular method of attack.

It must be remembered that a single operation, be it crystallisation, precipitation, fusion, partial solution, &c., can only separate a mixture of several bodies into two parts, just as the addition of a reagent only divides a mixture into two portions, a precipitate and a solution, and these divisions will be effected on different lines according to the reagent employed. We add, *e.g.* ammonia to a mixture, and at once get a separation into two parts. Or we add, say, oxalic acid to the same original solution, and we then split up the mixture into two other parts differently arranged.

Thus by crystallising didymium nitrate (in Auer's way) we divide the components into two parts. By fusing didymium nitrate we divide its components in a different way ; but so long as different methods of attack split up a body differently, it is evident that we have not yet got down to " bed rock."

Further, a compound molecule may easily act as an element. Take the case of didymium, which is certainly a compound, whether the products of Auer's operation be final or not. Didymium has a definite atomic weight ; it has well-defined salts, and has been subjected to the closest scrutiny by some of the ablest chemists in the world. I refer particularly to Clève's classical memoir. Still the compound molecule known as didymium was first too firmly held together to act otherwise than as an element, and as a seeming element it emerged from every trial. The simple operations to which it had been submitted in the preparation of its salts, and in its purification from other compound molecules, such as samarium and lanthanum, were not sufficient to split it up further. But subjected to a new method of attack it decomposes at once.

We have, in fact, a certain number of reagents, operations, processes, &c., in use. If a body resist all these and behave otherwise as a simple substance, we are apt to take it at its own valuation, and to call it an element. But for all that, it may, as we see, be compound, and as soon as a new and appropriate method of attack is devised we find it can be split up with comparative ease. Still, we must never forget that, however complex, it can hardly be resolved into more than two parts at one operation.

From considerations above laid down I do not feel in a position to recognise neodymium and praseodymium as elements. We need some criterion for an element which shall appeal to our reason more clearly than the old untrustworthy characteristic of having not as yet been decomposed, and to this point I must beg to call the special attention of my colleagues. It may be that whatever body gives only one absorption-band is an element, but we cannot conversely say that an element may be known by its giving only one absorption-band, since most of our elements give no bands at all !

Until these important and difficult questions can be decided, I have preferred to open what may be figuratively called a suspense account, wherein, as I have previously suggested, we may provisionally enter all these doubtful bodies as " meta-elements."

But these meta-elements may have more than a mere provisional value. Besides compounds, we have hitherto recognised merely ultimate atoms, or the aggregations of such atoms into simple molecules. But it becomes more and more probable that between the atom and the compound we have a gradation of molecules of different ranks, which, as we have seen, may pass for simple elementary bodies. It might be the easier plan, so soon as a constituent of these earths can be found to be chemically and spectroscopically distinguishable from its next of kin to give it a name, and to claim for it elemental rank ; but it seems to me the duty of a man of science to treat every subject, not in the manner which may earn for him the greatest temporary κυδος, but in that which will be of most service to Science.

If the study of the rare earths leads us to clearer views on the nature of the elements, neither my colleagues nor myself will, I am sure, regret the months spent in tedious and apparently wearisome fractionations. No one can be more conscious than myself how much ground is yet uncovered, and how many radical questions have received but very inadequate answers. But we can only work on, " unresting, unhasting," trusting that in the end our work will throw some white light upon this deeply interesting department of chemical physics.

EXPERIMENTS IN RADIO-ACTIVITY, AND THE PRODUCTION OF HELIUM FROM RADIUM.*

By Sir WILLIAM RAMSAY, K.C.B., F.R.S.,
and
FREDERICK SODDY, M.A.

1. *Experiments on the Radio-activity of the Inert Gases of the Atmosphere.*

OF recent years many investigations have been made by Elster and Geitel, Wilson, Strutt, Rutherford, Cooke, Allen, and others, on the spontaneous ionisation of the gases of the atmosphere, and on the excited radio-activity obtainable from it. It became of interest to ascertain whether the inert monatomic gases of the atmosphere bear any share in these phenomena. For this purpose a small electroscope contained in a glass tube of about 20 c.c. capacity, covered in the interior with tin-foil, was employed. After charging, the apparatus, if exhausted, retained its charge for thirty-six hours without diminution. Admission of air caused a slow discharge. In similar experiments with helium, neon, argon, krypton, and xenon, the last mixed with oxygen, the rate of discharge was proportional to the density and pressure of the gas. This shows that the gases have no special radio-activity of their own, and accords with the explanation already advanced by these investigators, that the discharging power of the air is caused by extraneous radio-activity.

Experiments were also made with the dregs left after liquefied air had nearly entirely evaporated, and again with the same result ; no increase in discharging power is produced by concentration of a possible radio-active constituent of the atmosphere.

2. *Experiments on the Nature of the Radio-active Emanation from Radium.*

The word emanation originally used by Boyle (" substantial emanations from the celestial bodies ") was resuscitated by Rutherford to designate definite substances of a gaseous nature, continuously produced from other substances. The term was also used by Russell (" emanation from hydrogen peroxide ") in much the same sense. If the adjective " radio-active " be added, the phenomenon of Rutherford is distinguished from the phenomena observed by Russell. In this section we are dealing with the emanation, or radio-active gas obtained from radium. Rutherford and Soddy investigated the chemical nature of the thorium emanation (*Phil. Mag.*, 1902, p. 580), and of the radium emanation (*Phil. Mag.*, 1903, p. 457), and came to the conclusion that these emanations are inert gases which withstand the action of reagents in a manner hitherto unobserved, except with the members of the argon family. This conclusion was arrived at because the emanations from thorium and radium could be passed without alteration over platinum and palladium black, chromate of lead, zinc dust, and magnesium powder, all at a red heat.

We have since found that the radium emanation withstands prolonged sparking with oxygen over alkali, and also, during several hours, the action of a heated mixture of magnesium powder and lime. The discharging power was maintained unaltered after this treatment, and inasmuch as a considerable amount of radium was employed, it was possible to use the self-luminosity of the gas as an optical demonstration of its persistence.

In an experiment in which the emanation mixed with oxygen had been sparked for several hours over alkali, a minute fraction of the total mixture was found to discharge an electroscope almost instantly. From the main quantity of the gas, the oxygen was withdrawn by ignited phosphorus, and no visible residue was left. When, however, another gas was introduced, so as to come into contact with the top of the tube, and then withdrawn, the emanation was found to be present in it in unaltered amount. It appears, therefore, that phosphorus burning in oxygen and sparking with oxygen have no effect upon the gas so far as can be detected by its radio-active properties.

* A Paper communicated to the Royal Society, July 28, 1903.

The experiments with magnesium-lime were more strictly quantitative. The method of testing the gas before and after treatment with the reagent was to take 1/2000 part of the whole mixed with air, and after introducing it into the reservoir of an electroscope to measure the rate of discharge. The magnesium-lime tube glowed brightly when the mixture of emanation and air was admitted, and it was maintained at a red heat for three hours. The gas was then washed out with a little hydrogen, diluted with air and tested as before. It was found that the discharging power of the gas had been quite unaltered by this treatment.

The emanation can be dealt with as a gas ; it can be extracted by aid of a Töpler pump ; it can be condensed in a U-tube surrounded by liquid air, and when condensed it can be " washed " with another gas, which can be pumped off completely, and which then possesses no luminosity and practically no discharging power. The passage of the emanation from place to place through glass tubes can be followed by the eye in a darkened room. On opening a stopcock between a tube containing the emanation and the pump, the slow flow through the capillary tube can be noticed ; the rapid passage along the wider tubes, the delay caused by the plug of phosphorus pentoxide, and the sudden diffusion into the reservoir of the pump. When compressed, the luminosity increased, and when the small bubble was expelled through the capillary it was exceedingly luminous. The peculiarities of the excited activity left behind on the glass by the emanation could also be well observed. When the emanation had been left a short time in contact with the glass, the excited activity lasts only for a short time ; but after the emanation has been stored a long time the excited activity decays more slowly.

The emanation causes chemical change in a similar manner to the salts of radium themselves. The emanation pumped off from 50 m.grms. of radium bromide after dissolving in water, when stored with oxygen in a small glass tube over mercury, turns the glass distinctly violet in a single night ; if moist, the mercury becomes covered with a film of the red oxide, but if dry, it appears to remain unattacked. A mixture of the emanation with oxygen produces carbon dioxide when passed through a lubricated stopcock.

3. *Occurrence of Helium in the Gases Evolved from Radium Bromide.*

The gas evolved from 20 m.grms. of pure radium bromide (which we are informed had been prepared three months) by its solution in water, and which consisted mainly of hydrogen and oxygen (*cf.* Giesel, *Ber.*, 1903, 347) was tested for helium, the hydrogen and oxygen being removed by contact with a red-hot spiral of copper wire, partially oxidised, and the resulting water vapour by a tube of phosphorus pentoxide. The gas issued into a small vacuum tube which showed the spectrum of carbon dioxide. The vacuum tube was in train with a small U-tube, and the latter was then cooled with liquid air. This much reduced the brilliancy of the CO_2 spectrum, and the D_3 line of helium appeared. The coincidence was confirmed by throwing the spectrum of helium into the spectroscope through the comparison prism, and shown to be at least within 0·5 of an Angström unit.

The experiment was carefully repeated in apparatus constructed of previously unused glass with 30 m.grms. of radium bromide, probably four or five months old, kindly lent us by Prof. Rutherford. The gases evolved were passed through a cooled U-tube on their way to the vacuum tube, which completely prevented the passage of carbon dioxide and the emanation. The spectrum of helium was obtained, and practically all the lines were seen, including those at 6677, 5876, 5016, 4932, 4713, and 4472. There were also present three lines of approximate wave-lengths, 6180, 5695, 5455, that have not yet been identified.

On two subsequent occasions, the gases evolved from both solutions of radium bromide were mixed, after four days' accumulation, which amounted to about 2·5 c.c. in each case, and were examined in a similar way. The D_3 line of helium could not be detected. It may be well to state the composition found for the gases continuously generated by a solution of radium, for it seemed likely that the large excess of hydrogen over the composition required to form water, shown in the analysis given by Bödlander (*Ber.*, *loc. cit.*) might be due to the greater solubility of the oxygen. In our analyses the gases were extracted with the pump, and the first gave 28·6, the second 29·2 per cent of oxygen. The slight excess of hydrogen is doubtless due to the action of the oxygen on the grease of the stopcocks, which has already been mentioned. The rate of production of these gases is about 0·5 c.c. per day for 50 m.grms. of radium bromide, which is over twice as great as that found by Bödlander.

4. *Production of Helium by the Radium Emanation.*

The maximum amount of the emanation obtained from 50 m.grms. of radium bromide was conveyed by means of oxygen into a U-tube cooled in liquid air, and the latter was then extracted by the pump. It was then washed out with a little fresh oxygen, which was again pumped off. The vacuum tube sealed on to the U-tube, after removing the liquid air, showed no trace of helium. The spectrum was apparently a new one, probably that of the emanation, but this has not yet been completely examined, and we hope to publish further details shortly. After standing from the 17th to the 21st inst., the helium spectrum appeared, and the characteristic lines were observed identical in position with those of a helium tube thrown into the field of vision at the same time. On the 22nd, the yellow, the green, the two blues, and the violet, were seen, and in addition the three new lines also present in the helium obtained from radium. A confirmatory experiment gave identical results.

We wish to express our indebtedness to the Research Fund of the Chemical Society for a part of the radium used in this investigation.

PROF. A. LADENBURG.

THE death occurred at Breslau, on August 15, of Dr. Albert Ladenburg, professor of chemistry in the University of Breslau. Dr. Ladenburg was born at Mannheim in 1842, and graduated as doctor of philosophy in 1863. In 1873 he accepted an invitation to take up a position as professor of chemistry and director of the laboratory at Kiel. In 1886 the honorary degree of doctor of medicine of Berne University was conferred on Dr. Ladenburg in recognition of his scientific investigations, and British and other societies, including the Pharmaceutical Society of Great Britain, also honoured him with honorary membership. He was also awarded the Hanbury gold medal for his services in the promotion of research on the chemistry of drugs. It was in 1889 that Dr. Ladenburg took up the post of professor of chemistry at Breslau, and he occupied the office with very great success.

Ladenburg's name is best known by his synthetic work on the production of homatropine. On splitting up atropine, tropic acid and tropine can be formed as derivatives; the latter Ladenburg combined with amygdalic acid to form a compound which is converted into oxy-toluyl-tropeine, or homatropine, an artificial alkaloid which, with its salts, has proved of the greatest service in ophthalmic surgery. His mathematical method of treating synthetic formulæ, and his prismatic benzene ring, place him in the first rank of chemists as a theorist; while as to his practical work, his list of communications to scientific societies and literature in this country and elsewhere includes articles on "The Valency of Nitrogen," on "Synthetic Alkaloids," on "The Relationship between Hyoscyamine and Atropine and the Conversion of the one Alkaloid into the other," on "Hyoscine," on "The Mydriatic Alkaloids occurring in Nature," on "The Synthesis of Coniine," and on "The History and Constitution of Atropine," in addition to the compilation with other collaborators of a dictionary ("Handwörterbuch der Chemie"), consisting of thirteen volumes dealing with inorganic and organic chemistry.

THE BRITISH ASSOCIATION AT PORTSMOUTH.

BY the time this issue reaches the readers of NATURE the eighty-first meeting of the British Association will have been inaugurated at Portsmouth, and, given fair weather conditions, we trust it will be a useful and enjoyable gathering. Judging from the number of distinguished men of science who have expressed their intention of being present, the meeting should be of importance as regards its scientific work, as well as successful from a social point of view.

The reception-room is the large Connaught Drill Hall, which appears to be ideal for that purpose. It gives under one roof a large reception hall with post office, telephone, &c., and a comfortably furnished reading and writing room for the members. In addition to this there is also a small room set apart for the use of ladies.

In point of view of numbers, the Portsmouth meeting may not reach that of Sheffield last year, but this is accounted for partly by the absence of any special industry attached to the town, and also may, to some extent, be due to the absence of any university or university college. Most of the accommodation available is, however, booked, and those who arrive late may have difficulty in finding quarters.

The meeting rooms are a little scattered, but this was unavoidable, and notices will be displayed making the routes to be taken to the various section-rooms easy to find.

In passing, mention may be made of a convenient plan for communication between members of the association. It is a box which will be placed in the reception-room, into which notes may be dropped addressed to other members. This box will be frequently cleared, and the notes delivered on request to those to whom they are written.

The pleasures of the meeting commence to-day (Thursday), when at 2.30 a party will be taken over the dockyard and battleships. A garden-party is to be given this afternoon by Sir John and Lady Brickwood at their beautiful residence in the town. In the evening the Mayor will give a reception at the South Parade Pier, which is the property of the Corporation.

On Friday afternoon there will be a special visit to the new filtration works of the Borough of Portsmouth Water Company, and Saturday will be entirely devoted to all-day excursions, including two to the Isle of Wight, and three drives in the South Downs, starting from Chichester, to which city there will be a special train. The drives are to (1) Kingly Vale, West Dean, and Goodwood; (2) Boxgrove Priory and Arundel Castle; (3) Bignor (with the Roman remains) and Parham Park.

On Sunday the Bishop of Winchester is to preach at the Portsea parish church, and on Tuesday the Mayor will entertain the members at a garden-party. In addition, the naval authorities have organised a naval display in Stokes' Bay, consisting of an attack by torpedo-boat destroyers and submarines. Visitors should not neglect a visit to the old *Victory*, one of the most interesting "links with the past" in existence, and a full description of which, written by Mr. W. L. Wyllie, R.A., will be found in an interesting little handbook to Portsmouth which will be presented to members.

INAUGURAL ADDRESS BY PROF. SIR. WILLIAM RAMSAY, K.C.B., PH.D., LL.D., D.SC., M.D., F.R.S., PRESIDENT.

IT is now eighty years since this Association first met at York, under the presidency of Earl Fitzwilliam. The object of the Association was then explicitly stated :—" To give a stronger impulse and a more systematic direction to scientific inquiry, to promote the intercourse of those who cultivate science in different parts of the British Empire with one another and with foreign philosophers, to obtain a more general attention to the objects of science and a removal of any disadvantages of a public kind which impede its progress."

In 1831 the workers in the domain of science were relatively few. The Royal Society, which was founded by Dr. Willis, Dr. Wilkins, and others, under the name of the " Invisible, or Philosophical College," about the year 1645, and which was incorporated in December, 1660, with the approval of King Charles II., was almost the only meeting-place for those interested in the progress of science ; and its Philosophical Transactions, begun in March, 1664–5, almost the only medium of publication. Its character was described in the following words of a contemporary poem :—

> " This noble learned Corporation
> Not for themselves are thus combined
> To prove all things by demonstration,
> But for the public good of the nation,
> And general benefit of mankind."

The first to hive off from the Royal Society was the Linnean Society for the promotion of botanical studies, founded in 1788 by Sir James Edward Smith, Sir Joseph Banks, and other Fellows of the Royal Society; in 1807 it was followed by the Geological Society; at a later date the Society of Antiquaries, the Chemical, the Zoological, the Physical, the Mathematical, and many other Societies were founded. And it was felt by those capable of forming a judgment that, as well expressed by Lord Playfair at Aberdeen in 1885, " Human progress is so identified with scientific thought, both in its conception and realisation, that it seems as if they were alternative terms in the history of civilisation." This is only an echo through the ages of an utterance of the great Englishman, Roger Bacon, who wrote in 1250 A.D.: " Experimental science has three great prerogatives over all other sciences: it verifies conclusions by direct experiment; it discovers truths which they could never reach; and it investigates the secrets of Nature, and opens to us a knowledge of the past and of the future."

The world has greatly changed since 1831; the spread of railways and the equipment of numerous lines of steamships have contributed to the peopling of countries at that time practically uninhabited. Moreover, not merely has travelling been made almost infinitely easier, but communication by post has been enormously expedited and cheapened; and the telegraph, the telephone, and wireless telegraphy have simplified as well as complicated human existence. Furthermore, the art of engineering has made such strides that the question " Can it be done? " hardly arises, but rather " Will it pay to do it? " In a word, the human race has been familiarised with the applications of science; and men are ready to believe almost anything if brought forward in its name.

Education, too, in the rudiments of science has been introduced into almost all schools; young children are taught the elements of physics and chemistry. The institution of a Section for Education in our Association (L) has had for its object the organising of such instruction, and much useful advice has been proffered. The problem is, indeed, largely an educational one; it is being solved abroad in various ways—in Germany and in most European States by elaborate Governmental schemes dealing with elementary and advanced instruction, literary, scientific, and technical; and in the United States and in Canada by the far-sightedness of the people: both employers and employees recognise the value of training and of originality, and on both sides sacrifices are made to ensure efficiency.

In England we have made technical education a local, not an Imperial, question; instead of half a dozen first-rate institutions of University rank, we have a hundred, in which the institutions are necessarily understaffed, in which the staffs are mostly overworked and underpaid; and the training given is that, not for captains of industry, but for workmen and foremen. " Efficient captains cannot be replaced by a large number of fairly good corporals." Moreover, to induce scholars to enter these institutions, they are bribed by scholarships, a form of pauperism practically unknown in every country but our own; and to crown the edifice, we test results by examinations of a kind not adapted to gauge originality and character (if, indeed, these can ever be tested by examination), instead of, as on the Continent and in America, trusting the teachers to form an honest estimate of the capacity and ability of each student, and awarding honours accordingly.

The remedy lies in our own hands. Let me suggest that we exact from all gainers of University scholarships an undertaking that, if and when circumstances permit, they will repay the sum which they have received as a scholarship, bursary, or fellowship. It would then be possible for an insurance company to advance a sum representing the capital value, viz. 7,464,931l., of the scholarships, reserving, say, twenty per cent. for non-payment, the result of mishap or death. In this way a sum of over six million pounds, of which the interest is now expended on scholarships, would be available for University purposes. This is about one-fourth of the sum of twenty-four millions stated by Sir Norman Lockyer at the Southport meeting as necessary to place our University education on

a satisfactory basis. A large part of the income of this sum should be spent in increasing the emoluments of the chairs; for, unless the income of a professor is made in some degree commensurate with the earnings of a professional man who has succeeded in his profession, it is idle to suppose that the best brains will be attracted to the teaching profession. And it follows that unless the teachers occupy the first rank, the pupils will not be stimulated as they ought to be.

Again, having made the profession of a teacher so lucrative as to tempt the best intellects in the country to enter it, it is clear that such men are alone capable of testing their pupils. The modern system of " external examinations," known only in this country, and answerable for much of its lethargy, would disappear; schools of thought would arise in all subjects, and the intellectual as well as the industrial prosperity of our nation would be assured. As things are, can we wonder that as a nation we are not scientific? Let me recommend those of my hearers who are interested in the matter to read a recent report on Technical Education by the Science Guild.

I venture to think that, in spite of the remarkable progress of science and of its applications, there never was a time when missionary effort was more needed. Although most people have some knowledge of the results of scientific inquiry, few, very few, have entered into its spirit. We all live in hope that the world will grow better as the years roll on. Are we taking steps to secure the improvement of the race? I plead for recognition of the fact that progress in science does not only consist in accumulating information which may be put to practical use, but in developing a spirit of prevision, in taking thought for the morrow; in attempting to forecast the future, not by vague surmise, but by orderly marshalling of facts, and by deducing from them their logical outcome; and chiefly in endeavouring to control conditions which may be utilised for the lasting good of our people. We must cultivate a belief in the " application of trained intelligence to all forms of national activity."

The Council of the Association has had under consideration the formation of a Section of Agriculture. For some years this important branch of applied science, borrowing as it does from botany, from physics, from chemistry, and from economics, has in turn enjoyed the hospitality of each of these sections, itself having been made a sub-section of one of these more definite sciences. It is proposed this year to form an Agricultural Section. Here there is need of missionary effort; for our visits to our colonies have convinced many of us that much more is being done for the farmer in the newer parts of the British Empire than at home. Agriculture is, indeed, applied botany, chemistry, entomology, and economics, and has as much right to independent treatment as has engineering, which may be strictly regarded as applied physics.

The question has often been debated whether the present method of conducting our proceedings is the one best adapted to gain our ends. We exist professedly " to give a stronger impulse and a more systematic direction to scientific inquiry." The Council has had under consideration various plans framed with the object of facilitating our work, and the result of its deliberations will be brought under your attention at a later date. To my mind, the greatest benefit bestowed on science by our meetings is the opportunity which they offer for friendly and unrestrained intercourse, not merely between those following different branches of science, but also with persons who, though not following science professionally, are interested in its problems. Our meetings also afford an opportunity for younger men to make the acquaintance of older men. I am afraid that we who are no longer in the spring of our lifetime, perhaps from modesty, perhaps through carelessness, often do not sufficiently realise how stimulating to a young worker a little sympathy can be; a few words of encouragement go a long way. I have in my mind words which encouraged me as a young man, words spoken by the leaders of Associations now long past—by Playfair, by Williamson, by Frankland, by Kelvin, by Stokes, by Francis Galton, by Fitzgerald, and many others. Let me suggest to my older scientific colleagues that they should not let such pleasant opportunities slip.

Since our last meeting the Association has to mourn the

loss by death of many distinguished members. Among these are :—

Dr. John Beddoe, who served on the Council from 1870 to 1875, has recently died at a ripe old age, after having achieved a world-wide reputation by his magnificent work in the domain of anthropology.

Sir Rubert Boyce, called away at a comparatively early age in the middle of his work, was for long a colleague of mine at University College, and was one of the staff of the Royal Commission on Sewage Disposal. The service he rendered science in combating tropical diseases is well known.

Sir Francis Galton died at the beginning of the year at the advanced age of eighty-nine. His influence on science has been characterised by Prof. Karl Pearson in his having maintained the idea that exact quantitative methods could —nay, must—be applied to many branches of science which had been held to be beyond the field of either mathematical or physical treatment. Sir Francis was General Secretary of this Association from 1863 to 1868 ; he was President of Section E in 1862, and again in 1872 ; he was President of Section H in 1885 ; but, although often asked to accept the office of President of the Association, his consent could never be obtained. Galton's name will always be associated with that of his friend and relative, Charles Darwin, as one of the most eminent and influential of English men of science.

Prof. Thomas Rupert Jones, also, like Galton, a member of this Association since 1860, and in 1891 President of the Geological Section, died in April last at the advanced age of ninety-one. Like Dr. Beddoe, he was a medical man with wide scientific interests. He became a distinguished geologist, and for many years edited the Quarterly Journal of the Geological Society.

Prof. Story Maskelyne, at one time a diligent frequenter of our meetings, and a member of the Council from 1874 to 1880, was a celebrated mineralogist and crystallographer. He died at the age of eighty-eight. The work which he did in the University of Oxford and at the British Museum is well known. In his later life he entered Parliament.

Dr. Johnstone Stoney, President of Section A in 1897, died on July 1, in his eighty-sixth year. He was one of the originators of the modern view of the nature of electricity, having given the name " electron " to its unit as far back as 1874. His investigations dealt with spectroscopy and allied subjects, and his philosophic mind led him to publish a scheme of ontology which, I venture to think, must be acknowledged to be the most important work which has ever been done on that difficult subject.

Among our corresponding members we have lost Prof. Bohr, of Copenhagen ; Prof. Brühl, of Heidelberg ; Hofrat Dr. Caro, of Berlin ; Prof. Fittig, of Strassburg ; and Prof. van 't Hoff, of Berlin. I cannot omit to mention that veteran of science Prof. Cannizzaro, of Rome, whose work in the middle of last century placed chemical science on the firm basis which it now occupies.

I knew all these men, some of them intimately ; and, if I have not ventured on remarks as to their personal qualities, it is because it may be said of all of them that they fought a good fight and maintained the faith that only by patient and unceasing scientific work is human progress to be hoped for.

It has been the usual custom of my predecessors in office either to give a summary of the progress of science within the past year or to attempt to present in intelligible language some aspect of the science in which they have themselves been engaged. I possess no qualifications for the former course, and I therefore ask you to bear with me while I devote some minutes to the consideration of ancient and modern views regarding the chemical elements. To many in my audience part of my story will prove an oft-told tale ; but I must ask those to excuse me, in order that it may be in some wise complete.

In the days of the early Greeks the word " element " was applied rather to denote a property of matter than one of its constituents. Thus, when a substance was said to contain fire, air, water, and earth (of which terms a childish game doubtless once played by all of us is a relic), it probably meant that they partook of the nature of the so-called elements. Inflammability showed the presence of concealed fire ; the escape of " airs " when some substances are heated or when vegetable or animal matter is distilled no doubt led to the idea that these airs were imprisoned in the matters from which they escaped ; hardness and permanence were ascribed to the presence of earth, while liquidity and fusibility were properties conveyed by the presence of concealed water. At a later date the " Spagyrics " added three " hypostatical principles " to the quadrilateral ; these were " salt," " sulphur," and " mercury." The first conveyed solubility, and fixedness in fire ; the second, inflammability ; and the third, the power which some substances manifest of producing a liquid, generally termed " phlegm," on application of heat, or of themselves being converted into the liquid state by fusion.

It was Robert Boyle, in his " Skeptical Chymist," who first controverted these ancient and mediæval notions, and who gave to the word " element " the meaning that it now possesses—the constituent of a compound. But in the middle of the seventeenth century chemistry had not advanced far enough to make his definition useful, for he was unable to suggest any particular substance as elementary. And, indeed, the main tenet of the doctrine of " phlogiston," promulgated by Stahl in the eighteenth century, and widely accepted, was that all bodies capable of burning or of being converted into a " calx," or earthy powder, did so in virtue of the escape of a subtle fluid from their pores ; this fluid could be restored to the " calces " by heating them with other substances rich in phlogiston, such as charcoal, oil, flour, and the like. Stahl, however false his theory, had at least the merit of having constructed a reversible chemical equation :—

$$\text{Metal} - \text{phlogiston} = \text{Calx} \; ; \quad \text{Calx} + \text{phlogiston} = \text{Metal}.$$

It is difficult to say when the first element was known to be an element. After Lavoisier's overthrow of the phlogistic hypothesis, the part played by oxygen, then recently discovered by Priestley and Scheele, came prominently forward. Loss of phlogiston was identified with oxidation ; gain of phlogiston with loss of oxygen. The scheme of nomenclature (" Méthode de Nomenclature chimique "), published by Lavoisier in conjunction with Guyton de Morveau, Berthollet, and Fourcroy, created a system of chemistry out of a wilderness of isolated facts and descriptions. Shortly after, in 1789, Lavoisier published his " Traité de Chimie," and in the preface the words occur : " If we mean by ' elements ' the simple and indivisible molecules of which bodies consist, it is probable that we do not know them ; if, on the other hand, we mean the last term in analysis, then every substance which we have not been able to decompose is for us an element ; not that we can be certain that bodies which we regard as simple are not themselves composed of two or even a larger number of elements, but because these elements can never be separated, or rather, because we have no means of separating them, they act, so far as we can judge, as elements ; and we cannot call them ' simple ' until experiment and observation shall have furnished a proof that they are so."

The close connection between " crocus of Mars " and metallic iron, the former named by Lavoisier " oxyde de fer," and similar relations between metals and their oxides, made it likely that bodies which reacted as oxides in dissolving in acids and forming salts must also possess a metallic substratum. In October, 1807, Sir Humphry Davy proved the correctness of this view for soda and potash by his famous experiment of splitting these bodies by a powerful electric current into oxygen and hydrogen, on the one hand, and the metals sodium and potassium on the other. Calcium, barium, strontium, and magnesium were added to the list as constituents of the oxides, lime, barytes, strontia, and magnesia. Some years later Scheele's " dephlogisticated marine acid," obtained by heating pyrolusite with " spirit of salt," was identified by Davy as in all likelihood elementary. His words are : " All the conclusions which I have ventured to make respecting the undecompounded nature of oxymuriatic gas are, I conceive, entirely confirmed by these new facts." " It has been judged most proper to suggest a name founded upon one of its obvious and characteristic properties, its colour, and to call it chlorine." The subsequent discovery of iodine by Courtois in 1812, and of bromine by Balard in 1826, led to the inevitable conclusion that fluorine, if isolated, should resemble the other halogens in properties, and much later, in the able hands of Moissan, this was shown to be true.

The modern conception of the elements was much strengthened by Dalton's revival of the Greek hypothesis of the atomic constitution of matter, and the assigning to each atom a definite weight. This momentous step for the progress of chemistry was taken in 1803; the first account of the theory was given to the public, with Dalton's consent, in the third edition of Thomas Thomson's " System of Chemistry " in 1807; it was subsequently elaborated in the first volume of Dalton's own " System of Chemical Philosophy," published in 1808. The notion that compounds consisted of aggregations of atoms of elements united in definite or multiple proportions, familiarised the world with the conception of elements as the bricks of which the Universe is built. Yet the more daring spirits of that day were not without hope that the elements themselves might prove decomposable. Davy, indeed, went so far as to write in 1811: " It is the duty of the chemist to be bold in pursuit; he must recollect how contrary knowledge is to what appears to be experience. . . . To inquire whether the elements be capable of being composed and decomposed is a grand object of true philosophy." And Faraday, his great pupil and successor, at a later date, 1815, was not behind Davy in his aspirations when he wrote: " To decompose the metals, to re-form them, and to realise the once absurd notion of transformation—these are the problems now given to the chemist for solution."

Indeed, the ancient idea of the unitary nature of matter was in those days held to be highly probable. For attempts were soon made to demonstrate that the atomic weights were themselves multiples of that of one of the elements. At first the suggestion was that oxygen was the common basis; and later, when this supposition turned out to be untenable, the claims of hydrogen were brought forward by Prout. The hypothesis was revived in 1842, when Liebig and Redtenbacher, and subsequently Dumas, carried out a revision of the atomic weights of some of the commoner elements, and showed that Berzelius was in error in attributing to carbon the atomic weight 12.25 instead of 12.00. Of recent years a great advance in the accuracy of the determinations of atomic weights has been made, chiefly owing to the work of Richards and his pupils, of Gray, and of Guye and his collaborators, and every year an international committee publishes a table in which the most probable numbers are given on the basis of the atomic weight of oxygen being taken as sixteen. In the table for 1911, of eighty-one elements, no fewer than forty-three have recorded atomic weights within one-tenth of a unit above or below an integral number. My mathematical colleague, Karl Pearson, assures me that the probability against such a condition being fortuitous is 20,000 millions to one.

The relation between the elements has, however, been approached from another point of view. After preliminary suggestions by Döbereiner, Dumas, and others, John Newlands in 1862 and the following years arranged the elements in the numerical order of their atomic weights, and published in *The Chemical News* of 1863 what he termed his law of octaves—that every eighth element, like the octave of a musical note, is in some measure a repetition of its forerunner. Thus, just as C on the third space is the octave of C below the line, so potassium, in 1863 the eighth known element numerically above sodium, repeats the characters of sodium, not only in its physical properties—colour, softness, ductility, malleability, &c.— but also in the properties of its compounds, which, indeed, resemble each other very closely. The same fundamental notion was reproduced at a later date, and independently, by Lothar Meyer and Dmitri Mendeléeff; and to accentuate the recurrence of such similar elements in *periods*, the expression " the periodic system of arranging the elements " was applied to Newlands' arrangement in octaves. As everyone knows, by help of this arrangement Mendeléeff predicted the existence of then unknown elements, under the names of eka-boron, eka-aluminium, and eka-silicon, since named *scandium, gallium*, and

germanium, by their discoverers, Cleve, Lecoq de Boisbaudran, and Winckler.

It might have been supposed that our knowledge of the elements was practically complete; that perhaps a few more might be discovered to fill the outstanding gaps in the periodic table. True, a puzzle existed, and still exists, in the classification of the " rare earths," oxides of metals occurring in certain minerals; these metals have atomic weights between 139 and 180, and their properties preclude their arrangement in the columns of the periodic table. Besides these, the discovery of the inert gases of the atmosphere, of the existence of which Johnstone Stoney's spiral curve, published in 1888, pointed a forecast, joined the elements like sodium and potassium, strongly electronegative, to those like fluorine and chlorine, highly electropositive, by a series of bodies electrically as well as chemically inert; and neon, argon, krypton, and xenon formed links between fluorine and sodium, chlorine and potassium, bromine and rubidium, and iodine and cæsium.

Including the inactive gases, and adding the more recently discovered elements of the rare earths, and radium, of which I shall have more to say presently, there are eighty-four definite elements, all of which find places in the periodic table if merely numerical values be considered. Between lanthanum, with atomic weight 139, and tantalum, 181, there are in the periodic table seventeen spaces; and although it is impossible to admit, on account of their properties, that the elements of the rare earths can be distributed in successive columns (for they all resemble lanthanum in properties), yet there are now fourteen such elements; and it is not improbable that other three will be separated from the complex mixture of their oxides by further work. Assuming that the metals of the rare earths fill these seventeen spaces, how many still remain to be filled? We will take for granted that the atomic weight of uranium, 238.5, which is the highest known, forms an upper limit not likely to be surpassed. It is easy to count the gaps; there are eleven.

But we are confronted by an *embarras de richesse*. The discovery of radio-activity by Henri Becquerel, of radium by the Curies, and the theory of the disintegration of the radio-active elements, which we owe to Rutherford and Soddy, have indicated the existence of no fewer than twenty-six elements hitherto unknown. To what places in the periodic table can they be assigned?

But what proof have we that these substances are elementary? Let us take them in order.

Beginning with radium, its salts were first studied by Madame Curie; they closely resemble those of barium— sulphate, carbonate, and chromate insoluble; chloride and bromide similar in crystalline form to chloride and bromide of barium; metal, recently prepared by Madame Curie, white, attacked by water, and evidently of the type of barium. The atomic weight, too, falls into its place; as determined by Madame Curie and by Thorpe, it is 89.5 units higher than that of barium; in short, there can be no doubt that radium fits the periodic table, with an atomic weight of about 226.5. It is an undoubted element.

But it is a very curious one. For it is *unstable*. Now, stability was believed to be the essential characteristic of an element. Radium, however, disintegrates—that is, changes into other bodies, and at a constant rate. If 1 gram of radium is kept for 1760 years, only half a gram will be left at the end of that time; half of it will have given other products. What are they? We can answer that question. Rutherford and Soddy found that it gives a condensable gas, which they named " radium emanation "; and Soddy and I, in 1903, discovered that, in addition, it evolves helium, one of the inactive series of gases, like argon. Helium is an undoubted element, with a well-defined spectrum; it belongs to a well-defined series. And radium emanation, which was shown by Rutherford and Soddy to be incapable of chemical union, has been liquefied and solidified in the laboratory of University College, London; its spectrum has been measured, and its density determined. From the density the atomic weight can be calculated, and it corresponds with that of a congener of argon, the whole series being: helium, 4; neon, 20; argon, 40; krypton, 83; xenon, 130; unknown, about 178; and niton (the name proposed for the emanation to recall its connection with its congeners and its phosphorescent properties), about 222·4. The formation of niton from radium would therefore be represented by the equation : radium (226·4)=helium (4)+niton (222·4).

Niton, in its turn, disintegrates, or decomposes, and at a rate much more rapid than the rate of radium ; half of it has changed in about four days. Its investigation, therefore, had to be carried out very rapidly, in order that its decomposition might not be appreciable while its properties were being determined. Its product of change was named by Rutherford " radium A," and it is undoubtedly deposited from niton as a metal, with simultaneous evolution of helium ; the equation would therefore be :

$$niton (222·4)=helium (4)+radium A (218·4).$$

But it is impossible to investigate radium A chemically, for in three minutes it has half changed into another solid substance, radium B, again giving off helium. This change would be represented by the equation :

$$radium A (218·4)=helium (4)+radium B (214·4).$$

Radium B, again, can hardly be examined chemically, for in twenty-seven minutes it has half changed into radium C^1. In this case, however, no helium is evolved ; only atoms of negative electricity, to which the name " electrons " has been given by Dr. Stoney, and these have minute weight which, although approximately ascertainable, at present has defied direct measurement. Radium C^1 has a half-life of 19·5 minutes, too short, again, for chemical investigation ; but it changes into radium C^2, and in doing so each atom parts with a helium atom, hence the equation :

$$radium C^1 (214·4)=helium (4)+radium C^2 (210·4).$$

In 2·5 minutes radium C^2 is half gone, parting with electrons, forming radium D. Radium D gives the chemist a chance, for its half-life is no less than sixteen and a half years. Without parting with anything detectable, radium D passes into radium E, of which the half-life period is five days ; and, lastly, radium E changes spontaneously into radium F, the substance to which Madame Curie gave the name " polonium," in allusion to her native country, Poland. Polonium, in its turn, is half changed in 140 days, with loss of an atom of helium, into an unknown metal, supposed to be possibly lead. If that be the case, the equation would run :

$$polonium (210·4)=helium (4)+lead (206·4).$$

But the atomic weight of lead is 207·1, and not 206·4 ; however, it is possible that the atomic weight of radium is 227·1, and not 226·4.

We have another method of approaching the same subject. It is practically certain that the progenitor of radium is uranium, and that the transformation of uranium into radium involves the loss of three α particles, that is, of three atoms of helium. The atomic weight of helium may be taken as one of the most certain ; it is 3·994, as determined by Mr. Watson in my laboratories. Three atoms would therefore weigh 11·98, practically 12. There is, however, still some uncertainty in the atomic weight of uranium ; Richards and Merigold make it 239·4, but the general mean, calculated by Clarke, is 239·0. Subtracting 12 from these numbers, we have the values 227·0, and 227·4 for the atomic weight of radium. It is as yet impossible to draw any certain conclusion.

The importance of the work, which will enable a definite and sure conclusion to be drawn, is this : For the first time, we have accurate knowledge as to the descent of some of the elements. Supposing the atomic weight of uranium to be certainly 239, it may be taken as proved that, in losing three atoms of helium, radium is produced, and, if the change consists solely in the loss of the three atoms of helium, the atomic weight of radium must necessarily be 227. But it is known that β rays, or electrons, are also parted with during this change ; and electrons have weight. How many electrons are lost is unknown ; therefore, although the weight of an electron is approximately known, it is impossible to say how much to allow for in estimating the atomic weight of radium. But it is possible to solve this question indirectly by determining exactly the atomic weights of radium and of uranium ; the difference between the atomic weight of radium *plus* 12, *i.e.* plus the weight of three atoms of helium, and that of uranium,

will give the weight of the number of electrons which escape. Taking the most probable numbers available, viz. 239·4 for uranium and 226·8 for radium, and adding 12 to the latter, the weight of the escaping electrons would be 0·6.

The correct solution of this problem would in great measure clear up the mystery of the irregularities in the periodic table, and would account for the deviations from Prout's Law, that the atomic weights are multiples of some common factor or factors. I also venture to suggest that it would throw light on allotropy, which in some cases, at least, may very well be due to the loss or gain of electrons, accompanied by a positive or negative heat-change. Incidentally, this suggestion would afford places in the periodic table for the somewhat overwhelming number of pseudo-elements the existence of which is made practically certain by the disintegration hypothesis. Of the twenty-six elements derived from uranium, thorium, and actinium, ten, which are formed by the emission of electrons alone, may be regarded as allotropes or pseudo-elements ; this leaves sixteen, for which sixteen or seventeen gaps would appear to be available in the periodic table, provided the reasonable supposition be made that a second change in the length of the periods has taken place. It is, above all things, certain that it would be a fatal mistake to regard the existence of such elements as irreconcilable with the periodic arrangement, which has rendered to systematic chemistry such signal service in the past.

Attention has repeatedly been drawn to the enormous quantity of energy stored up in radium and its descendants. That, in its emanation, niton is such that if what it parts with as heat during its disintegration were available, it would be equal to three and a half million times the energy available by the explosion of an equal volume of detonating gas—a mixture of one volume of oxygen with two volumes of hydrogen. The major part of this energy comes, apparently, from the expulsion of particles (that is, of atoms of helium) with enormous velocity. It is easy to convey an idea of this magnitude in a form more realisable by giving it a somewhat mechanical turn. Suppose that the energy in a ton of radium could be utilised in thirty years, instead of being evolved at its invariable slow rate of 1760 years for half-disintegration, it would suffice to propel a ship of 15,000 tons, with engines of 15,000 horse-power, at the rate of 15 knots an hour for thirty years—practically the lifetime of the ship. To do this actually requires a million and a half tons of coal.

It is easily seen that the virtue of the energy of the radium consists in the small weight in which it is contained ; in other words, the radium-energy is in an enormously concentrated form. I have attempted to apply the energy contained in niton to various purposes ; it decomposes water, ammonia, hydrogen chloride, and carbon dioxide each into its constituents ; further experiments on its action on salts of copper appeared to show that the metal copper was converted partially into lithium, a metal of the sodium column ; and similar experiments, of which there is not time to speak, indicate that thorium, zirconium, titanium, and silicon are degraded into carbon ; for solutions of compounds of these, mixed with niton, invariably generated carbon dioxide, while cerium, silver, mercury, and some other metals gave none. One can imagine the very atoms themselves, exposed to bombardment by enormously quickly moving helium atoms, failing to withstand the impacts. Indeed, the argument *à priori* is a strong one ; if we know for certain that radium and its descendants decompose spontaneously, evolving energy, why should not other more stable elements decompose when subjected to enormous strains?

This leads to the speculation whether, if elements are capable of disintegration, the world may not have at its disposal a hitherto unsuspected source of energy. If radium were to evolve its stored-up energy at the same rate that gun-cotton does, we should have an undreamt-of explosive ; could we control the rate we should have a useful and potent source of energy, provided always that a sufficient supply of radium were forthcoming. But the supply is certainly a very limited one ; and it can be safely affirmed that the production will never surpass half an ounce a year. If, however, the elements which we have been used to consider as permanent are capable of changing with evolution of energy, if some form of catalyser could be discovered which would usefully increase their almost inconceivably slow rate of change, then it is not too much to say that the whole future of our race would be altered.

The whole progress of the human race has indeed been due to individual members discovering means of concentrating energy and of transforming one form into another. The carnivorous animals strike with their paws and crush with their teeth; the first man who aided his arm with a stick in striking a blow discovered how to concentrate his small supply of kinetic energy; the first man who used a spear found that its sharp point in motion represented a still more concentrated form; the arrow was a further advance, for the spear was then propelled by mechanical means; the bolt of the crossbow, the bullet shot forth by compressed hot gas, first derived from black powder, later from high explosives, all these represent progress. To take another sequence: the preparation of oxygen by Priestley applied energy to oxide of mercury in the form of heat; Davy improved on this when he concentrated electrical energy into the tip of a thin wire by aid of a powerful battery, and isolated potassium and sodium.

Great progress has been made during the past century in effecting the conversion of one form of energy into others with as little useless expenditure as possible. Let me illustrate by examples: A good steam engine converts about one-eighth of the potential energy of the fuel into useful work; seven-eighths are lost as unused heat and useless friction. A good gas engine utilises more than one-third of the total energy in the gaseous fuel; two-thirds are uneconomically expended. This is a universal proposition; in order to effect the conversion from one form of energy into another, some energy must be expended uneconomically. If A is the total energy which it is required to convert, if B is the energy into which it is desired to convert A, then a certain amount of energy, C, must be expended to effect the conversion. In short, $A = B + C$. It is eminently desirable to keep C, the useless expenditure, as small as possible; it can never equal zero, but it can be made small. The ratio of C to B (the economic coefficient) should therefore be as large as is attainable.

The middle of the nineteenth century will always be noted as the beginning of the golden age of science, the epoch when great generalisations were made, of the highest importance on all sides, philosophical, economic, and scientific. Carnot, Clausius, Helmholtz, Julius Robert Mayer abroad, and the Thomsons, Lord Kelvin and his brother James, Rankine, Tait, Joule, Clerk Maxwell, and many others at home, laid the foundations on which the splendid structure has been erected. That the latent energy of fuel can be converted into energy of motion by means of the steam engine is what we owe to Newcomen and Watt; that the kinetic energy of the fly-wheel can be transformed into electrical energy was due to Faraday, and to him, too, we are indebted for the re-conversion of electrical energy into mechanical work; and it is this power of work which gives us leisure, and which enables a small country like ours to support the population which inhabits it.

I suppose that it will be generally granted that the Commonwealth of Athens attained a high-water mark in literature and thought which has never yet been surpassed. The reason is not difficult to find; a large proportion of its people had ample leisure, due to ample means; they had time to think and time to discuss what they thought. How was this achieved? The answer is simple: each Greek Freeman had, on an average, at least five helots who did his bidding, who worked his mines, looked after his farm, and, in short, saved him from manual labour. Now we in Britain are much better off; the population of the British Isles is in round numbers 45 millions; there are consumed in our factories at least 50 million tons of coal annually, and "it is generally agreed that the consumption of coal per indicated horse-power per hour is, on an average, about 5 lb." (Royal Commission on Coal Supplies, Part I.). This gives seven million horse-power per year. How many man-power are equal to a horse-power? I have arrived at an estimate

thus: A Bhutanese can carry 230 lb. *plus* his own weight, in all 400 lb., up a hill 4000 feet high in eight hours; this is equivalent to about one twenty-fifth of a horse-power; seven million horse-power are therefore about 175 million man-power. Taking a family as consisting, on the average, of five persons, our 45 millions would represent nine million families, and dividing the total man-power by the number of families, we must conclude that each British family has, on the average, nearly twenty "helots" doing his bidding, instead of the five of the Athenian family. We do not appear, however, to have gained more leisure thereby; but it is this that makes it possible for the British Isles to support the population which it does.

We have in this world of ours only a limited supply of stored-up energy, in the British Isles a very limited one—namely, our coalfields. The rate at which this supply is being exhausted has been increasing very steadily for the last forty years, as anyone can prove by mapping the data given on p. 27, table D, of the General Report of the Royal Commission on Coal Supplies (1906). In 1870 110 million tons were mined in Great Britain, and ever since the amount has increased by three and a third million tons a year. The available quantity of coal in the proved coalfields is very nearly 100,000 million tons; it is easy to calculate that if the rate of working increases as it is doing, our coal will be completely exhausted in 175 years. But, it will be replied, the rate of increase will slow down. Why? It has shown no sign whatever of slackening during the last forty years. Later, of course, it must slow down, when coal grows dearer owing to approaching exhaustion. It may also be said that 175 years is a long time; why, I myself have seen a man whose father fought in the '45 on the Pretender's side, nearly 170 years ago! In the life of a nation 175 years is a span.

This consumption is still proceeding at an accelerated rate. Between 1905 and 1907 the amount of coal raised in the United Kingdom increased from 236 to 268 million tons, equal to six tons per head of the population, against three and a half tons in Belgium, two and a half tons in Germany, and one ton in France. Our commercial supremacy and our power of competing with other European nations are obviously governed, so far as we can see, by the relative price of coal; and when our prices rise, owing to the approaching exhaustion of our supplies, we may look forward to the near approach of famine and misery.

Having been struck some years ago with the optimism of my non-scientific friends as regards our future, I suggested that a committee of the British Science Guild should be formed to investigate our available sources of energy. This Guild is an organisation, founded by Sir Norman Lockyer after his tenure of the Presidency of this Association, for the purpose of endeavouring to impress on our people and their Government the necessity of viewing problems affecting the race and the State from the standpoint of science; and the definition of science in this, as in other connections, is simply the acquisition of knowledge, and orderly reasoning on experience already gained and on experiments capable of being carried out, so as to forecast and control the course of events, and, if possible, to apply this knowledge to the benefit of the human race.

The Science Guild has enlisted the services of a number of men, each eminent in his own department, and each has now reported on the particular source of energy of which he has special knowledge.

Besides considering the uses of coal and its products, and how they may be more economically employed, in which branches the Hon. Sir Charles Parsons, Mr. Dugald Clerk, Sir Boverton Redwood, Dr. Beilby, Dr. Hele-Shaw, Prof. Vivian Lewes and others have furnished reports, the following sources of energy have been brought under review: the possibility of utilising the tides; the internal heat of the earth; the winds; solar heat; water-power; the extension of forests, and the use of wood and peat as fuels; and, lastly, the possibility of controlling the undoubted, but almost infinitely slow, disintegration of the elements, with the view of utilising their stored-up energy.

However interesting a detailed discussion of these possible sources of energy might be, time prevents my dwelling on them. Suffice it to say that the Hon. R. J. Strutt has shown that in this country, at least, it would be impracticable to attempt to utilise terrestrial heat from bore-holes; others have deduced that from the tides, the winds, and water-power small supplies of energy are no doubt obtainable, but that, in comparison with that derived from the combustion of coal, they are negligible; nothing is to be hoped for from the direct utilisation of solar heat in this temperate and uncertain climate, and it would be folly to consider seriously a possible supply of energy in a conceivable acceleration of the liberation of energy by atomic change. It looks utterly improbable, too, that we shall ever be able to utilise the energy due to the revolution of the earth on her axis, or to her proper motion round the sun.

Attention should undoubtedly be paid to forestry and to the utilisation of our stores of peat. On the Continent, the forests are largely the property of the State; it is unreasonable, especially in these latter days of uncertain tenure of property, to expect any private owner of land to invest money in schemes which would at best only benefit his descendants, but which, under our present trend of legislation, do not promise even that remote return. Our neighbours and rivals, Germany and France, spend annually 2,200,000l. on the conservation and utilisation of their forests; the net return is 6,000,000l. There is no doubt that we could imitate them with advantage. Moreover, an increase in our forests would bring with it an increase in our water-power, for without forest land rain rapidly reaches the sea, instead of distributing itself so as to keep the supply of water regular, and so more easily utilised.

Various schemes have been proposed for utilising our deposits of peat: I believe that in Germany the peat industry is moderately profitable; but our humid climate does not lend itself to natural evaporation of most of the large amount of water contained in peat, without which processes of distillation prove barely remunerative.

We must therefore rely chiefly on our coal reserve for our supply of energy, and for the means of supporting our population; and it is to the more economical use of coal that we must look in order that our life as a nation may be prolonged. We can economise in many ways: By the substitution of turbine engines for reciprocating engines, thereby reducing the coal required per horse-power from 4 to 5 lb. to $1\frac{1}{2}$ or 2 lb.; by the further replacement of turbines by gas engines, raising the economy to 30 per cent. of the total energy available in the coal, that is, lowering the coal consumption per horse-power to 1 or $1\frac{1}{4}$ lb.; by creating the power at the pit-mouth, and distributing it electrically, as is already done in the Tyne district. Economy can also be effected in replacing "beehive" coke ovens by recovery ovens; this is rapidly being done; and Dr. Beilby calculates that in 1909 nearly six million tons of coal, out of a total of sixteen to eighteen millions, were coked in recovery ovens, thus effecting a saving of two to three million tons of fuel annually. Progress is also being made in substituting gas for coal or coke in metallurgical, chemical, and other works. But it must be remembered that for economic use gaseous fuel must not be charged with the heavy costs of piping and distribution.

The domestic fire problem is also one which claims our instant attention. It is best grappled with from the point of view of smoke. Although the actual loss of thermal energy in the form of smoke is small—at most less than a half per cent. of the fuel consumed—still the presence of smoke is a sign of waste of fuel and careless stoking. In works, mechanical stokers, which ensure regularity of firing and complete combustion of fuel, are more and more widely replacing hand-firing. But we are still utterly wasteful in our consumption of fuel in domestic fires. There is probably no single remedy applicable; but the introduction of central heating, of gas fires, and of grates which permit of better utilisation of fuel will all play a part in economising our coal. It is open to argument whether it might not be wise to hasten the time when smoke is no more by imposing a sixpenny fine for each offence; an instantaneous photograph could easily prove the offence to have been committed, and the imposition of the fine might be delayed until three warnings had been given by the police.

Now I think that what I wish to convey will be best expressed by an allegory. A man of mature years, who has surmounted the troubles of childhood and adolescence without much disturbance to his physical and mental state, gradually becomes aware that he is suffering from loss of blood; his system is being drained of this essential to life and strength. What does he do? If he is sensible he calls in a doctor, or perhaps several, in consultation; they ascertain the seat of the disease, and diagnose the cause. They point out that while consumption of blood is necessary for healthy life, it will lead to a premature end if the constantly increasing drain is not stopped. They suggest certain precautionary measures; and if he adopts them he has a good chance of living at least as long as his contemporaries; if he neglects them his days are numbered.

That is our condition as a nation. We have had our consultation in 1903; the doctors were the members of the Coal Commission. They showed the gravity of our case, but we have turned a deaf ear.

It is true that the self-interest of coal consumers is slowly leading them to adopt more economical means of turning coal into energy. But I have noticed, and frequently publicly announced, a fact which cannot but strike even the most unobservant. It is this: When trade is good, as it appears to be at present, manufacturers are making money; they are overwhelmed with orders, and have no inclination to adopt economies which do not appear to them to be essential, and the introduction of which would take thought and time, and which would withdraw the attention of their employees from the chief object of the business—how to make the most of the present opportunities. Hence improvements are postponed. When bad times come, then there is no money to spend on improvements; they are again postponed until better times arrive.

What can be done?

I would answer: Do as other nations have done and are doing; take stock annually. The Americans have a permanent Commission initiated by Mr. Roosevelt, consisting of three representatives from each State, the sole object of which is to keep abreast with the diminution of the stores of natural energy, and to take steps to lessen its rate. This is a non-political undertaking, and one worthy of being initiated by the ruler of a great country. If the example is followed here the question will become a national one.

Two courses are open to us: first, the *laissez-faire* plan of leaving to self-interested competition the combating of waste; or second, initiating legislation which, in the interest of the whole nation, will endeavour to lessen the squandering of our national resources. This legislation may be of two kinds; penal, that is, imposing a penalty on wasteful expenditure of energy supplies; and helpful, that is, imparting information as to what can be done, advancing loans at an easy rate of interest to enable reforms to be carried out, and insisting on the greater prosperity which would result from the use of more efficient appliances.

This is not the place, nor is there the time, to enter into detail; the subject is a complicated one, and it will demand the combined efforts of experts and legislators for a generation; but if it be not considered with the definite intention of immediate action, we shall be held up to the deserved execration of our not very remote descendants.

The two great principles which I have alluded to in an earlier part of this address must not, however, be lost sight of; they should guide all our efforts to use energy economically. Concentration of energy in the form of electric current at high potential makes it possible to convey it for long distances through thin, and therefore comparatively inexpensive, wires; and the economic coefficient of the conversion of mechanical into electrical, and of electrical into mechanical, energy is a high one; the useless expenditure does not much exceed one-twentieth part of the energy which can be utilised. These considerations would point to the conversion at the pit-mouth of the energy of the fuel into electrical energy, using as an intermediary turbines, or preferably gas engines, and distributing the electrical energy to where it is wanted. The use of gas engines may, if desired, be accompanied by the production of half-distilled coal, a fuel which burns nearly without smoke, and one which is suitable for domestic fires, if it is found too difficult to displace them and to induce our population to adopt the more efficient and economical systems of domestic heating which are used in America and on the Continent. The increasing use of gas for factory, metallurgical, and chemical purposes points to the gradual concentration of works near the coal mines in order that the laying-down of expensive piping may be avoided.

An invention which would enable us to convert the energy of coal directly into electrical energy would revolutionise our ideas and methods, yet it is not unthinkable. The nearest practical approach to this is the Mond gas-battery, which, however, has not succeeded, owing to the imperfection of the machine.

In conclusion, I would put in a plea for the study of pure science, without regard to its applications. The discovery of radium and similar radio-active substances has widened the bounds of thought. While themselves, in all probability, incapable of industrial application, save in the domain of medicine, their study has shown us to what enormous advances in the concentration of energy it is permissible to look forward, with the hope of applying the knowledge thereby gained to the betterment of the whole human race. As charity begins at home, however, and as I am speaking to the *British* Association for the Advancement of Science, I would urge that our first duty is to strive for all which makes for the permanence of the British Commonweal, and which will enable us to transmit to our posterity a heritage not unworthy to be added to that which we have received from those who have gone before.

———

Index

Figures in bold type indicate articles written by the author under whose name the entry appears